Lecture Notes in Computer Science 2676

Edited by G. Goos, J. Hartmanis, and J. van Leeuwen

Springer
Berlin
Heidelberg
New York
Barcelona
Hong Kong
London
Milan
Paris
Tokyo

Ricardo Baeza-Yates   Edgar Chávez
Maxime Crochemore (Eds.)

Combinatorial Pattern Matching

14th Annual Symposium, CPM 2003
Morelia, Michoacán, Mexico, June 25-27, 2003
Proceedings

 Springer

Series Editors

Gerhard Goos, Karlsruhe University, Germany
Juris Hartmanis, Cornell University, NY, USA
Jan van Leeuwen, Utrecht University, The Netherlands

Volume Editors

Ricardo Baeza-Yates
Universidad de Chile
Depto. de Ciencias de la Computación
Blanco Encalada 2120
Santiago 6511224, Chile
E-mail: rbaeza@dcc.uchile.cl

Edgar Chávez
Universidad Michoacana
Escuela de Ciencias Físico-Matemáticas
Edificio "B", ciudad universitaria
Morelia, Michoacán, Mexico
E-mail: elchavez@fismat.umich.mx

Maxime Crochemore
Université de Marne-la-Vallée
77454 Marne-la-Vallée CEDEX 2, France
E-mail: Maxime.Crochemore@univ-mlv.fr

Cataloging-in-Publication Data applied for

A catalog record for this book is available from the Library of Congress.

Bibliographic information published by Die Deutsche Bibliothek
Die Deutsche Bibliothek lists this publication in the Deutsche Nationalbibliografie;
detailed bibliographic data is available in the Internet at <http://dnb.ddb.de>.

CR Subject Classification (1998): F.2.2, I.5.4, I.5.0, I.7.3, H.3.3, E.4, G.2.1

ISSN 0302-9743
ISBN 3-540-40311-6 Springer-Verlag Berlin Heidelberg New York

Springer-Verlag Berlin Heidelberg New York
a member of BertelsmannSpringer Science+Business Media GmbH

http://www.springer.de

© Springer-Verlag Berlin Heidelberg 2003
Printed in Germany

Typesetting: Camera-ready by author, data conversion by PTP-Berlin GmbH
Printed on acid-free paper   SPIN: 10927533   06/3142   5 4 3 2 1 0

Preface

The papers contained in this volume were presented at the 14th Annual Symposium on Combinatorial Pattern Matching, held June 25–27, 2003 at the *Centro Cultural Universitario* of the *Universidad Michoacana*, in Morelia, Michoacán, Mexico. They were selected from 57 abstracts submitted in response to the call for papers. In addition, there were invited lectures by Vladimir Levenshtein, from the University of Bergen, Norway, and Ian Munro, from the University of Waterloo, Canada.

Combinatorial Pattern Matching (CPM) addresses issues of searching and matching strings and more complicated patterns such as trees, regular expressions, graphs, point sets, and arrays, in various formats. The goal is to derive nontrivial combinatorial properties of such structures and to exploit these properties in order to achieve superior performance for the corresponding computational problems. Another important goal is to analyze and pinpoint the properties and conditions under which searches cannot be performed efficiently.

Over the past decade a steady flow of high quality-research on this subject has changed a sparse set of isolated results into a full-fledged area of algorithmics. This area is continuing to grow even further due to the increasing demand for speed and efficiency that stems from important applications such as the World Wide Web, computational biology, computer vision, and multimedia systems. These involve requirements for information retrieval in heterogeneous databases, data compression, and pattern recognition. The objective of the annual CPM gathering is to provide an international forum for research in combinatorial pattern matching and related applications.

The first 13 meetings were held in Paris, London, Tucson, Padova, Asilomar, Helsinki, Laguna Beach, Aarhus, Piscataway, Warwick, Montreal, Jerusalem, and Fukuoka, over the years 1990–2002. After the first meeting, a selection of papers appeared in Volume 92 of *Theoretical Computer Science*. Selected papers of the 10th meeting appeared as a special issue of the *Journal of Discrete Algorithms*. Selected papers of the 12th meeting will appear in a special issue of *Discrete Applied Mathematics*. The proceedings of the 3rd to 13th meetings appeared as Volumes 644, 684, 807, 937, 1075, 1264, 1448, 1645, 1848, 2089, and 2373 of the Springer LNCS series.

The general organization and orientation of the CPM conferences is coordinated by a steering committee composed of Alberto Apostolico (*Universities of Padova and Purdue*), Maxime Crochemore (*University of Marne-la-Vallée and King's College London*), Zvi Galil (*Columbia University*), and Udi Manber (*Amazon*). The conference chair was Edgar Chávez (*University Michoacana*).

April 2003   R. Baeza-Yates, E. Chávez, and M. Crochemore

Program Committee

Ricardo Baeza-Yates, co-chair, *Univ. of Chile*
Edgar Chávez, *Univ. of Michoacán, Mexico*
Richard Cole, *New York University, USA*
Maxime Crochemore, co-chair, *Univ. of Marne-la-Vallée, France*
Rafaelle Giancarlo, *Univ. of Palermo, Italy*
Roberto Grossi, *Univ. of Pisa, Italy*
Dan Gusfield, *U.C. Davis, USA*
Costas Iliopoulos, *King's College London, UK*
Joao Paulo Kitajima, *Alellyx Applied Genomics, Brazil*
Gad Landau, *Univ. of Haifa, Israel*
Thierry Lecroq, *Univ. of Rouen, France*
Udi Manber, *Amazon, USA*
Gonzalo Navarro, *Univ. of Chile*
Wojciech Plandowski, *Warsaw University, Poland*
Marie-France Sagot, *INRIA Rhône-Alpes, France*
Cenk Sahinalp, *Case Western Reserve Univ., USA*
Jeanette Schmidt, *Incyte, USA*
Ayumi Shinohara, *Kyushu University, Japan*
Kaizhong Zhang, *Univ. of Western Ontario, Canada*

Local Organization

Local arrangements and the conference Web site were coordinated by Edgar Chávez. Organizational help was provided by the School of Physics and Mathematics, University of Michoacán.

Sponsoring Institutions

The conference was sponsored by the *Consejo Nacional de Ciencia y Tecnología (CONACyT)* and the *Universidad Michoacana*.

Table of Contents

Multiple Genome Alignment: Chaining Algorithms Revisited

Mohamed Ibrahim Abouelhoda[1] and Enno Ohlebusch[2]

[1] Faculty of Technology, University of Bielefeld, P.O. Box 10 01 31, 33501 Bielefeld, Germany. mibrahim@techfak.uni-bielefeld.de
[2] Faculty of Computer Science, University of Ulm, Albert-Einstein-Allee, 89069 Ulm, Germany. eo@informatik.uni-ulm.de

Abstract. Given n fragments from $k > 2$ genomes, we will show how to find an optimal chain of colinear non-overlapping fragments in time $O(n \log^{k-2} n \log \log n)$ and space $O(n \log^{k-2} n)$. Our result solves an open problem posed by Myers and Miller because it reduces the time complexity of their algorithm by a factor $\frac{\log^2 n}{\log \log n}$ and the space complexity by a factor $\log n$. For $k = 2$ genomes, our algorithm takes $O(n \log n)$ time and $O(n)$ space.

1 Introduction

Given the continuing improvements in high-throughput genomic sequencing and the ever-expanding sequence databases, new advances in software tools for post-sequencing functional analysis are being demanded by the biological scientific community. Whole genome comparisons have been heralded as the next logical step toward solving genomic puzzles, such as determining coding regions, discovering regulatory signals, and deducing the mechanisms and history of genome evolution. However, before any such detailed analyses can be addressed, methods are required for comparing such large sequences. If the organisms under consideration are closely related, then global alignments are the strategy of choice. Although there is an immediate need for "reliable and automatic software for aligning three or more genomic sequences" [12], currently only the software tool *MGA* [10] solves the problem of aligning multiple complete genomes. This is because all previous multiple alignment algorithms were designed for comparing single protein sequences or DNA sequences containing a single gene, and are incapable of producing long alignments. In order to cope with the shear volume of data, *MGA* uses an anchor-based method that is divided into three phases: (1) computation of fragments (regions in the genomes that are similar—in *MGA* these are multiple maximal exact matches), (2) computation of an optimal chain of colinear non-overlapping fragments: these are the anchors that form the basis of the alignment, (3) alignment of the regions between the anchors.

This paper is concerned with algorithms for solving the combinatorial chaining problem of the second phase; see Fig. 1. Note that every genome alignment tool has to solve the chaining problem somehow, but the algorithms differ from

R. Baeza-Yates et al. (Eds.): CPM 2003, LNCS 2676, pp. 1–16, 2003.

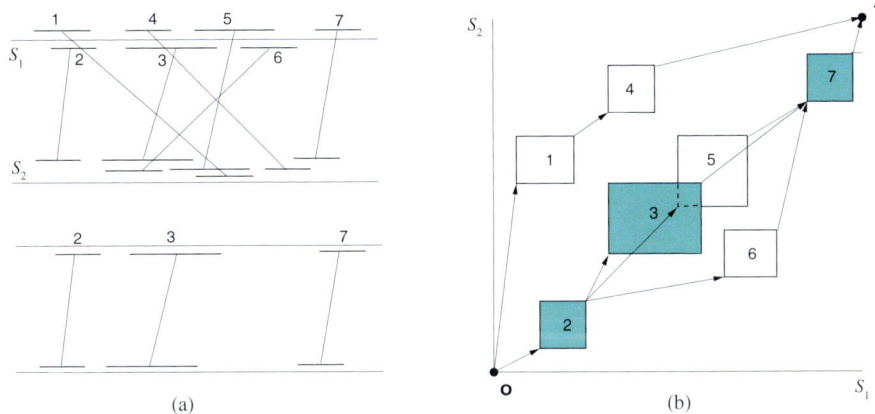

Fig. 1. Given a set of fragments (upper left figure), an optimal chain of colinear non-overlapping fragments (lower left figure) can be computed, e.g., by computing an optimal path in the graph in (b) (in which not all edges are shown).

tool to tool; see, e.g., [3,13,17]. Chaining algorithms are also useful in other bioinformatics applications such as comparing restriction maps [9] or solving the exon assembly problem which is part of eucaryotic gene prediction [6].

A well-known solution to the chaining problem consists of finding a maximum weight path in a weighted directed acyclic graph; see, e.g., [10]. However, the running time of this chaining algorithm is quadratic in the number n of fragments. This can be a serious drawback if n is large. To overcome this obstacle, *MGA* currently uses a variant of an algorithm devised by Zhang et al. [20], but without taking gap costs into account. This algorithm takes advantage of the geometric nature of the chaining problem. It constructs an optimal chain using orthogonal range search based on kd-trees, a data structure known from computational geometry. As is typical with kd-tree methods, however, no rigorous analysis of the running time of the algorithm is known; cf. [15].

Another chaining algorithm, devised by Myers and Miller [15], falls into the category of *sparse dynamic programming* [5]. Their algorithm is based on the line-sweep paradigm, and uses orthogonal range search supported by *range trees* instead of kd-trees. It is the only chaining algorithm for $k > 2$ sequences that runs in sub-quadratic time $O(n \log^k n)$, "but the result is a time bound higher by a logarithmic factor than what one would expect" [4]. In particular, for $k = 2$ sequences it is one log-factor slower than previous chaining algorithms [5,14], which require only $O(n \log n)$ time. In the epilog of their paper [15], Myers and Miller wrote: "We thought hard about trying to reduce this discrepancy but have been unable to do so, and the reasons appear to be fundamental" and "To improve upon our result appears to be a difficult open problem." In this paper, we solve this problem. Surprisingly, we can not only reduce the time and space complexities by a log-factor but actually improve the time complexity by a factor $\frac{\log^2 n}{\log \log n}$. In essence, this improvement is achieved by (1) a combination of fractional cascading [19] with the efficient priority queues of [18,8], which yields a more efficient search than on ordinary range trees, and (2) by incorporating gap costs into the weight of fragments, so that it is enough to determine a maximum function value over a semi-dynamic set (instead of a dynamic set). In related work, Baker and Giancarlo [1] have shown how to efficiently compute a longest common subsequence from fragments, which is a variant of our problem, but their algorithm is restricted to two sequences.

2   Basic Concepts and Definitions

For any point $p \in \mathbb{R}^k$, let $p.x_1, p.x_2, \ldots, p.x_k$ denote its coordinates. If $k = 2$, the coordinates of p will also be written as $p.x$ and $p.y$. A hyper-rectangle (called hyperrectangular domain in [16]) is the Cartesian product of intervals on distinct coordinate axes. A hyper-rectangle $[l_1 \ldots h_1] \times [l_2 \ldots h_2] \times \ldots \times [l_k \ldots h_k]$ (with $l_i < h_i$ for all $1 \leq i \leq k$) will also be denoted by $R(p, q)$, where $p = (l_1, \ldots, l_k)$ and $q = (h_1, \ldots, h_k)$ are its two extreme corner points. In the problem we consider, all points are given in advance (off-line). Therefore, it is possible to map the points into $\mathbb{N}^k$, called the *rank space*; see, e.g., [2]. Every point $(x_1, x_2, \ldots, x_k)$ is mapped to point $(r_1, r_2, \ldots, r_k)$, where r_i, $1 \leq i \leq k$, is the index (or rank) of point p in a list which is sorted in ascending order w.r.t. dimension x_i. This transformation takes $O(kn \log n)$ time and $O(n)$ space because one has to sort the points k times. Thus, we can assume that the points are already transformed to the rank space.

For $1 \leq i \leq k$, S_i denotes a string of length $|S_i|$. In our application, S_i is the DNA sequence of a genome. $S_i[l_i \ldots h_i]$ is the substring of S_i starting at position l_i and ending at position h_i. An *exact fragment* (or *multiple exact match*) f consists of two k-tuples $beg(f) = (l_1, l_2, \ldots, l_k)$ and $end(f) = (h_1, h_2, \ldots, h_k)$ such that $S_1[l_1 \ldots h_1] = S_2[l_2 \ldots h_2] = \ldots = S_k[l_k \ldots h_k]$, i.e., the substrings are identical. It is maximal, if the substrings cannot be simultaneously extended to the left and to the right in every S_i. If mismatches are allowed in the substrings, then we speak of a *gap-free fragment*. If one further allows insertions and deletions (so that the substrings may be of unequal length), we will use the general term *fragment*. Many algorithms have been developed to efficiently compute all kinds of fragments (e.g., [10,11]), and the algorithms presented here work for arbitrary fragments.

A fragment f of k genomes can be represented by a hyper-rectangle in $\mathbb{R}^k$ with the two extreme corner points $beg(f)$ and $end(f)$, where each coordinate of the points is non-negative. In the following, the words *number of genomes* and *dimension* will thus be used synonymously. With every fragment f, we associate a weight $f.weight \in \mathbb{R}$. This weight can, for example, be the length of the fragment (in case of gap-free fragments) or its statistical significance.

In what follows, we will often identify the point $beg(f)$ or $end(f)$ with the fragment f. For example, if we speak about the score of a point $beg(f)$ or $end(f)$, we mean the score of the fragment f. For ease of presentation, we consider the points $\mathbf{0} = (0, \ldots, 0)$ (the origin) and $\mathbf{t} = (|S_1|, \ldots, |S_k|)$ (the terminus) as fragments with weight 0. For these fragments, we define $beg(0) = \perp$, $end(0) = 0$, $0.score = 0$, $beg(t) = t$, and $end(t) = \perp$.

Definition 1. *We define the relation $\ll$ on the set of fragments by $f \ll f'$ if and only if $end(f).x_i < beg(f').x_i$ for all $1 \leq i \leq k$. If $f \ll f'$, then we say that f precedes f'. We further define $0 \ll f \ll t$ for every fragment f with $f \neq 0$ and $f \neq t$.*

Definition 2. *A* chain *of colinear non-overlapping fragments (or chain for short) is a sequence of fragments $f_1, f_2, \ldots, f_\ell$ such that $f_i \ll f_{i+1}$ for all $1 \leq i < \ell$. The score of C is $Score(C) = \sum_{i=1}^{\ell-1}(f_i.weight - g(f_{i+1}, f_i))$, where $g(f_{i+1}, f_i)$ is the cost of connecting fragment f_i to f_{i+1} in the chain. We will call this cost* gap cost.

Given a set of n fragments and a *gap cost* function g, the *fragment-chaining problem* is to determine a chain of maximum score (called optimal chain in the following) starting at the origin 0 and ending at terminus t. A direct solution to this problem is to construct a weighted directed acyclic graph $G = (V, E)$, where the set V of vertices consists of all fragments (including 0 and t) and the set of edges E is characterized as follows: There is an edge $f \rightarrow f'$ with weight $f'.weight - g(f', f)$ if $f \ll f'$; see Fig. 1(b). An optimal chain of fragments, starting at the origin 0 and ending at terminus t, corresponds to a path with maximum score from vertex 0 to vertex t in the graph. Because the graph is acyclic, such a path can be computed as follows. Let $f'.score$ be defined as the maximum score of all chains that start at 0 and end at f'. $f'.score$ can be expressed by the recurrence: $0.score = 0$ and

$$f'.score = f'.weight + \max\{f.score - g(f', f) : f \ll f'\} \qquad (1)$$

A dynamic programming algorithm based on this recurrence takes $O(|V| + |E|)$ time provided that computing *gap costs* takes constant time. Because $|V| + |E| \in O(n^2)$, computing an optimal chain takes quadratic time and linear space. This graph-based solution works for any number of genomes and for any kind of gap cost. As explained in Section 1, however, the time bound can be improved by considering the geometric nature of the problem. In order to present our result systematically, we first give a chaining algorithm that neglects gap costs. Then we will modify this algorithm in two steps, so that it can deal with certain gap costs.

3   The Chaining Algorithm without Gap Cost

3.1   The Chaining Algorithm

Because our algorithm is based on orthogonal range search for maximum, we have to recall two notions. Given a set S of points in $\mathbb{R}^k$ with associated score, a *range query* (RQ) asks for all the points of S that lie in a hyper-rectangle $R(p, q)$, while a *range maximum query* (RMQ) asks for a point of maximum score in $R(p, q)$. In the following, RMQ will also denote a procedure that takes two points p and q as input and returns a point of maximum score in the hyper-rectangle $R(p, q)$.

Lemma 3. *Suppose that the gap cost function g is the constant function 0. If $RMQ(0, beg(f'))$ returns the end point of fragment f, then $f'.score = f'.weight + f.score$.*

Proof. This follows immediately from recurrence (1).

We will further use the line-sweep paradigm to construct an optimal chain. Suppose that the start and end points of the fragments are sorted w.r.t. their x_1 coordinate. Then, processing the points in ascending order of their x_1 coordinate simulates a line (plane or hyper-plane in higher dimensions) that sweeps the points w.r.t. their x_1 coordinate. If a point has already been scanned by the sweeping line, it is said to be *active*; otherwise it is said to be *inactive*. During the sweeping process, the x_1 coordinates of the active points are smaller than the x_1 coordinate of the currently scanned point s. According to Lemma 3, if s is the start point of fragment f', then an optimal chain ending at f' can be found by a RMQ over the set of active end points of fragments. Since $p.x_1 < s.x_1$ for every active end point p, the RMQ need not take the first coordinate into account. In other words, the RMQ is confined to the range $R(0, (s.x_2, \ldots, s.x_k))$, so that the dimension of the problem is reduced by one. To manipulate the point set during the sweeping process, we need a semi-dynamic data structure D that stores the end points of fragments and efficiently supports the following two operations: (1) activation and (2) RMQ over the set of active points. The following algorithm is based on such a data structure D, which will be defined later.

Algorithm 4 *k-dimensional chaining of n fragments*

Sort all start and end points of the n fragments in ascending order w.r.t. their x_1 coordinate and store them in the array points; *because we include the end point of the origin and the start point of the terminus, there are $2n + 2$ points. Store all end points of the fragments (ignoring their x_1 coordinate) as inactive in the $(k - 1)$-dimensional data structure D.*
for $i := 1$ **to** $2n + 2$
    if points$[i]$ *is the start point of fragment f'* **then**
        $q := RMQ(0, (\text{points}[i].x_2, \ldots, \text{points}[i].x_k))$
        determine the fragment f with $end(f) = q$
        $f'.prec := f$
        $f'.score := f'.weight + f.score$
    else $\backslash \star$ points$[i]$ *is end point of a fragment* $\star \backslash$
        activate $(\text{points}[i].x_2, \ldots, \text{points}[i].x_k)$ *in D*

In the algorithm, $f'.prec$ denotes a field that stores the preceding fragment of f' in a chain. It is an immediate consequence of Lemma 3 that Algorithm 4 finds an optimal chain. The complexity of the algorithm depends of course on how the data structure D is implemented. In the following subsection, we will outline an implementation of D that supports RMQ with activation in time $O(n \log^{d-1} n \log \log n)$ and space $O(n \log^{d-1} n)$, where d is the dimension and n is the number of points. Because in our chaining problem $d = k-1$, finding an optimal chain by Algorithm 4 takes $O(n \log^{k-2} n \log \log n)$ time and $O(n \log^{k-2} n)$ space.

3.2 Answering RMQ with Activation Efficiently

In the following, we assume the reader to be familiar with *range trees*. An introduction to this well-known data structure can, for example, be found in [16, pp. 83-88]. Given a set S of n d-dimensional points, its range tree can be built in $O(n \log^{d-1} n)$ time and space and it supports range queries RQ(p,q) in $O(\log^d n + z)$ time, where z is the number of points in the hyper-rectangle $R(p,q)$. The technique of *fractional cascading* [19] saves one log-factor in answering range queries.[1] We briefly describe this technique because we want to modify it to answer RMQ$(0,q)$ with activation efficiently. For ease of presentation, we consider the case $d = 2$. In this case, the range tree is a binary search tree (called *x-tree*) of binary search trees (called *y-trees*). In fractional cascading, the *y-trees* are replaced with arrays (called *y-arrays*) as follows. Let $v.L$ and $v.R$ be the left and right child nodes of a node $v \in x$-*tree* and let A_v denote the *y-array* of v. That is, A_v contains all the points in the leaf list of v sorted in ascending order w.r.t. their y coordinate. Every element $p \in A_v$ has two downstream pointers: The left pointer *Lptr* and the right pointer *Rptr*. The left pointer *Lptr* points to an element q_1 of $A_{v.L}$, where q_1 is either p itself or the rightmost element in A_v that precedes p and also occurs in $A_{v.L}$. In an implementation, *Lptr* is the index with $A_{v.L}[Lptr] = q_1$. Analogously, the right pointer *Rptr* points to an element q_2 of $A_{v.R}$, where q_2 is either p itself or the rightmost element in A_v that precedes p and also occurs in $A_{v.R}$. Fig. 2 shows an example of this structure.

Locating all the points in a rectangle $R(0,(h_1,h_2))$ is done in two stages. In the first stage, a binary search is performed over the *y-array* of the root node of the *x-tree* to locate the rightmost point p_{h_2} such that $p_{h_2}.y \in [0 \ldots h_2]$. Then, in the second stage, the *x-tree* is traversed (while keeping track of the downstream pointers) to locate the rightmost leave p_{h_1} such that $p_{h_1}.x \in [0 \ldots h_1]$. During the traversal of the *x-tree*, we identify a set of nodes which we call *maximum splitting nodes*. A maximum splitting node is either a node on the path from the root to p_{h_1} such that the points in its leaf list are within $[0 \ldots h_1]$ or it is a left child of a node on the path satisfying the same condition. The set of maximum splitting nodes is the smallest set of nodes $v_1, \ldots, v_\ell \in x$-*tree* such that $\biguplus_{j=1}^\ell A_{v_j} = $ RQ$(0,(h_1,\infty))$.[2] In other words, $P := \biguplus_{j=1}^\ell A_{v_j}$ contains every point $p \in S$ such that $p.x \in [0 \ldots h_1]$. However, not every point $p \in P$ satisfies $p.y \in [0 \ldots h_2]$. Here, the downstream pointers come into play. As already mentioned, the downstream pointers are followed while traversing the *x-tree*, and to follow one pointer takes constant time. If we encounter a maximum splitting node v_j, then the element e_j, to which the last downstream pointer points, partitions the list A_{v_j} as follows: Every e that is strictly to the right of e_j is not in $R(0,(h_1,h_2))$, whereas all other elements of A_{v_j} lie in $R(0,(h_1,h_2))$. For this reason, we will call the element e_j the *splitting* element. It is easy to see that the number of maximum splitting nodes is $O(\log n)$. Moreover, we can find all of them and the splitting elements of their *y-arrays* in $O(\log n)$ time; cf. [19].

---

[1] In the same construction time and using the same space as the original range tree.
[2] $\uplus$ denotes disjoint union.

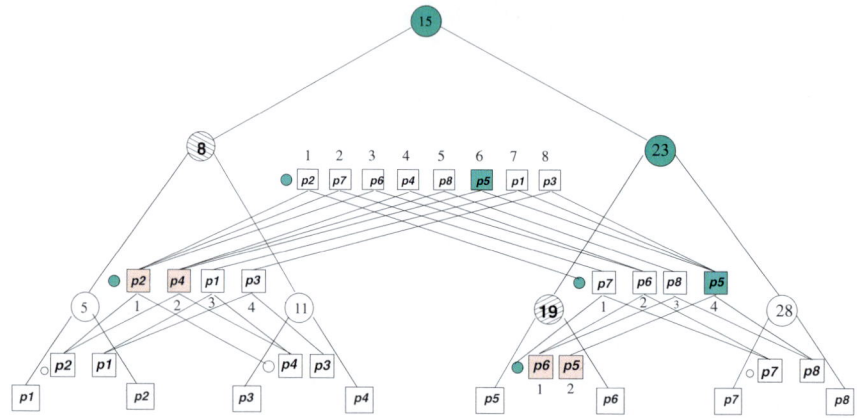

Fig. 2. Fractional cascading: colored nodes are the visited ones. Hatched nodes are the maximum splitting nodes. The small circles refer to NULL pointers. In this example, $p_{h_1} = p_6$ and $p_{h_2} = p_5$. The colored elements of the y-arrays of the maximum splitting nodes are the points in the query rectangle, which is shown in Fig. 3. The numerical value in every internal node is the x coordinate that separates the points in its left subtree from those occurring in its right subtree.

Therefore, the range tree with fractional cascading supports 2-dimensional range queries in $O(\log n + z)$ time. For dimension $d > 2$, it takes time $O(\log^{d-1} n + z)$.

In order to answer $\mathtt{RMQ}(0, q)$ with activation efficiently, we will further enhance every y-array that occurs in the fractional cascading data structure with a priority queue as described in [18,8]. Each of these queues is (implicitly) constructed over the rank space of the points in the y-array (note that the y-arrays are sorted w.r.t. the y dimension). The rank space of the points in the y-array consists of points in the range $[0 \ldots m]$, where m is the size of the y-array. The priority queue supports the operations $insert(r)$, $delete(r)$, $predecessor(r)$

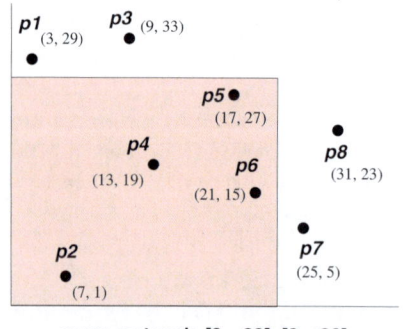

query rectangle [0 .. 22]x[0 .. 28]

Fig. 3. Query rectangle for the example of Fig. 2.

(gives the largest element $\leq r$), and $successor(r)$ (gives the smallest element $> r$) in time $O(\log \log m)$, where r is an integer in the range $[0 \ldots m]$. Algorithm 5 shows how to activate a point in the range tree and Algorithm 6 answers a RMQ.

Algorithm 5 *Activation of a point q in the data structure D*

> $v :=$ *root node of the x-tree*
> *find the rank (index) r of q in A_v by a binary search*
> **while** $(v \neq \perp)$
>> **if** $(A_v[r].score > A_v[predecessor(r)].score)$ **then**
>>> *insert(r) into the priority queue attached to A_v*
>>> **while**$(A_v[r].score > A_v[successor(r)].score)$
>>>> *delete($successor(r)$) from the priority queue attached to A_v*
>> **if** $(A_v[r] = A_{v.L}[A_v[r].Lptr])$ **then**
>>> $r := A_v[r].Lptr$
>>> $v := v.L$
>> **else**
>>> $r := A_v[r].Rptr$
>>> $v := v.R$

Note that in the outer while-loop of Algorithm 5, the following invariant is maintained: If $0 \leq i_1 < i_2 < \ldots < i_\ell \leq m$ are the entries in the priority queue attached to A_v, then $A_v[i_1].score < A_v[i_2].score < \ldots < A_v[i_\ell].score$.

Algorithm 6 *RMQ$(0, q)$ in the data structure D*

> $v :=$ *root node of the x-tree*
> $max\_score := -\infty$
> $max\_point := \perp$
> *find the rank (index) r of the rightmost point p with $p.y \in [0 \ldots q.y]$ in A_v*
> **while** $(v \neq \perp)$
>> **if** $(v.xmax \leq q.x)$ **then** \⋆ *v is a maximum splitting node* ⋆\
>>> $tmp := predecessor(r)$ *in the priority queue of A_v*
>>> $max\_score := \max\{max\_score, A_v[tmp].score\}$
>>> **if** $(max\_score = tmp.score)$ **then** $max\_point := A_v[tmp]$
>> **else if** $(v.xkey \leq q.x)$ **then** \⋆ *v.L is a maximum splitting node* ⋆\
>>> $tmp := predecessor(A_v[r].Lptr)$ *in the priority queue of $A_{v.L}$*
>>> $max\_score := \max\{max\_score, A_{v.L}[tmp].score\}$
>>> **if** $(max\_score = tmp.score)$ **then** $max\_point := A_{v.L}[tmp]$
>>> $r := A_v[r].Rptr$
>>> $v := v.R$
>> **else**
>>> $r := A_v[r].Lptr$
>>> $v := v.L$

In Algorithm 6, we assume that every node v has a field $v.xmax$ such that $v.xmax = \max\{p.x \mid p \in A_v\}$. Furthermore, $v.xkey$ is an x-coordinate (computed during the construction of D) that separates the points occurring in $A_{v.L}$

(or equivalently, in the leaf list of $v.L$) from those occurring in $A_{v.R}$ (or equivalently, in the leaf list of $v.R$). Algorithm 6 gives pseudocode for answering RMQ$(0, q)$, but we would also like to describe the algorithm on a higher level. In essence, Algorithm 6 locates all maximal splitting nodes $v_1, \ldots, v_\ell$ in D for the hyper-rectangle $R(0, q)$. For any v_j, $1 \leq j \leq \ell$, let the r_jth element be the splitting element in A_{v_j}. We have seen that $\uplus_{j=1}^{\ell} A_{v_j}$ contains every point $p \in S$ such that $p.x \in [0 \ldots q.x]$. Now if r_j is the index of the splitting element of A_{v_j}, then all points $A_{v_j}[i]$ with $i \leq r_j$ are in $R(0, q)$, whereas all other elements $A_{v_j}[i]$ with $i > r_j$ are not in $R(0, q)$. Since Algorithm 5 maintains the above-mentioned invariant, the element with highest score in the priority queue of A_{v_j} that lies in $R(0, q)$ is $q_j = predecessor(r_j)$ (if r_j is in the priority queue of A_{v_j}, then $q_j = r_j$ because $predecessor(r_j)$ gives the largest element $\leq r_j$). Algorithm 6 then computes $max\_score := \max\{A_{v_j}[q_j].score \mid 1 \leq j \leq \ell\}$ and returns $max\_point = A_{v_i}[q_i]$, where $A_{v_i}[q_i].score = max\_score$.

Because the number of maximum splitting nodes is $O(\log n)$ and any of the priority queue operations takes $O(\log \log n)$ time, answering a 2-dimensional RMQ takes $O(\log n \log \log n)$ time. The total complexity of activating n points is $O(n \log n \log \log n)$ because every point occurs in at most $\log n$ priority queues and hence there are at most $n \log n$ delete operations.

Theorem 7. *Given $k > 2$ genomes and n fragments, an optimal chain (without gap costs) can be found in $O(n \log^{k-2} n \log \log n)$ time and $O(n \log^{k-2} n)$ space.*

Proof. In Algorithm 4, the points are first sorted w.r.t. their first dimension and the RMQ with activation is required only for $d = k - 1$ dimensions. For $d \geq 2$ dimensions, the preceding data structure is implemented for the last two dimensions of the range tree, which yields a data structure D that requires $O(n \log^{d-1} n)$ space and $O(n \log^{d-1} n \log \log n)$ time for n RMQ and n activation operations. Consequently, one can find an optimal chain in $O(n \log^{k-2} n \log \log n)$ time and $O(n \log^{k-2} n)$ space.

In case $k = 2$, the data structure D is simply a priority queue over the rank space of all points. But the transformation to the rank space and the sorting procedure in Algorithm 4 require $O(n \log n)$ time, and thus dominate the overall time complexity of Algorithm 4. To sum up, Algorithm 4 takes $O(n \log n)$ time and $O(n)$ space for $k = 2$.

4   Incorporating Gap Costs

In the previous section, fragments were chained without penalizing the gaps in between them. In this section we modify the algorithm, so that it can take gap costs into account.

4.1   Gap Costs in the L_1 Metric

We first handle the case in which the cost for the gap between two fragments is the distance between the end and start point of the two fragments in the L_1

```
ACCXXXX____AGG                 ACCXXXXAGG
ACC____YYYAGG                  ACCYYY_AGG
```

Fig. 4. Alignments based on the fragments ACC and AGG w.r.t. gap cost g_1 (left) and g_∞ (right), where X and Y are anonymous characters.

metric. For two points $p, q \in \mathbb{R}^k$, this distance is defined by

$$d_1(p, q) = \sum_{i=1}^{k} |p.x_i - q.x_i|$$

and for two fragments $f \ll f'$ we define $g_1(f', f) = d_1(beg(f'), end(f))$. If an alignment of two sequences S_1 and S_2 shall be based on fragments and one uses this gap cost, then the characters between the two fragments are *deleted/inserted*; see left side of Fig. 4.

The problem with gap costs in our approach is that a RMQ does not take the cost $g(f', f)$ from recurrence (1) into account, and if we would explicitly compute $g(f', f)$ for every pair of fragments with $f \ll f'$, then this would yield a quadratic time algorithm. Thus, it is necessary to express the gap costs implicitly in terms of weight information attached to the points. We achieve this by using the *geometric cost* of a fragment f, which we define in terms of the terminus point t as $gc(f) = d_1(t, end(f))$.

Lemma 8. *Let f, $\tilde{f}$, and f' be fragments such that $f \ll f'$ and $\tilde{f} \ll f'$. Then we have $\tilde{f}.score - g_1(f', \tilde{f}) > f.score - g_1(f', f)$ if and only if the inequality $\tilde{f}.score - gc(\tilde{f}) > f.score - gc(f)$ holds.*

Proof.

$$\tilde{f}.score - g_1(f', \tilde{f}) > f.score - g_1(f', f)$$
$$\Leftrightarrow \tilde{f}.score - \sum_{i=1}^{k}(beg(f').x_i - end(\tilde{f}).x_i) > f.score$$
$$- \sum_{i=1}^{k}(beg(f').x_i - end(f).x_i)$$
$$\Leftrightarrow \tilde{f}.score - \sum_{i=1}^{k}(t.x_i - end(\tilde{f}).x_i) > f.score - \sum_{i=1}^{k}(t.x_i - end(f).x_i)$$
$$\Leftrightarrow \tilde{f}.score - gc(\tilde{f}) > f.score - gc(f)$$

The second equivalence follows from adding $\sum_{i=1}^{k} beg(f').x_i$ to and subtracting $\sum_{i=1}^{k} t.x_i$ from both sides of the inequality. Fig. 5 illustrates the lemma for $k = 2$.

Because t is fixed, the value $gc(f)$ is known in advance for every fragment f. Therefore, Algorithm 4 needs only two slight modifications to take gap costs into account. In order to apply Lemma 8, we set $0.score = -g_1(t, 0)$. Moreover, in Algorithm 4 we replace the statement $f'.score := f'.weight + f.score$ with

$$f'.score := f'.weight - gc(f') + f.score + gc(f) - g_1(f', f)$$

We subtract $gc(f')$ in view of Lemma 8. Furthermore, we have to add $gc(f)$ to compensate for the subtraction of this value when the score of fragment f was

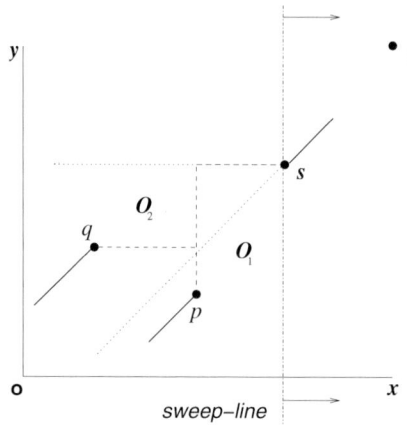

Fig. 5. Points p and q are active end points of the fragments f and $\tilde{f}$. The start point s of fragment f' is currently scanned by the sweeping line and t is the terminus point.

computed. This modified algorithm maintains the following invariant: If the end point of a fragment f is active, then $f.score$ stores the maximum score of all chains (with g_1 gap costs taken into account) that start at 0 and end at f minus the geometric cost of f.

4.2   The Sum-of-Pair Gap Cost

For clarity of presentation, we first treat the case $k = 2$ because the general case $k > 2$ is rather involved.

The case $k = 2$: For two points $p, q \in \mathbb{R}^2$, we write $\Delta_{x_i}(p, q) = |p.x_i - q.x_i|$, where $i \in \{1, 2\}$. We will sometimes simply write Δ_{x_1} and Δ_{x_2} if their arguments can be inferred from the context. The sum-of-pair distance of two points $p, q \in \mathbb{R}^2$ depends on the parameters ϵ and λ and was defined by Myers and Miller [15] as follows:

$$d(p, q) = \begin{cases} \epsilon \Delta_{x_2} + \lambda(\Delta_{x_1} - \Delta_{x_2}) \text{ if } \Delta_{x_1} \geq \Delta_{x_2} \\ \epsilon \Delta_{x_1} + \lambda(\Delta_{x_2} - \Delta_{x_1}) \text{ if } \Delta_{x_2} \geq \Delta_{x_1} \end{cases}$$

However, we rearrange these terms and derive the following equivalent definition:

$$d(p, q) = \begin{cases} \lambda \Delta_{x_1} + (\epsilon - \lambda)\Delta_{x_2} \text{ if } \Delta_{x_1} \geq \Delta_{x_2} \\ (\epsilon - \lambda)\Delta_{x_1} + \lambda \Delta_{x_2} \text{ if } \Delta_{x_2} \geq \Delta_{x_1} \end{cases}$$

For two fragments f and f' with $f \ll f'$, we define $g(f', f) = d(beg(f'), end(f))$. Intuitively, $\lambda > 0$ is the cost of aligning an anonymous character with a gap position in the other sequence, while $\epsilon > 0$ is the cost of aligning two anonymous characters. For $\lambda = 1$ and $\epsilon = 2$, this gap cost coincides with the g_1 gap cost, whereas for $\lambda = 1$ and $\epsilon = 1$, this gap cost corresponds to the L_∞ metric. (The gap cost of connecting two fragments $f \ll f'$ in the L_∞ metric is defined by $g_\infty(f', f) = d_\infty(beg(f'), end(f))$, where $d_\infty(p, q) = \max_{i \in [1..k]} |p.x_i - q.x_i|$ for

$p, q \in \mathbb{R}^k$.) Following [15,20], we demand that $\lambda > \frac{1}{2}\epsilon$ because otherwise it would always be best to connect fragments entirely by gaps as in the L_1 metric. So if an alignment of two sequences S_1 and S_2 shall be based on fragments and one uses the sum-of-pair gap cost with $\lambda > \frac{1}{2}\epsilon$, then the characters between the two fragments are *replaced* as long as possible and the remaining characters are *deleted* or *inserted*; see right side of Fig. 4.

In order to compute the score of a fragment f' with $beg(f') = s$, the following definitions are useful. The *first quadrant* of a point $s \in \mathbb{R}^2$ consists of all points $p \in \mathbb{R}^2$ with $p.x_1 \leq s.x_1$ and $p.x_2 \leq s.x_2$. We divide the first quadrant of s into regions O_1 and O_2 by the straight line $x_2 = x_1 + (s.x_2 - s.x_1)$. O_1, called the *first octant* of s, consists of all points p in the first quadrant of s satisfying $\Delta_{x_1} \geq \Delta_{x_2}$ (i.e., $s.x_1 - p.x_1 \geq s.x_2 - p.x_2$), these are the points lying below or on the straight line $x_2 = x_1 + (s.x_2 - s.x_1)$; see Fig. 5. The *second octant* O_2 consists of all points q satisfying $\Delta_{x_2} \geq \Delta_{x_1}$ (i.e., $s.x_2 - q.x_2 \geq s.x_1 - q.x_1$), these are the points lying above or on the straight line $x_2 = x_1 + (s.x_2 - s.x_1)$. Then $f'.score = f'.weight + \max\{v_1, v_2\}$, where $v_i = \max\{f.score - g(f', f) : f \ll f'$ and $end(f)$ lies in octant $O_i\}$, for $i \in \{1, 2\}$.

However, our chaining algorithms rely on RMQ, and these work only for orthogonal regions, not for octants. For this reason, we will make use of the *octant-to-quadrant* transformations of Guibas and Stolfi [7]. The transformation $T_1 : (x_1, x_2) \mapsto (x_1 - x_2, x_2)$ maps the first octant to a quadrant. More precisely, point p is in the first octant of point s if and only if $T_1(p)$ is in the first quadrant of $T_1(s)$.[3] Similarly, for the transformation $T_2 : (x_1, x_2) \mapsto (x_1, x_2 - x_1)$, point q is in the second octant of point s if and only if $T_2(q)$ is in the first quadrant of $T_2(s)$. By means of these transformations, we can apply the same techniques as in the previous sections. We just have to define the geometric cost properly. The following lemma shows how one has to choose the geometric cost gc_1 for points in the first octant O_1. An analogous lemma holds for points in the second octant.

Lemma 9. *Let f, $\tilde{f}$, and f' be fragments such that $f \ll f'$ and $\tilde{f} \ll f'$. If $end(f)$ and $end(\tilde{f})$ lie in the first octant of $beg(f')$, then $\tilde{f}.score - g(f', \tilde{f}) > f.score - g(f', f)$ if and only if $\tilde{f}.score - gc_1(\tilde{f}) > f.score - gc_1(f)$, where $gc_1(\tilde{f}) = \lambda\Delta_{x_1}(t, end(\tilde{f})) + (\epsilon - \lambda)\Delta_{x_2}(t, end(\tilde{f}))$ for any fragment $\tilde{f}$.*

Proof. Similar to the proof of Lemma 8.

In Section 4.1, we dealt with the geometric cost gc by modifying the field $f.score$. This is not possible here because we have to take two different geometric costs gc_1 and gc_2 into account. To cope with this problem, we need two data structures D_1 and D_2, where D_i stores the set of points

$$\{T_i(end(f).x_2, \ldots, end(f).x_k) \mid f \text{ is a fragment}\}$$

---

[3] Observe that the transformation may yield points with negative coordinates, but it is easy to overcome this obstacle by an additional transformation (a translation). Hence we will skip this minor problem.

If we encounter the end point of fragment f' in Algorithm 4, then we activate point $T_1(end(f').x_2, \ldots, end(f').x_k)$ in D_1 with priority $f'.score - gc_1(f')$ and point $T_2(end(f').x_2, \ldots, end(f').x_k)$ in D_2 with priority $f'.score - gc_2(f')$. If we encounter the start point of fragment f', then we launch two RMQ, namely RMQ$(0, T_1(beg(f').x_2, \ldots, beg(f').x_k))$ in the data structure D_1 and analogously RMQ$(0, T_2(beg(f').x_2, \ldots, beg(f').x_k))$ in D_2. If the first RMQ returns $T_1(end(f_1))$ and the second returns $T_2(end(f_2))$, then f_i is a fragment of highest priority in D_i such that $T_i(end(f_i).x_2, \ldots, end(f_i).x_k) < T_i(beg(f').x_2, \ldots, beg(f').x_k)$, where $1 \leq i \leq 2$. Because a point p is in the octant O_i of point $beg(f')$ if and only if $T_i(p)$ is in the first quadrant of $T_i(beg(f'))$, it follows that f_i is a fragment such that its priority $f_i.score - gc_i(f_i)$ is maximal in octant O_i. Therefore, according to Lemma 9, the value $v_i = f_i.score - g(f', f_i)$ is maximal in octant O_i. Hence, if $v_1 > v_2$, then we set $f'.prec = f_1$ and $f'.score := f'.weight + v_1$. Otherwise, we set $f'.prec = f_2$ and $f'.score := f'.weight + v_2$.

The case $k > 2$: In this case, the sum-of-pair gap cost is defined for fragments $f \ll f'$ by

$$g_{sop}(f', f) = \sum_{0 \leq i < j \leq k} g(f'_{i,j}, f_{i,j})$$

where $f'_{i,j}$ and $f_{i,j}$ are the two-dimensional fragments consisting of the ith and jth component of f' and f, respectively. For example, in case of $k = 3$, let $s = beg(f')$ and $p = end(f)$ and assume that $\Delta_{x_1}(s, p) \geq \Delta_{x_2}(s, p) \geq \Delta_{x_3}(s, p)$. In this case, we have $g_{sop}(f', f) = 2\lambda\Delta_{x_1} + \epsilon\Delta_{x_2} + (\epsilon - \lambda)2\Delta_{x_3}$ because $g(f'_{1,2}, f_{1,2}) = \lambda\Delta_{x_1} + (\epsilon - \lambda)\Delta_{x_2}$, $g(f'_{1,3}, f_{1,3}) = \lambda\Delta_{x_1} + (\epsilon - \lambda)\Delta_{x_3}$, and $g(f'_{2,3}, f_{2,3}) = \lambda\Delta_{x_2} + (\epsilon - \lambda)\Delta_{x_3}$. By contrast, if $\Delta_{x_1} \geq \Delta_{x_3} \geq \Delta_{x_2}$, then the equality $g_{sop}(f', f) = 2\lambda\Delta_{x_1} + (\epsilon - \lambda)2\Delta_{x_2} + \epsilon\Delta_{x_3}$ holds.

In general, each of the $k!$ permutations π of $1, \ldots, k$ yields a hyper-region R_π defined by $\Delta_{x_{\pi(1)}} \geq \Delta_{x_{\pi(2)}} \geq \ldots \geq \Delta_{x_{\pi(k)}}$ in which a specific formula for $g_{sop}(f', f)$ holds. That is, in order to obtain the score of a fragment f', we must compute $f'.score = f'.weight + \max\{v_\pi \mid \pi$ is a permutation of $1, \ldots, k\}$, where

$$v_\pi = \max\{f.score - g_{sop}(f', f) : f \ll f' \text{ and } end(f) \text{ lies in hyper-region } R_\pi\}$$

Because our RMQ-based approach requires orthogonal regions, each of these hyper-regions R_π of s must be transformed into the *first hyper-corner* of some point $\tilde{s}$. The first hyper-corner of a point $\tilde{s} \in \mathbb{R}^k$ is the k-dimensional analogue to the first quadrant of a point in $\mathbb{R}^2$. It consists of all points $p \in \mathbb{R}^k$ with $p.x_i \leq \tilde{s}.x_i$ for all $1 \leq i \leq k$ (note that there are 2^k hyper-corners). We describe the generalization of the *octant-to-quadrant* transformations for the case $k = 3$. The extension to the case $k > 3$ is obvious. There are 3! hyper-regions, hence 6 transformations:

$$\Delta_{x_1} \geq \Delta_{x_2} \geq \Delta_{x_3} : T_1(x_1, x_2, x_3) = (x_1 - x_2, x_2 - x_3, x_3)$$
$$\Delta_{x_1} \geq \Delta_{x_3} \geq \Delta_{x_2} : T_2(x_1, x_2, x_3) = (x_1 - x_3, x_2, x_3 - x_2)$$
$$\Delta_{x_2} \geq \Delta_{x_1} \geq \Delta_{x_3} : T_3(x_1, x_2, x_3) = (x_1 - x_3, x_2 - x_1, x_3)$$
$$\Delta_{x_2} \geq \Delta_{x_3} \geq \Delta_{x_1} : T_4(x_1, x_2, x_3) = (x_1, x_2 - x_3, x_3 - x_1)$$
$$\Delta_{x_3} \geq \Delta_{x_1} \geq \Delta_{x_2} : T_5(x_1, x_2, x_3) = (x_1 - x_2, x_2, x_3 - x_1)$$
$$\Delta_{x_3} \geq \Delta_{x_2} \geq \Delta_{x_1} : T_6(x_1, x_2, x_3) = (x_1, x_2 - x_1, x_3 - x_2)$$

In what follows, we will focus on the particular case where π is the identity permutation. The hyper-region corresponding to the identity permutation will be denoted by R_1 and its transformation by T_1. The other permutations are numbered in an arbitrary order and are handled similarly.

Lemma 10. *Point $p \in \mathbb{R}^k$ is in hyper-region R_1 of point s if and only if $T_1(p)$ is in the first hyper-corner of $T_1(s)$, where $T_1(x_1, x_2, \ldots, x_k) = (x_1 - x_2, x_2 - x_3, \ldots, x_{k-1} - x_k, x_k)$.*

Proof. $T_1(p)$ is in the first hyper-corner of $T_1(s)$

$$
\begin{aligned}
&\Leftrightarrow & T_1(s).x_i &\geq T_1(p).x_i & &\text{for all } 1 \leq i \leq k \\
&\Leftrightarrow & s.x_i - s.x_{i+1} &\geq p.x_i - p.x_{i+1} \text{ and } s.x_k \geq p.x_k & &\text{for all } 1 \leq i < k \\
&\Leftrightarrow & (s.x_1 - p.x_1) &\geq (s.x_2 - p.x_2) \geq \ldots \geq (s.x_k - p.x_k) \\
&\Leftrightarrow & \Delta_{x_1}(s, p) &\geq \Delta_{x_2}(s, p) \geq \ldots \geq \Delta_{x_k}(s, p)
\end{aligned}
$$

The last statement holds if and only if p is in hyper-region R_1 of s.

For each hyper-region R_j, we compute the corresponding geometric cost $gc_j(f)$ of every fragment f. Note that for every index j a k-dimensional analogue to Lemma 9 holds. Furthermore, for each transformation T_j, we keep a data structure D_j that stores the transformed end points $T_j(end(f))$ of all fragments f. Algorithm 11 generalizes the 2-dimensional chaining algorithm described above to k dimensions.

Algorithm 11 *k-dim. chaining of n fragments w.r.t. the sum-of-pair gap cost*

Sort all start and end points of the n fragments in ascending order w.r.t. their x_1 coordinate and store them in the array points; *because we include the end point of the origin and the start point of the terminus, there are $2n + 2$ points.*
for $j := 1$ **to** $k!$
    apply transformation T_j to the end points of the fragments and store the resulting points (ignoring their x_1 coordinate) as inactive in the $(k - 1)$-dimensional data structure D_j
for $i := 1$ **to** $2n + 2$
    if points$[i]$ *is the start point of fragment f'* **then**
        $maxRMQ := -\infty$
        for $j := 1$ **to** $k!$
            $q := $ RMQ$(0, T_j($points$[i].x_2, \ldots, $points$[i].x_k))$ *in D_j*
            determine the fragment f_q with $T_j(end(f_q)) = q$
            $maxRMQ := \max\{maxRMQ, f_q.score - g_{sop}(f', f_q)\}$
            if $f_q.score - g_{sop}(f', f_q) = maxRMQ$ **then** $f := f_q$
        $f'.prec := f$
        $f'.score := f'.weight + maxRMQ$
    else   * points$[i]$ *is end point of a fragment f'* *\
        for $j := 1$ **to** $k!$
            activate $T_j($points$[i].x_2, \ldots, $points$[i].x_k))$ in D_j with priority $f'.score - gc_j(f')$

For every start point $beg(f')$ of a fragment f', Algorithm 11 searches for a fragment f in the first hyper-corner of $beg(f')$ such that $f.score - g_{sop}(f', f)$ is maximal. This entails $k!$ RMQ because the first hyper-corner is divided into $k!$ hyper-regions. Analogously, for every end point $end(f')$ of a fragment f', Algorithm 11 performs $k!$ activation operations. Therefore, the total time complexity of Algorithm 11 is $O(k!\, n \log^{k-2} n \log \log n)$ and its space requirement is $O(k!\, n \log^{k-2} n)$. This result improves the running time of Myers and Miller's algorithm [15] by a factor $\frac{\log^2 n}{\log \log n}$ and the space requirement by one log-factor.

5   Conclusions

In this paper, we have presented a line-sweep algorithm that solves the fragment chaining problem of multiple genomes. For $k > 2$ genomes, our algorithm takes

- $O(n \log^{k-2} n \log \log n)$ time and $O(n \log^{k-2} n)$ space without gap costs,
- $O(n \log^{k-2} n \log \log n)$ time and $O(n \log^{k-2} n)$ space for gap costs in the L_1 metric,
- $O(k!\, n \log^{k-2} n \log \log n)$ time and $O(k!\, n \log^{k-2} n)$ space for the sum-of-pair gap cost.

For $k = 2$, it takes $O(n \log n)$ time and $O(n)$ space for any of the above-mentioned gap costs.

This solves the open problem of reducing the time complexity of Myers and Miller's [15] chaining algorithm. Specifically, our algorithm reduces the time complexity of their algorithm by a factor $O(\frac{\log^2}{\log \log n})$ and the space complexity by a log factor. Myers and Miller did not provide an implementation of their chaining algorithm, but we are currently implementing ours. To find the chaining algorithm that performs best in practice, we are planning to conduct experiments that compare the running times of various chaining algorithms, including the algorithms that are based on kd-trees.

It is worth-mentioning that the longest common subsequence (LCS) from fragments problem can also be solved within our framework. This generalizes the algorithm of [1] to more than two sequences.

References

1. B.S. Baker and R. Giancarlo. Longest common subsequence from fragments via sparse dynamic programming. In *Proc. 6th European Symposium on Algorithms*, LNCS 1461, pp. 79–90, 1998.
2. B. Chazelle. A functional approach to data structures and its use in multidimensional searching. *SIAM Journal on Computing*, 17(3):427–462, 1988.
3. A.L. Delcher, A. Phillippy, J. Carlton, and S.L. Salzberg. Fast algorithms for large-scale genome alignment and comparison. *Nucleic Acids Res.*, 30(11):2478–2483, 2002.
4. D. Eppstein. http://www.ics.uci.edu/~eppstein/pubs/p-sparsedp.html.

5. D. Eppstein, R. Giancarlo, Z. Galil, and G.F. Italiano. Sparse dynamic programming. I:Linear cost functions; II:Convex and concave cost functions. *Journal of the ACM*, 39:519–567, 1992.
6. M.S. Gelfand, A.A. Mironov, and P.A. Pevzner. Gene recognition via spliced sequence alignment. *Proc. Nat. Acad. Sci.*, 93:9061–9066, 1996.
7. L.J. Guibas and J. Stolfi. On computing all north-east nearest neighbors in the L_1 metric. *Information Processing Letters*, 17(4):219–223, 1983.
8. D.B. Johnson. A priority queue in which initialization and queue operations take $O(\log \log D)$ time. *Mathematical Systems Theory*, 15:295–309, 1982.
9. D. Joseph, J. Meidanis, and P. Tiwari. Determining DNA sequence similarity using maximum independent set algorithms for interval graphs. *Proc. 3rd Scandinavian Workshop on Algorithm Theory*, LNCS 621, pp. 326–337, 1992.
10. M. Höhl, S. Kurtz, and E. Ohlebusch. Efficient multiple genome alignment. *Bioinformatics*, 18:S312–S320, 2002.
11. M.-Y. Leung, B.E. Blaisdell, C. Burge, and S. Karlin. An efficient algorithm for identifying matches with errors in multiple long molecular sequences. *Journal of Molecular Biology*, 221:1367–1378, 1991.
12. W. Miller. Comparison of genomic DNA sequences: Solved and unsolved problems. *Bioinformatics*, 17:391–397, 2001.
13. B. Morgenstern. A space-efficient algorithm for aligning large genomic sequences. *Bioinformatics* 16:948–949, 2000.
14. E.W. Myers and X. Huang. An $O(n^2 \log n)$ restriction map comparison and search algorithm. *Bulletin of Mathematical Biology*, 54(4):599–618, 1992.
15. E.W. Myers and W. Miller. Chaining multiple-alignment fragments in sub-quadratic time. *Proc. 6th ACM-SIAM Symposium on Discrete Algorithms*, pp. 38–47, 1995.
16. F.P. Preparata and M.I. Shamos. *Computational geometry: An introduction.* Springer-Verlag, New York, 1985.
17. S. Schwartz, Z. Zhang, K.A. Frazer, A. Smit, C. Riemer, J. Bouck, R. Gibbs, R. Hardison, and W. Miller. PipMaker—A web server for aligning two genomic DNA sequences., *Genome Research*, 4(10):577–586, 2000.
18. P. van Emde Boas. Preserving order in a forest in less than logarithmic time and linear space. *Information Processing Letters*, 6(3):80–82, 1977.
19. D.E. Willard. New data structures for orthogonal range queries. *SIAM Journal of Computing*, 14:232–253, 1985.
20. Z. Zhang, B. Raghavachari, R. Hardison, and W. Miller. Chaining multiple-alignment blocks. *Journal of Computational Biology*, 1:51–64, 1994.

Two-Dimensional Pattern Matching with Rotations[*]

Amihood Amir[1,2][**], Ayelet Butman[1][***], Maxime Crochemore[3,4][†],
Gad M. Landau[5,6][‡], and Malka Schaps[1][§]

[1] Bar-Ilan University
[2] Georgia Tech
[3] University of Marne-La-Vallée
[4] King's College London
[5] University of Haifa
[6] Polytechnic University

Abstract. The problem of pattern matching with rotation is that of finding all occurrences of a two-dimensional pattern in a text, in all possible rotations. We prove an upper and lower bound on the number of such different possible rotated patterns. Subsequently, given an $m \times m$ array (pattern) and an $n \times n$ array (text) over some finite alphabet Σ, we present a new method yielding an $O(n^2 m^3)$ time algorithm for this problem.

Keywords: Design and analysis of algorithms, two-dimensional pattern matching, rotation.

[*] Part of this research was conducted while the first and fourth authors were visiting the University of Marne-La-Vallée supported by Arc-en-Ciel/Keshet, French-Israeli Scientific and Technical Cooperation Program.

[**] Department of Computer Science, Bar-Ilan University, 52900 Ramat-Gan, Israel, (972-3)531-8770, `amir@cs.biu.ac.il`. Partially supported by NSF grant CCR-01-04494, ISF grant 282/01, and an Israel-France exchange scientist grant funded by the Israel Ministry of Science.

[***] Department of Computer Science, Bar-Ilan University, 52900 Ramat-Gan, Israel, (972-3)531-8408, `ayelet@cs.biu.ac.il`.

[†] Institut Gaspard-Monge, University of Marne-La-Vallée, and King's College London; partially supported by CNRS, NATO Science Programme grant PST.CLG.977017, and by Arc-en-Ciel/Keshet, French-Israeli Scientific and Technical Cooperation Program.

[‡] Department of Computer Science, Haifa University, Haifa 31905, Israel, phone: (972-4) 824-0103, FAX: (972-4) 824-9331; Department of Computer and Information Science, Polytechnic University, Six MetroTech Center, Brooklyn, NY 11201-3840; email: `landau@poly.edu`; partially supported by NSF grants CCR-9610238 and CCR-0104307, by NATO Science Programme grant PST.CLG.977017, by the Israel Science Foundation grants 173/98 and 282/01, by the FIRST Foundation of the Israel Academy of Science and Humanities, by IBM Faculty Partnership Award, and by Arc-en-Ciel/Keshet, French-Israeli Scientific and Technical Cooperation Program.

[§] Department of Mathematics, Bar-Ilan University, Ramat-Gan, Israel, (972-3)531-8408, `mschaps@macs.biu.ac.il`.

R. Baeza-Yates et al. (Eds.): CPM 2003, LNCS 2676, pp. 17–31, 2003.
© Springer-Verlag Berlin Heidelberg 2003

1   Introduction

One of the main motivation for research in two-dimensional pattern matching is the problem of searching aerial photographs. The problem is a basic one in computer vision, but it was felt that pattern matching can not be of any use in its solution. Such feelings were based on the belief that pattern matching algorithms are only good for *exact matching* whereas in reality one seldom expects to find an exact match of the pattern. Rather, it is interesting to find all text locations that "approximately" match the pattern. The types of differences that make up these "approximations" are:

1. *Local Errors* – introduced by differences in the digitization process, noise, and occlusion (the pattern partly obscured by another object).
2. *Scale* – size difference between the image in the pattern and the text.
3. *Rotation* – The pattern image appearing in the text in a different angle.

Several of these approximation types have been handled in the past [2,4,5,7, 14,17].

The problem of pattern matching with rotation is that of finding all occurrences of a two-dimensional pattern in a text, in all possible rotations. There was no known efficient solution for this problem, even though many researchers were thinking about it for over a decade. Part of the difficulty lay in the lack of a rigorous definition to capture the concept of rotation for a discrete pattern.

The first breakthrough came quite recently. Fredriksson and Ukkonen [11] gave an excellent combinatorial definition of rotation. They resorted to a geometric interpretation of text and pattern and provided the following definition.

Let P be a two-dimensional $m \times m$ array and T be a two-dimensional $n \times n$ array over some finite alphabet Σ. The array of *unit pixels* for T consists of n^2 unit squares, called *pixels* in the real plane R^2. The corners of the pixel $T[i, j]$ are $(i - 1, j - 1), (i, j - 1), (i - 1, j)$, and (i, j). Hence the pixels of T form a regular $n \times n$ array covering the area between $(0, 0), (n, 0), (0, n)$, and (n, n). The *center* of each pixel is the geometric center point of its square location. Each pixel $T[i, j]$ is identified with the value from Σ that the original text had in that position. We say that the pixel has a *color* from Σ. See Figure 1 for an example of the grid and pixel centers of a 7×7 text.

The array of pixels for pattern P is defined similarly. A different treatment is necessary for patterns with odd sizes and for patterns with even sizes. For simplicity's sake we assume throughout the rest of this paper that the pattern is of size $m \times m$ and m is even. The *rotation pivot* of the pattern is its exact center, the point $(\frac{m}{2}, \frac{m}{2}) \in R^2$. See Figure 2 for an example of the rotation pivot of a 4×4 pattern P.

Consider now a rigid motion (translation and rotation) that moves P on top of T. Consider the special case where the translation moves the grid of P precisely on top of the grid of T, such that the grid lines coincide.

Assume that the rotation pivot of P is at location (i, j) on the text grid. The pattern is now rotated, centered at (i, j), creating an angle α between the x-axes of T and P. P is said to be at *location* $((i, j), \alpha)$ *over* T. Pattern P is said to

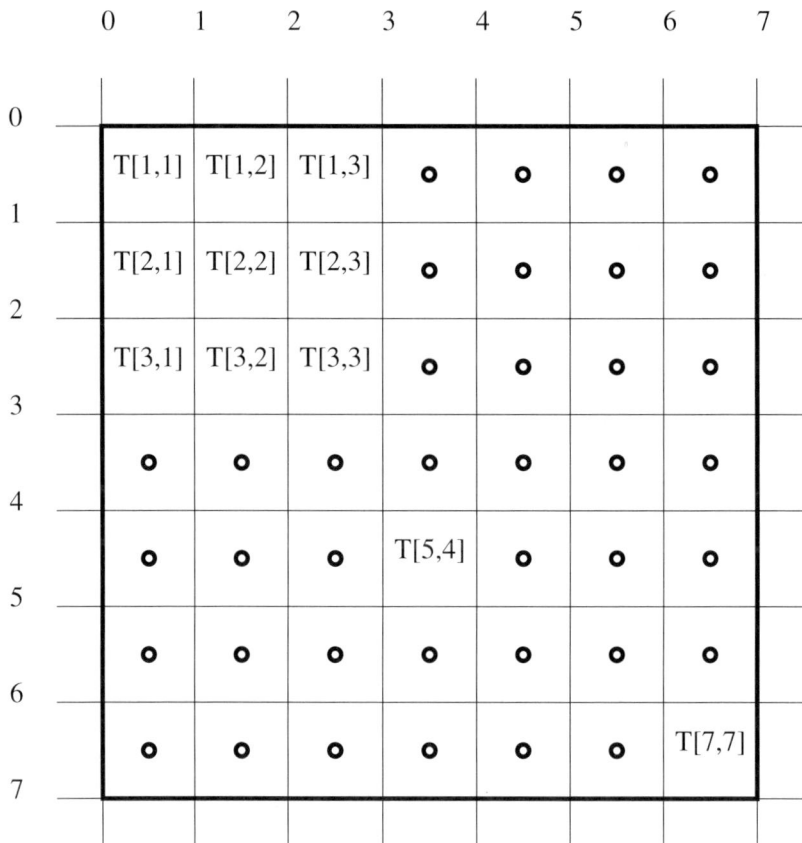

Fig. 1. The text grid and pixel centers of a 7 × 7 text.

have an *occurrence* at location $((i, j), \alpha)$ if the center of each pixel in P has the same color as the pixel of T under it. When the center of a pattern pixel is over a vertical (horizontal) border between text pixels, the color of the text pixel left (below) to the border is chosen. See Figure 3 for an example of 4 × 4 pattern P at location $((3, 4), 45°)$ on top of 8 × 8 text T.

Fredriksson and Ukkonen [11] give a rotation invariant filter for two-dimensional matching. Their algorithm performs well in reality. Combinatorially, though, their algorithm has a worst case time complexity of $O(n^2 m^5)$.

In a subsequent paper, Fredriksson, Navarro and Ukkonen [9] discuss an index for two-dimensional matching involving rotation. They give an indexing scheme for the rotated search problem. Their scheme allows a rotated search in expected time $O(n^2 m^3)$, but the worst case is $O(n^2 m^5)$.

Further improvements were done recently by the same authors [10]. They give fast filtering algorithms for seeking a 2-dimensional pattern in a 2-dimensional text allowing any rotation of the pattern. They consider the cases of exact and

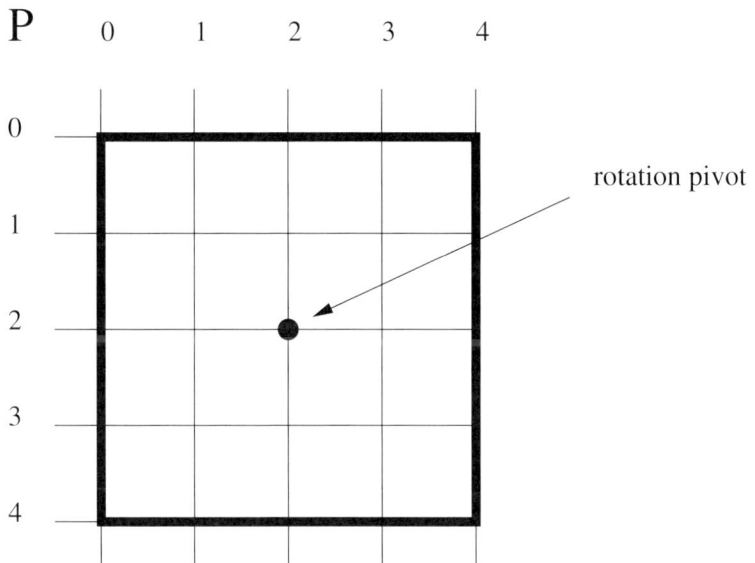

Fig. 2. The rotation pivot of a 4×4 pattern P.

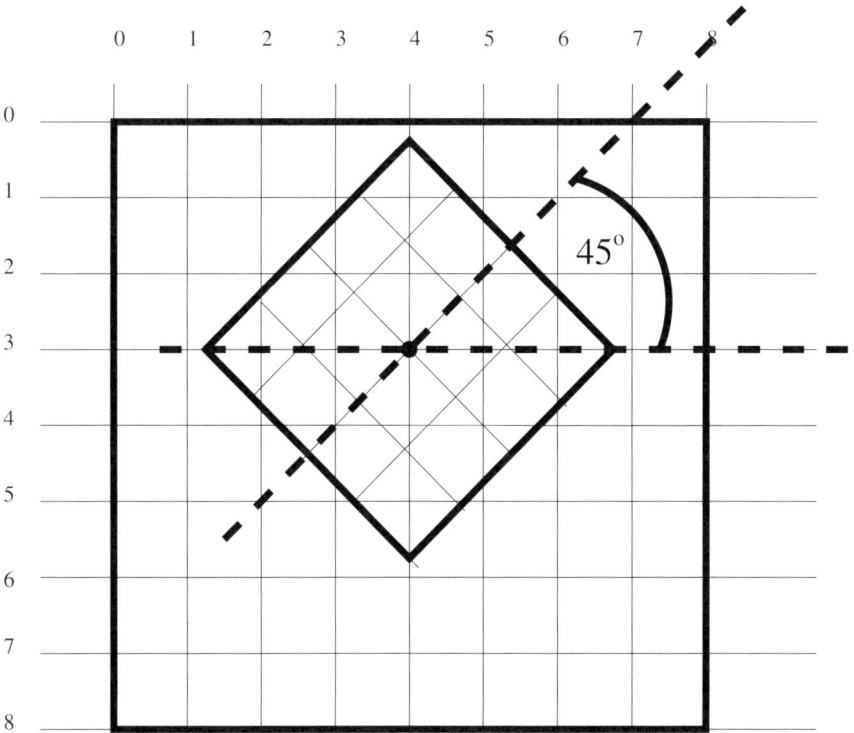

Fig. 3. 4×4 pattern P at location $((3, 4), 45^o)$ over 8×8 text T.

approximate matching, improving the previous results. The time complexity of their exact matching algorithm is $O((n^2m^3))$.

Fredriksson, Navarro and Ukkonen [9] give two possible definitions for rotation. One is as described above and the second is, in some way, the opposite. P is placed *under* the text T. More precisely, assume that the rotation pivot of P is under location (i, j) on the text grid. The pattern is now rotated, centered at (i, j), creating an angle α between the x-axes of T and P. P is said to be at *location* $((i, j), \alpha)$ *under* T. Pattern P is said to have an *occurrence* at location $((i, j), \alpha)$ if the center of each pixel in T has the same color as the pixel of P under it.

While the two definitions of rotation, "over" and "under", seem to be quite similar, they are not identical. For example, in the "pattern over text" model there exist angles for which two pattern pixel centers may find themselves in the same text pixel, or, alternately, there are angles where a text pixel does not have in it a center of a pattern pixel, but all text pixels around it have centers of pattern pixels. Figure 4 shows an example where text location $[3, 5]$ does not have any pattern pixel center in it, but $[2, 5]$ and $[3, 6]$ have pattern pixel centers. In the "pattern under text" model the center of the text pixel is the one to decide the color of the pixel, then all contiguous text pixels that have pattern area under them define a rotated pattern, thus it is impossible to have a rotated pattern with "don't cares" surrounded by symbols.

Although all our results apply to both definitions of rotation, to avoid confusion our paper will henceforth deal only with the rotation of pattern over the text.

This paper considers the worst-case behavior of the rotation problem. We analyze the simple strategy of searching for all different patterns that represent possible rotations, that had been hitherto the basis of practically all rotated searching algorithms. The main contribution of this paper is proving an upper and lower bound on the number of different possible rotated patterns. Subsequently, by using appropriate data structures, we present two deterministic algorithms whose worst-case complexity is $O(n^2m^3)$. The first algorithm is similar to the one derived from [12]. The second algorithm achieves the same upper bounds using another method. We present both algorithms since we believe that the problem can be solved more efficiently. Some of the methods we present may prove fruitful in further improving the solution for this problem.

The rotation problem is an interesting one since it brings together two research areas that deal with similar problems by using different means – pattern matching and computational geometry. Recent geometry papers have made use of pattern matching techniques to solve geometric problems see e.g. [18,16]). We believe this is the first pattern matching paper that makes heavy use of geometry to solve pattern matching problems. A related problem considering similarity between images is considered in [6].

This paper is organized as follows. First we present the dictionary matching solution. In Section 3 we give a tight bound on the number of different patterns that represent all possible rotations of a given pattern. This leads to a simple

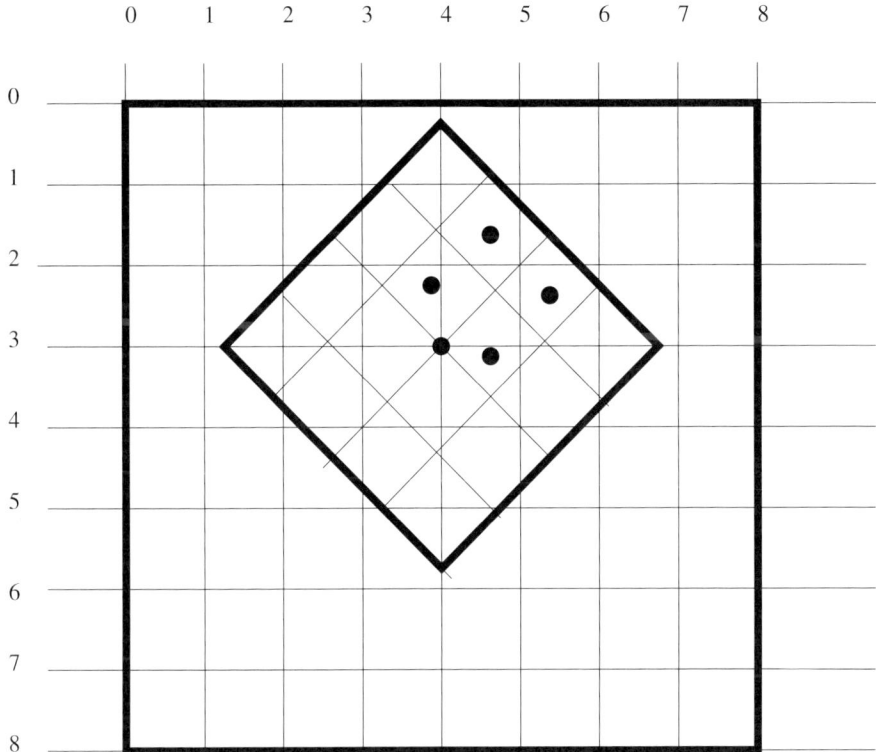

Fig. 4. Text pixel $T[3,5]$ has no pattern pixel over it.

$O(n^2 m^3 \log m)$ algorithm for the rotated matching problem. In the following section we present two different ways of improving the time to $O(n^2 m^3)$. Finally, we conclude with ideas for further improvements.

2   The Dictionary-Matching Solution

An immediate natural idea for solving the rotation problem is to use *dictionary matching* solutions. That is, construct a dictionary of different 2-dimensional arrays that represent all possible rotations of the pattern, one array for each rotation (see Figure 5). Then search for them in the text.

We slightly modify the definition given in [11] for *rotated occurrences of P in T*.

Definition 1. *Let P be an $m \times m$ pattern with m even. Place P on the real plane R^2 in the following manner:*

1. *The rotation pivot coincides with the origin $(0,0)$ in R^2.*
2. *Every pattern element corresponds to a 2×2 square.*

1	2	3	4
5	6	7	8
9	10	11	12
13	14	15	16

5a

	1,2	3	
5	6	7	4,8
13,9	10	11	12
	14	15,16	

5b

5	1	2	3
9	6	7	4
13	10	11	8
14	15	16	12

5c

		1		
5		2	3	
	9	6	7	4
13		10	11	8
	14	15		12
		16		

5d

5	2	3	3
5	6	7	8
9	10	11	12
14	14	15	12

5e

	1	2	
9	6	7	4
13	10	11	8
	15	16	

5f

Fig. 5. An example of some possible 2-dimensional arrays that represent one pattern. Fig 5a – the original pattern. Figures 5b-d are computed in the "pattern over the text" model. Fig 5b – a representation of the pattern rotated by 19^0 (the color of a cell with two points of the pattern is the color of the leftmost point). Fig 5c – Pattern rotated by 21^0. Fig 5d – Pattern rotated by 26^0. Figures 5e-f are computed in the "pattern under the text" model. Fig 5e – Pattern rotated by 17^0. Fig 5f – Pattern rotated by 26^0.

The segments $x = 2i$, $i = -\frac{m}{2}, ..., 0, ..., \frac{m}{2}$, $-m \le x \le m$ and $y = 2j$, $j = -\frac{m}{2}, ..., 0, ..., \frac{m}{2}$, $-m \le y \le m$ are the pattern grid.

The 2×2 squares whose corners, clockwise from the top left, namely $(2i, 2j), (2i, 2(j+1)), (2(i+1), 2(j+1))$ and $(2(i+1), 2j)$, $i, j = -\frac{m}{2}, ..., 0, ..., \frac{m}{2} - 1$, are the pattern grid cells.

The points $(2i + 1, 2j + 1)$, $i, j = -\frac{m}{2}, ..., 0, ..., \frac{m}{2} - 1$ are the pattern pixel centers.

The color of pattern pixel centered at $(2i+1, 2j+1)$ is $P[i + \frac{m}{2} + 1, j + \frac{m}{2} + 1]$.

Example: The pattern in Figure 6 is of size 8×8. The pattern grid is the solid (even valued) lines. The pattern pixel centers are the intersection of the odd lines.

Definition 2. *Let P^c be the lattice whose points are the pattern pixel centers. The color of lattice point p is the color of the pattern pixel center that coincides with p. Let P^c_α be the lattice P^c rotated around the origin by angle α. Color an R^2 grid cell by the color of the lattice point in that grid cell, if only one such point exists. If there is a R^2 grid cell with two lattice points of P^c_α with different colors in it, then the color of the grid cell is the color of the leftmost lattice point in that grid cell. If there are no such lattice points then color the grid cell by the don't care or wildcard symbol ϕ that matches every character in the alphabet.*

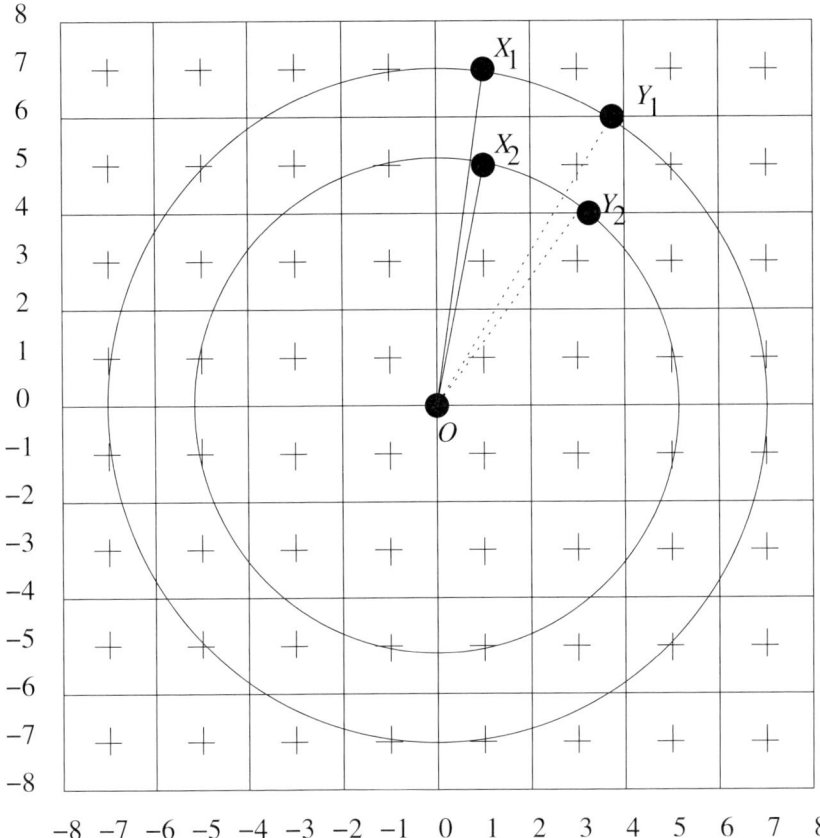

Fig. 6. Points X_1 and X_2 each have coprime integer coordinates. Orbits of them under rotation around point O cross an horizontal line at points Y_1 and Y_2 respectively. Then, $\measuredangle X_1 O X_2 \neq \measuredangle Y_1 O Y_2$ by Claim 3.

Consider the smallest square in R^2 centered at $(0,0)$ that includes all non-ϕ colored pixels, and construct a matrix of the pixel colors of that square. This 2-dimensional array represents the pattern P rotated by angle α (see Figure 5).

While there exist efficient two-dimensional dictionary-matching algorithms (e.g. [3,13]), none of them works with don't cares. The only known algorithm for efficiently solving string matching with don't cares is the Fischer-Paterson algorithm [8]. The Fischer-Paterson algorithm finds all occurrences of a string of length m in a string of length n, both possibly having occurrences of the don't care symbol, in time $O(n \log m)$ for finite alphabet. Unfortunately, this method does not generalize to dictionary matching.

Therefore, an immediate suggestion for a rotation algorithm is the following.
Preprocessing: Construct a data base of all the different 2-dimensional arrays $(P_i, \ i = 1, ..., k)$ that represent all possible pattern rotations.

Text scanning: Let T be the input text (an $n \times n$ array).
For every P_i in the data base do:
   Find all occurrences of P_i in T.
Time: The algorithm's running time is $O(kn^2 \log m)$. In the next section we
prove that the number of different pattern rotations k is $\Theta(m^3)$. Thus the running
time of the above algorithm becomes $O(n^2 m^3 \log m)$.

3   The Number of Different Rotations

Let *pattern* $P = P[1..m, 1..m]$ be a two-dimensional $m \times m$ array and let *text*
$T = T[1..n, 1..n]$ be an $n \times n$ array over some finite alphabet Σ. We assume that
$m < n$.

   We want to prove that the upper and lower bounds on the number of legal
pattern rotations is $\Theta(m^3)$.

   It is clear that the upper bound is $O(m^3)$ since we have m^2 different points
in the pattern, each creates no more than m different angles (whenever its orbit
crosses an odd coordinate).

   It suffices to prove that the order of the different rotations is $\Omega(m^3)$, and
that will establish a tight bound of $\Theta(m^3)$.

   We will restrict ourselves to points (x, y) in the first quadrant $(x, y \geq 0)$
whose coordinates x and y are *relatively prime*, or *co-prime*, i.e., the greatest
common divisor of x and y is 1. We prove that the number of different rotations
of just the first quadrant points with coprime coordinates is $\Omega(m^3)$.

Theorem 1. *There are $\Omega(m^3)$ different rotations of an $m \times m$ pattern.*

Proof: We will show that, for every two different first quadrant points X_1
and X_2 having coprime coordinates, it is impossible that they cross a grid
line at the same angle. The cardinality of the set $\{(n_0, m_0) \mid 1 \leq n_0, m_0 \leq$
m, and n_0, m_0 are coprime$\}$ is: $\frac{6m^2}{\pi^2} + o(m \log m)$. This is a direct corollary of
[theorem 330, [15]]. There are $\Theta(m^2)$ such points. Lemma 1 shows that $\Omega(m^2)$
of these co-prime points cross the grid $\Omega(m)$ times. Therefore, there are $\Omega(m^3)$
different rotations.

Lemma 1. *There are $\Theta(m^2)$ first quadrant points with co-prime coordinates
each of which crosses $\Omega(m)$ gridpoints.*

Proof: Let $X = (n_0, m_0)$, where n_0, m_0 are co-prime. Denote by d the number
of even vertical and horizontal grid lines intersected by a circle whose center is
the origin and whose radius is $\sqrt{n_0^2 + m_0^2}$.
Claim: $d \geq 4 \max\{|n_0|, |m_0|\}$.

   Recall that X is a first quadrant point, thus we may assume that $n_0, m_0 \geq 0$.
On the circle whose center is the origin and on whose circumference lies the
point $X = (n_0, m_0)$, also lie the points $(-n_0, m_0), (n_0, -m_0)$, and $(-n_0, -m_0)$.

   Because of symmetry, the number of horizontal grid lines intersected by the
circle is equal to the number of vertical grid lines, therefore it is sufficient to

prove that the number of intersecting horizontal grid lines is no greater than $2\max\{|n_0|, |m_0|\}$.

Without loss of generality, assume that $m_0 \geq n_0$. The arc that begins at (n_0, m_0) and ends at $(n_0, -m_0)$ on our circle, intersects the grid lines whose y-axis values lie between $-m_0$ and m_0 (see Figure 7). Since we are interested only in the horizontal lines with an even y-axis, we get that the horizontal grid lines intersected by the circle have values: $-m_0+1, ..., -4, -2, 0, 2, 4, ..., m_0-1$. Thus the total number of intersecting even horizontal grid lines is $2((m_0-1)/2)+1 = 2m_0$.

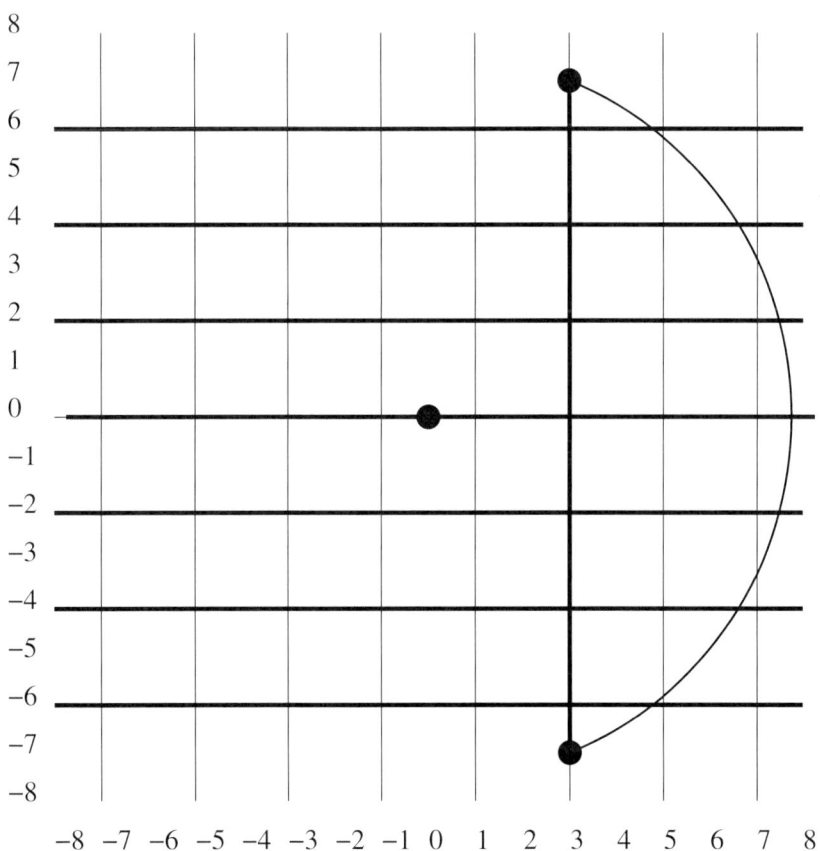

Fig. 7. The arc that begins at $(7,3)$ and ends at $(7,-3)$, intersects the grid lines whose y-axis values lie between 7 and -7.

Consider now the $6m^2/\pi^2$ points with co-prime coordinates in the first quadrant. No more than $m^2/4$ of them have both x and y coordinates smaller than $m/2$. Thus, at least

$$\frac{6m^2}{\pi^2} - \frac{m^2}{4} = (\frac{24 - \pi^2}{4\pi^2})m^2 = \Theta(m^2)$$

lie on circles that intersect at least $2m^2$ even grid lines.   □

Assume, then, that X_1 and X_2 each have coprime integer coordinates. Assume also that if $X_1 = (c, s)$ then $X_2 \neq (s, c)$. This second assumption reduces the number of pairs under consideration by a half, but it is still $\Theta(m^2)$. Rotation moves X_1 to Y_1 and X_2 to Y_2. Assume that both Y_1 and Y_2 just crossed a horizontal text grid line. This means that both Y_1, Y_2 have coordinates of which one is an even integer and one is of the form $\sqrt{2n}$, where n is odd. The reason for this is the following.

Let $X_1 = (c, s)$, and $Y_1 = (c', s')$. Note that c, s are odd integers and therefore can be denoted by $c = 2k_1 + 1, s = 2k_2 + 1$. s' is even so it can be denoted by $s' = 2l_1$. Using the Pythagorean Theorem we get $(c')^2 = (2k_1 + 1)^2 + (2k_2 + 1)^2 - 4l_1^2 = 4k_1^2 + 4k_1 + 1 + 4k_2^2 + 4k_2 + 1 - 4l_1^2 = 4(k_1^2 + k_1 + k_2^2 + k_2 - l_1) + 2$. Therefore, $(c')^2$ is even. It can be denoted by $(c')^2 = 2l_2$. Substituting $(c')^2$ with $2l_2$ we get: $4(k_1^2 + k_1 + k_2^2 + k_2 - l_1) + 2 = 2l_2$. Therefore, $2(k_1^2 + k_1 + k_2^2 + k_2 - l_1) + 1 = l_2$, consequently, l_2 is odd.

Claim. $\angle X_1 O Y_1 \neq \angle X_2 O Y_2$.

Proof: It suffices to show that $\angle X_1 O X_2 \neq \angle Y_1 O Y_2$. Consider the three possibilities:

- case 1: Both Y_1, Y_2 have a horizontal coordinate as an even integer coordinate.
  Denote,

$$X_1 = (c_1, s_1)$$

$$X_2 = (c_2, s_2)$$

$$Y_1 = (c'_1, s'_1)$$

$$Y_2 = (c'_2, s'_2)$$

then, by the formula $\sin(\alpha - \beta) = \sin\alpha\cos\beta - \cos\alpha\sin\beta$, we get

$$\sin \angle X_1 O X_2 = \frac{s_1 c_2 - s_2 c_1}{r_1 r_2}$$

$$\sin \angle Y_1 O Y_2 = \frac{s'_1 c'_2 - s'_2 c'_1}{r_1 r_2}$$

where, $r_1 = \sqrt{c_1^2 + s_1^2}$, $r_2 = \sqrt{c_2^2 + s_2^2}$.
Assume $\sin \angle X_1 O X_2 = \sin \angle Y_1 O Y_2$
then, $s_1 c_2 - s_2 c_1 = s'_1 c'_2 - s'_2 c'_1$ where, $s'_1, s'_2 \in 2\mathbb{Z}$,
$c'_1 = l_1\sqrt{2m_1}$, $c'_2 = l_2\sqrt{2m_2}$ s.t. m_1, m_2 are odd and square free (i.e. they do not have a square factor).
$d' = s'_1 c'_2 - s'_2 c'_1 \in \mathbb{Z}[\sqrt{2m_1}, \sqrt{2m_2}]$
Now, either $m_1 = m_2$ or not. In both cases we will reach a contradiction.

- If $m_1 \neq m_2$ then $d' \notin \mathbb{Z}$ since, $\sqrt{2m_1}$, $\sqrt{2m_2}$ have non zero coefficients and are linearly independent. This contradicts the fact that $d' \in \mathbb{Z}$ since $d' = s_1 c_2 - s_2 c_1$.
- If $m_1 = m_2$ then $d' \in \mathbb{Z}$ iff $d' = 0$. In that case either X_1 and X_2 are the same point, or $c_1 = s_2$ and $s_1 = c_2$, or one of them does not have coprime coordinates. Both latter cases contradict the assumption.
  - case 2: Both Y_1, Y_2 have an even vertical coordinate: exchange s'_i with c'_i.
  - case 3: One has an even vertical coordinate and one has an even horizontal coordinate: replace cos with sin.

This ends the proof. □

4   Getting Rid of the log Factor

We provide a new method, called "Invalid-interval" for solving the problem of pattern matching with rotation in time $O(n^2 m^3)$. This improves on the immediate $O(n^2 m^3 \log m)$ solution of Section 2. The aim is to present a new methodological approach, because the resulting complexity is the same as the method of the next section that can be derived easily from [12]. For the sake of completeness the latter is presented first.

4.1   Invalid-Intervals Method

Definition: The *relevant text pixels for P at location* $((i,j), \alpha)$ are all text pixels that are under at least one pattern pixel center. The *relevant text pixels for P at text location* (i,j) are all relevant text pixels for P at location $((i,j), \alpha)$, for all $\alpha \in [0, 2\pi)$.

Let P be centered at some fixed text location (i,j), and let $\langle (x_P, y_P), (x_T, y_T) \rangle$ be a pair where (x_P, y_P) is a pattern location and (x_T, y_T) is a relevant text pixel of P at location (i,j). The *angle interval of pair* $\langle (x_P, y_P), (x_T, y_T) \rangle$ is the interval $[\alpha, \beta], 0 \leq \alpha < \beta \leq 2\pi$, where α is the smallest angle for which the center of pattern pixel (x_P, y_P) is on text pixel (x_T, y_T) at location $((i,j), \alpha)$, and β is the largest angle for which the center of pattern pixel (x_P, y_P) is on text pixel (x_T, y_T) at location $((i,j), \beta)$.

Claim. Let P be centered at some fixed text location (i,j). There are $O(m^3)$ pairs $\langle (x_P, y_P), (x_T, y_T) \rangle$, where (x_P, y_P) is a pattern location and (x_T, y_T) is a relevant text pixel of P at location (i,j), that have non-empty angle intervals.

Proof: For a fixed text location there are $O(m^2)$ relevant text pixels. Therefore the total number of pairs is $O(m^4)$. However, some pattern elements cannot match some text pixels in a rotation centered at fixed text location (i,j). During a complete 2π radian rotation, the center of every pattern pixel moves through at most $O(m)$ text pixels. Since there are m^2 different pattern pixels, this means that the number of pairs that generate non-empty angle intervals is $O(m^3)$. □

The preprocessing in the *invalid-intervals* method is also structural. Fix a text location, as the center of rotations. Compute the angle interval for each one of the $O(m^3)$ pairs. In the text scanning phase, for every text location as center, we are guaranteed that every interval where the pattern symbol does not equal the text symbol in its pair, cannot have a match. Any angle not covered by an invalid interval is an angle where there is a rotated match.

Our algorithm's pattern preprocessing is then:

Preprocessing

    For every pattern pixel center do:
        Generate the formula of the circle created by that center during a
        2π rotation.
        For every text pixel intersected by the circle, mark the angle interval.
    endFor

end Preprocessing

Implementation Remarks: The angle precision is given by the problem definition. The preprocessing can accommodate any precision that is at least $\log m$ bits since the exact angle is not necessary for the preprocessing. The only necessary information is differentiating between different angles. For this, a precision of $\log m$ bits is sufficient since there are $O(m^3)$ different angles. In fact, we designate the angles by integers in $\{1, ..., 2m^3\}$ (with an attached conversion table of appropriate precision).

Time for Preprocessing: The preprocessing running time can be made $O(m^3)$.

The text scanning involves ascertaining which are the invalid angle intervals and computing the union of these intervals. Any remaining angle is a rotated match. The invalid intervals are declared via a brute-force check. The union is computed by sorting and merging the intervals.

Text Scanning

    For every text pixel do:
        For every pair $\langle (x_P, y_P), (x_T, y_T) \rangle$ do:
            If $P[x_P, y_P] \neq T[x_T, y_T]$ then mark the interval $[\alpha, \beta]$ of pair
            $\langle (x_P, y_P), (x_T, y_T) \rangle$ as **invalid**.
        endFor
        Bucket Sort the invalid intervals by starting angle.
        Merge overlapping intervals into a single larger interval.
        If the interval $[0, 2\pi]$ is achieved, then there is no rotated match
        at text pixel. Otherwise, all angles outside the merged intervals
        are rotated matches.
    endFor

end Text Scanning

Implementation Remarks: Recall that our angles are denoted by numbers in $\{1, ..., 2m^3\}$, thus bucket sort is appropriate in the above algorithm.

Time for Text Scanning: For each of the n^2 text pixels the algorithm makes $O(m^3)$ comparisons. Bucket sorting the resulting $O(m^3)$ invalid intervals is done in time $O(m^3)$, and merging is done in linear time, for a total of $O(m^3)$ per text pixel. The total algorithm time is, therefore, $O(n^2m^3)$.

5 Future Work

We proved an upper and lower bound on the number of such different possible rotated patterns, and we have presented a new strategy to search for rotated two-dimensional patterns in arrays. The time complexity of our algorithms is not entirely satisfactory and seems to leave room for further improvements. Our opinion is based on the fact that each position in the text array is examined independently of other positions. We have been unable to take into account the information collected at one position to accelerate the test at next positions, as it is classical in most pattern matching methods. However, we believe that this is possible.

References

1. K. Abrahamson. Generalized string matching. *SIAM J. Comp.*, 16(6):1039–1051, 1987.
2. A. Amir, A. Butman, and M. Lewenstein. Real scaled matching. *Information Processing Letters*, 70(4):185–190, 1999.
3. A. Amir and M. Farach. Two dimensional dictionary matching. *Information Processing Letters*, 44:233–239, 1992.
4. A. Amir and G. Landau. Fast parallel and serial multidimensional approximate array matching. *Theoretical Computer Science*, 81:97–115, 1991.
5. A. Apostolico and Z. Galil (editors). *Pattern Matching Algorithms*. Oxford University Press, 1997.
6. R. Baeza-Yates and G. Valiente. An image similarity measure based on graph matching. In *Proc. of the 7th Symposium on String Processing and Information Retrieval (SPIRE'2000)*, pages 28–38. I.E.E.E. CS Press, 2000.
7. M. Crochemore and W. Rytter. *Text Algorithms*. Oxford University Press, 1994.
8. M.J. Fischer and M.S. Paterson. String matching and other products. *Complexity of Computation, R.M. Karp (editor), SIAM-AMS Proceedings*, 7:113–125, 1974.
9. K. Fredriksson, G. Navarro, and E. Ukkonen. An index for two dimensional string matching allowing rotations. In *Proc. IFIP International Conference on Theoretical Computer Science (IFIP TCS)*, volume 1872 of *LNCS*, pages 59–75. Springer, 2000.
10. K. Fredriksson, G. Navarro, and E. Ukkonen. Optimal Exact and Fast Approximate Two Dimensional Pattern Matching Allowing Rotations. In *Proceedings of the 13th Annual Symposium on Combinatorial Pattern Matching (CPM 2002)*, volume 2373 of *LNCS*, pages 235–248. Springer, 2002.
11. K. Fredriksson and E. Ukkonen. A rotation invariant filter for two-dimensional string matching. In *Proc. 9th Annual Symposium on Combinatorial Pattern Matching (CPM 98)*, pages 118–125. Springer, LNCS 1448, 1998.
12. K. Fredriksson and E. Ukkonen. Combinatorial methods for approximate pattern matching under rotations and translations in 3D arrays. In *Proc. of the 7th Symposium on String Processing and Information Retrieval (SPIRE'2000)*, pages 96–104. I.E.E.E. CS Press, 2000.

13. R. Giancarlo and R. Grossi. On the construction of classes of suffix trees for square matrices: Algorithms and applications. *Information and Computation*, 130(2):151–182, 1996.

14. Dan Gusfield. *Algorithms on Strings, Trees, and Sequences: Computer Science and Computational Biology.* Cambridge University Press, 1997.

15. G. H. Hardy and E. M. Wright. *An Introduction to the Theory of Numbers.* Oxford at the Clarendon Press, fifth edition, 1979.

16. P. Indyk, R. Motwani, and S. Venkatasubramanian. Geometric matching under noise: Combinatorial bounds and algorithms. In *Proc. 10th ACM-SIAM Symposium on Discrete Algorithms (SODA)*, pages 354–360, 1999.

17. G. M. Landau and U. Vishkin. Pattern matching in a digitized image. *Algorithmica*, 12(3/4):375–408, 1994.

18. L. Schulman and D. Cardoze. Pattern matching for spatial point sets. *Proc. 39th IEEE FOCS*, pages 156–165, 1998.

An Improved Algorithm for Generalized Comparison of Minisatellites

Behshad Behzadi and Jean-Marc Steyaert

LIX, Ecole Polytechnique, Palaiseau cedex 91128, France
{Behzadi, Steyaert}@lix.polytechnique.fr

Abstract. One of the most important objects in genetic mapping and forensic studies are minisatellites. They consist of a heterogeneous tandem array of short repeat units called variants. The evolution of minisatellites is realized by tandem duplication and tandem deletion of variants. Jeffrey et al. proposed a method to obtain the sequence of variants, called maps. Bérard and Rivals designed the first algorithm of comparison of two minisatellite maps under an evolutionary model including deletion, insertion, mutation, amplification and contraction. The complexity of this first algorithm was $O(n^4)$ in time and $O(n^3)$ in space where n is the size of the maps. In this paper we propose a more efficient algorithm using the same evolutionary model which is $O(n^3)$ in time and $O(n^2)$ in space. Our algorithm with this better efficiency can even solve generalized and more refined models.

1 Introduction

Comparing sequences is a long-addressed problem in computer science as well as in biology. Numerous algorithms have been designed starting from `diff` in Unix and ending (for the moment) at the subquadratic algorithm of Crochemore et al. (see [8]). Our interest in this paper is devoted to a structured comparison of sequences when complex operations can be use to transform strings. These notions intervene naturally in the algorithmic study and treatment of minisatellites — a very important concept in biology. These genomic subsequences are commonly used to understand the dynamic of mutations in particular for inter-allelic gene conversion-like processes at autosomal loci [3,4]. Jobling et al. [2] have characterized the Y-specific locus MSY1, a haploid minisatellite, which is composed of 48 to 114 copies of a repeat unit of length 25, rich in AT and predicted to form stable hairpin structures. These sequences are of great interest since they constitute markers for Y chromosome diversity: therefore they allow to trace male descendence proximity in populations.

In order to built phylogenetic trees based on the proximity of minisatellites one has to compute precisely, and rapidly if possible, the distance between such sequences, taking into account their structure and their specific mutation rules. Basically, these sequences are characterized by the fact that, in addition to the classical mutations, insertions and deletions, they quite often modify their structure by duplicating — triplicating at times — a subsequence of length 10

R. Baeza-Yates et al. (Eds.): CPM 2003, LNCS 2676, pp. 32–41, 2003.

to 140, thus producing what is called tandem repeats: tandem deletion is also considered.

Modelling minisatellite evolution is therefore necessary in order to provide biologists with a rigorous tool for comparing sequences and establishing likely conclusions as to their proximity. Bérard and Rivals [1] have proposed a combinatorial algorithm to solve the edit distance problem for minisatellites: they consider the five operations mentioned in the previous paragraph, with symmetric costs for each type of operations and design a $O(n^4)$ algorithm. They use this algorithm to compute a phylogenetic tree for a set of Y chromosomes which were first studied and analyzed by Jobling et al. [2,5,6]. In fact, this approach lacks flexibility since all mutations have the same cost, irrespective of their frequencies, and similarly for duplication/deletion. Their algorithm cannot be adapted to a more subtle model where the different possible variants (in tandem repeats) could have different costs. We show that it is possible to take into account this generalized model, and we design an algorithm which runs in time $O(n^3)$, thus being more efficient even in a more involved context. This algorithm is based on dynamic programming and has been run on the same data set.

In Section 2, we describe the combinatorial model for minisatellite evolution, and we state the problem in its full generality.

Section 3 is devoted to two technical lemmas which are essential to prove the correctness of the algorithm. Intuitively, these lemmas state that there exists a canonical way of ordering the sequence of operations which transform a minisatellite into another, and that it is sufficient to compare the costs of all possible "normal forms".

In Section 4, we then make explicit the algorithm which computes the best possible cost for the evolution from one minisatellite to another.

Finally in Section 5, we comment on the efficiency of the method.

2   Evolutionary Model and Description of the Problem

The symbols are the elements from an alphabet Σ. In this paper we will use the letters x, y, z,... for the symbols in Σ and s, t, u, v,... for strings[1] over Σ. The empty string is denoted by ϵ. We will denote by $s[i]$ the symbol in position i of the string s (the first symbol of a string s is $s[1]$). The substring of s starting at position i and ending at position j is denoted by $s[i..j] = s[i]s[i+1]\ldots s[j]$.

In the evolutionary model, five elementary operations are considered for a string. These operations are mutation (replacement), insertion, deletion, amplification and contraction. The first three are the well-known string edit distance operations (see for example [7]). The last two are new operations which are significant in the study of the evolution of minisatellite strings. Amplification of a symbol x in a string s amounts to repeating this symbol after one of its occurrences in s. A p-plication of a symbol x which is an amplification of *order p*

---

[1] Throughout the paper we use the word string to designate what biologists call sequences or maps [1]. The word sequence will refer to a sequence of operations on a string.

amounts to $p-1$ times repeat symbol x after the initial symbol x. Conversely, the p-contraction of a symbol x means to delete $p-1$ consecutive symbol x in condition that the symbol just before them is also an x. Given two strings s and t, there are infinitely many sequences of elementary operations which transform the string s into the string t. Among this infinity, some evolution sequences are more likely; in order to identify them, we introduce a cost function for each elementary operation depending on the symbols involved in the operation: $I(x)$ and $D(x)$ are the costs of an insertion or a deletion of symbol x. $M(x,y)$ is the cost of the replacement of symbol x by symbol y in the string. For $p>1$, $A_p(x)$ is the cost of a p-plication of symbol x in the string and finally $C_p(x)$ is the cost of a p-contraction of a symbol x. We suppose that $A_p(x) > A_{p-1}(x)$ and $C_p(x) > C_{p-1}(x)$ for any $p>1$. In the symmetric model $I(x) = D(x)$ and $C_p(x) = A_p(x)$ but we will present the algorithms for the generalized (and not necessarily symmetric) model. We suppose that the mutation cost function satisfies the triangle inequality property: $M(x,y) + M(y,z) \geq M(x,z)$ for all different x, y, z in $\Sigma \cup \{\epsilon\}$. In the case that one of the three symbols is ϵ, $M(x, \epsilon)$ is replaced by either $D(x)$ or $C_p(x) - C_{p-1}(x)$ and $M(\epsilon, x)$ is replaced by either $I(x)$ or $A_p(x) - A_{p-1}(x)$ where $p>1$ and $C_1(x) = A_1(x) = 0$. In addition, $M(x,x) = 0$ for any symbol x and all other values of all of the cost functions are strictly greater than zero. These hypotheses do not reduce the generality of our statements.

A *transformation* of s into t amounts to applying a sequence of operations on s transforming it into t. When s is transformed into t by a sequence of operations we write $s \xrightarrow{*} t$ and when s is transformed into t in one elementary operation we use the notation $s \to t$. The cost of a transformation is the sum of the costs of its operations. The *transformation distance* from s into t is the minimum cost for a possible transformation from s into t. The transformation which gives this minimum is called *optimal transformation* (it is not necessarily unique). Our objective in this paper is to find this distance between two strings and one of their optimal transformations. In the next section we will study the optimal transformation properties.

It will be convenient to add an extra special symbol $\$$ to the alphabet and to consider that the value of all the functions with $\$$ as one of their variables is ∞. Whenever we are asked to find the transformation distance between strings s and t, we will compute the optimal transformation of $\$s$ into $\$t$. By our assumption these two values are equal. This is a way to forbid any insertion at the beginning of strings. (So from now without loss of generality we suppose that the insertions are allowed only after symbols.)

3   Optimal Transformation Properties

Any given string can be transformed into any other string in the evolutionary system. Each transformation applies a sequence of operations on a string s and the result will be a string t. We will call this sequence, *transformation sequence* from s into t. Note that if we inverse this sequence and replace the deletions and contractions respectively by insertions and amplifications and vice versa, we will have a transformation sequence from t to s. In a transformation of s into t, each symbol of s will generate a substring of t: this substring can possibly be the empty string. The symbols of s which generate a non empty substring of t are called *generative symbols* and the other symbols are called *vanishing* symbols. Note that the same symbol can be generative and vanishing in two different transformation sequences of s into t.

Let us study the transformation of one symbol x to a non-empty string s: this transformation is called *generation* of s from x. Some generations are less expensive than others: in particular let us consider the generations which use only mutations, amplifications and insertions. These generations are called *non-decreasing* generations because the length of the string during the generation is never decreased. We will show that the optimal generation of any string s from any symbol x is a non-decreasing generation. It is convenient to represent a non-decreasing generation by a tree. For any non-decreasing generation sequence we construct the tree using the following rules:

1) The root of the tree has label x.
2) For any k-plication of a symbol y in the sequence we add k new nodes with label y as children of the corresponding node with label y.
3) The insertion of a letter z after a symbol y is shown by adding two children to the corresponding node y which have labels y and z from left to right.
4) The mutation of a symbol y into z is represented by a single child with label z for the node with label y.

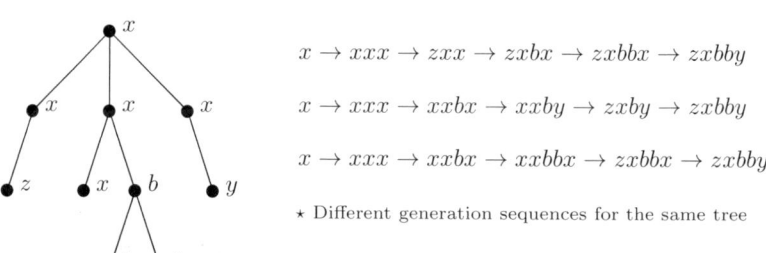

$$x \to xxx \to zxx \to zxbx \to zxbbx \to zxbby$$

$$x \to xxx \to xxbx \to xxby \to zxby \to zxbby$$

$$x \to xxx \to xxbx \to xxbbx \to zxbbx \to zxbby$$

$\star$ Different generation sequences for the same tree

Generation Cost $= A_3(x) + A_2(b) + I(b) + M(x,z) + M(x,y)$

Fig. 1. The tree representation of a non-decreasing generation

Note that two different generation sequences can have the same tree representation. They differ by the order of operations but their costs are the same. In other words for a given tree there are several generation sequences but all these different generation sequences have the same cost (see Fig. 1). An interesting sequence that we can construct for a given tree is the sequence which is obtained by a *left depth first search* of the tree where we visit the children of a node from left to right; This special sequence has the property that the operations are applied to the string from left to right. We now prove the following lemma on which the algorithm depends:

Lemma 1. *(The generation lemma):*
The optimal generation of a non-empty string s from a symbol x can be achieved by a non-decreasing generation.

Proof: Let $\mathcal{S}$ be a sequence of operations of length N which transforms a symbol x into a string s. This transformation is made explicit step by step by $N + 1$ strings: $x = s^0 \rightarrow s^1 \rightarrow \ldots \rightarrow s^N = s$ where s^i is obtained from s^{i-1} by applying a single operation. Now we show that if there is a contraction or deletion operation in this transformation (generation) sequence then we can construct another transformation sequence of lower cost. Let h be the smallest non-negative integer such that s^{h+1} is obtained from s^h by a deletion or contraction. In the transformation sequence $x = s^0 \rightarrow s^1 \rightarrow \ldots \rightarrow s^h \rightarrow s^{h+1}$ there is only one contraction or deletion operation which is the last operation. The generation of the s^h from x in this transformation is a non-decreasing generation and it can be represented by a generation tree $\mathcal{T}$. We consider two separate cases.

The first case is when $s^h \rightarrow s^{h+1}$ is the deletion of a symbol z; this z is a leaf of the tree. If this leaf is not the only child of its father then we eliminate it from the tree. The relations $A_p(z) > A_{p-1}(z)$ and $I(y) + D(z) \geq M(z, y)$ guarantee that the new tree cost is not higher than the first one. On the other hand, if z is the unique child of its father we eliminate this node with its fathers until arriving to a node with more than one child. By the triangle inequality property of the mutation costs and the above relations the remaining tree is a generation tree for $x \overset{*}{\rightarrow} s^{h+1}$ without deletion nor contraction and which is not more expensive than the initial one.

The second case to be considered is when $s^h \rightarrow s^{h+1}$ is a p-contraction z. This means that there are p leaves in the tree which are consecutive in a left-to-right traversal of the tree. Now by the same method we eliminate the last z of these p leaves. Applying a $(p - 1)$-contraction to the remaining tree yields s^{h+1} with a cost not higher than the initial generation. So we have a transformation sequence from x to s^{h+1} whose cost is not higher and with at most one contraction as its last operation. By repeating the process we can decrease the order of this last contraction without increasing the cost. We can construct a transformation sequence from x to s^{h+1} which has neither deletions nor contractions and costs no more than the initial generation which had ultimately one deletion or contraction in the last position. If we replace this new generation of s^{h+1} from x in the first generation sequence we will get a new generation sequence whose cost is not higher and whose number of contractions or deletions is reduced. By induction we obtain a generation sequence which does not use contraction nor deletion and costs less than the first generation sequence which completes the proof. □

Similarly it is possible to show that *reduction* of a non-empty string s to a single symbol x only uses mutations, deletions and contractions. As discussed before in a transformation of a string s into a string t the symbols of s can be either generative or vanishing. The previous lemma shows that the generative symbols only use insertions, amplifications and mutations. The following lemma states an important property of an optimal transformation of a string s into a string t.

Lemma 2. *(The independency of contractions):*
In an optimal transformation of a string s to a string t any contraction operation can be done before any generation.

Proof. Suppose the statement is not true: then there is an optimal transformation of the string s to string t in which at least one of the contractions should be done after some operations of a generation tree. Without loss of generality we can suppose that this contraction is a 2-contraction of a symbol y which should be done after a part of transformation of a generative symbol x. Thus the rightmost leaf of the tree rooted at x should be y at some moment during the generation. This process is shown in the left-hand side transformation of Figure 2. The right-hand side transformation uses the initial vanishing symbol y in s to generate w and in the generation starting from x it does not generate the last y in the tree therefore there is no need to use the 2-contraction of y. Note that the substring u is transformed to ϵ in both cases. Both transformations transform xuy into vw. The right-hand side transformation is equivalent and less expensive which completes the proof.   □

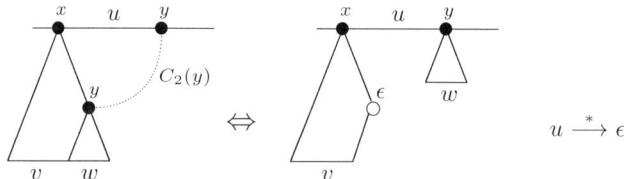

Fig. 2. The schema of the proof of lemma 2: the right-hand side transformation is equivalent and less expensive.

4   The General Algorithm

The two lemmas of the last section show that the optimal transformation of a string s into a string t can be considered as a union of some independent substring reductions followed by some substring generations. More precisely if m and n are respectively the lengths of s and t, there are two sequences of positions $1 = i_1 < i_2 < \ldots < i_l = m + 1$ and $1 = j_1 < j_2 < \ldots < j_l = n + 1$ such that $i_1, i_2, \ldots, i_{l-1}$

are the generative positions in the string s: $s[i_k]$ generates $t[j_k..(j_{k+1}-1)]$ for any $1 \leq k < l$. Before the generations, the substrings $s[i_k..(i_{k+1}-1)]$ are reduced to $s[i_k]$ for all $k < l$. In other words all the deletions and contractions can be done before all insertions and amplifications. Thus the optimal transformation has two main phases: firstly the reduction of the vanishing substrings and then the generations (see Fig. 3). The algorithm consists of two parts. In the preprocessing part, we compute the costs of generation of substrings of t and the costs of reduction of substrings of s. In the second phase we determine the optimal transformation distance by dynamic programming.

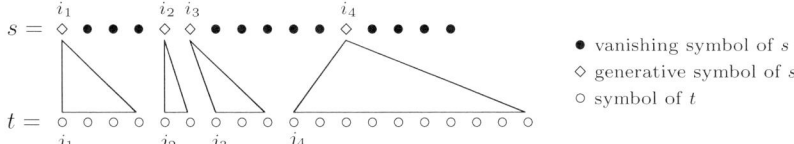

Fig. 3. Generative and vanishing symbols can be transformed in two phases.

4.1 Substring Generation Costs: Dynamic Programming

As a preprocessing part we calculate the minimal cost necessary to generate any given substring in the target string t from any given symbol x. Generating a substring $t[i..j]$ from a symbol x amounts to applying a sequence of operations on x transforming it into $t[i..j]$. Let $mc[i, j, p, x]$ be the minimum generation cost for generating $t[i..j]$ from a symbol x among all possible generations of this substring starting with a p-plication (see Fig. 4). In other words in the tree representation of this generation, the root has p children with the same label.

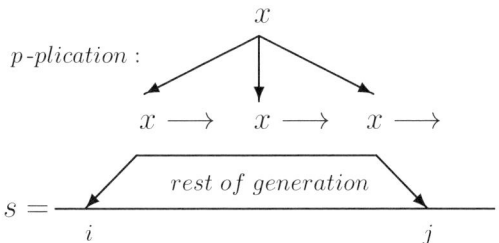

Fig. 4. The illustration of $mc[i, j, p, x]$

Let $G[i, j, x]$ be the minimum generation cost for the substring $t[i..j]$ starting from a symbol x among all generations which do not start by a mutation. $T[i, j, x]$ is the minimum generation cost of the substring $t[i..j]$ from the symbol x. Then we have the following recurrence relations system, described in Fig. 5.

Initialization:
$$\forall i \; \forall x \in \Sigma \; \forall p > 1 \; : \; T[i,i,x] = M(x,t[i]) \quad \text{and} \quad mc[i,i,p,x] = \infty$$

Recurrence:
(1)
$$\forall i < j \; \forall x \in \Sigma \; : \; mc[i,j,2,x] = A_2(x) + \min_{i<k\leq j}\{T[i,k-1,x] + T[k,j,x]\}$$

(2)
$$\forall p > 2 \; \forall i < j \; \forall x \in \Sigma \; :$$
$$p \leq j-i+1 : \; mc[i,j,p,x] = A_p(x) - A_{p-1}(x) + \min_{i<k\leq j}\{T[i,k-1,x] + mc[k,j,p-1,x]\}$$
$$p > j-i+1 : \; mc[i,j,p,x] = \infty$$

(3)
$$\forall i < j \; \forall x \in \Sigma : G[i,j,x] = \min \begin{cases} mc[i,j,p,x] & \forall p > 1 \\ T[i+1,j,y] + I(y) & \text{if } t[i] = x \; \forall y \in \Sigma \end{cases}$$

(4)
$$\forall i < j \; \forall x \in \Sigma : T[i,j,x] = \min_{y \in \Sigma}\{G[i,j,y] + M(x,y)\}$$

Fig. 5. The recurrence relations for substrings generation costs

Proposition 1 *The relations of Fig. 5 determine correctly the minimum cost of generation of substring $t[i..j]$ from x in $T[i,j,x]$.*

Proof: The proof is done by induction on the length of the substring $t[i..j]$. After the initialization part $T[i,j,x]$ is correctly computed for all substrings of length 1 where $i = j$. Now consider i and j with $j - i = l > 0$. The root of the tree representation of the optimal generation sequence which generates $t[i..j]$ from x, together with its children correspond to the first operation of this generation sequence. If this first operation is an amplification of order 2 the x will be replaced by two x's; the first x generates $t[i..(k-1)]$ and the second one generates $t[k..j]$ for some $i < k \leq j$ which are considered in equations (1). In equations (2) an amplification of order p is considered as a union of a 2-plication followed by a $(p-1)$-plication where the difference of the costs are correctly considered. In equations (3) the minimum generation cost among all possible generations starting by an amplification or an insertion is computed in $G[i,j,x]$. By lemma 1, $G[i,j,x]$ is correctly computed. In equations (4) the minimum of the value computed in $G[i,j,x]$ and the value of the other generations which start by a mutation is written in $T[i,j,x]$. By the triangle inequality property of mutation costs there cannot be two consecutive mutations on the same symbol, therefore equations (4) compute correctly the optimal answer in $T[i,j,x]$.   □

The time complexity of this preprocessing part is $O(n^3|\Sigma|\rho)$ where $|\Sigma|$ is the number of symbols in the alphabet Σ and ρ is the maximum possible p for a p-plication. The space complexity is $O(n^2|\Sigma|\rho)$. By the same method the substring reduction cost can be computed. Let $S[i,j]$ be the minimum cost of reduction of the substring $s[i..j]$ into $s[i]$. This table can be determined by a similar recurrence system in time $O(n^3)$. We will use the two tables T and S in our main algorithm for determining the transformation distance.

4.2   Transformation Distance: Dynamic Programming

Once we have computed the tables T and S, we use another dynamic programming algorithm to calculate the transformation distance between the two strings s and t. Let $TD[i, j]$ be the transformation distance between $s[1..i]$ and $t[1..j]$. Then the following recurrence relations hold for TD:

Initialization:
$$TD[0, 0] = 0 \quad \text{and} \quad TD[0, j] = \infty \quad \forall j > 0$$

Recurrence:
$$\forall i > 0 \quad TD[i, j] = \min \begin{cases} TD[i-1, l] + T[l+1, j, s[i]] & \forall\, 0 \leq l < j \\ TD[k, j] + S[k, i] & \forall\, 0 < k < i \end{cases}$$

Fig. 6. Transformation distance

Proposition 2 *The recurrence relations of Fig. 6 determine correctly the transformation distance between $s[1..i]$ and $t[1..j]$ in $TD[i, j]$ for all $i \leq m$ and $j \leq n$.*

Proof: We prove the proposition by induction on i. For $i = 0$, the starting string $s[1..0] = \epsilon$; thus we cannot have any operation on this string, because we supposed that the insertion can be done only after some symbols. This proves that the initialization part determines correctly the values for $i = 0$. Suppose the table TD is correctly filled for all $i' < q$ and let $i = q$. In an optimal transformation of $s[1..i]$ into $t[1..j]$, $s[i]$ is either generative and generates a proper suffix of $t[1..j]$ (denoted by $t[(l+1)..j]$) or is a vanishing symbol which is reduced together with some of its previous symbols to $s[k]$ for some $k < i$ (the substring $s[(k+1)..i]$ is eliminated). In both cases, the problem is reduced to the problem of determining $TD[i', q]$ for some $i' < i$ which is already correctly computed by the induction hypothesis. The proposition is true for $i = q$ and this completes the proof. □

The complexity of this main part of the algorithm is $O(n^3)$ in time and $O(n^2)$ in space. So the total complexity of the algorithm is $O(n^3|\Sigma|\rho)$ in time and $O(n^2|\Sigma|\rho)$ in space.

5   Remarks, Conclusion, and Future Work

In this paper we gave a generalized algorithm for the comparison of the minisatellite sequences. In practice there are a small number of variants ($|\Sigma|$ is small) for minisatellites. The order of an amplification in the real minisatellites is believed not to be more than 3 or 4. Therefore the complexity of the algorithm for the real minisatellite data is $O(n^3)$ which is efficient for the real biological data. On the set of data (690 strings) proposed by Jobling and previously processed by Bérard and Rivals, a direct implementation of the algorithm on a PC took 2 hours of computation for 690×345 comparisons.

The hypotheses on the cost function values are not really restrictive for the problem. The triangle inequality condition can be released by a preprocessing in which we replace the mutation cost of x into y by the lowest cost of a path of mutations from x to y. We use our algorithm and we replace each mutation in the optimal found answer by its corresponding path computed in the preprocessing.

The important point is that our hypotheses are very weak and should apply to most situations issued from the biological observations.

As a last remark, it is also possible to adapt a simplified version of the algorithm to solve the problem of "compatible arches" computation in the $O(n^4)$ Bérard-Rivals algorithm, thus giving a $O(n^3)$ solution (independent of the size of the alphabet).

References

1. Bérard, S., Rivals, E.: Comparison of Minisatellites. Proceedings of the 6th Annual International Conference on Research in Computational Molecular Biology. ACM Press, (2002).
2. Jobling, M.A., Bouzekri, N., Taylor, P.G.: Hypervariable digital DNA codes for human paternal lineages: MVR-PCR at the Y-specific minisatellite, MSY1(DYF155S1). Human Molecular Genetics, Vol. 7,No. 4. (1998)643–653
3. Bouzekri, N., Taylor, P.G., Hammer M.F, Jobling, M.A.: Novel mutation processes in the evolution of haploid minisatellites, MSY1: array homogenization without homogenization. Human Molecular Genetics,Vol. 7, No. 4. (1998)655–659
4. Jeffreys, A.J., Tamaki, K., Macleod, A., Monckton, D.G., Neil, D.L and Armour, J.A.L: Complex gene conversion events in germline mutation at human minisatellites. Nature Genetics, 6. (1994)136–145
5. Brión, M., Cao, R., Salas, A., Lareu M.V., Carracedo A.: New Method to Measure Minisatellite Variant Repeat Variation in Population Genetic Studies. American Journal of Human Biology, Vol. 14.(2002) 421–428
6. Elemento, O., Gascuel, O., Lefranc, M.-P.: Reconstructing the duplication history of tandemly repeated genes. Molecular Biology and Evolution,vol 19(3). (2002) 278–288
7. Sankoff, D. and Kruskal, J.B: Time Warps, String Edits and Macromolecules: The Theory and Practice of Sequence Comparison. Addison-Wesley. (1983)
8. Crochemore, M., Landau, G. M., Ziv-Ukelson, M.: A sub-quadratic sequence alignment algorithm for unrestricted cost matrices. SODA'2002. ACM-SIAM. (2002)679–688.

Optimal Spaced Seeds for Hidden Markov Models, with Application to Homologous Coding Regions

Broňa Brejová, Daniel G. Brown, and Tomáš Vinař

School of Computer Science, University of Waterloo, Waterloo ON N2L 3G1 Canada
{bbrejova, browndg, tvinar}@math.uwaterloo.ca

Abstract. We study the problem of computing optimal spaced seeds for detecting sequences generated by a Hidden Markov model. Inspired by recent work in DNA sequence alignment, we have developed such a model for representing the conservation between related DNA coding sequences. Our model includes positional dependencies and periodic rates of conservation, as well as regional deviations in overall conservation rate. We show that, for hidden Markov models in general, the probability that a seed is matched in a region can be computed efficiently, and use these methods to compute the optimal seed for our models. Our experiments on real data show that the optimal seeds are substantially more sensitive than the seeds used in the standard alignment program BLAST, and also substantially better than those of PatternHunter or WABA, both of which use spaced seeds. Our results offer the hope of improved gene finding due to fewer missed exons in DNA/DNA comparison, and more effective homology search in general, and may have applications outside of bioinformatics.

1 Introduction

Consider a probabilistic process P that generates finite binary sequences. We wish to compute the probability that a given pattern s matches the random sequence, for a particular definition of pattern matching. Then, given that the pattern s comes from a small class of possible patterns, we wish to find the pattern s^* with the highest probability of matching the random sequence A.

Our form of pattern matching is based on *spaced seeds*. In this framework, one chooses a seed, represented by a binary string $s = (s_1, \ldots, s_k)$. In the seed, ones intuitively correspond to places in the random sequence which must also be ones, while zeros correspond to wildcards. Formally, for the seed s to match the randomly generated sequence A, there must be a position i in A such that $a_i \geq s_1, a_{i+1} \geq s_2, \ldots, a_{i+k-1} \geq s_k$.

Here, we wish to identify the optimal seed s^* when the random process is a hidden Markov model (HMM). For such models, we show that one can identify the probability that a given seed detects a sequence generated from an HMM efficiently. When k is small, one can examine all seeds efficiently to find the best

R. Baeza-Yates et al. (Eds.): CPM 2003, LNCS 2676, pp. 42–54, 2003.

one. We use our methods to find optimal seeds for finding homologous protein coding regions in long DNA sequences. The optimal seed for a reasonable HMM for this problem is more effective than seeds used by existing methods.

A launching point for our work is the recent study of DNA sequence homology by Ma et al. [8], who showed that spaced seeds can be useful in increasing the probability of detecting related DNA regions. The process of identifying related DNA sequences, large scale similarity search, is one of the most time consuming processes in bioinformatics, so procedures that improve it are needed. Ma et al.'s local alignment program PatternHunter, because of its use of spaced seeds tuned for a particular model, is more successful at detecting homologous regions, and thus is more likely to build useful local alignments between two sequences.

The model used by Ma et al., $PH(N, p)$, assumes that correct local alignments are of length N, and that in comparing two homologous sequences, the probability that they are the same at position i is p. Keich et al. [5] give a dynamic programming algorithm for computing the optimal seed in this simple probabilistic model. In $PH(64, .7)$, the seed of size less than or equal to 18, with 10 1's, that is most likely to detect homologous sequences is 111001001001010111.

Traditional programs for sequence alignment, like BLASTN [1], use in effect a seed of the form 1111111111, and only detect homologous sequences with a set of contiguous exact matches. The expected number of matches of the BLAST seed to a sequence generated by $PH(64, .7)$ is quite close to the expected number of matches of the PatternHunter seed, but the probability that an alignment has any match to the seed is substantially higher with PatternHunter's seed.

In Section 2, we review HMMs, and in Section 3, we consider the question of identifying the best seeds when the random process is more complicated, in particular an HMM. If the sequence A is given by a known HMM, we give an efficient algorithm that computes the seed s^* most likely to match A, given a distribution on the lengths of the sequence A.

One application of these techniques is increasing the probability of detecting homologous protein coding regions in DNA sequences. Protein coding regions form most significant local alignments between distantly related species, and are also specifically important. Alignments between homologous coding regions can, for example, aid in gene finding [7]. Alignments between homologous protein coding regions, however, are very poorly modeled by PH, as it does not take into account basic characteristics of protein alignments, such as their three-periodicity and the inhomogeneity of conservation rates within alignments.

In Section 4, we give an HMM modeling conservation within alignments of coding regions, and in Section 5, we use compute the optimal seeds under this model, using data, based on curated alignments between the human genome and the mouse and *Drosophila* (fruit fly) genomes. Comparisons between the seeds' predicted sensitivity and their actual sensitivity show that the HMM does more accurately model the probability that a seed is useful in practice than do simpler models, and similarly, we find that the seeds which our algorithms predict as having the highest probability of success are, in practice, optimal. In particular, the best seed with 10 ones and length less than or equal to 18, according to the algorithm, is the second best in our testing set, and detects 85% of alignments. This compares with 45% for the BLAST seed, 59% for the PatternHunter seed, and 82% for the seed 11011011011011, used by the program WABA [6].

We note that our methods were developed contemporaneously with other work by Buhler et al. [3], who extend Keich et al.'s algorithm [5] to the case of multiple seeds used for the same model. However, they consider neither HMMs, nor probabilistic models specifically developed for coding regions.

2   Probabilistic Models of Local Alignments

Local alignments can be represented as binary sequences, where 1 represents a match and 0 a mismatch. We model only ungapped alignments, as seed hits are found only in gapless regions. We can characterize a particular type of local alignment by giving a probabilistic model defining a probability distribution over all binary strings of a given length. Then, given a particular probabilistic model, we can find a seed s^* from a class of seeds C maximizing the probability of a hit in an alignment sampled from this model. For us, the class C is all seeds with W ones and length at most M. To avoid redundancy, we require that the first and last positions of a seed are 1. The parameter W is the *weight* of a seed.

Ma et al. [8] introduced the concept of optimal seeds in PatternHunter. PatternHunter's model $PH(N, p)$ represents a similarity region of length N, where each position is a match independently with probability p. Formally, it is a sequence of N independent Bernoulli random variables $X_0, X_1, \ldots, X_{N-1}$, with $\Pr(X_i = 1) = p$ for each i. PatternHunter's seed optimizes this model with parameters $N = 64$ and $p = 0.7$. Keich et al. [5] give a dynamic programming algorithm to compute the probability that a given seed of length M has at least one hit in an alignment sampled from $PH(N, p)$. The running time is $O(2^{M-W} M(M^2 + N))$ for each seed. One computes the probability for each seed in C and chooses the best.

Here, we explore the problem of finding the optimal seed under more general class of probabilistic models, namely hidden Markov models (HMMs). HMMs model dependencies between adjacent positions in the sequence and have successfully modeled biological sequences in a number of applications [4]. Here, we show that HMMs can capture several important properties of DNA alignments of protein coding regions.

An HMM is a simple generative process defined on a collection of states and transitions between the states. First, a starting state is chosen according to initial probabilities associated with each state. In each step, a single character is generated according to the emission probability table associated with the current state. Then, a new state is chosen according to the transition probabilities for the current state. The probability of a string in the model equals the probability that the string is generated by this process.

In the next section, we present an algorithm for computing the probability that a seed has a hit in a string sampled from an HMM. Our algorithm is an extension of the algorithm presented by Keich et al. [5]. In particular, the *PH* probabilistic model is equivalent to a single state HMM, and in this special case, our algorithm is identical to the one presented by Keich et al.

We also extend the probabilistic model so that the length of the alignment is also a random variable, chosen from some fixed distribution of alignment lengths. The probability of a seed match in such an extended model is the weighted average of the probabilities for models with fixed lengths.

3 Finding Optimal Seeds for HMMs

Assume that we are given a seed Q of length M and weight W and an HMM characterizing alignments. We will compute the probability that Q has at least one hit in a length N alignment.

Let S be the set of states of the HMM. Each step of the generative process consists of emission of one character in the current state, followed by transition to the next state. Let $String(s, x)$ be the event that the HMM generates string x, assuming that it starts in state s. Let $State(s, t, i)$ be the probability that after i steps the HMM will be in state t, assuming that it starts in state s. Notice that $\Pr(State(s, 0, s)) = 1$ and $\sum_{x \in \{0,1\}^i} \Pr(String(s, x)) = 1$, for all i.

We will proceed by dynamic programming, where the subproblem is defined as follows: for any $i \leq N$, binary string x of length at most $\min\{i, M\}$, and $s \in S$, the subproblem $A_{i,x,s}$ is the probability that the string generated by the HMM in i steps starting in state s contains a match of the seed Q, provided that x is a prefix of the generated string. The probability that Q has at least one hit in the entire alignment then equals the sum of $A_{N,\lambda,s}$ over all states s, weighted by the initial probabilities of all states (λ denotes the empty string).

Let $matches(x, Q)$ be the event that string $x1^{M-|x|}$ is a hit for seed Q. That is, when we align the starts of x and Q, x has 1 at all positions where Q has 1. If $matches(x, Q)$, we say that x is a matching string. Let $suffix(x, Q)$ be the longest suffix z of x such that z is a matching string.

The probability $A_{i,x,s}$ can be computed by the following recurrent formula:

$$
A_{i,x,s} = \begin{cases}
0 & \text{if } i < M & \text{(A)} \\
1 & \text{if } |x| = M \text{ and } matches(x, Q) & \text{(B)} \\
\sum_{t \in S} p_t \cdot A_{i-|y|,z,t} & \begin{array}{l} \text{if not } matches(x, Q), \text{ where} \\ x = yz, z = suffix(x, Q), \\ p_t = \Pr[State(s, t, |y|) \mid String(s, x)] \end{array} & \text{(C)} \\
\begin{array}{l} p \cdot A_{i,x1,s} \\ +(1 - p) \cdot A_{i,x0,s} \end{array} & \begin{array}{l} \text{if } |x| < M \text{ and } matches(x, Q), \text{ where} \\ p = \Pr[String(s, x1) \mid String(s, x)] \end{array} & \text{(D)}
\end{cases}
$$

Case (A) recognizes that seeds of length M cannot have hits in shorter regions. In case (B), string x guarantees a hit. In case (C), since the string x does not match Q, a hit will not start at the first position of the alignment, so only following positions need to be considered. In particular, the longest matching suffix z of string x corresponds to the first possible position where a hit can occur. We need to consider all possible states t in which the HMM can start to generate suffix z. Finally, case (D) provides a formula for combining subproblems where the last $|x|+1$ characters are fixed into a subproblem where only $|x|$ characters are fixed.

This recurrent formula can be used to compute probabilities $A_{i,x,s}$ in order of increasing i and shrinking x. Keich et al. [5] show how to efficiently organize the computation of a similar recurrent formula so that the running time is $O(M2^{M-W}(M^2 + N))$ for each of the $\binom{M-2}{W-2}$ possible seeds (under the assumption that numbers of size between 0 and 2^{M-W} can be manipulated in constant time). We apply their methods to our modified algorithm yielding the following Lemma.

Lemma 1. *The probability that a length M, weight W seed Q has a hit in an alignment of length N generated by a given σ-state HMM can be computed in $O(M2^{M-W}(M^2 + \sigma^3 + \sigma^2 N))$ time.*

Proof. To complete the dynamic programming algorithm we need to demonstrate how to compute the probabilities needed in cases (C) and (D). First, the standard Forward algorithm for HMMs [9] can be used to compute values $B_{s,t,x}$ and $C_{s,x}$ defined as follows:

$$B_{s,t,x} = \Pr(String(s,x) \text{ and } State(s,t,|x|))$$

$$C_{s,x} = \Pr(String(s,x)) = \sum_{t \in S} B_{s,t,x}$$

The conditional probabilities required can now be computed using these two formulas:

$$\Pr(State(s,t,|y|) \mid String(s,\overbrace{yz}^{x})) = \frac{B_{s,t,y} \cdot C_{t,z}}{C_{s,x}} \qquad (1)$$

$$\Pr(String(s,x1) \mid String(s,x)) = \frac{\sum_t B_{s,t,x} \cdot C_{t,1}}{C_{s,x}} \qquad (2)$$

Let $\Gamma = \{x, x0 \mid x \text{ is a matching string}, |x| \leq M\}$. The number of strings in Γ is $O(M2^{M-W})$. Observe that to compute values $A_{N,\Lambda,s}$, we do not need to compute values of $A_{i,x,s}$ for all x, but it is sufficient to compute the values for $x \in \Gamma$.

A value of $suffix(x0, Q)$ can be easily computed in $O(M^2)$. Then we compute values of $B_{s,t,x}$ for each pair of states s and t and for each $x \in \Gamma$. Using the fact that Γ is closed under prefix operation this can be done in $O(\sigma^3 M2^{M-W})$ time. It is possible to prove that only values of $B_{s,t,x}$ for $x \in \Gamma$ are needed to compute probabilities from formulas (1) and (2) for all $x \in \Gamma$. In particular, notice that if $x \in \Gamma$ and $x = yz$ where $z = suffix(x, Q)$, then both y and z are also in Γ.

Once all auxiliary tables are computed, each value $A_{i,x,s}$ can be computed in $O(\sigma)$ time. This gives total time $O(M2^{M-W}(M^2 + \sigma^3 + \sigma^2 N))$. □

4   Modeling Local Alignments in Protein Coding Regions

Now that we have demonstrated that the notion of optimal seeds can be extended to more general probabilistic models, namely HMMs, we apply this approach to the problem of detecting local alignments of homologous protein coding regions. These have specific properties not modeled well by the simple *PH* model. Also, because protein coding regions are very important, local alignment programs must perform well on these regions of genomes.

Protein coding regions encode an organism's proteins in its genome. A protein's amino acid sequence is encoded in DNA as a sequence of triplet codons, each representing one amino acid according to the genetic code. However, the code is redundant: some amino acids have multiple encodings. In particular, third positions in codons can often mutate "silently," not altering the amino acid, and even similar proteins can differ greatly in corresponding coding regions in DNA.

In higher organisms, a protein's coding DNA is often split into disjoint regions, called exons, separated by long stretches of non-coding DNA, called introns. Exons are cut out and spliced together during protein production. Here, we assume that the locations of the protein coding regions is unknown prior to local alignment. Because of the existence of introns, finding corresponding coding regions is difficult, since a protein alignment may correspond to several short alignments of exonic DNA, separated by long intron gaps.

Some local alignment programs, such as BLASTN [1] and PatternHunter [8], use seeds not reflecting properties of protein coding regions. Thus, they are not able to detect some related pairs of coding sequences. By contrast, Kent and Zahler [6] use the seed 110110110 in their alignment program, WABA. This seed represents the intuition that coding DNA has 3-periodic structure, and the third position of a codon is less conserved than the first two. Kent and Zahler report a substantial improvement over BLAST in sensitivity in detecting homologous coding sequences. We obtain better seeds for protein coding regions by modeling properties of alignments in these regions by probabilistic models. We describe three models increasing in complexity and accuracy.

Three-periodicity. The most obvious property of alignments in coding regions is their three-periodicity and that some of the positions of a codon are less conserved than others. A simple extension of the *PH* model, which we call $M^{(3)}$, encapsulates this idea. Model $M^{(3)}(N, p_0, p_1, p_2)$ represents a region of length N, where the probability of a position being a match depends on its relative codon position, but alignment positions are still independent. Formally, it is a sequence of N independent Bernoulli random variables $X_0, X_1, \ldots, X_{N-1}$ where $\Pr(X_i = 1) = p_{i \bmod 3}$. This model can be expressed as a simple 3-state HMM, as depicted in Figure 1, and we can use our algorithm to compute the probability that a seed matches an alignment sampled from this model.

Table 1 shows the parameters of the model $M^{(3)}$ estimated from our training set of alignments between human and fruit fly coding regions and between human and mouse coding regions (more details about the data sets can be found in

Table 1. Parameters of model $M^{(3)}$ estimated from our training set.

Data set	p_0	p_1	p_2
human/mouse	0.82	0.87	0.61
human/fruit fly	0.67	0.77	0.40

Section 5). In both cases, the third position is much less conserved than the others, and the first position is somewhat less conserved than the second.

Dependencies within codon. Model $M^{(3)}$ models the different conservation levels within codons arising from the redundancy of the genetic code. However, dependencies exist among positions within codons. These arise in part from the differing prevalence of some amino acids and their corresponding codons. To model these dependencies, we use a new model, $M^{(8)}$.

The model $M^{(8)}(n, p_{000}, p_{001}, \ldots, p_{111})$ represents a region of n codons ($N = 3n$ nucleotides). Each codon has conservation pattern $x \in \{0,1\}^3$ with probability p_x; the sum of $p_{000}, p_{001}, \ldots, p_{111}$ is 1. In this model, positions within one codon have arbitrary dependencies, yet individual codons are independent. Formally, this model is a sequence of n independent triples of Bernoulli random variables $(X_0, X_1, X_2), \ldots, (X_{3n-3}, X_{3n-2}, X_{3n-1})$, such that $\Pr(X_{3i} = a, X_{3i+1} = b, X_{3i+2} = c) = p_{abc}$. The HMM shown in Figure 1 represents model $M^{(8)}$.

Table 2 demonstrates the existence of dependencies within codon. The first row of the table shows the probabilities of all triplets as estimated by model $M^{(3)}$, under the assumption that codon positions are independent. The second row shows the probabilities of all triplets in model $M^{(8)}$. In particular, triplets 000 and 111 occur in the data more often than expected based on model $M^{(3)}$.

Inhomogeneity of alignments. The previous models assume that alignments have roughly the same conservation rate throughout their length. In fact, the pattern of conservation of a typical alignment is highly non-uniform. Many alignments include short, highly conserved regions surrounded by less well conserved regions. (This is unsurprising, as highly conserved regions are more likely to be functional parts of the proteins [10].)

We address this problem with a new HMM. In this model, regions with high and low conservation can alternate. This model is an HMM consisting of 4 copies of the HMM for $M^{(8)}$ shown in Figure 1, each copy corresponding to a different level of conservation. To allow transitions between different conservation levels, we add transitions from all copies of both states labeled by a star in Figure 1 to all copies of both states labeled by two stars. In our experiments this model will be called *HMM*.

5   Experimental Data

Data and methods. We have conducted our experiments on two datasets of protein coding region alignments: human vs. fruit fly and human vs. mouse, each

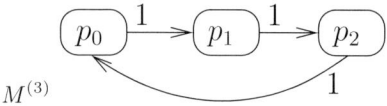

$M^{(3)}$

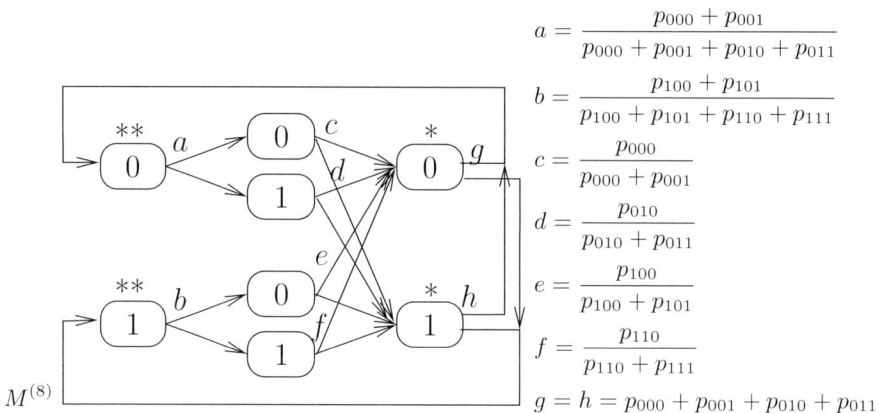

$M^{(8)}$

$$a = \frac{p_{000} + p_{001}}{p_{000} + p_{001} + p_{010} + p_{011}}$$

$$b = \frac{p_{100} + p_{101}}{p_{100} + p_{101} + p_{110} + p_{111}}$$

$$c = \frac{p_{000}}{p_{000} + p_{001}}$$

$$d = \frac{p_{010}}{p_{010} + p_{011}}$$

$$e = \frac{p_{100}}{p_{100} + p_{101}}$$

$$f = \frac{p_{110}}{p_{110} + p_{111}}$$

$$g = h = p_{000} + p_{001} + p_{010} + p_{011}$$

Fig. 1. HMM representation of models $M^{(3)}(N, p_0, p_1, p_2)$ and $M^{(8)}(n, p_{000}, \ldots, p_{111})$. Each state is labeled with emission probability of '1'. In the HMM for $M^{(3)}$, each transition has probability 1. In the HMM for $M^{(8)}$, each state has two outgoing transitions. One of them has the transition probability shown in the picture; the other has probability such that they add to 1.

treated separately. First, we found statistically significant interspecies protein alignments. Then, we filtered these so that each protein occurs in at most one alignment. Next, we split the resulting sets into testing and training halves. The protein sequences were then mapped to their coding sequences in the genomes. In this way, we transformed alignments of pairs of proteins into alignments of pairs of DNA sequences. Note that one protein alignment can yield several DNA alignments because the coding regions for each protein can be interrupted by non-coding introns. We call the DNA alignments in this set *fragments*. We discarded weakly conserved and very short fragments because they cannot be detected by alignment programs using the spaced seed method.

The fragments may be gapped alignments. However, to find a fragment by an alignment program using the spaced seed method, the seed hit must be within an ungapped region. Thus, we broke gapped fragments in the training set into ungapped fragments, and again discarded weak and short fragments. In the testing set, we have kept gapped fragments, because a single hit is sufficient to discover the entire fragment by the spaced seed method.

This process ensures that our data sets contain biologically meaningful nucleotide alignments of protein coding regions. Only coding regions of related proteins are aligned, codon boundaries are always correctly aligned, and alignments do not extend to non-coding parts of the genome.

Table 2. Comparison of probabilities of conservation patterns within a codon in models $M^{(3)}$ and $M^{(8)}$. Parameters of models estimated from human/fruit fly alignments.

Model	p_{000}	p_{001}	p_{010}	p_{011}	p_{100}	p_{101}	p_{110}	p_{111}
$M^{(3)}$	0.05	0.03	0.15	0.10	0.09	0.06	0.31	0.20
$M^{(8)}$	0.11	0.04	0.12	0.06	0.06	0.03	0.32	0.27

The initial data set consisted of all human, fruit fly and mouse proteins from SWISSPROT database [2], release 40.38, for which a correctly annotated coding regions could be found in the GenBank database. We created the initial alignments by BLASTP 2.0.8 [1] using E-value threshold 10^{-30}. The resulting set contained 339 human vs. fruit fly protein alignments and 675 human vs. mouse protein alignments. In the final data sets we filtered out gapped fragments with alignment score less than 16 (scoring scheme: match +1, mismatch −1, gap opening −5, gap extension −1) and ungapped fragments with less than 10 matches. The human vs. fruit fly dataset contains 972 ungapped fragments in the training set and 810 gapped fragments in the testing set. The human vs. mouse dataset contains 2171 ungapped fragments in the training set and 1660 gapped fragments in the testing set.

We used the training set of ungapped fragments to estimate the parameters of PH, $M^{(3)}$, $M^{(8)}$, and HMM. We estimated the parameters of PH, $M^{(3)}$, and $M^{(8)}$ by counting frequencies of corresponding conservation patterns, while we estimated the parameters of the HMM model by the Baum-Welch algorithm [4]. We estimated the length distribution of alignments from the set of ungapped fragments as well.

In our experiments we studied all 24310 seeds of weight 10 and length at most 18. We used the dynamic programming algorithm to compute the probability of a hit in a random alignment for each seed under each of the models.

To evaluate the seed performance, we have computed for each seed how many gapped alignments from the testing set contain a hit of the seed and can be thus potentially found by an alignment program using this seed. The results of these experiments follow.

Optimal seeds and their performance. Here, we present the optimal seeds under different probabilistic models of alignments, and we compare their sensitivity on the testing data set as well as sensitivity when using the seeds with alignment program PatternHunter.

Table 3 lists the seeds we selected for further examination, and presents the sensitivity of each seed on both testing sets. We also evaluated the seeds with the program PatternHunter and we list the fraction of testing fragments overlapping an alignment reported by PatternHunter.

On the human vs. fruit fly data set, the seed DATA-OPT has the highest sensitivity on the testing set. It was not discovered by any model. Models HMM and $M^{(8)}$ both discovered the seed HMM-OPT, which has almost the same performance as DATA-OPT. The seed M3-OPT is optimal under model $M^{(3)}$, but its performance is lower than that of the other seeds. We also include seeds used by WABA, PatternHunter and BLAST for comparison. Out of these, only WABA performs comparably with M3-OPT. We also include the seed globally worst on the testing set.

Results are similar for the human vs. mouse data set. Here, models $M^{(8)}$ and $M^{(3)}$ found the seed DATA-OPT, which is again optimal for the testing data. Model HMM reported the seed HMM-OPT again. In this case, there is much less difference between the best and the worst seeds, since alignments between human and mouse are much more conserved than alignments between drosophila and mouse, and are thus easier to detect by any seed. Still, there is a clear separation between performance of seeds tailored to detecting alignments in coding regions and other seeds.

Table 3. Performance of selected seeds on both testing sets. Columns labeled Hit indicate how many fragments in the testing set have a hit of the seed. Columns labeled PH indicate how many of these fragments overlap an alignment discovered by the PatternHunter program using the seed.

Seed		Fruit fly		Mouse	
		Hit	PH	Hit	PH
11011011000011011	DATA-OPT	86%	85%	94%	92%
11011000011011011	HMM-OPT	85%	85%	94%	92%
11001011001011011	M3-OPT	82%	76%	93%	90%
11011011011011	WABA	81%	79%	92%	90%
111001001001010111	PH-OPT	59%	57%	87%	86%
1111111111	BLAST	45%	43%	82%	81%
10101010101010101011	WORST	38%	39%	80%	79%

In both data sets, the seeds that take into account the 3-periodic structure of the coding regions (WABA, M3-OPT, HMM-OPT and DATA-OPT), have higher sensitivity than other seeds. Further, the optimal seeds under models HMM and $M^{(8)}$ have significantly better performance than the WABA seed, currently used for local alignments in coding regions [6]. By using these seeds we were able to reduce the number false negatives by 29% in the fruit fly set and by 20% in the mouse set.

This increase in sensitivity does not come at a cost of greatly increased running time. The running time of PatternHunter increased by at most 3% compared to the original PH-OPT seed.

Our models as predictors of seed performance. For further validation, we studied how well the sensitivity predicted under each of the models corresponds to the sensitivity measured on the testing data set. Good models should assign higher probability to seeds which perform better in practice, and the probability predicted by the model should correspond to sensitivity on real data.

As illustrated in Figure 2, the predicted probability increases with the sensitivity of the seed on testing data in models $M^{(3)}$, $M^{(8)}$, and HMM. This is in contrast to the model *PH* where there is no clear correspondence between predicted and real sensitivity. Models *HMM* and $M^{(8)}$ exhibit better quality of ordering among the top seeds as well as among the worst seeds. In addition to that, *HMM* is clearly the best predictor of the real sensitivity; in fact, the correlation between the estimated sensitivity by *HMM* and the actual sensitivity is the highest, at $r^2 = 0.9687$. Sensitivity is consistently underpredicted in all models, since the training set consisted of ungapped fragments, which are slightly shorter than the gapped fragments in the testing set.

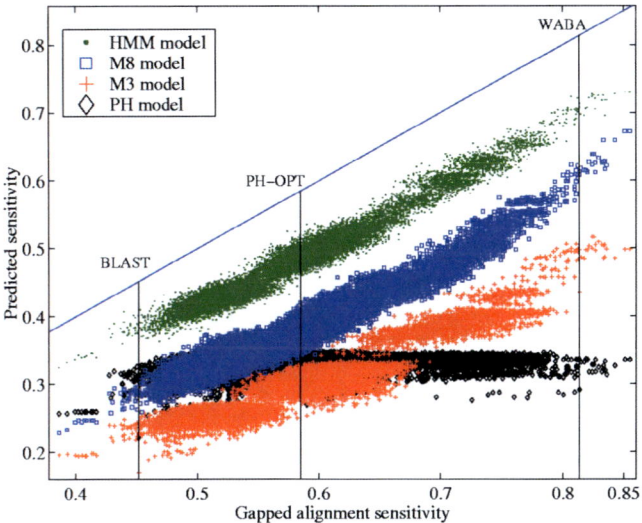

Fig. 2. Sensitivity of all seeds on the testing set versus their predicted sensitivity under different probabilistic models. Each dot represents one seed under one model. Vertical lines indicate sensitivity of the WABA, PH-OPT and BLAST seeds.

Table 4 further demonstrates the ordering capabilities of each of the models. Models *HMM* and $M^{(8)}$ correctly identified the top 3 seeds, while the seed classified as the best in $M^{(3)}$ model ranks 15th on the testing data. Also, there is no clear correspondence between ranks in *PH* model and rank on the testing data. The BLAST seed (which is currently the most widely used) is among the worst seeds in any of the considered models.

6   Conclusion

We have given an algorithm to compute the optimal spaced seeds for detecting sequences generated by a hidden Markov model. By extending work of Keich et al.[5], we have given an efficient algorithm for computing the probability that a

Table 4. Ranks of seeds under different probabilistic models. The table shows the rank of each of the chosen seeds in the testing data set as well as under each of the considered probabilistic models.

Seed	Testing data Rank	Sensitivity	HMM	$M^{(8)}$	$M^{(3)}$	PH	Name
11011011000011011	1	0.855	2	2	17	9746	DATA-OPT
11011000011011011	2	0.851	1	1	16	9746	HMM-OPT
11000011011011011	3	0.843	3	3	42	19124	
11011011001011001	4	0.836	15	15	15	23212	
10110110000110101	5	0.835	42	18	37	13208	
11001011001011011	15	0.824	4	6	1	17945	M3-OPT
11011011011011	22	0.814	17	27	129	24187	WABA
111001001001010111	11258	0.585	10427	10350	3254	1	PH-OPT
1111111111	24270	0.451	24285	24233	24310	24310	BLAST
101010101010101011	24310	0.386	24310	24310	24298	24306	WORST

sequence generated by such a model matches a given seed; then, by examining all such seeds, one can find the best.

We have adapted our methods to the subject of detecting homologous coding sequences in DNA. Our HMM for local alignments of these regions models their 3-periodic nature, dependence among positions within triplets, and inhomogeneity in conservation. This HMM is much better at predicting the sensitivity of a given seed than more naive models. In particular, in our experiments, the 3 best seeds according to this model were, in fact, the best overall.

Also, these optimal seeds are substantially more sensitive to alignments than seeds used by existing sequence alignment methods, while alignments based on them require essentially no additional time. Our experiments show that they substantially outperform the naive BLAST seed, and the PatternHunter seed, developed for a different model. They also outperform the simple 3-periodic seed used in WABA, matching 20-29% of the alignments missed by that seed.

Other questions for future work remain open. First, is it possible to develop similar seed-finding algorithms for other probabilistic models more complicated than HMMs, while still keeping the reasonable runtime? Alternatively, can one approximate the probability of a seed match without the expensive dynamic programming we have described? Also, do other pattern-finding applications of this sort exist, where one searches for patterns in the output of HMMs, and if so, how readily can these fit within our framework? Finally, can still more accurate models for coding alignments be developed, or can similar techniques can be used to study patterns of conservations in sequence elements other than protein coding regions?

Acknowledgements. We would like to thank Bioinformatics Solutions, Inc. (http://www.BioinformaticsSolutions.com/) for providing us with a version of PatternHunter customized to our needs and Ming Li for providing access to the manuscript [5].

References

1. S. F. Altschul, W. Gish, W. Miller, E. W. Myers, and D. J. Lipman. Basic local alignment search tool. *Journal of Molecular Biology*, 215(3):403–410, 1990.
2. A. Bairoch and R. Apweiler. The SWISS-PROT protein sequence database and its supplement TrEMBL in 2000. *Nucleic Acids Research*, 28(1):45–48, 2000.
3. J. Buhler, U. Keich, and Y. Sun. Designing seeds for similarity search in genomic dna. In *Proceedings of the 7th Annual International Conference on Computational Biology (RECOMB)*, 2003. To appear.
4. R. Durbin, S. Eddy, A. Krogh, and G. Mitchison. *Biological sequence analysis*. Cambridge University Press, 1998.
5. U. Keich, M. Li, B. Ma, and J. Tromp. On spaced seeds. Unpublished.
6. W. J. Kent and A. M. Zahler. Conservation, regulation, synteny, and introns in a large-scale C. briggsae-C. elegans genomic alignment. *Genome Research*, 10(8):1115–1125, 2000.
7. I. Korf, P. Flicek, D. Duan, and M. R. Brent. Integrating genomic homology into gene structure prediction. *Bioinformatics*, 17 Suppl 1:S140–8, 2001.
8. B. Ma, J. Tromp, and M. Li. PatternHunter: faster and more sensitive homology search. *Bioinformatics*, 18(3):440–445, March 2002.
9. L. R. Rabiner. A tutorial on Hidden Markov models and selected applications in speech recognition. *Proceedings of the IEEE*, 77(2):257–285, 1989.
10. Z. Yang. Maximum-likelihood estimation of phylogeny from DNA sequences when substitution rates differ over sites. *Molecular Biology and Evolution*, 10(6):1396–1401, 1993.

Fast Lightweight Suffix Array Construction and Checking

Stefan Burkhardt[1]⋆ and Juha Kärkkäinen[1]⋆

Max-Planck-Institut für Informatik
Stuhlsatzenhausweg 85, 66123 Saarbrücken, Germany
{stburk,juha}@mpi-sb.mpg.de

Abstract. We describe an algorithm that, for any $v \in [2, n]$, constructs the suffix array of a string of length n in $\mathcal{O}(vn + n \log n)$ time using $\mathcal{O}(v + n/\sqrt{v})$ space in addition to the input (the string) and the output (the suffix array). By setting $v = \log n$, we obtain an $\mathcal{O}(n \log n)$ time algorithm using $\mathcal{O}(n/\sqrt{\log n})$ extra space. This solves the open problem stated by Manzini and Ferragina [ESA '02] of whether there exists a lightweight (sublinear extra space) $\mathcal{O}(n \log n)$ time algorithm. The key idea of the algorithm is to first sort a sample of suffixes chosen using mathematical constructs called difference covers. The algorithm is not only lightweight but also fast in practice as demonstrated by experiments. Additionally, we describe fast and lightweight suffix array checkers, i.e., algorithms that check the correctness of a suffix array.

1 Introduction

The suffix array [21,9], a lexicographically sorted array of the suffixes of a string, has numerous applications, e.g., in string matching [21,9], genome analysis [1] and text compression [5]. In many cases, the construction of the suffix array is a bottleneck. As suffix arrays are often generated for very long strings, both space and time requirement matter.

In a typical situation, a string of length n occupies n bytes and its suffix array $4n$ bytes of space. The suffix array can be constructed using little extra space, but then the worst-case running time is $\Omega(n^2)$. All fast construction algorithms with runtime guarantee $\mathcal{O}(n \log n)$ or better require at least $4n$ bytes of extra space, which almost doubles the space requirement. Manzini and Ferragina [22] asked whether it is possible to achieve $\mathcal{O}(n \log n)$ runtime using sublinear extra space. In this paper, we answer the question positively describing an algorithm with $\mathcal{O}(n \log n)$ worst-case runtime that uses only $\mathcal{O}(n/\sqrt{\log n})$ extra space.

Previous Work. In addition to time and space requirements, suffix array construction algorithms differ in their alphabet model. The models of interest, in order from the most restrictive to the most general, are constant alphabet (the

---

⋆ Partially supported by the Future and Emerging Technologies programme of the EU under contract number IST-1999-14186 (ALCOM-FT).

R. Baeza-Yates et al. (Eds.): CPM 2003, LNCS 2676, pp. 55–69, 2003.

size of the alphabet is bounded by a constant), integer alphabet (characters are integers in a range of size $n^{\mathcal{O}(1)}$), and general alphabet (only character comparisons are allowed).

Previous suffix array construction algorithms can be classified into four main categories.

The algorithms in the first category compute the suffix array from the suffix tree in linear time. The classical suffix tree construction algorithms [28,23,26] work in linear time for constant alphabets. Farach's algorithm [8] achieves linear time for integer alphabets. The drawback of these algorithms is their space requirement. The most space efficient implementation by Kurtz [18] uses $8n$–$14n$ bytes of space in total, but this comes at the cost of limiting the maximum string length to 135 million characters.

The second category, direct linear time construction algorithms, has appeared very recently [16,17,11]. It is not yet clear what the space requirements of practical implementations are but all of them appear to require at least $4n$ bytes of extra space and likely more. One of these algorithms [11] is, in fact, closely related to the present algorithm as described in Section 9. All of these algorithms support integer alphabets.

The algorithms in the third category are based on the doubling algorithm of Karp, Miller and Rosenberg [13]. These algorithms sort the suffixes initially by their first characters and then double the significant prefix length in each further pass. The first algorithm in this category was by Manber and Myers [21], but the Larsson-Sadakane algorithm [19] is considered to be the best in practice. Both algorithms run in $\mathcal{O}(n \log n)$ time, need $4n$ bytes of extra space, and support general alphabets.

The final category consists of algorithms based on sorting the suffixes as independent strings. For most real world inputs, string sorting is very fast [19] and it needs little extra space. Furthermore, Itoh and Tanaka [10] as well as Seward [24] reduce the number of suffixes to be sorted to about half by taking advantage of the correlation between suffixes starting at consecutive positions. However, the worst case running time of these algorithms is $\Omega(n^2)$ and they can also be very slow on real world inputs with long repeats [22]. Basic string sorting supports general alphabets but the Itoh–Tanaka, Seward, and Manzini–Ferragina (below) heuristics assume a constant alphabet.

The string sorting based algorithms are attractive in practice since they are both the fastest in most cases and use little space, but the possibility of quadratic running time is unacceptable for many applications. The solution adopted by the `bzip2` compression package [25] implementing Seward's algorithm uses the Larsson–Sadakane algorithm as a fallback when the primary algorithm takes too much time. However, in such cases, the algorithm is slow and, more importantly, needs $4n$ bytes of extra space. The algorithms of Manzini and Ferragina [22] sort the suffixes using only their first ℓ characters and apply fallback heuristics to the groups of suffixes that remain unsorted. The algorithms use little extra space (less than $0.03n$ bytes), are fast even for many difficult cases, but the worst case running times remain $\Omega(n^2)$.

Our Contribution. We describe an algorithm that runs in $\mathcal{O}(vn + n \log n)$ time using $\mathcal{O}(v + n/\sqrt{v})$ extra space for any $v \in [2, n]$. Thus, the choice of v offers a space–time tradeoff. Moreover, setting $v = \log n$ gives an $\mathcal{O}(n \log n)$ time, $\mathcal{O}(n/\sqrt{\log n})$ extra space algorithm, the first $\mathcal{O}(n \log n)$ time algorithm using sublinear extra space. The algorithm is alphabet-independent, i.e., it supports general alphabets. Note that $\mathcal{O}(n \log n)$ time is optimal under general alphabets.

The key to the result are mathematical constructs called *difference covers* (see Section 2). Difference covers are a powerful tool that can be used to obtain a number of other results, too. We give a brief glimpse to such results in Section 9. Difference covers have also been used for VLSI design [15] and distributed mutual exclusion [20,7].

Experiments show that the algorithm is also practical. A straightforward implementation using less than n bytes of extra space is competitive with the Larsson-Sadakane algorithm [19], which is considered the best among the good worst case algorithms. The algorithm of Manzini and Ferragina [22], probably the fastest string sorting based implementation available, is significantly faster and slightly more space efficient on real world data but is unusably slow on worst case data. We believe that an improved implementation of our algorithm can bring its running time close to that of Manzini–Ferragina even on real world data (see Section 6).

Verifying the correctness of an implementation of an algorithm is a difficult, often impossible task. A more modest but still useful guard against incorrect implementation is a *result checker* [4,27] that verifies the output of a computation. A *suffix array checker* verifies the correctness of a suffix array. The trivial suffix array checker has an $\mathcal{O}(n^2)$ worst case running time. In Section 8, we describe some simple, fast and lightweight checkers. These checkers are not directly related to our main result, but we found them useful during the implementation of the difference cover algorithm.

2   Basic Idea

Our algorithm is based on the following simple observation. Suppose we want to compare two suffixes S_i and S_j (i.e., suffixes starting at positions i and j) to each other. If we can find another pair S_{i+k} and S_{j+k}, called an *anchor pair*, of suffixes whose relative order is already known, then at most the first k characters of S_i and S_j need to be compared. Some of the fallback heuristics of Manzini and Ferragina [22] rely on this observation. Unlike their methods, our algorithm gives a guarantee of finding an anchor pair with a small offset k for any pair of suffixes. This is achieved by sorting first a *sample* of suffixes, and then finding the anchor pairs among the sample.

The question is how to choose such a sample.

An initial attempt might be to take every vth suffix for some v. However, for any pair of suffixes S_i and S_j such that $i \not\equiv j \pmod{v}$, there would be no anchor pairs at all. A second try could be a *random* sample of size $n/\sqrt{v}$, which makes expected distance to an anchor pair $\mathcal{O}(v)$. The problem is that there are no time and space efficient algorithms for sorting an arbitrary set of suffixes. The alternatives are basically general string sorting and full suffix array construction, neither of which is acceptable in our case.

Our solution is based on using difference covers. A *difference cover* D modulo v is a set of integers in the range $[0, v)$ such that for all $i \in [0, v)$, there exist $j, k \in D$ such that $i \equiv k - j \pmod{v}$. For example, $D = \{1, 2, 4\}$ is a difference cover modulo 7:

$$
\begin{array}{c|ccccccc}
i & 0 & 1 & 2 & 3 & 4 & 5 & 6 \\
\hline
k - j & 1 - 1 & 2 - 1 & 4 - 2 & 4 - 1 & 1 - 4 & 2 - 4 & 1 - 2
\end{array}
$$

The sample we use contains all suffixes whose starting positions are in D modulo v. The key properties of the sample are:

- The size of the sample is $\mathcal{O}(n/\sqrt{v})$ (see Section 3).
- The sample guarantees that an anchor pair is found within distance v (see Section 4).
- The periodicity of the sample (with period v) makes it possible to sort the sample efficiently (see Section 5).

3   Tools

Let $s[0, n)$ be a string of length n over a general alphabet, i.e., we assume that the characters can be compared in constant time but make no other assumptions on the alphabet. Let S_i, $i \in [0, n)$, denote the suffix $s[i, n)$. A set of suffixes is v-*ordered* if they are ordered by their first v characters, i.e., as if the suffixes longer than v were truncated to length v.

Lemma 1. *A set of m suffixes (represented by an array of their starting positions) can be v-ordered in $\mathcal{O}(vm + m \log m)$ time using $\mathcal{O}(\log m)$ extra space.*

Proof. Use the multikey quicksort algorithm of Bentley and Sedgewick [3] that can sort m strings of total length M in time $\mathcal{O}(M + m \log m)$. Only the recursion stack requires non-constant extra space.

The suffix array $\mathrm{SA}[0, n)$ of s is a permutation of $[0, n)$ specifying the lexicographic order of the suffixes of s, i.e., for all $0 \leq i < j < n$, $S_{\mathrm{SA}[i]} < S_{\mathrm{SA}[j]}$. The inverse suffix array $\mathrm{ISA}[0, n)$ is the inverse permutation of SA, i.e., $\mathrm{ISA}[i] = j$ if and only if $\mathrm{SA}[j] = i$. The inverse suffix array constitutes a *lexicographic naming* of the suffixes allowing constant time comparisons: $S_i \leq S_j$ if and only if $\mathrm{ISA}[i] \leq \mathrm{ISA}[j]$. The suffix array and its inverse can be computed quickly with the algorithms of Manber and Myers [21] or Larsson and Sadakane [19].

Lemma 2 ([21,19]). *The suffix array SA and the inverse suffix array ISA of a string s of length n can be computed in $\mathcal{O}(n \log n)$ time and $\mathcal{O}(n)$ space.*

A difference cover D modulo v is a set of integers in the range $[0, v)$ such that for all $i \in [0, v)$, there exists $j, k \in D$ such that $i \equiv k - j \pmod{v}$. It is easy to see that a difference cover modulo v must be of size at least $\sqrt{v}$. A simple method for generating small difference covers is described by Colbourn and Ling [7]:

Lemma 3 ([7]). *For any v, a difference cover modulo v of size $\leq \sqrt{1.5v} + 6$ can be computed in $\mathcal{O}(\sqrt{v})$ time.*

For any integers i, j, let $\delta(i, j)$ be an integer $k \in [0, v)$ such that $(i+k) \bmod v$ and $(j + k) \bmod v$ are both in D.

Lemma 4. *Given a difference cover D modulo v, a data structure allowing constant time evaluation of $\delta(i, j)$ for any i and j can be computed in $\mathcal{O}(v)$ time and space.*

Proof. Build a lookup table d such that, for all $h \in [0, v)$, both $d[h]$ and $(d[h] + h) \bmod v$ are in D, and implement δ as $\delta(i, j) = (d[(j-i) \bmod v] - i) \bmod v$. Then $i + \delta(i, j) = d[(j - i) \bmod v] \in D$ and $j + \delta(i, j) = d[(j - i) \bmod v] + (j - 1) \in D$ $\pmod{v}$.

For the difference cover $D = \{1, 2, 4\}$ modulo 7, we have

h	0	1	2	3	4	5	6
$d[h]$	1	1	2	1	4	4	2

For example, let $i = 2$ and $j = 12$. Then we have $k = \delta(2, 12) = (d[10 \bmod 7] - 2) \bmod 7 = (1 - 2) \bmod 7 = 6$, and $(i + \delta(i, j)) \bmod 7 = (2 + 6) \bmod 7 = 1 \in D$ and $(j + \delta(i, j)) \bmod 7 = (12 + 6) \bmod 7 = 4 \in D$.

4   The Algorithm

The algorithm consists of the following phases:

Phase 0. Choose $v \geq 2$, a difference cover D modulo v with $|D| = \Theta(\sqrt{v})$, and compute the associated function δ.

Phase 1. Sort the suffixes whose starting position modulo v is in D.

Phase 2. Construct SA by exploiting the fact that, for any $i, j \in [0, n - v]$, the relative order of the suffixes starting at $i + \delta(i, j)$ and $j + \delta(i, j)$ is already known.

The implementation of Phase 0 was already explained in the previous section. Phase 1 is described in detail in the next section. In this section, we concentrate on Phase 2.

Let D be a difference cover modulo v. A D-sample D_n is the set $\{i \in [0, n] \mid i \bmod v \in D\}$. Let $m = |D_n| \leq (n/v + 1)|D|$. A lexicographic naming of the D-sample suffixes is a function $\ell : D_n \mapsto [0, m)$ satisfying $\ell(i) \leq \ell(j)$ if and only if $S_i \leq S_j$ (see Fig. 1). Phase 1 produces the function ℓ as its output. In the next section, we prove the following lemma.

Lemma 5. *A data structure allowing constant time evaluation of $\ell(i)$ for any $i \in D_n$ can be computed in $\mathcal{O}(\sqrt{v}n + (n/\sqrt{v})\log(n/\sqrt{v}))$ time using $\mathcal{O}(v + n/\sqrt{v})$ space in addition to the string s.*

Input:

i	0	1	2	3	4	5	6	7	8	9	10	11	12	13	14	15	16	17	18	19	20	21	22	23	24	25
$s[i]$	a		r	o	s	e		i	s		a		r	o	s	e		i	s		a		r	o	s	e

Result of phase 1:

$i \in D_{26}$	1	2		4				8	9		11					15	16		18				22	23		25
$\ell(i)$	3	8		11				10	0		2					5	1		9				7	6		4

Result of step 2.1:

i	0	1	2	3	4	5	6	7	8	9	10	11	12	13	14	15	16	17	18	19	20	21	22	23	24	25
$SA^7[i]$	9	19	6	16	21	1	11	20	0	10	25	5	15	7	17	23	3	13	22	2	12	8	18	24	4	14

Output:

i	0	1	2	3	4	5	6	7	8	9	10	11	12	13	14	15	16	17	18	19	20	21	22	23	24	25
$SA[i]$	19	9	16	6	21	11	1	20	10	0	25	15	5	17	7	23	13	3	22	12	2	18	8	24	14	4

Fig. 1. The difference cover algorithm with the difference cover $D = \{1, 2, 4\}$ modulo 7 applied to the string "a rose is a rose is a rose"

The essence of Phase 2 is the observation that the following comparisons are equivalent for all $i, j \in [0, n - v]$:

$$S_i \leq S_j$$
$$\langle s[i, i + v), S_{i+\delta(i,j)}\rangle \leq \langle s[j, j + v), S_{j+\delta(i,j)}\rangle$$
$$\langle s[i, i + v), \ell(i + \delta(i,j))\rangle \leq \langle s[j, j + v), \ell(j + \delta(i,j))\rangle$$

Thus we can implement Phase 2 as follows.

Step 2.1. v-order the suffixes using multikey quicksort.

Step 2.2. For each group of suffixes that remains unsorted, i.e., shares a prefix of length v, complete the sorting with a comparison based sorting algorithm using $\ell(i + \delta(i,j))$ and $\ell(j + \delta(i,j))$ as keys when comparing suffixes S_i and S_j.

The operation of the algorithm is illustrated in Fig. 1. As an example of Step 2.2, consider the ordering of suffixes S_2 and S_{12}, which were not sorted in Step 2.1 because they share the prefix "rose is". We compute $\delta(2, 12) = 6$ and compare $\ell(2 + 6) = 10$ and $\ell(12 + 6) = 9$ to find out that S_{12} is smaller than S_2.

Lemma 6. *If δ and ℓ can be evaluated in constant time, the suffix array $SA[0, n)$ of a string $s[0, n)$ can be constructed in $\mathcal{O}(vn + n \log n)$ time using $\mathcal{O}(\log n)$ space in addition to s, SA, and the space needed for δ and ℓ.*

Proof. The time complexity follows from Lemma 1. To keep the space requirement within $\mathcal{O}(\log n)$, Step 2.2 is performed for an unsorted group immediately when it is formed during Step 2.1. Then the only extra space is needed for the stack.

Combining Lemmas 3, 4, 5, and 6 we obtain the main result.

Theorem 1. *For any positive integer v, the suffix array of a string of length n can be constructed in $\mathcal{O}(vn + n \log n)$ time using $\mathcal{O}(v + n/\sqrt{v})$ space in addition to the string and the suffix array.*

By choosing $v = \log n$, we get:

Corollary 1. *The suffix array of a string of length n can be constructed in $\mathcal{O}(n \log n)$ time using $\mathcal{O}(n/\sqrt{\log n})$ extra space.*

5 Sorting the Sample

The remaining part is to show how to sort the D-sample suffixes. More precisely, we want to compute the lexicographic naming function ℓ. Let $D = \{d_0, \dots, d_{k-1}\}$ with $d_0 < d_1 < \dots < d_{k-1} < d_k = v$, and let h be such that $d_h \leq n \bmod v < d_{h+1}$. Then $D_n = \{d_0, \dots, d_{k-1}, d_0 + v, \dots, d_{k-1} + v, d_0 + 2v, \dots, d_0 + \lfloor n/v \rfloor v, \dots, d_h + \lfloor n/v \rfloor v\}$. Let μ be the mapping $D_n \mapsto [0, m)$: $\mu(d_i + jv) = \lfloor n/v \rfloor i + \min(i, h) + j$, i.e., it maps each sequence $d_i, d_i + v, d_i + 2v, \dots$ to consecutive positions (see Fig. 2). To compute $\mu(k)$, for $k \in D_n$, we need to find i and j such that $k = d_i + jv$. This is done by computing $j = \lfloor k/v \rfloor$, $d_i = k \bmod v$, and using a lookup table to compute i from d_i. Thus, the function μ can be evaluated in constant time after an $\mathcal{O}(v)$ time and space preprocessing.

Let ℓ^v be a lexicographic naming of the D-sample suffixes based on their first v characters, i.e., $\ell^v(i) \leq \ell^v(j)$ if and only if $s[i, \min(i+v, n)) \leq s[j, \min(j+v, n))$. Let $s'[0, m)$ be the string defined by $s'[\mu(i)] = \ell^v(i)$ for $i \in D_n$. Note that, for all $i \in (n - v, n] \cap D_n$, the character $s'[\mu(i)] = \ell^v(i)$ is unique and acts as a separator.

Let S'_i, $i \in [0, m)$, denote the suffixes of s', and let SA$'$ and ISA$'$ be the suffix array of s' and its inverse. The following inequalities are all equivalent for all $i, j \in D_n$.

$$\ell(i) \leq \ell(j)$$
$$S_i \leq S_j$$
$$s[i, n) \leq s[j, n)$$
$$s[i, i + v) \cdot s[i + v, i + 2v) \cdots \leq s[j, j + v) \cdot s[j + v, j + 2v) \cdots$$
$$\ell^v(i) \cdot \ell^v(i + v) \cdots \leq \ell^v(j) \cdot \ell^v(j + v) \cdots$$
$$s'[\mu(i)] \cdot s'[\mu(i) + 1] \cdots \leq s'[\mu(j)] \cdot s'[\mu(j) + 1] \cdots$$
$$S'_{\mu(i)} \leq S'_{\mu(j)} \qquad \text{(Note: separators in } s')$$
$$\text{ISA}'[\mu(i)] \leq \text{ISA}'[\mu(j)]$$

Input:

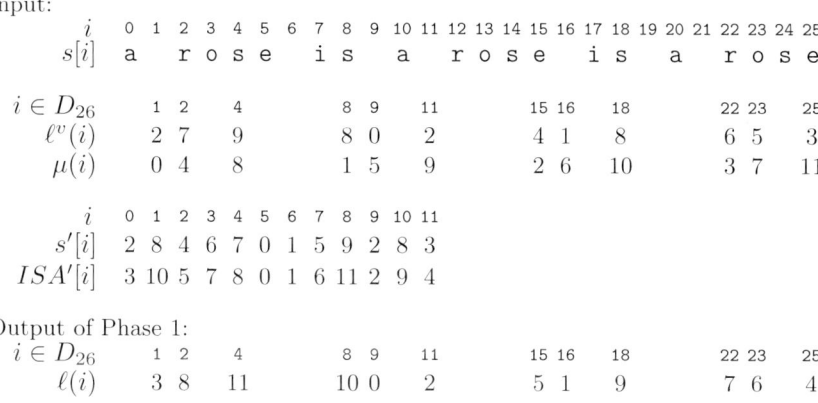

Fig. 2. Phase 1 of the algorithm with the difference cover $D = \{1, 2, 4\}$ modulo 7 applied to the string "a rose is a rose is a rose"

Thus we can implement $\ell(i)$ as $\text{ISA}'[\mu(i)]$.

The inverse suffix array ISA' is computed as follows:

Step 1.1. v-order the D-sample suffixes using multikey quicksort.
Step 1.2. Compute $\ell^v(i)$ for all $i \in D_n$ by traversing the D-sample suffixes in lexicographic order and construct s' by setting $s'[\mu(i)] = \ell^v(i)$.
Step 1.3. Compute ISA' using Manber-Myers or Larsson-Sadakane algorithm.

Proof (of Lemma 5). For D of size $\mathcal{O}(\sqrt{v})$, $m = \mathcal{O}(n/\sqrt{v})$. Thus, the first step takes $\mathcal{O}(vm + m \log m) = \mathcal{O}(\sqrt{v}n + (n/\sqrt{v})\log(n/\sqrt{v}))$ time (Lemma 1). The second step requires $\mathcal{O}(m)$ and the third $\mathcal{O}(m \log m)$ time (Lemma 2). The space requirements are $\mathcal{O}(v)$ for implementing μ, $\mathcal{O}(m) = \mathcal{O}(n/\sqrt{v})$ for s', SA', and ISA', and $\mathcal{O}(\log m)$ for a stack.

6   Implementation

We implemented the difference cover algorithm in C++ using difference covers modulo powers of two for fast division (available through http://www.mpi-sb.mpg.de/~juha/publications.html). The difference covers up to modulo 64 are from [20]. The difference covers modulo 128 and 256 we found using an exhaustive search algorithm. For larger v, we computed the difference covers using the algorithm in [7] (see Lemma 3). In addition to the string and the suffix array, the only non-constant data structures in the implementation are an array storing the difference cover, two lookup tables of size v to implement δ and μ, and an array of size about $4n|D|/v$ bytes for the inverse suffix array ISA'. During Phase 1, another array of size $4n|D|/v$ is needed for SA' but we use (a part of) the suffix array for this. The sizes of the difference cover D and the ISA' array (the latter in bytes) for different values of v are:

v	4	8	16	32	64	128	256	512	1024	2048
$|D|$	3	4	5	7	9	13	20	28	40	58
$|ISA'|$	$3n$	$2n$	$1.25n$	$0.88n$	$0.56n$	$0.41n$	$0.31n$	$0.22n$	$0.16n$	$0.11n$

We used our own implementations of multikey quicksort [3] for v-ordering, and Larsson-Sadakane [19] for ISA' construction.

There are several possibilities for further improving the running time. For real world data, the majority of the construction time is spent in Step 2.1 for v-ordering the suffixes. The Manzini–Ferragina algorithm [22] also spends it most of its time on v-ordering the suffixes but, as shown by the experiments in the next section, they are much faster. Their implementation of v-ordering uses several additional optimization techniques that could also be used in our algorithm. The most important of these is probably Seward's "pointer copying" heuristic [24].

7 Experiments

We evaluated the performance of our implementation for typical inputs and for bad cases, texts with very high average LCP (longest common prefix). We compared our implementation using a difference cover modulo 32 to Larsson's and Sadakane's implementation of their algorithm (http://www.cs.lth.se/~jesper/qsufsort.c) and to that of Manzini and Ferragina (http://www.mfn.unipmn.it/~manzini/papers/ds.tgz).

We ran the experiments on a single processor of a Sun-Fire-15000 and compiled with g++ -O3 respectively gcc -O3. The Solaris operating system provides a virtual file, /proc/self/as, the size of which equals the total amount of memory a process occupies (including code, stack and data). We monitored this file during our runs and report the largest value.

For the real-world experiments we selected text files from the Canterbury Corpus (http://corpus.canterbury.ac.nz/index.html), the tarfile of gcc 3.0, the Swissprot protein database (Version 34) and a random (symmetric Bernoulli) string. Table 1 shows alphabet size $|\Sigma|$, average LCP and maximum LCP for these files. The results can be found in Table 2. The algorithm by Manzini and Ferragina outperforms the other two (which are of comparable speeds) by roughly a factor of 3. It also uses the least memory, followed closely by our implementation which requires about 17% more space. The algorithm of Larsson and Sadakane requires between 52 and 59% more space than that of Manzini and Ferragina.

To evaluate performance for difficult cases we created four artificial strings with 50 million characters each. The first contains solely the letter A. This is a frequently used worst-case text, but as our results indicate, it is far from the worst case for some algorithms. Therefore we also created three others consisting of a single random (symmetric Bernoulli) 'seed' string which is repeated until 50 million characters are reached. We used seed strings of length 20, 1000 and 500 000. Statistics for these texts are presented in Table 3 followed by results in Table 4. Apart from the not-so worst case string consisting only of the letter

Table 1. Real-world text files sorted by average LCP

Text	$|\Sigma|$	Characters	average LCP	maximum LCP
random string	26	50 000 000	4.84	10
King James Bible (*bible.txt*)	63	4 047 392	13.97	551
E. coli genome (*E.coli*)	4	4 638 690	17.38	2 815
CIA World Fact Book (*world192.txt*)	94	2 473 400	23.01	559
Swissprot V34 Protein Database	66	109 617 186	89.08	7 373
gcc 3.0 source code tarfile	150	86 630 400	8 603.21	856 970

Table 2. Runtimes and space consumption of Larsson-Sadakane(LS), Manzini-Ferragina(MF) and our difference cover algorithm(DC32) for real world files

Text	LS		MF		DC32	
	Time[sec]	Space[MB]	Time[sec]	Space[MB]	Time[sec]	Space[MB]
random	227.46	401.0	106.06	251.9	275.60	295.4
Bible	11.98	33.3	2.12	21.6	5.91	25.5
E. coli	13.56	38.1	2.05	24.6	7.20	28.9
World	4.11	20.8	0.99	13.7	3.81	16.2
Swissprot	996.10	877.9	292.66	551.2	1126.80	645.7
gcc 3.0	528.63	694.0	298.89	439.1	577.60	510.6

Table 3. Text files with long common prefixes

Text	$|\Sigma|$	Characters	average LCP	maximum LCP
length 500000 random, repeated	26	50 000 000	24 502 500.5	49 500 000
length 1000 random, repeated	26	50 000 000	24 999 000.5	49 999 000
length 20 random, repeated	15	50 000 000	24 999 980.5	49 999 980
$\{A\}^{5\cdot10^7}$	1	50 000 000	24 999 999.5	49 999 999

Table 4. Runtimes and space consumption of Larsson-Sadakane(LS), Manzini-Ferragina(MF) and our difference cover algorithm(DC32) for worst-case strings. For the first three databases the algorithm of Manzini and Ferragina did not finish within 24 hours, so we broke off the tests (marked with a -)

Text	LS		MF		DC32	
	Time[sec]	Space[MB]	Time[sec]	Space[MB]	Time[sec]	Space[MB]
500000 char repeats	1310.82	401.0	-	-	618.91	295.4
1000 char repeats	1221.71	401.0	-	-	1460.26	295.4
20 char repeats	1118.49	401.0	-	-	808.36	295.4
$\{A\}^{5\cdot10^7}$	160.13	401.0	8.30	251.9	343.44	295.4

A, the speed of our implementation (DC32) is comparable to that of Larsson and Sadakane (between 53% faster and 19% slower) but it saves 26% space. For these instances, the algorithm of Manzini and Ferragina had runtimes of more than one day (an estimate based on experiments with smaller datasets resulted in a time of 2 weeks).

We also studied the effect of the choice of the difference cover on the performance of the algorithm. For texts with low or moderate average LCP, large covers result in lower runtimes and space requirements. However, for texts with high average LCP, there exists a tradeoff between space and time consumption. Figure 3 illustrates this with the results of experiments using difference covers modulo 4 to 1024 applied to a random (symmetric Bernoulli) string with 50 million characters and a high-LCP string of the same length (length 1000 random, repeated). It clearly displays the different behaviour of our algorithm for low and high average LCP texts. In addition to the better performance of large covers for texts with low or moderate LCP, it shows that moderately sized covers are relatively good overall performers.

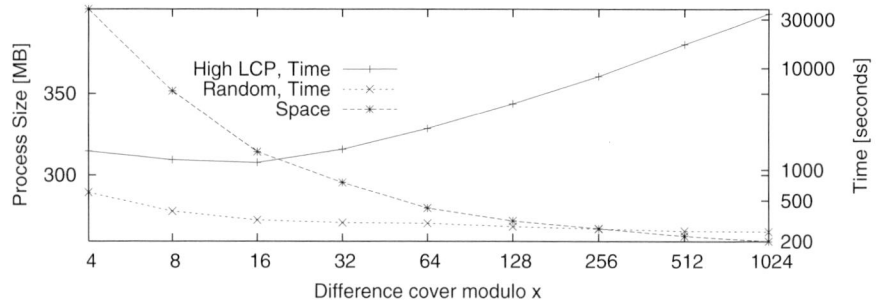

Fig. 3. Influence of the difference cover on performance

8   Suffix Array Checkers

A suffix array checker is an algorithm that takes a string and its suffix array and verifies the correctness of the suffix array. In this section, we describe suffix array checkers based on the following fact.

Theorem 2. *An array* $\mathrm{SA}[0,n)$ *of integers is the suffix array of a string* $s[0,n)$ *if and only if the following conditions are satisfied:*

1. *For all* $i \in [0,n)$, $\mathrm{SA}[i] \in [0,n)$.
2. *For all* $i \in [1,n)$, $s[\mathrm{SA}[i-1]] \leq s[\mathrm{SA}[i]]$.
3. *For all* $i \in [1,n)$ *such that* $s[\mathrm{SA}[i-1]] = s[\mathrm{SA}[i]]$ *and* $\mathrm{SA}[i-1] \neq n-1$, *there exists* $j,k \in [0,n)$ *such that* $\mathrm{SA}[j] = \mathrm{SA}[i-1]+1$, $\mathrm{SA}[k] = \mathrm{SA}[i]+1$ *and* $j < k$.

We omit the proof here.

The conditions 1 and 2 are easy to check in linear time. For checking condition 3, we can find the values j and k by $j = \text{ISA}[\text{SA}[i-1]+1]$ and $k = \text{ISA}[\text{SA}[i]+1]$ if the inverse suffix array ISA is available. Since ISA can be computed from SA (or checked against SA) in linear time, we obtain a simple linear time checker.

The drawback of the above algorithm is the extra space required for the inverse suffix array. Therefore, we offer lightweight alternatives based on the following alternative formulation of Condition 3:

For all characters $c \in \Sigma$: If $\text{SA}[a,b]$ contains the suffixes starting with the character c, then $\text{SA}[a]+1, \text{SA}[a+1]+1, \ldots, \text{SA}[b]+1$ occur in SA in this order (but not consecutively in general), except that the first entry $\text{SA}[a]+1$ is missing when $c = s[n-1]$.

For a given character, checking the condition in linear time is trivial. This leads to an $\mathcal{O}(\sigma_2 n)$ time, $\mathcal{O}(1)$ extra space algorithm, where σ_2 is the number of distinct characters that occur at least twice in s (no checking is needed for characters that occur only once).

The following algorithm does the checking for all the characters in one pass.

```
1    for i := n − 1, . . . , 0 do A(s[SA[i]]) = i
2    A(s[n − 1])++                          // skip SA[i] = n − 1
3    for k := 0, . . . , n − 1 do
4        if SA[k] > 1 then
5            c := s[SA[k] − 1]
6            check that SA[A(c)] + 1 = SA[k]
7            A(c)++
```

The time and space complexity of the algorithm depends on the implementation of A. The algorithm performs $\mathcal{O}(n)$ accesses to A and everything else runs in linear time and constant extra space. Simple implementations include a lookup table, a hash table and a sorted array.

For large alphabets, the space requirement can approach the simpler inverse suffix array checker. A simple optimization, already mentioned above, is to include only characters that occur at least twice. Furthermore, the σ_2 characters can be split into smaller groups and the lines 3–7 of the above algorithm performed separately for each group. Limiting group size to $d < \sigma_2$, the sorted array implementation runs in $\mathcal{O}((\sigma_2 n/d) \log d)$ time and $\mathcal{O}(d)$ extra space. Setting $d = n/\sqrt{\log n}$ to match the space requirement of our construction algorithm, we obtain an $\mathcal{O}(n \log \sigma_2 + \sigma_2 \log^{1.5} n)$ time checker.

A straightforward sorted array implementation (to make it comparison based as the construction algorithm is) never took more than half the construction time and usually much less.

9   Discussion

We have described a suffix array construction algorithm that combines fast worst case running times with small space consumption. The key idea behind the algorithm is using a sample of suffixes based on a difference cover. There are several further possibilities for extensions and variations of the algorithm. For example, the algorithm can be extended to compute the longest common prefix (LCP) of each pair of adjacent suffixes with the same time and extra space limits. There is a linear time LCP algorithm [14] but it requires $4n$ bytes of extra space.

Another application is sorting a *subset* of suffixes. Some special kinds of subsets can be sorted efficiently [12,2,6,17]. Also, the technique of Section 5 can be used when the set of starting positions is periodic. However, the previous alternatives for sorting an *arbitrary* subset of m suffixes of a string of length n are string sorting, with $\Omega(nm)$ worst case running time, and full suffix array construction, with $\Omega(n)$ extra space requirement. Phase 2 of the difference cover algorithm works just as well for an arbitrary subset of suffixes, giving an algorithm that runs in $\mathcal{O}(\sqrt{v}n + (n/\sqrt{v})\log(n/\sqrt{v}) + vm + m\log m)$ time and $\mathcal{O}(v + n/\sqrt{v})$ extra space. For example, if $m = n/\log n$, by choosing $v = \log^2 n$ we obtain an $\mathcal{O}(n\log n)$ time and $\mathcal{O}(n/\log n) = \mathcal{O}(m)$ extra space algorithm.

The basic idea of using difference covers can also be realized in ways that are quite different from the present one. This is demonstrated by the *linear time* (but not lightweight) algorithm in [11]. To illustrate the similarities and differencies of the two algorithms, we give a brief description (which is quite different from the one in [11]) of the linear time algorithm:

Phase 0. Choose $v = 3$ and $D = \{1, 2\}$. Then the D-sample contains two thirds of the suffixes.

Phase 1. Perform the same steps as in Phase 1 of the present algorithm but do Step 1.1 by radix sort and Step 1.3 by a recursive call.

Phase 2(a). Perform Phase 2 on the remaining one third of the suffixes (not on all suffixes) using radix sort for both steps. Note that using radix sort in Step 2.2 is possible because $\delta(i, j) = 1$ for all $i, j \equiv 0 \pmod 3$.

Phase 2(b). Merge the two sorted groups of suffixes. Comparisons of suffixes are done as in Phase 2 of the present algorithm.

References

1. M. I. Abouelhoda, S. Kurtz, and E. Ohlebusch. The enhanced suffix array and its applications to genome analysis. In *Proc. 2nd Workshop on Algorithms in Bioinformatics*, volume 2452 of *LNCS*, pages 449–463. Springer, 2002.

2. A. Andersson, N. J. Larsson, and K. Swanson. Suffix trees on words. *Algorithmica*, 23(3):246–260, 1999.

3. J. L. Bentley and R. Sedgewick. Fast algorithms for sorting and searching strings. In *Proc. 8th Annual Symposium on Discrete Algorithms*, pages 360–369. ACM, 1997.

4. M. Blum and S. Kannan. Designing programs that check their work. *J. ACM*, 42(1):269–291, Jan. 1995.

5. M. Burrows and D. J. Wheeler. A block-sorting lossless data compression algorithm. Technical Report 124, SRC (digital, Palo Alto), May 1994.
6. R. Clifford. Distributed and paged suffix trees for large genetic databases. In *Proc. 14th Annual Symposium on Combinatorial Pattern Matching*. Springer, 2003. This volume.
7. C. J. Colbourn and A. C. H. Ling. Quorums from difference covers. *Inf. Process. Lett.*, 75(1–2):9–12, July 2000.
8. M. Farach. Optimal suffix tree construction with large alphabets. In *Proc. 38th Annual Symposium on Foundations of Computer Science*, pages 137–143. IEEE, 1997.
9. G. Gonnet, R. Baeza-Yates, and T. Snider. New indices for text: PAT trees and PAT arrays. In W. B. Frakes and R. Baeza-Yates, editors, *Information Retrieval: Data Structures & Algorithms*. Prentice-Hall, 1992.
10. H. Itoh and H. Tanaka. An efficient method for in memory construction of suffix arrays. In *Proc. 6th Symposium on String Processing and Information Retrieval*, pages 125–136. IEEE, 1999.
11. J. Kärkkäinen and P. Sanders. Simple linear work suffix array construction. In *Proc. 13th International Conference on Automata, Languages and Programming*. Springer, 2003. To appear.
12. J. Kärkkäinen and E. Ukkonen. Sparse suffix trees. In *Proc. 2nd Annual International Conference on Computing and Combinatorics*, volume 1090 of *LNCS*, pages 219–230. Springer, 1996.
13. R. M. Karp, R. E. Miller, and A. L. Rosenberg. Rapid identification of repeated patterns in strings, trees and arrays. In *Proc. 4th Annual Symposium on Theory of Computing*, pages 125–136. ACM, 1972.
14. T. Kasai, G. Lee, H. Arimura, S. Arikawa, and K. Park. Linear-time longest-common-prefix computation in suffix arrays and its applications. In *Proc. 12th Annual Symposium on Combinatorial Pattern Matching*, volume 2089 of *LNCS*, pages 181–192. Springer, 2001.
15. J. Kilian, S. Kipnis, and C. E. Leiserson. The organization of permutation architectures with bused interconnections. *IEEE Transactions on Computers*, 39(11):1346–1358, Nov. 1990.
16. D. K. Kim, J. S. Sim, H. Park, and K. Park. Linear-time construction of suffix arrays. In *Proc. 14th Annual Symposium on Combinatorial Pattern Matching*. Springer, 2003. This volume.
17. P. Ko and S. Aluru. Linear time construction of suffix arrays. In *Proc. 14th Annual Symposium on Combinatorial Pattern Matching*. Springer, 2003. This volume.
18. S. Kurtz. Reducing the space requirement of suffix trees. *Software – Practice and Experience*, 29(13):1149–1171, 1999.
19. N. J. Larsson and K. Sadakane. Faster suffix sorting. Technical report LU-CS-TR:99–214, Dept. of Computer Science, Lund University, Sweden, 1999.
20. W.-S. Luk and T.-T. Wong. Two new quorum based algorithms for distributed mutual exclusion. In *Proc. 17th International Conference on Distributed Computing Systems*, pages 100–106. IEEE, 1997.
21. U. Manber and G. Myers. Suffix arrays: A new method for on-line string searches. *SIAM J. Comput.*, 22(5):935–948, Oct. 1993.
22. G. Manzini and P. Ferragina. Engineering a lightweight suffix array construction algorithm. In *Proc. 10th Annual European Symposium on Algorithms*, volume 2461 of *LNCS*, pages 698–710. Springer, 2002.
23. E. M. McCreight. A space-economic suffix tree construction algorithm. *J. ACM*, 23(2):262–272, 1976.

24. J. Seward. On the performance of BWT sorting algorithms. In *Proc. Data Compression Conference*, pages 173–182. IEEE, 2000.
25. J. Seward. The bzip2 and libbzip2 official home page, 2002. `http://sources.redhat.com/bzip2/`.
26. E. Ukkonen. On-line construction of suffix trees. *Algorithmica*, 14(3):249–260, 1995.
27. H. Wasserman and M. Blum. Software reliability via run-time result-checking. *J. ACM*, 44(6):826–849, Nov. 1997.
28. P. Weiner. Linear pattern matching algorithm. In *Proc. 14th Symposium on Switching and Automata Theory*, pages 1–11. IEEE, 1973.

Distributed and Paged Suffix Trees for Large Genetic Databases

Raphaël Clifford and Marek Sergot

Department of Computing, Imperial College, London.
raphael@clifford.net, m.sergot@imperial.ac.uk

Abstract. We present two new variants of the suffix tree which allow much larger genome sequence databases to be handled efficiently. The method is based on a new linear time construction algorithm for "sparse" suffix trees, which are subtrees of the whole suffix tree. The new data structures are called the *paged suffix tree* (PST) and the *distributed suffix tree* (DST). Both tackle the memory bottleneck by constructing subtrees of the full suffix tree independently and are designed for single processor and distributed memory parallel computing environments (e.g. Beowulf clusters), respectively. The standard operations on suffix trees of biological importance are shown to be easily translatable to these new data structures. While none of these operations on the DST require inter-process communication, many have optimal expected parallel running times.

1   Introduction

The suffix tree is the key data structure of computational pattern matching which allows a multitude of sophisticated operations to be performed efficiently (see e.g. [3,12]). In the field of bioinformatics this includes whole genome alignment [6], analysis of repetitive elements [17], and fast protein classification [7], amongst many others. However, the main obstacle to more widespread acceptance of these methods remains that of memory use. Suffix trees have high memory overheads, and the poor memory locality, both of their construction and of querying algorithms, make disk-based implementations highly problematic.

Generally, the existing approaches to tackling this memory bottleneck can be divided into two categories. On the one hand, there are those that attempt to improve the implementation of the suffix tree itself, at the cost of limiting the maximum problem size (see e.g. [16]). On the other hand, related data structures have been developed which have lower memory overheads at the cost of either restricting the range of queries that can be performed or increasing their time complexity (e.g. suffix arrays [18], level-compressed tries [2] and suffix cactuses [14], sparse suffix trees [15]).

To tackle significantly larger problem sizes (e.g. data that is hundreds of times larger than available RAM) a disk-based scheme would be desirable. However, as a result of the poor locality mentioned above, most existing applications of disk-based schemes assume that the order of node traversal to be performed at

R. Baeza-Yates et al. (Eds.): CPM 2003, LNCS 2676, pp. 70–82, 2003.

query time is known in advance. Ferragina and Grossi have studied the efficient external memory construction of string indices in general [8,9,10]. To the authors' knowledge, these methods have not yet been applied to large-scale bioinformatics problems.

We present here two new data structures for problems of intermediate size— that is, problems larger than can be handled by existing suffix tree/array methods but small enough that the input can be stored entirely in real memory—a range of at least an order of magnitude. To give some indication, the new methods allow us to store and analyse the whole human genome, perform cross species pattern matching on all available bacterial genomes at once, or search a large EST database, using a small cluster of standard PC's. The data structures are termed the *distributed suffix tree* (DST) and the *paged suffix tree* (PST). They are both based on a new extension of Ukkonen's suffix tree construction algorithm [19] which allows subtrees of a suffix tree to be constructed efficiently in space proportional to the size of the resultant data structure and not the whole suffix tree. This enables a suffix tree to be either distributed over a number of computing nodes (and queried in parallel) or for a single node to compute independent subtrees successively, querying each in turn. By effectively splitting the input string lexicographically (not into contiguous substrings) we show that all the most popular biologically inspired operations on suffix trees exhibit optimal or near optimal parallel speedups. Furthermore problems which would previously have been impossible to solve due to their size can now be tackled efficiently, either in parallel or serial and with modest hardware requirements.

The DST and PST construction algorithms have been implemented in C on an 8 processor distributed memory parallel computer, increasing by a factor of 7.65 the size of the largest database that could be indexed. Exact set matching and repeat finding procedures for random data have also been implemented and performed on the DST. The results, not shown here for space reasons, showed substantial speedups (with average efficiencies in excess of 90% and 99%, respectively) and exhibited good scalability, confirming the theoretical analysis [5]. For systematically biased genetic data, preliminary results show that simple load balancing schemes can successfully increase the parallel efficiency of biological operations to close to 90%.

The method is simple to apply. Almost any current bioinformatic technique that relies on suffix trees can be modified to take advantage of DSTs or PSTs, greatly extending the range of problem sizes that can be tackled. Also, complex or time consuming queries, such as the preliminary stages of matching all ESTs against the human genome, can be performed with optimal or near optimal efficiency in parallel. In the next section we first describe the new data structures and then present the construction algorithms. We then present the expected time efficiencies of a sample of operations on the DST and show empirical results using a snapshot of the sequencing of human chromosomes 21, 22, and X.

Construction algorithms for two previous types of (sparse) suffix tree have been considered in [15] and [1]. The first work considers suffix trees that contain only *evenly spaced* suffixes of the text. This allows pattern matching to be performed on a single, smaller, suffix tree at the cost of increasing the running time of the query. The definition of suffix links proposed there is not suitable for the class of sparse suffix trees we consider as it depends on properties of evenly spaced suffixes that do not hold in general. However, the construction algorithm presented here can be viewed as an extension of that work. The method in [1] uses a quite different word-oriented approach that relies on delimiters between "words" in the text. These delimiters may not overlap and so the method can not be directly applied to the problem of constructing subtrees of a suffix tree.

2   Distributed and Paged Suffix Trees

A suffix tree of input string t is a compacted trie of the suffixes of t. We define a *sparse suffix tree* (SST) of input string t to be a compacted trie of a subset of the suffixes of t. Here, we are interested in the special case where all the suffixes in this subset start with the same prefix z and assume from now on that all SSTs are of this type. Both *distributed suffix trees* (DST) and *paged suffix trees* (PST) are simply collections of SSTs defined in this way. The most efficient use for a PST is for applications where the different SSTs can be queried successively and independently. In that case the SSTs will be constructed on a single processor only as they are needed and then discarded before the next one is required. We concentrate on the DST from now on to simplify the explanation and return to the PST in Section 2.4. The SSTs in a PST are defined in exactly the same way as they are for a DST.

Usually, a single SST will be held at each computing node and the union of the path labels of the leaves of these SSTs will be the full set of suffixes of t. In other words, every suffix of t will be represented by exactly one SST at exactly one of the computing nodes. An example DST and the corresponding standard suffix tree are given in Figures 1 and 2.

In this case the prefixes for the 6 different SSTs are *aa*, *ac*, *ca*, *cc*, *a$* and *$*. Each SST has been connected to a central root node. The *sparse suffix links* will be explained below but the most important feature is that in the standard suffix tree the suffix links can point the full width of the tree. In the DST the new links point only to nodes that are within the same SST. This allows the SSTs to be constructed independently without any inter-process communication.

2.1   Preliminaries

We call the input string t and assume that $n = |t|$ throughout this paper. The characters of t are drawn from an ordered alphabet Σ and we let $\sigma = |\Sigma|$. $t[i, j]$ represents the substring of t starting at position i and terminating at position j, inclusively. A suffix of t is said to be *repeated* if it occurs at least once as a non-suffix substring of t. We also need to be able to specify which suffixes of t are to be included in an SST. This is done by considering a short prefix string z, and a set of start positions, V_z, for those suffixes of the input which have the string z as a prefix. V_z is also called the *valid set*. We say that a substring s of

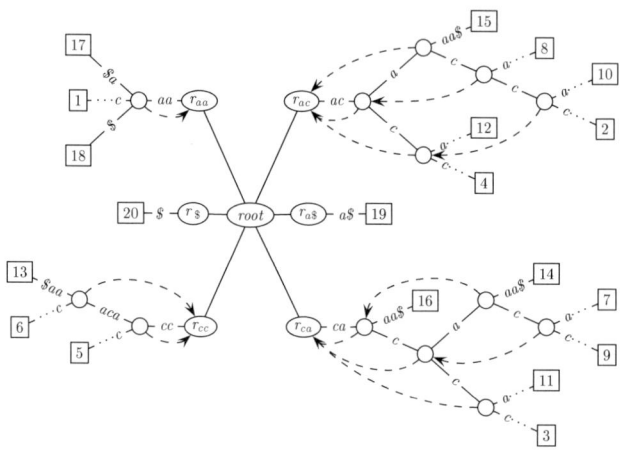

Fig. 1. The SSTs for $aacacccacacaccacaaa\$$ with their respective root nodes labelled $r_{aa}, r_{ac}, r_{ca}, r_{cc}, r_{a\$}$ and $r_\$$. The sparse suffix links for the valid sets $V_{aa}, V_{ac}, V_{ca}, V_{cc}, V_{a\$}$ and $V_\$$ are marked with dashed arrows. Note that the final suffixes, $a\$$ and $\$$, are included but typically will not be used.

t is *valid* if z is a prefix of s and that an interval $I[i, j]$ is *valid* (with respect to V_z and t) if $i \in V_z$. We say that s is a *valid suffix for* for $I[i, j]$ if $s = t[k, j]$ for some $k \in V_z$ and $k > i$. Note that it is possible that a valid suffix for an interval may be shorter than the fixed prefix z, depending on the characters that follow directly in the input string.

Let z be a string of length greater than or equal to 1. We say (t, V_z) is an *input pair* if V_z is the valid set. We write $sst(t, V_z)$ for the sparse suffix tree of t using the valid set V_z. For a sparse suffix tree $T = sst(t, V_z)$ and an arbitrary set of strings S, we say that $T' = T$ *augmented by* S if T' is the sparse suffix tree of the valid suffixes of the input pair (t, V_z) and the strings in S.

We call the concatenation of the edge labels on the path from the *root* to some position in the SST (either a node, or somewhere on an edge between two nodes) the *path label* of that position. To avoid confusion between strings and nodes, we use the notation $\overline{w}$ to label a node whose path label is w. A string which is a path label of some node is called *nodal*. A string which corresponds to any path label in the tree is said to *occur* in the tree.

Example 1. Consider the input

$$s = agacagc\$ \quad \text{with} \quad z = ag$$

The valid set is $V_z = \{1, 5\}$ and $sst(t, V_z)$ has two leaves labelled $\overline{agacagc}$ and $\overline{agc}$ and one branching node labelled $\overline{ag}$. The root has only one child.

2.2   Building the DST

We are able to construct the DST of string t in $O(n)$ time in parallel with no communication overheads apart from the one-off cost of sending the input to

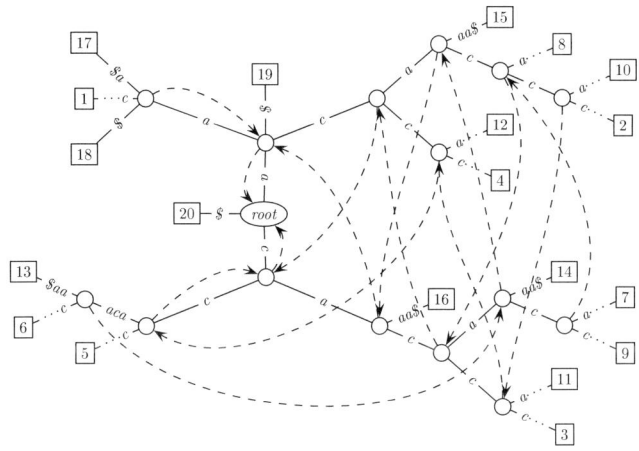

Fig. 2. The standard suffix tree of *aacacccacacaccacaaa*$ with standard suffix links. This is for comparison with the merged tree in Figure 1. See the text for further explanation.

the different nodes. This communication can be achieved on most modern local area networks by broadcasting the data to all nodes simultaneously, making the entire construction time $O(n)$. The time complexity can be achieved trivially, of course, by simply constructing the full suffix tree at each computing node using Ukkonen's algorithm [19] and then pruning it to remove unwanted nodes and edges. However, the assumption we make is that this will not be possible for large problems due to memory constraints and so a novel construction method is required.

To construct the DST in linear time we show how to construct an SST in linear time and simply run the algorithm in parallel for the different prefixes that are chosen. The resulting algorithm uses space at each node which is proportional to the size of the SST constructed, as required.

Sparse Suffix Links

Suffix links play a critical role in the linear time construction of suffix trees. However, the standard definition is not sufficient for SSTs as, in general, a substring of t and its longest suffix may have different prefixes. Moreover, suffixes that are shorter than the length of the prefix z have to be considered separately. Online construction proceeds by reading in one character at a time from the input and stepping down in the tree until there is a mismatch. At this point a new edge or node is inserted and a jump is made in the tree to a new position, from which the process is continued. To perform this jump we need to consider the longest suffix of the current position which might require an additional edge or node in the tree, either using the current character or when more data is read in. Suffix links are used to perform this traversal and it is a fundamental property that if the current position is a node then there will be a corresponding node to which the jump should be made. A new definition of suffix links is required.

Definition 1. *Consider an input pair (t, V_z) and sparse suffix tree $T = sst(t, V_z)$. Let aw be nodal in T and v be the longest repeated suffix of aw that occurs in T. A sparse suffix link or ssl is an unlabelled edge from $\overline{aw}$ to the root if $|v| < |z|$ and from $\overline{aw}$ to $\overline{v}$, otherwise.*

The following proposition shows that sparse suffix links are always well defined in a sparse suffix tree.

Proposition 1. *Consider an input pair (t, V_z) and sparse suffix tree $T = sst(t, V_z)$. Let aw be nodal in T and v be the longest repeated suffix of aw that occurs in T. If $|v| \geq |z|$ then v is nodal and therefore the sparse suffix link from $\overline{aw}$ to $\overline{v}$ is well defined.*

Proof. If aw is nodal then there must be at least two occurrences of aw in t, each with a different character directly to its right. v must occur as a suffix of both occurrences. Therefore there are two occurrences of v in t with different characters directly to their right. If $|v| \geq |z|$ then v must be valid (as v occurs in T). This implies that v is nodal in T, as required.   □

We must also ensure that the total number of steps required by the construction algorithm is linear and that the correct positions can be visited by the sparse suffix links. A full proof is provided in [5]. Here we show which suffixes need to be inspected at each stage of construction and argue that this can be done efficiently.

The demonstration is an extension of that provided in [11] for standard suffix trees. First we need some further definitions. The valid set V_z is always with respect to input string t. When the valid set is applied to a prefix p of t, any start positions that are greater than $|p|$ are simply discarded.

Definition 2. *Consider input pair (p, V_z) and a valid repeated suffix s of p. Denote the set of such suffixes by $R(p, V_z)$. We define this set to include the empty string, ϵ. Let $\alpha(p)$ be the longest suffix in $R(p, V_z)$.*

Example 2. Consider input string $t = aabaaa$ with prefix $p = aabaa$ and $V_{aa} = \{1, 4, 5\}$. $\alpha(p) = aa$ and $R(p, V_{aa}) = \{aa, \epsilon\}$.

It is an important property of $R(p, V_z)$ that all its elements must, by definition, either have length zero or be at least as long as the prefix string z.

Theorem 1. *Consider the input pair (t, V_z) with p, a proper prefix of t. Let a be the character in t that directly follows the prefix p. Consider also the set, S, of valid suffixes, sa, for $I[1, |pa|]$ such that*

$$|\alpha(p)a| \geq |sa| > |\alpha(pa)|$$

and $s \in R(p, V_z)$. Then $sst(pa, V_z) = sst(p, V_z)$ augmented by S.

Proof. Let sa be a valid suffix for $I[1, |pa|]$. We need to insert this new suffix into the tree if and only if $s \in R(p, V_z)$ but $sa \notin R(pa, V_z)$ ($s = \epsilon$ is a special case). In this case sa corresponds to a leaf in $sst(pa, V_z)$ but s does not correspond to a leaf in $sst(p, V_z)$ so a new node is necessary.

1. If $|sa| > |\alpha(p)a|$ then $s \notin R(p, V_z)$. But s is valid and thus corresponds to a leaf in $sst(p, V_z)$. In such a case sa will correspond to the same leaf in $sst(pa, V_z)$ by the implicit growing of the corresponding open edge. No action is needed.
2. If $|\alpha(p)a| \geq |sa| > |\alpha(pa)|$ then $sa \notin R(pa, V_z)$. To determine whether any action is needed we consider the first part of the inequality which gives us $|\alpha(p)| \geq |s|$. In this case $s \in R(p, V_z)$ and therefore a new leaf $\overline{sa}$ must be added.
3. If $|\alpha(pa)| \geq |sa|$ then either $sa \in R(pa, V_z)$ or none of the non-suffix occurrences of sa is valid. Recall that either $|s| \geq |z|$ or $|s| = \epsilon$. Consider the two cases:
   a) $|s| \geq |z|$. sa is a valid suffix for $I[1, |pa|]$ and therefore z is a prefix of sa. As sa is a suffix of $\alpha(pa)$ it now follows that $sa \in R(pa, V_z)$. No action is needed.
   b) $s = \epsilon$. sa is simply the first character of z and must therefore occur in p (as $\alpha(pa)$ has non-zero length). Therefore no action is needed.

So we need only consider case 2 to insert any new leaves required. Therefore only suffixes, sa, satisfying $|\alpha(p)a| \geq |sa| > |\alpha(pa)|$ where $s \in R(p, V_z)$ need to be inserted into $sst(p, V_z)$ as required. □

The positions in the tree of successive elements of $R(p, V_z)$ can be visited using sparse suffix links in an analogous way to the way positions of successive suffixes are visited in a standard suffix tree. The total number of these suffixes is $O(n)$ and, assuming that $|z|$ is constant, the total time taken to visit all their positions is also $O(n)$. Therefore, the whole construction algorithm runs in $O(n)$ time.

Theorem 2. *For a input pair (t, V_z) with $|z| \geq 1$, $sst(t, V_z)$ can be constructed in $O(n)$ time, where $n = |t|$.*

Proof. There are two main differences between our SST construction algorithm and the suffix tree construction algorithms of [19] and [15]. The first is that we use ssl's instead of suffix links and the second is that we must apply a different rule if we follow an ssl to the root. If we ignore this extra rule for a moment, the running time follows closely the reasoning that is presented in the previous papers and so we do not describe it further here. However, if an ssl is followed to the root then the position of the longest repeated prefix of the current suffix may be anywhere between the root and its child node. The position of the next suffix which must be inserted into the tree can trivially be found however, by simply stepping down in the tree, one character at a time. The characters of the next valid suffix in t are used to step down until a mismatch is found. As the next valid suffix must, by definition, have z as a prefix no mismatch will be found before the first child node is reached. Each time an ssl is followed the index of the current suffix being considered increases. This index is never decreased, so the total number of ssl's that are followed is $O(n)$. Therefore, the total number of one character steps taken down the tree is $O(n|z|)$. Assuming that $|z|$ is bounded above by a small constant, the total running time for the algorithm is therefore $O(n)$ as required.   □

2.3   Experimental Results

To test the sparse suffix tree construction algorithm we ran a series of experiments using random binary data and different prefix lengths. A fixed prefix of the desired length was chosen for each test. For the first test the prefix is set to "a", for the second "aa" and so on. As there is no inter-node communication requred in DST construction, the running times shown reflect the parallel running time for computing all the SSTs of a DST on a cluster of computers (excluding the time to broadcast the input to the different processors). The construction times for a PST can be calculted by multiplying the running time for each prefix by the number of prefixes of that length. 20 bytes/valid suffix were required for our implementation. More sophisticated implementations such as those described in [11] could significantly reduce this number. The purpose here is to compare the new and old data structures using the same implementation techniques.

The timings for different length prefixes are presented in Figure 3. The implementation is in C and was run on a 512MB 800MHz AMD system running Linux 2.2.19. The construction of the complete suffix tree slows drastically for inputs larger than 24.4 million characters. This is when real memory is exhausted. With prefix "aa" the maximum size is 47.8 million. With prefixes "aa" and "aaa" it grows to 94.4 million and 186.7 million respectively. So, using 8 SSTs and a binary alphabet, we are able to construct a DST or PST for problems approximately 7.65 times larger than before.

2.4   Operations on the DST and PST

Gusfield [12] provides what can be regarded as a canonical list of major bioinformatic techniques on suffix trees. The algorithms can be broadly classified into three categories according to how they perform on a DST. The first category is that of exact pattern matching algorithms which have very fast serial solutions. The only obstacle to their practical use is the size of the suffix tree that must be computed. A DST will allow a much larger text to be indexed and each search will require only one processor to perform a computation. This is in contrast to a segmentation of the text into contiguous sections which would require all processors to perform calculations for each query. The second category consists of operations such as the calculation of matching statistics [4] which can be performed on a DST but for which there is little parallel speedup. The third category are the algorithms which both benefit from being able to be run on larger data sets and which show optimal or near optimal speedup on a DST. It

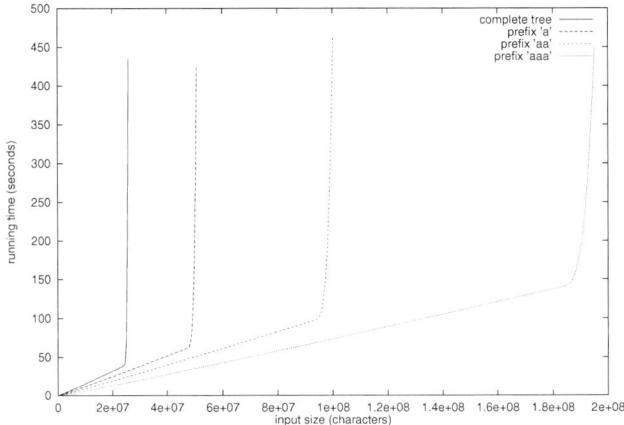

Fig. 3. A comparison of sparse suffix tree construction times using a binary alphabet. The sharp upturn in each line indicates the point at which real memory was exhausted.

is perhaps surprising that all commonly used bioinformatic operations fall into either the first or the third category—that is, they perform traversals of the tree which can easily be translated to a DST without incurring any communication overheads. There is, of course, another class of algorithms outside this list which would require inter-process comunication if run on a DST. It is likely that these algorithms would also be impractical to run on a PST as only one SST will be in real memory at any given time. It is an open question which of them can be translated into efficient parallel algorithms on a DST.

We now summarise the analysis of five representative problems from the first and third categories described above. They are Longest Common Substring, Exact Local Matching (LCS, ELM), Maximal Repeat Finding, All Pairs Suffix-Prefix [13] and Exact Set Matching. Full descriptions along with their serial solutions using suffix trees can be found in Gusfield [12] and elsewhere. All five problems can clearly be solved using a PST with the running time complexity, excluding the time to construct the SSTs, exactly the same as for the serial case. We now examine their solutions on a DST.

Because we are interested in average case and not worst case analysis we make the commonly used assumption that the input characters are independent and uniformly distributed (i.u.d.). In practice, this assumption may not hold, of course; load balancing for systematically biased data is discussed at the conclusion.

We suppose that there are k computing nodes and assume for simplicity that $k = \sigma^{|z|}$, where σ is the alphabet size and z is the fixed prefix. Table 1 compares the expected running times for the solution of these five problems using the fastest serial method (based on a standard suffix tree) and a parallel method (based on distributed suffix trees). The derivation of these results is sketched briefly below.

Table 1. Post-construction average time complexities for five different problems using standard and distributed suffix trees with k computing nodes. r is the number of strings for the All Pairs Suffix-Prefix problem and the number of patterns for Exact Set Matching.

Problem	Expected Running Time	
	Standard ST (Serial)	Distributed ST (Parallel)
LCS and ELM	$O(n)$	$O(n/k)$
Maximal Repeat Finding	$O(n)$	$O(n/k)$
All Pairs Suffix-Prefix	$O(n + r^2)$	$O((n + r^2)/k)$
Exact Set Matching	$O(r \log n)$	$O((r \log n)/k)$

Longest Common Substring and Exact Local Matching

The serial solutions to these problems perform a full traversal of the suffix tree. As both the longest common substring and an exact local match of two strings will have the same prefix by definition, the problem can be solved by simply applying the serial solution to the SSTs in the DST in parallel. The average running time is therefore determined by the expectation of the size of the largest SST. Assuming i.u.d. characters in the input, this can be shown to be

$$\mathbb{E}\left[\max_z(|V_z|)\right] = \frac{n}{k} + o(n), \qquad \text{as } n \to \infty.$$

Maximal Repeat Finding

The serial solution to this problem also performs a full traversal of the suffix tree. Any pair of maximal repeats will also have the same prefix by definition and so can be found in a single SST. Therefore the average running time is determined as above.

All Pairs Suffix-Prefix

Definition 3. *Given two strings S_i and S_j, any suffix of S_i that matches a prefix of S_j is called a* suffix-prefix match *of S_i, S_j. Given a set of strings $S = \{S_1, \ldots, S_r\}$ of total length n, the* all-pairs suffix-prefix problem *is the problem of finding, for each ordered pair S_i, S_j in S, the longest suffix-prefix match of S_i, S_j.*

The serial solution to this problem performs a depth-first search of a generalised suffix tree of the input strings recording each suffix-prefix match in one of r stacks. Using a generalised DST the same problem can be solved in parallel in optimal expected time. A full traversal must be made at each SST and each suffix-prefix match has to be recorded. Call r_z the number of strings in the input that have z as a prefix. There can be no more than $r_z r$ suffix-prefix matches at each SST and therefore the running time at each SST is $O(|V_z| + r_z r)$. The time to completion is the maximum of these times and it can shown that

$$\mathbb{E}\left[\max_z(|V_z| + r_z r)\right] = \frac{n + r^2}{k} + o(n + r^2), \qquad \text{as } n, r \to \infty.$$

Exact Set Matching

In the exact set matching problem there is a set $\mathcal{P}$ consisting of r patterns to be matched against a text t of length n. For simplicity the patterns are assumed to be of equal length m and, to ensure that the number of matches found by each string is bounded, we let $m = \log_\sigma(n/c_0)$. The running time calculation is complicated by the fact that the number of patterns sent to the relevant SST and the size of the associated SST are both subject to random variation. The following bound on the expected running time can be proved.

Theorem 3. *Suppose the characters in the input text are i.u.d with alphabet size σ and that there are r patterns each with length $m = \log_\sigma(n/c_0)$. Assume the characters in each of the patterns are also i.u.d. with alphabet size σ and let R be the maximum running time of the computing nodes in a DST using prefixes of equal length. Then for suitable constants λ and γ,*

$$\mathbb{E}(R) \leqslant \frac{r}{k}(\lambda m + \gamma c_0) + o(r \log n), \qquad as\ n, r \to \infty.$$

3   Discussion

The algorithms presented here were implemented and tested on random data. For systematically biased biological data the SSTs may not be as balanced as with random data. This is likely to decrease the work done at some computing nodes at the expense of increasing it at others, thereby reducing the overall parallel efficiency. Using a snapshot of the sequences available for human chromosomes 21 and 22 combined and that of chromosome X we were able to estimate the parallel efficiencies for maximal repeat finding and exact local matching on a DST. We found that the efficiencies were 90 and 82 percent for 4 computing nodes and 72 and 61 percent for 16 nodes. In order to increase these figures we introduced a simple load balancing scheme. An example of how it worked using 16 computing nodes follows.

Instead of considering the 16 different prefixes of length 2 for DNA data we consider all 64 prefixes of length 3. The number of substrings of the input which start with each prefix was counted and, using this information, the set of 64 prefixes was partitioned into 16 subsets as evenly as possible using a simple greedy heuristic. Each computing node was then associated with one of the 16 subsets and a variant of the SSTs described above was constructed. Instead of associating one prefix with each SST, the new structure has a set of prefixes associated with it. So, an SST represented all the suffixes in the input which started with any of the prefixes in the given set. In this way the size of the SSTs was evened out considerably, thereby increasing the overall efficiency. For example, we were able to increase the parallel efficiency of maximal repeat finding on chromosome X using 16 computing nodes from 61 to 89 percent. Simple load balancing schemes for the other problems listed above gave similar improvements in efficiency for real genetic data.

Acknowledgements. This work was undertaken while the first author was a PhD student at Imperial College London. We gratefully acknowledge his support through an EPSRC studentship. We should also like to thank Costas Iliopoulos, Wojtek Rytter, and the anonymous reviewers for a number of valuable suggestions.

References

[1] A. Andersson, N. Larsson, Jesper, and K. Swanson. Suffix trees on words. In *Proceedings of the 7th Annual Symposium on Combinatorial Pattern Matching*, LNCS 1075, pages 102–115. Springer-Verlag, 1996.

[2] A. Andersson and S. Nilsson. Improved behaviour of tries by adaptive branching. *Information Processing Letters*, 46:293–300, 1993.

[3] A. Apostolico. The myriad virtues of subword trees. In A. Apostolico and Z. Galil, editors, *Combinatorial Algorithms on Words*, volume F12 of *NATO ASI Series*, pages 85–96. Springer-Verlag, 1985.

[4] W. I. Chang and E. L. Lawler. Sublinear expected time approximate string matching and biological applications. *Algorithmica*, 12:327–344, 1994.

[5] R. Clifford. *Indexed strings for large-scale genomic analysis*. PhD thesis, Imperial College of Science Technology and Medicine, London, April 2001.

[6] A. Delcher, S. Kasif, R. Fleischmann, J. Peterson, O. White, and S. Salzberg. Alignment of whole genomes. *Nucleic Acids Research*, 27(11):2369–2376, 1999.

[7] B. Dorohonceanu and C. Nevill-Manning. Accelerating protein classification using suffix trees. In *Proceedings of the 8th International Conference on Intelligent Systems for Molecular Biology (ISMB)*, pages 126–133, 2000.

[8] P. Ferragina and R. Grossi. A fully-dynamic data structure for external substring search. In *Proceedings of the 27th Annual ACM Symposium on Theory of Computing*, pages 693–702, Las Vegas, Nevada, 1995.

[9] P. Ferragina and R. Grossi. Fast string searching in secondary storage: Theoretical developments and experimental results. In *Proceedings of the Seventh Annual Symposium on Discrete Algorithms*, pages 373–382, Atlanta, Georgia, 1996.

[10] P. Ferragina and R. Grossi. The string B-Tree: a new data structure for string search in external memory and its applications. *Journal of the ACM*, 46(2):238–280, 1999.

[11] R. Giegerich and S. Kurtz. From Ukkonen to McCreight and Weiner: A unifying view of linear-time suffix tree construction. *Algorithmica*, 1997.

[12] D. Gusfield. *Algorithms on strings, trees and sequences. Computer Science and Computational Biology*. Cambridge University Press, 1997.

[13] D. Gusfield, G. M. Landau, and D. Schieber. An efficient algorithm for the all pairs suffix-prefix problem. *Information Processing Letters*, 41:181–185, 1992.

[14] J. Kärkkäinen. Suffix cactus : a cross between suffix tree and suffix array. In Z. Galil and E. Ukkonen, editors, *Proceedings of the 6th Annual Symposium on Combinatorial Pattern Matching*, LNCS 937, pages 191–204. Springer-Verlag, 1995.

[15] J. Kärkkäinen and E. Ukkonen. Sparse suffix trees. In *COCOON '96, Hong Kong*, LNCS 1090, pages 219–230. Springer-Verlag, 1996.

[16] S. Kurtz. Reducing the space requirement of suffix trees. Report 98-03. Technical report, Technische Fakultat, Universität Bielefeld, 1998.

[17] S. Kurtz and C. Schleiermacher. Reputer: Fast computation of maximal repeats in complete genomes. *Bioinformatics*, 15(5):426–427, 1999.

[18] U. Manber and G. Myers. Suffix arrays: a new method for on-line string searches. In *Proceedings of the 1st Annual ACM-SIAM Symposium on Discrete Algorithms*, 1990.

[19] E. Ukkonen. On-line construction of suffix-trees. *Algorithmica*, 14:249–260, 1995.

Analysis of Tree Edit Distance Algorithms

Serge Dulucq[1] and Hélène Touzet[2]

[1] LaBRI – Université Bordeaux I
33 405 Talence cedex, France
Serge.Dulucq@labri.fr
[2] LIFL – Université Lille 1
59 655 Villeneuve d'Ascq cedex, France
Helene.Touzet@lifl.fr

Abstract. In this article, we study the behaviour of dynamic programming methods for the tree edit distance problem, such as [4] and [2]. We show that those two algorithms may be described in a more general framework of *cover strategies*. This analysis allows us to define a new tree edit distance algorithm, that is optimal for cover strategies.

1 Introduction

One way of comparing two ordered trees is by measuring their edit distance: the minimum cost to transform one tree into another using elementary operations. This problem has several application areas. Examples include comparison of hierarchically structured data [1], such as XML documents, or alignment of RNA secondary structures in computational biology [3]. There are currently two main algorithms for solving the tree edit distance: Zhang-Shasha [4] and Klein [2]. Both algorithms use the dynamic programming paradigm, and they may be seen as an extension of the basic string edit distance algorithm. The difference lies in the set of subforests that are involved in the decomposition. This leads to different complexities: Zhang-Shasha is in n^4 in worst case, as Klein is in $n^3 \log(n)$. However, this does not mean that Klein's approach is strictly better than Zhang-Shasha's. The performance of each algorithm depends on the shape of the two trees to be compared. For example, for the pair of trees (A, B) below Klein's approach needs 84 distinct recursive calls, whereas the Zhang-Shasha's approach needs only 72 distinct recursive calls.

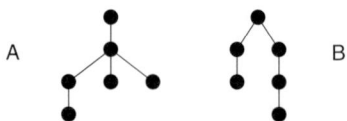

The purpose of this paper is to present a general analysis of dynamic programming for tree edit distance algorithms. For that, we consider a class of tree decompositions, which involves Zhang-Shasha and Klein algorithms. We study the complexity of those decompositions by counting the exact number of distinct

R. Baeza-Yates et al. (Eds.): CPM 2003, LNCS 2676, pp. 83–95, 2003.
© Springer-Verlag Berlin Heidelberg 2003

recursive calls for the underlying corresponding algorithm. As a corollary, this gives the number of recursive calls for Zhang-Shasha, that was a known result [4], and for Klein, that was not known. In the last section, we take advantage of this analysis to define a new edit distance algorithm for trees, which improves Zhang-Shasha and Klein algorithms with respect to the number of recursive calls.

2   Edit Distance for Trees and Forests

Definition 1 (Trees and forests). *A* tree *is a node (called the root) connected to an ordered sequence of disjoint trees. Such a sequence is called a* forest. *We write* $l(A_1 \circ \cdots \circ A_n)$ *for the tree composed of the node l connected to the sequence of trees* $A_1, \ldots, A_n$.

This definition assumes that trees are ordered trees, and that the nodes are labeled. Trees with labeled edges can by handled similarly. In the sequel, we may use the word *forest* for denoting both forests and trees, a tree being a sequence reduced to a single element.

Notation 1 *Let F be a forest.*

- $|F|$ *denotes the number of nodes of the forest* F,
- $\mathcal{SF}(F)$ *is the set of all subforests of* F,
- $F(i)$, *i is a node of* F, *denotes the subtree of* F *rooted at* i,
- $deg(i)$ *is the degree of* i, *that is the number of children of* i.

As usual, the edit distance relies on three elementary edit operations: the *substitution*, which consists in replacing a label of a node by another label, the *insertion* of a node, and the *deletion* of a node. Each edit operation is assigned a cost: c_s, c_i and c_d denote the costs for (respectively) substituting a label, inserting and deleting a node.

Definition 2 (Edit distance). *Let F and G be two forests. The* edit distance *between F and G, denoted* $d(F, G)$, *is the minimal cost of edit operations needed to transform F into G.*

Before investigating algorithms for trees, we recall some basic facts about edit distance for strings. The usual well-known dynamic programming algorithm proceeds by solving the more general problem of measuring the distance for all pairs of suffixes of the two strings. This is based on the following recursive relationship:

$$d(xu, yv) = \min \begin{cases} c_d(x) + d(u, yv) \\ c_i(y) + d(xu, v) \\ c_s(x, y) + d(u, v) \end{cases}$$

where u and v are strings, x and y are alphabet symbols. It is likely to develop an alternative algorithm by constructing the distance for all pairs of prefixes.

$$d(ux, vy) = \min \begin{cases} c_d(x) + d(u, vy) \\ c_i(y) + d(ux, v) \\ c_s(x, y) + d(u, v) \end{cases}$$

For strings, these two approaches are of course equivalent, since the number of pairs of prefixes equals the number of pairs of suffixes. Let us now come to our main concern, edit distance for trees. As for strings, the distance can be computed with dynamic programming techniques. The first decomposition applies necessarily to the roots of the trees:

$$d(l(F), l'(F')) = \min \begin{cases} c_d(l) + d(F, l'(F')) \\ c_i(l') + d(l(F), F') \\ c_s(l, l') + d(F, F') \end{cases}$$

We now have to solve the problem for forests. Since a forest is a sequence, two directions of decomposition are allowed. This gives the following recursive relationships.

$$d(l(F) \circ T, l'(F') \circ T') = \min \begin{cases} c_d(l) + d(F \circ T, l'(F') \circ T') \\ c_i(l') + d(l(F) \circ T, F' \circ T') \\ d(l(F), l'(F')) + d(T, T') \end{cases} \quad (1)$$

or alternatively

$$d(T \circ l(F), T' \circ l'(F')) = \min \begin{cases} c_d(l) + d(T \circ F, T' \circ l'(F')) \\ c_i(l') + d(T \circ l(F), T' \circ F') \\ d(l(F), l'(F')) + d(T, T') \end{cases} \quad (2)$$

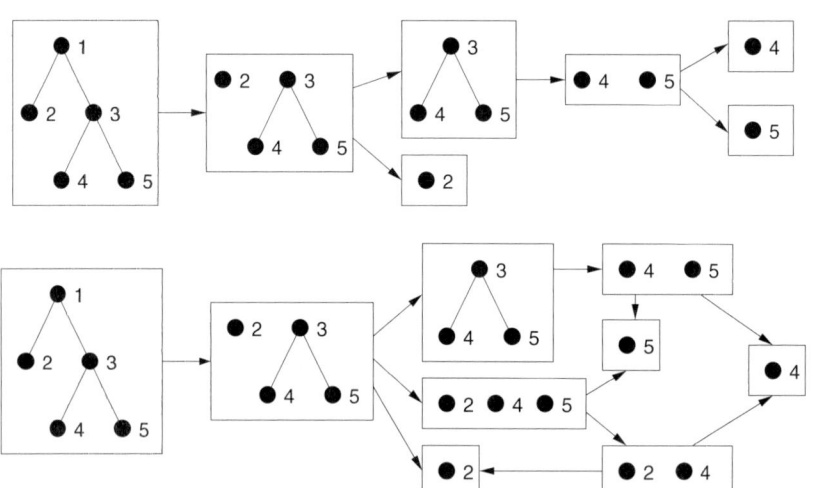

Fig. 1. These two graphics show two strategies of decomposition for a five-node tree: in the first case, all decompositions are left decompositions (according to Equation 1), and in the second case, all decompositions are right decompositions (according to Equation 2). This gives respectively 7 and 9 subforests.

We call decomposition according to Equation 1 a *left* decomposition and decomposition according to Equation 2 a *right* decomposition. In the sequel, we use the word *direction* to indicate that the decomposition is left or right. The difference with string edit distance is that the choice between 1 and 2 may lead to different numbers of recursive calls. Figure 1 shows such an example. We will see in this article that it is advantageous to alternate both schemes. An alternation of left and right decompositions gives rise to a *strategy*.

Definition 3 (Strategy). *Let F and G be two forests. A strategy is a mapping from $\mathcal{SF}(F) \times \mathcal{SF}(G)$ to $\{left, right\}$.*

Each strategy is associated with a specific set of recursive calls. We name *relevant forests* the forests that are involved in these recursive calls. This terminology is adapted from [2].

Definition 4 (Relevant Forests). *Let (F, F') be a pair of forests provided with a strategy ϕ. The set $\mathcal{RF}_\phi(F, F')$ of relevant forests is defined as the least subset of $\mathcal{SF}(F) \times \mathcal{SF}(F')$ such that if the decomposition of (F, F') meets the pair (G, G'), then (G, G') belongs to $\mathcal{RF}_\phi(F, F')$.*

Proposition 1. *Let (F, F') be a pair of forests provided with a strategy ϕ. The set $\mathcal{RF}_\phi(F, F')$ of relevant forests satisfies:*

- *if F and F' are empty forests, then $\mathcal{RF}_\phi(F, F') = \emptyset$,*
- *if $\phi(F, F') = left$, $F = l(G) \circ T$ and F' is empty, then $\mathcal{RF}_\phi(F, F')$ is $\{(F, F')\} \cup \mathcal{RF}_\phi(G \circ T, F')$,*
- *if $\phi(F, F') = right$, $F = T \circ l(G)$ and F' is empty, then $\mathcal{RF}_\phi(F, F')$ is $\{(F, F')\} \cup \mathcal{RF}_\phi(T \circ G, F')$,*
- *if $\phi(F, F') = left$, $F' = l'(G') \circ T'$ and F is empty, then $\mathcal{RF}_\phi(F, F')$ is $\{(F, F')\} \cup \mathcal{RF}_\phi(F, G' \circ T')$,*
- *if $\phi(F, F') = right$, $F' = T' \circ l'(G')$ and F is empty, then $\mathcal{RF}_\phi(F, F')$ is $\{(F, F')\} \cup \mathcal{RF}_\phi(F, T' \circ G')$,*
- *if $\phi(F, F') = left$, $F = l(G) \circ T$ and $F' = l'(G') \circ T'$, then $\mathcal{RF}_\phi(F, F')$ is $\{(F, F')\} \cup \mathcal{RF}_\phi(G \circ T, F') \cup \mathcal{RF}_\phi(F, G' \circ T') \cup \mathcal{RF}_\phi(l(G), l'(G')) \cup \mathcal{RF}_\phi(T, T')$,*
- *if $\phi(F, F') = right$, $F = T \circ l(G)$, and $F' = T' \circ l'(G')$, then $\mathcal{RF}_\phi(F, F')$ is $\{(F, F')\} \cup \mathcal{RF}_\phi(T \circ G, F') \cup \mathcal{RF}_\phi(F, T' \circ G') \cup \mathcal{RF}_\phi(l(G), l'(G')) \cup \mathcal{RF}_\phi(T, T')$.*

Proof. By induction on $|F| + |F'|$, using equations 1 and 2.   □

We write $\mathcal{RF}_\phi(F)$ and $\mathcal{RF}_\phi(F')$ to denote the projection of $\mathcal{RF}_\phi(F, F')$ on $\mathcal{SF}(F)$ and $\mathcal{SF}(F')$ respectively. The index ϕ in $\mathcal{RF}_\phi$ will be omitted when it is clear from the context. Finally, we denote `#relevant` the number of relevant forests.

Lemma 1. *Given a tree $A = l(A_1 \circ \cdots \circ A_n)$, for any strategy we have*

$$\texttt{\#relevant}(A) \geq |A| - |A_i| + \texttt{\#relevant}(A_1) + \cdots + \texttt{\#relevant}(A_n)$$

where $i \in [1..n]$ is such that the size of A_i is maximal.

Proof. Let $F = A_1 \circ \cdots \circ A_n$. By Proposition 1, we have $\mathcal{RF}(A) = \{A\} \cup \mathcal{RF}(F)$. So it is sufficient to prove that $\texttt{\#relevant}(F) \geq |F| - |A_i| + \texttt{\#relevant}(A_1) + \cdots + \texttt{\#relevant}(A_n)$. The proof is by induction on n. If $n = 1$, then the result is direct. If $n > 1$, assume that a left operation is applied to F. Let l, F_1, T such that $A_1 = l(F_1)$ and $T = A_2 \circ \cdots \circ A_n$. We have $\mathcal{RF}(F) = \{F\} \cup \mathcal{RF}(A_1) \cup \mathcal{RF}(T) \cup \mathcal{RF}(F_1 \circ T)$. It is possible to prove by induction on $|F_1| + |T|$ that the number of relevant forests of $F_1 \circ T$ containing both nodes of F_1 and nodes of T is greater than $\min\{|F_1|, |T|\}$. Therefore $\texttt{\#relevant}(F) \geq 1 + \texttt{\#relevant}(A_1) + \texttt{\#relevant}(T) + \min\{|F_1|, |T|\}$. Let $j \in [2..n]$ such that A_j has the maximal size among $\{A_2, \ldots, A_n\}$. By induction hypothesis for T, $\texttt{\#relevant}(F)$ is greater than

$$1 + \texttt{\#relevant}(A_1) + \cdots + \texttt{\#relevant}(A_n) + |T| - |A_j| + \min\{|F_1|, |T|\}.$$

It remains to verify that

$$1 + |T| - |A_j| + \min\{|F_1|, |T|\} \geq |F| - |A_i|.$$

If $|F_1| \leq |T|$, then $1 + |T| + \min\{|F_1|, |T|\} = |F|$. Since $|A_j| \leq |A_i|$, it follows that $1 + |T| + \min\{|F_1|, |T|\} - |A_j| \geq |F| - |A_i|$. If $|T| < |F_1|$, then $i = 1$ and $|F| - |A_i| = |T|$, that implies $1 + |T| - |A_j| + \min\{|F_1|, |T|\} \geq |F| - |A_i|$. $\square$

This Lemma yields a lower bound for the number of relevant forests.

Lemma 2. *For every natural number n, there exists a tree A of size n such that for any strategy, $\texttt{\#relevant}(A)$ has a lower bound in $O(n \log(n))$.*

Proof. For each complete balanced binary tree T_n of size n, we prove by induction on n that $\texttt{\#relevant}(T_n) \geq (n+1)\log_2(n+1)/2$ using Lemma 1. $\square$

This gives a $O(n^2 \log^2(n))$ bound for the total number of pairs of relevant forests.

3   Cover Strategies

In this section, we define a main family of strategies that we call *cover strategies*. The idea comes from the following remark. Assume that $l(F) \circ T$ is a relevant forest for a given strategy. Suppose that the direction of $l(F) \circ T$ is left. This decomposition generates three relevant forests: $l(F)$, T and $F \circ T$. The forest T is a subforest of $F \circ T$. An opportune point of view is then to eliminate in priority nodes of F in $F \circ T$, so that $F \circ T$ and T share relevant forests as most as possible. We make this intuitive property more formal by defining *covers*.

Definition 5 (Cover). *Let F be a forest. A cover r of F is a mapping from F to $F \cup \{right, left\}$ satisfying for each node i in F*

 - *if $deg(i) = 0$ or $deg(i) = 1$, then $r(i) \in \{right, left\}$,*
 - *if $deg(i) > 1$, then $r(i)$ is a child of i.*

In the first case, $r(i)$ is called the direction *of i, and in the latter case, $r(i)$ is called the* favorite child *of i.*

Definition 6 (Cover Strategy). *Given a pair of trees (A, B) and a cover r for A, we associate a unique strategy ϕ as follows.*

- *if $deg(i) = 0$ or $deg(i) = 1$, then $\phi(A(i), G) = r(i)$, for each forest G of B.*
- *if $A(i)$ is of the form $l(A_1 \circ \ldots \circ A_n)$ with $n > 1$, then let $p \in \{1, \ldots, n\}$ such that the favorite child $r(i)$ is the root of A_p. For each forest G of B, we define*

$$\phi(A(i), G) = right \text{ whenever } p = 1, \text{ left otherwise,}$$
$$\phi(T \circ A_p \circ \cdots \circ A_n, G) = left, \text{ for each forest } T \text{ of } A_1 \circ \cdots \circ A_{p-1},$$
$$\phi(A_p \circ T, G) = right, \text{ for each forest } T \text{ of } A_{p+1} \circ \cdots \circ A_n.$$

The tree A is called the cover tree*. A* strategy is a cover strategy *if there exists a cover associated to it.*

The family of cover strategies includes Zhang-Shasha and Klein algorithms. The Zhang-Shasha algorithm is the decomposition strategy uniquely based on Equation 1. It corresponds to the cover strategy

- for any node of degree 0 or 1, the direction is *left*,
- for any other node, the favorite child is the rightmost child.

Klein algorithm may be described as the cover strategy

- for any node of degree 0 or 1, the direction is *left*,
- for any other node, the favorite child is the root of the heaviest subtree.

The aim is now to study the number of relevant forests for a cover strategy. This task is divided into two steps. First, we compute the number of strategy for the cover tree alone. This intermediate result will indirectly contribute to our final objective.

Lemma 3. *Let F and G be two forests. For any strategy ϕ, for any nodes i of F and j of G, $(F(i), G(j))$ is an element of $\mathcal{RF}_\phi(F, G)$.*

Proof. By induction on the depth of i in F and of j in G.   □

Lemma 4. *Let F be a forest, T a non empty forest and l a labeled node.*

- *Let ϕ be a cover strategy for $l(F) \circ T$ such that $\phi(l(F) \circ T) = left$, let $k = |F|$. We write $F_1, \ldots, F_k$ for denoting the k subforests of F corresponding to the successive left decompositions of F: F_1 is F, and each F_{i+1} is obtained from F_i by a left deletion.*

$$\mathcal{RF}(l(F) \circ T) = \{l(F) \circ T, F_1 \circ T, \ldots, F_k \circ T\} \cup \mathcal{RF}(l(F)) \cup \mathcal{RF}(T).$$

– Let ϕ be a cover strategy for $T \circ l(F)$ such that $\phi(T \circ l(F)) = right$, let $k = |F|$. We write $G_1, \ldots, G_k$ for denoting the k subforests of F corresponding to the successive right decompositions of F: G_1 is F, and each G_{i+1} is obtained from G_i by a right deletion.

$$\mathcal{RF}(T \circ l(F)) = \{T \circ l(F), T \circ G_1, \ldots, T \circ G_k\} \cup \mathcal{RF}(l(F)) \cup \mathcal{RF}(T).$$

Proof. We show the first claim, the proof of the other one being symmetrical. By Proposition 1,

$$\mathcal{RF}(l(F) \circ T) = \{l(F) \circ T\} \cup \mathcal{RF}(l(F)) \cup \mathcal{RF}(T) \cup \mathcal{RF}(F \circ T).$$

Let's have a closer look at $F \circ T$. We establish that

$$\mathcal{RF}(F \circ T) = \{F_1 \circ T\} \cup \cdots \cup \{F_k \circ t\} \cup_{i \in F} \mathcal{RF}(F(i)) \cup \mathcal{RF}(T)$$

The proof is by recurrence on k. If $k = 1$, then $F = F_1$ and $\mathcal{RF}(F \circ T) = \{F_1 \circ T\} \cup \mathcal{RF}(T)$. If $k > 1$, let l', F' and T' such that $F = l'(F') \circ T'$. We have

$$\mathcal{RF}(F \circ T) = \{F \circ T\} \cup \mathcal{RF}(l'(F')) \cup \mathcal{RF}(F \circ T' \circ T) \cup \mathcal{RF}(T' \circ T).$$

On the other hand $F = F_1$, $F' \circ T' = F_2$ and $T' = F_{|l'(F')|+1}$. Since ϕ is a cover strategy, the direction for $F \circ T$ and for successive subforests containing T is left. We apply the induction hypothesis to T' and $F' \circ T'$ to get the expected result.

We come back to $\mathcal{RF}(l(F) \circ T)$. For $\mathcal{RF}(l(F) \circ T)$, we finally get

$$\{l(F) \circ T\} \cup \{F_1 \circ T\} \cup \cdots \cup \{F_k \circ T\} \cup_{i \in F} \mathcal{RF}(F(i)) \cup \mathcal{RF}(l(F)) \cup \mathcal{RF}(T).$$

According to Lemma 3, for each node i in F, $\mathcal{RF}(Fi))$ is included in $\mathcal{RF}(l(F))$. It follows that

$$\mathcal{RF}(l(F) \circ T) = \{l(F) \circ T\} \cup \{F_1 \circ T\} \cup \cdots \cup \{F_k \circ T\} \cup \mathcal{RF}(l(F)) \cup \mathcal{RF}(T).$$

$\square$

The first consequence of this Lemma is that cover strategies can reach the $n \log(n)$ lower bound of Lemma 2. We give a criterion for that. Note that this criterion is fulfilled by Klein strategy.

Lemma 5. *Let F be a forest provided with a cover strategy ϕ, such that for each relevant forest $A \circ T \circ B$ (A and B are trees, T is a forest)*

– *if $|B| > |A \circ T|$, then $\phi(A \circ T \circ B) = left$,*
– *if $|A| > |T \circ B|$, then $\phi(A \circ T \circ B) = right$,*

then `#relevant`$(F) \leq |F| \log_2(|F| + 1)$.

Proof. The proof is by induction on $|F|$. If F is a tree, the result is direct. Otherwise assume $\phi(F) = left$. Let $F = A \circ T \circ B$, $n = |A|$ and $m = |T \circ B|$. By Lemma 4, `#relevant`$(F) = n + $`#relevant`$(A) + $`#relevant`$(T \circ B)$. By induction hypothesis for A and $T \circ B$, it follows that `#relevant`$(F) \leq n + n \log_2(n+1) + m \log_2(m+1)$. Since $\phi(F) = left$, $n \leq m$ and so `#relevant`$(F) \leq (n + m) \log_2(n + m + 1)$.

$\square$

Another application of Lemma 4 is that it is possible to know the exact number of relevant forests for the cover tree.

Lemma 6. *Let* $A = l(A_1 \circ \ldots \circ A_n)$ *be a cover tree such that* $n = 1$ *or the root of* A_j *is the favorite child.*

$$\texttt{\#relevant}(A) = |A| - |A_j| + \texttt{\#relevant}(A_1) + \cdots + \texttt{\#relevant}(A_n).$$

Proof. It is sufficient to establish that $\mathcal{RF}(A)$ equals

$$\{A\} \cup \{F_1 \circ A_j \circ \cdots \circ A_n, \ldots, F_k \circ A_j \circ \cdots \circ A_n\} \cup \{G_1, \ldots, G_h\} \cup_i \mathcal{RF}(A_i)$$

where k is the size of $A_1 \circ \cdots \circ A_{j-1}$, h is the size of $A_{j+1} \circ \cdots \circ A_n$, F_1 is $A_1 \circ \cdots \circ A_{j-1}$, and each F_{i+1} is obtained from F_i by a left deletion, G_1 is $A_{j+1} \circ \cdots \circ A_n$ and each G_{i+1} is obtained from G_i by a right deletion. The proof is by induction on the size of A. If $|A| = 1$, then $\mathcal{RF}(A)$ is $\{A\}$. If $|A| > 1$, by Proposition 1, $\mathcal{RF}(A)$ is $\{A\} \cup \mathcal{RF}(A_1 \circ \cdots \circ A_n)$. Repeated applications of Lemma 4 yield the expected result. □

As a corollary of Lemma 6 and Lemma 1, we know that Klein algorithm is a strategy that minimizes the number of relevant forests for the cover tree, and Lemma 5 implies that the number of relevant forests for the cover tree is in $n \log(n)$.

After the study of the number of relevant forests for the cover tree, we are now able to look for the total number of relevant forests for a pair of trees. Given a pair of trees (A, B) provided with a cover for A, it appears that all relevant forests of A fall within three categories:

(α) those that are compared with all *rightmost* forests of B,
(β) those that are compared with all *leftmost* forests of B,
(δ) those that are compared with all *special* forests of B.

Definition 7 (Special, Rightmost and Leftmost Forests). *Let A be a tree.*

– *The set of special forests of A is the least set containing all relevant forests of A: F is a special forest of A if there exists a strategy ϕ such that $F \in \mathcal{RF}_\phi(A)$;*
– *the set of rightmost forests is the set of relevant forests wrt the strategy Left (Zhang-Shasha);*
– *the set of leftmost forests is the set of relevant forests wrt the strategy Right.*

We write $\texttt{\#special}(A)$, $\texttt{\#right}(A)$ *and* $\texttt{\#left}(A)$ *for the number of special, rightmost and leftmost forests of A.*

The task is to assign the proper category (α), (β) or (δ) to each relevant forest of A. We have this immediate result.

Lemma 7. *Let A be a cover tree, and let F be a relevant forest of A.*

– *if the direction of F is right, then F is at least compared with all leftmost forests of B,*
– *if the direction of F is left, then F is at least compared with all rightmost forests of B.*

Proof. By induction on $|A| - |F|$. □

For subtrees whose roots are *free nodes*, the category is entirely determined by the direction.

Definition 8 (Free Node). *Let A be a cover tree. A node i is free if i is the root of A, or if its parent is of degree greater than one and i is not the favorite child.*

Lemma 8. *Let i be a free node of A.*

1. *if the direction of i is left, then $A(i)$ is (α),*
2. *if the direction of i is right, then $A(i)$ is (β).*

Proof. We establish the first claim. Assume there exists a free node i with direction left, such that $A(i)$ is not (α). Applying Lemma 7, it means that $A(i)$ is compared with a forest that is not a rightmost forest. It is then possible to consider G, the largest forest of B such that $(A(i), G)$ belongs to $\mathcal{RF}(A, B)$ and G is not a rightmost forest. G is not the whole tree B, since B is a particular rightmost forest. So there are four possibilities for the generation of $(A(i), G)$.

– If $(A(i), G)$ is generated by an insertion: since the direction of i is *left*, there exists a node l and two forests H and P such that $G = H \circ P$ and $(A(i), l(H) \circ P)$ is in $\mathcal{RF}(A, B)$. Since the size of $l(H) \circ P$ is greater than the size of G, $l(H) \circ P$ is a rightmost forest, that implies that $G = H \circ P$ is a also rightmost forest.

– If $(A(i), G)$ is generated by a deletion: there exists a node l such that either $l \circ A(i)$, $A(i) \circ l$ or $l(A(i))$ is a relevant forest. In the two first cases, this would imply that i is a favorite child, and in the third case, that the degree of the parent of i is 1. In all cases, this contradicts the hypothesis that i is a free node.

– If $(A(i), G)$ is generated by a substitution, being the matching part of the substitution: this would imply that G is a tree, that contradicts the hypothesis that G is not a rightmost forest.

– If $(A(i), G)$ is generated by a substitution, being the remaining part of the substitution: $(A(i), G)$ should be obtained from a relevant pair of the form $(A' \circ A(i), B' \circ G)$ or $(A(i) \circ A', G \circ B')$, where A' and B' are subtrees of A and B respectively. In both cases, this contradicts the hypothesis that i is not a favorite child. □

For nodes that are not free and for forests, the situation is more complex. It is then necessary to take into account the category of the parent too. The following lemma establishes that those nodes inherit the relevant forests of B from their parents.

Lemma 9. *Let F be a relevant forest of A that is not a tree. Let i be the lower common ancestor of the set of nodes of F and let j be the favorite child of i.*

1. *if F is a rightmost forest whose leftmost tree is not $A(j)$, then F has the same category as $A(i)$,*
2. *if F is a leftmost forest, then F has the same category as $A(i)$,*
3. *otherwise F is (δ).*

Proof. The proof of the two first cases is similar to the proof of Lemma 8. We give a detailed proof of the third case. By definition of cover strategies, the first tree of F is $A(j)$ and the direction of F is right. Assume there is a special forest G of B such that (F, G) is not a relevant pair. We suppose that G is of maximal size.

   – If G is a rightmost forest. Since F is not a leftmost forest, j is not the leftmost child of i, that implies that the direction of $A(i)$ is *left*. Lemma 7 implies that $(A(i), G)$ is a relevant pair. The pair (F, G) is obtained from $(A(i), G)$ by successive left deletions until j and right deletion until F.

   – If G is not a rightmost forest. There exists a node l such that $G \circ l$ is a special forest. By construction of G, $(F, G \circ l)$ is a relevant pair. Since the direction of F is right, a right deletion gives (F, G).   □

Lemma 10. *Let A be a cover tree, i be a node of A that is not free, and j be the parent of i.*

– *if the direction of i is left, if i is the rightmost child of j and $A(j)$ is (α), then $A(i)$ is (α),*
– *if the direction of i is right, if i is the leftmost child of j and $A(j)$ is (β), then $A(i)$ is (β),*
– *otherwise $A(i)$ is (δ).*

Proof. Similar to proof of Lemma 9.   □

We introduce notations for classifying nodes of A according to their inheritance.

Notation 2 *Let i be a node of A, let j be the parent of i (if i is not the root).*

   $Free(A(i))$ *: cardinality of $\mathcal{RF}(A, B) \cap (A(i), B)$ if i is free*

   $Right(A(i))$ *: cardinality of $\mathcal{RF}(A, B) \cap (A(i), B)$ if $A(j)$ is (α)*

   $Left(A(i))$ *: cardinality of $\mathcal{RF}(A, B) \cap (A(i), B)$ if $A(j)$ is (β)*

   $All(A(i))$ *: cardinality of $\mathcal{RF}(A, B) \cap (A(i), B)$ if $A(j)$ is (δ).*

With this notation, `#relevant`(A, B) equals $Free(A)$. We are now able to formulate the main result of this section, which gives the total number of relevant forests for a cover strategy.

Theorem 1. *Let (A, B) be a pair of trees, A being a cover tree.*

1. *If A is reduced to a single node whose direction is right*

$$Free(A) = \ Left(A) \ = \text{\#left}(B)$$
$$All(A) \ = Right(A) = \text{\#special}(B)$$

2. If A is reduced to a single node whose direction is left

$$Free(A) = Right(A) = \text{\#right}(B)$$
$$All(A) \ = \ Left(A) \ = \text{\#special}(B)$$

3. If $A = l(A')$ and the direction of l is right

$$Free(A) = \ Left(A) \ = \text{\#left}(B) + Right(A')$$
$$All(A) \ = Right(A) = \text{\#special}(B) + All(A')$$

4. If $A = l(A')$ and the direction of l is left

$$Free(A) = Right(A) = \text{\#right}(B) + Left(A')$$
$$All(A) \ = \ Left(A) \ = \text{\#special}(B) + All(A')$$

5. If $A = l(A_1 \circ \cdots \circ A_n)$ and the favorite child is the leftmost child

$$Free(A) = \ Left(A) \ = \textstyle\sum_{i>1} Free(A_i) + Left(A_1) + \text{\#left}(B)(|A| - |A_1|)$$
$$All(A) \ = Right(A) = \textstyle\sum_{i>1} Free(A_i) + All(A_1) + \text{\#special}(B)(|A| - |A_1|)$$

6. If $A = l(A_1 \circ \cdots \circ A_n)$ and the favorite child is the rightmost child

$$Free(A) = Right(A) = \textstyle\sum_{i<n} Free(A_i) + Right(A_n) + \text{\#right}(B)(|A| - |A_n|)$$
$$All(A) \ = \ Left(A) \ = \textstyle\sum_{i<n} Free(A_i) + All(A_n) + \text{\#special}(B)(|A| - |A_n|)$$

7. If $A = l(A_1 \circ \cdots \circ A_n)$ and the favorite child is A_j with $1 < j < n$
$$Free(A) = \textstyle\sum_{i \neq j} Free(A_i) + All(A_j) + \text{\#right}(B)(1 + |A_1 \circ \cdots \circ A_{j-1}|)$$
$$+ \text{\#special}(B)|A_j \circ \cdots \circ A_n|$$
$$Right(A) = Free(A)$$
$$All(A) = Left(A) = \textstyle\sum_{i \neq j} Free(A_i) + All(A_j) + \text{\#special}(B)(|A| - |A_j|)$$

Proof. By induction on the size of A, using Lemmas 6, 9, 8 and 10.   □

Lemma 11. *For Zhang-Shasha algorithm,*

$$\text{\#relevant}(A, B) = \text{\#right}(A) * \text{\#right}(B).$$

Proof. In Theorem 1, it appears that the cases *2*, *4* and *6* are the only useful cases. Applying Lemma 6, it follows that $\text{\#relevant}(A, B) = \text{\#right}(A) * \text{\#right}(B)$ (by induction on the size of A).   □

We get the symmetrical result for the symmetrical strategy: for nodes of degree 0 or 1, the direction is *right*, and for other nodes, the favorite child is the leftmost child. In this case, $\text{\#relevant}(A, B) = \text{\#left}(A) * \text{\#left}(B)$.

For Theorem 1 to be effective, it remains to evaluate the values of $\text{\#right}$, $\text{\#left}$ and $\text{\#special}$.

Lemma 12. *Let A be a tree.*

$$\texttt{\#right}(A) = \sum(|A(i)|, i \in A) - \sum(|A(j)|, j \text{ is a rightmost child})$$
$$\texttt{\#left}(A) = \sum(|A(i)|, i \in A) - \sum(|A(j)|, j \text{ is a leftmost child})$$

Proof. Lemma 6 gives

$$\texttt{\#right}(A) = \texttt{\#left}(A) = 1, \text{if the size of } A \text{ equals } 1$$
$$\texttt{\#right}(l(A_1, \dots, A_n)) = \sum_i \texttt{\#right}(A_i) + |A| - |A_n|, \text{otherwise.}$$
$$\texttt{\#left}(l(A_1, \dots, A_n)) = \sum_i \texttt{\#left}(A_i) + |A| - |A_1|, \text{otherwise.}$$

Induction on the size of A concludes the proof. □

Lemma 13. *Let F be a forest.* $\texttt{\#special}(F) = \frac{|F|(|F|+3)}{2} - \sum_{i \in F} |F(i)|.$

Proof. The proof is by induction on the size of F. If $|F| = 0$, then $\texttt{\#special}(F) = 0$, which is consistent with the Lemma. If $F > 0$, then $F = l(T) \circ P$, where P and T are (possibly empty) subforest of F. There are two kinds of special subforests of F to be considered:

1. those containing the node l: there are $|P| + 1$ such subforests,
2. those not containing the node l: there are $\texttt{\#special}(T \circ P)$ such subforests.

    It follows that $\texttt{\#special}(F) = |P| + 1 + \texttt{\#special}(T \circ P)$. The induction hypothesis applied to $T \circ P$, whose size is $|F| - 1$, ensures.

$$\begin{aligned}
\texttt{\#special}(F) &= |P| + 1 + \frac{(n-1)(n+2)}{2} - \sum_{i \in T \circ P} |t \circ P(i)| \\
&= n - |l(t)| + 1 + \frac{(n-1)(n+2)}{2} - \sum_{i \in T \circ p} |T \circ P(i)| \\
&= n + 1 + \frac{(n-1)(n+2)}{2} - \sum_{i \in F} |F(i)| \\
&= \frac{n(n+3)}{2} - \sum_{i \in F} |F(i)|
\end{aligned}$$

This concludes the proof. □

4   A New Optimal Cover Strategy

In this last section, we show that Theorem 1 makes it possible to design an optimal cover strategy. An optimal cover strategy is of course a strategy that minimizes the total number of relevant forests. The algorithm is as follows. For any pair of trees (A, B), define four dynamic programming structures Right, Left, Free, and All of size $|A|$. The definition of Right, Left, Free, and All is directly borrowed from Theorem 1. The only difference is that the favorite child is not known. So at each step, the favorite child is chosen to be the child that minimizes the number of relevant forests. For instance, if $A = l(A_1 \circ \cdots \circ A_n)$ then

$$Free(A) = \sum_{i \geq 1} Free(A_i) + \min \begin{cases} Left(A_1) - Free(A_1) + \texttt{\#left}(B) * (|A| - |A_1|) \\ All(A_j) - Free(A_j) + \texttt{\#special}(B)|A_j \circ \cdots \circ A_n| \\ \quad + \texttt{\#right}(B)(1 + |A_1 \circ \cdots \circ A_{j-1}|), \quad 1 < j < n \\ Right(A_n) - Free(A_n) + \texttt{\#right}(B)(|A| - |A_n|) \end{cases}$$

The optimal cover is then built up using tracing back from $Free(A)$. The size of each table is $|A|$. As for the time complexity, the time computation for a cell of a node i is proportional to the degree of i. So the overal time computation is in $O(\sum_{i \in A} deg(i))$, that is in $O(|A|)$. Figure 2 shows an example of optimal cover.

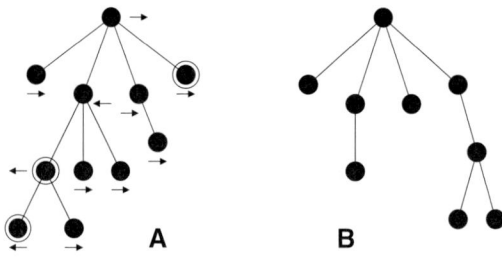

Fig. 2. This figure shows an optimal cover for A built up for the pair (A, B). For each node, the direction, left or right, is indicated with an arrow. For each node of arity greater than 1, the favorite child is the circled node. This cover yields 340 relevant forests. For the two Zhang-Shasha strategies, we get 405 and 350 relevant forests, and for the Klein strategy, we get 391 relevant forests.

By construction, this algorithm involves less relevant forests than Zhang-Shasha and Klein strategies. In particular it is in $n^3 \log(n)$. But we do not claim that this is the best algorithm, since we do not take into account the cost of the preprocessing in comparison with the expected gain of number of relevant forests. This is still an open question.

References

1. S. Chawathe, "Comparing hierarchical data in external memory" *Proceedings of the Twenty-fifth International Conference on Very Large Data Bases* (1999), Edinburgh, Scotland, p. 90–101.
2. P. Klein, "Computing the edit-distance between unrooted ordered trees" *Proceeding of 6th European Symposium on Algorithms* (1998), p. 91–102.
3. B. Shapiro and K. Zhang, "Comparing multiple RNA secondary structures using tree comparisons", *Comput. Appl. Biosciences*, Vol.4, 3 (1988), p. 387–393.
4. K. Zhang and D. Shasha, "Simple fast algorithms for the editing distance between trees and related problems" *SIAM Journal of Computing*, Vol 18-6, (1989), p. 1245–1262.

An Exact and Polynomial Distance-Based Algorithm to Reconstruct Single Copy Tandem Duplication Trees

Olivier Elemento[1,2] and Olivier Gascuel[1]

[1] Département d'Informatique Fondamentale et Applications, LIRMM, 161 rue Ada, 34392, Montpellier, France, http://www.lirmm.fr/w3ifa/MAAS/, {elemento, gascuel}@lirmm.fr
[2] IMGT, the international ImMunoGeneTics database, http://imgt.cines.fr, Laboratoire d'Immunogénétique Moléculaire, LIGM, Université Montpellier II, UPR CNRS 1142, IGH, 141 rue de la Cardonille, 34396, Montpellier Cedex 5, France

Abstract. The problem of reconstructing the duplication tree of a set of tandemly repeated sequences which are supposed to have arisen by unequal recombination, was first introduced by Fitch (1977), and has recently received a lot of attention. In this paper, we deal with the restricted problem of reconstructing single copy duplication trees. We describe an exact and polynomial distance based algorithm for solving this problem, the parsimony version of which has previously been shown to be NP-hard (like most evolutionary tree reconstruction problems). This algorithm is based on the minimum evolution principle, and thus involves selecting the shortest tree as being the correct duplication tree. After presenting the underlying mathematical concepts behind the minimum evolution principle, and some of its benefits (such as consistency), we provide a new recurrence equation to estimate the tree length using ordinary least-squares, given a matrix of pairwise distances between the copies. We then show how this equation naturally forms the dynamic programming framework on which our algorithm is based, and provide an implementation in $O(n^3)$ time and $O(n^2)$ space, where n is the number of copies.

1 Introduction

Tandemly repeated DNA sequences consist of two or more adjacent copies of a stretch of DNA. They arise from tandem duplication, in which a sequence of DNA (which may itself contain several copies) is transformed into two adjacent and identical sequences. Since copies are then free to evolve independently and are likely to undergo additional mutation events, they become approximate over time. Unequal recombination is widely viewed as the predominant biological mechanism responsible for the production of tandemly repeated sequences [1, 2,3,4,5,6], at least when the basic repeated motif is large (*i.e.* corresponds to minisatellites or genes). The problem of reconstructing the duplication history of tandemly repeated sequences was first considered by Fitch in 1977 [3]. It has

R. Baeza-Yates et al. (Eds.): CPM 2003, LNCS 2676, pp. 96–108, 2003.

not received much attention until recently, probably due to the lack of available repeated sequence data, and also because there has been no dedicated computer program available to reconstruct duplication histories. With the sequencing of human genome, and the ongoing sequencing of several other species, this problem has gained a lot of attention. The reason is that fast and efficient methods for reconstructing the duplication history of these tandemly repeated sequences would be important tools to study genome evolution. They should provide deeper insights into the processes, dynamics and mechanisms of gene duplications, which is considered as one of the major evolutionary forces at the genome level [1].

Most of the recent studies have been devoted to repeated sequences generated by single copy duplication events [7,8,9,10]. Indeed, the mechanism of unequal recombination allows simultaneous duplication of several copies, but there is now evidence [3,5,6,7,8] that single copy duplications are predominant over multiple copies duplications, at least with tandemly repeated genes. For example, one of the most famous tandemly arranged gene families, the Antennapedia (*antp*)-class homeobox genes, is assumed to have arisen through repetitive single copy duplications [11].

The series of duplications that has given rise to tandemly repeated sequences can be represented by way of a "duplication tree", which we formally describe below. A duplication tree which only contains single copy duplications is simply called a "single copy duplication tree". Reconstructing optimal single copy duplication trees has been shown to be NP-hard within a parsimony framework [10], and several authors described approximation algorithms. Benson and Dong [7] developed a greedy algorithm for reconstructing single copy duplication trees, based on the parsimony criterion. Using simulations, they showed that their algorithm performs better than approximation algorithms based on minimum ordered spanning trees, which themselves guarantee a performance ratio of 2. More recently, Tang et al. [8] described a dynamic programming algorithm within a parsimony framework, based on the lifting technique [12], for the same problem. Then, they showed that the cost of the obtained tree is at most twice higher than the cost of the most parsimonious tree. Later, Tang et al. [9] and Jaitly et al. [10] independently described polynomial time approximation schemes (PTAS) for the single copy problem (within the same parsimony framework), also obtained using the lifting technique combined with local optimization and dynamic programming.

In this paper, we present an exact and polynomial $O(n^3)$ time and $O(n^2)$ space algorithm for reconstructing the optimal single copy duplication tree, where n is the number of copies. Our algorithm is based on the minimum evolution principle [13,14], and uses as input a matrix of pairwise evolutionary distances, calculated from a set of ordered nucleotide or protein sequences. The minimum evolution principle involves selecting the tree with shortest least-squares length estimate as being the correct tree. Due to the use of this principle, our reconstruction algorithm is consistent [14,15], as opposed to parsimony methods which were shown inconsistent by Felsenstein [16]. The content of this paper is organized as follows. First we describe the duplication model, *i.e.* the characteristics of the mathematical objects we aim at reconstructing. Then we describe the minimum evolution framework on which our algorithm is based and provide a novel recurrence equation for estimating the length of any given tree, from a distance matrix between copies. Using this equation, we describe a dynamic programming algorithm to solve the single copy duplication tree problem under the minimum evolution principle.

2   Duplication Model

Assuming unequal recombination as the sole mechanism responsible in generating the copies, Fitch [3], and more recently Tang et al. [8,9] and Elemento et al. [5,6] independently introduced the following duplication model. A duplication history (Figure 1(a) and 1(b)) is a rooted tree with n labelled and ordered leaves denoted as $(1,2,3,...,n)$, in which internal nodes correspond to duplication events. In a real duplication history, the time intervals between consecutive duplications are completely known, and the internal nodes are ordered from top to bottom according to the moment they occurred in the course of evolution. However, in the absence of molecular clock (which is almost always the case), both the order between the duplication events of two different lineages and the root location are impossible to recover from the sequences. In this case, we are only able to infer a *duplication tree* (Figure 1(c)), *i.e.* an unrooted tree with ordered leaves, whose topology is compatible with at least one duplication history. Recovering the position of the root can sometimes be achieved, through the use of rooting procedures (outgroups, midpoint, etc.), and creates *a rooted duplication tree* (Figure 1(d)).

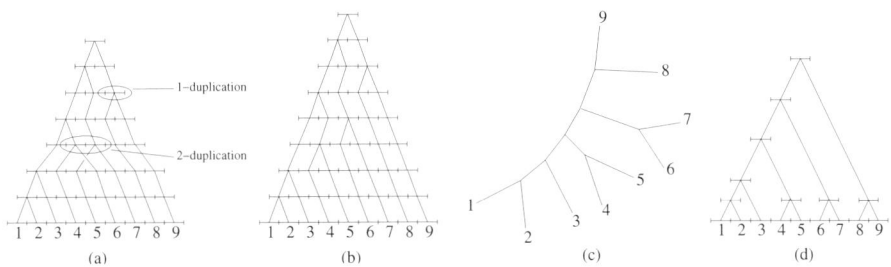

Fig. 1.   (a) Duplication history, (b) single copy duplication history, (c) single copy duplication tree and (d) single copy rooted duplication tree

A duplicated fragment may only contain a single copy, in which case we say that the duplication event is a 1-duplication (or a single copy duplication). It may also contains 2, 3 or k copies, in which case we call the duplication event a 2-, 3- or k-duplication. When a rooted duplication tree only contains 1-duplication events (such as Figure 1(d)), it is analoguous to a binary search tree. Consequently, the number of distinct single copy rooted duplication trees is equal to the number of binary search trees, which is given by the Catalan recursion [17] :

$$C_n = \sum_{k=1}^{n-1} C_k C_{n-k} = \frac{(2n)!}{n!(n+1)!} \sim \frac{4^n}{\sqrt{\pi} n^{3/2}}.$$

It was shown in [3] and later in [6] that the general duplication model constraints the root of a duplication tree to be located somewhere in the tree along the path between the most distant copies on the locus, but (generally) not anywhere due to the presence of multiple duplications. It is easy to see that a single copy duplication tree can be rooted anywhere along the path between the two most distant copies. Suppose that we systematically root duplication trees on the rightmost branch, $i.e.$ the branch associated with label n. In this situation, the left subtree is a single copy rooted duplication tree with $n-1$ leaves. Therefore, the number of distinct single copy unrooted duplication trees with n leaves is simply equal to C_{n-1}. Since this number is exponential (see above), exploring the landscape of single copy duplication trees in search for the optimal one cannot be done using a trivial algorithm.

A single copy rooted duplication tree $X_{1,n}$, whose leaves are labelled with the ordered set of copies $(1, 2, 3, ..., n)$, is obtained by combining two rooted subtrees $X_{1,p}$ and $X_{p+1,n}$ whose leaves are labelled with the ordered sets $(1, 2, 3, ..., p)$ and $(p + 1, p + 2, ..., n)$, respectively. Identically, a single copy unrooted duplication tree $X_{1,n}$ is obtained by combining three rooted subtrees $X_{1,p}$, $X_{p+1,q}$ and $X_{q+1,n}$ with $1 \leq p < q < n$. In the rest of this paper, the ordered set $(p, p + 1, ..., q)$ is denoted as $[p, q]$, while, depending on the context, $X_{p,q}$ refers to a rooted tree on $[p, q]$ or to $[p, q]$ itself.

3   Minimum Evolution Principle and Least-Squares Tree Length Estimation

3.1   The Minimum Evolution Principle

The minimum evolution (ME) principle [13,14] involves selecting the shortest tree as being the tree which best explains the observed sequences. The tree length is equal to the sum of branch lengths and branch lengths are estimated by minimizing a least-squares criterion. The problem of inferring optimal phylogenies (*i.e.* without restriction to duplication trees) within the ME principle is suspected to be NP-hard, but not classified, to the best of our knowledge. Nonetheless, the ME principle forms the basis of several phylogenetic reconstruction methods, generally based on greedy heuristics. Among them is the popular Neighbor-Joining (NJ) algorithm [18]. Starting from a star tree, NJ iteratively agglomerates external pairs of taxa so as to minimize the tree length at each step.

Assuming that we have consistent evolutionary distance estimators which converge towards the true distance as the length of the sequences increases, the

ME principle combined with ordinary least-squares (OLS) tree length estimation is statistically consistent [14,15]. Statistical consistency is an essential property in phylogenetic reconstruction, since it ensures that, for the given method, the probability of recovering the correct topology increases with sequence length. Consistent methods are in contrast with inconsistent ones, *e.g.* parsimony in some cases [16], which may converge towards a wrong tree as the amount of data increases.

In this section, we introduce a new recurrence equation for estimating the length of a tree topology using OLS, given a matrix of pairwise evolutionary distances between copies. As we shall see, this equation forms the basis of our reconstruction algorithm.

3.2   Notation

Δ is a matrix of pairwise evolutionary distances between copies, and δ_{ij} is the distance in Δ between copy i and copy j; Υ is an unrooted tree topology, and T represents a valued tree with topology Υ. T induces a matrix of pairwise tree distances between copies, which we denote Δ^T. In this matrix, δ_{ij}^T denotes the length of the tree path linking copy i and copy j. The sum of the branch lengths of T is denoted as $L(T)$. As shown in Figure 2, we consider in the rest of this section that T is composed of three non-intersecting subtrees A, B and C. These subtrees are linked together by three branches whose lengths are a, b and c. A is the union of two subtrees A_1 and A_2, and in turn A_1 is the union of two subtrees A_{11} and A_{12}. Two branches with lengths a_1 and a_2 link the root of A to the roots of A_1 and A_2, respectively. In the remainder of this paper, we call R the subset of leaves that do not belong to A (*i.e.* $R = B \bigcup C$).

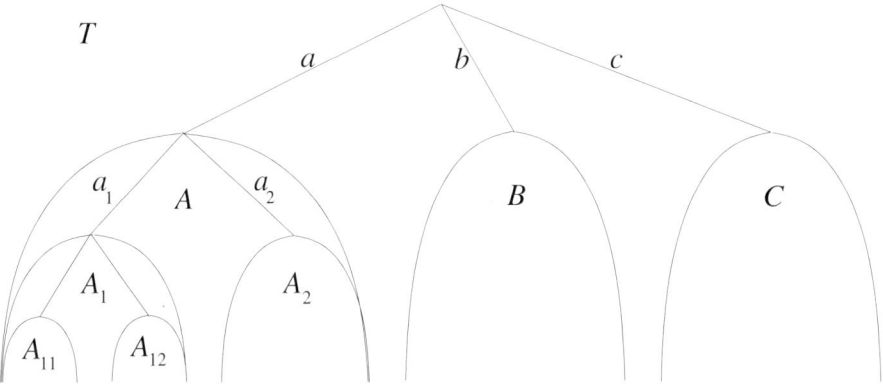

Fig. 2. Unrooted tree T, composed of three subtrees A, B and C.

Let X be any subtree of T, and $\overline{X}$ be the average distance in T between the root of X and its leaves. Δ_{XY} and Δ_{XY}^T are the average distances between the leaves of two non-intersecting subtrees X and Y, in the distance matrices Δ and Δ^T, respectively :

$$\Delta_{XY} = \frac{1}{|X||Y|} \sum_{i \in X, j \in Y} \delta_{ij}$$

$$\Delta^T_{XY} = \frac{1}{|X||Y|} \sum_{i \in X, j \in Y} \delta^T_{ij}.$$

Given a topology Υ and a distance matrix Δ, the OLS branch length estimation of T is obtained by minimizing the following sum of squares:

$$\sum_{i,j \in T} (\delta^T_{ij} - \delta_{ij})^2.$$

3.3 OLS Tree Length Expression

Theorem: Let the branches of T be estimated by OLS. Then:

$$L(T) = (L(A) - \overline{A}) + (L(B) - \overline{B}) + (L(C) - \overline{C}) \tag{1}$$
$$+ \tfrac{1}{2}(\Delta_{AB} + \Delta_{AC} + \Delta_{BC}).$$

Moreover, $(L(A) - \overline{A})$ is recursively obtained in the following way:
(a) if A is a leaf, then $(L(A) - \overline{A}) = 0$,
(b) otherwise, $(L(A) - \overline{A})$ is given by:

$$(L(A) - \overline{A}) = (L(A_1) - \overline{A_1}) + (L(A_2) - \overline{A_2}) \tag{2}$$
$$+ \tfrac{1}{2}\Delta_{A_1 A_2} + \tfrac{1}{2}\left(\frac{|A_2| - |A_1|}{|A|}\right)\Delta_{A_1 R} + \tfrac{1}{2}\left(\frac{|A_1| - |A_2|}{|A|}\right)\Delta_{A_2 R},$$

and the same applies to $(L(B) - \overline{B})$ and $(L(C) - \overline{C})$, by symmetry.

Proof: Using Figure 2, we see that:

$$L(T) = L(A) + L(B) + L(C) + a + b + c. \tag{3}$$

It has been shown that the average distance between two non-intersecting subtrees X and Y is preserved between Δ and Δ^T, when these subtrees are adjacent to a common ternary node (*i.e.* A and B, A and C or B and C in Figure 2), and when branch lengths of T are estimated by OLS [19,20]. Using this property, Δ_{AB}, Δ_{AC} and Δ_{BC} can be expressed in the following way :

$$\begin{align}
(i) \quad & \Delta_{AB} = \overline{A} + a + b + \overline{B}, \\
(ii) \quad & \Delta_{AC} = \overline{A} + a + c + \overline{C}, \\
(iii) \quad & \Delta_{BC} = \overline{B} + b + c + \overline{C},
\end{align} \tag{4}$$

and Equation (1) is obtained by combining Equations (3) and (4). Identically, the length of A is equal to the sum of its branch lengths:

$$L(A) = L(A_1) + L(A_2) + a_1 + a_2, \tag{5}$$

while $\overline{A}$ is given by:

$$\overline{A} = \frac{|A_1|}{|A|}(a_1 + \overline{A_1}) + \frac{|A_2|}{|A|}(a_2 + \overline{A_2}). \tag{6}$$

a_1 and a_2 are obtained by rewriting Equation (4) for A_1, A_2 and R, in place of A, B and C; solving this linear system, we obtain:

$$a1 = \frac{1}{2}\Delta_{A_1 A_2} + \frac{1}{2}\Delta_{A_1 R} - \frac{1}{2}\Delta_{A_2 R} - \overline{A_1},$$

$$a2 = \frac{1}{2}\Delta_{A_1 A_2} + \frac{1}{2}\Delta_{A_2 R} - \frac{1}{2}\Delta_{A_1 R} - \overline{A_2}. \tag{7}$$

Equation (2) is finally obtained by substracting (6) to (5), and replacing a_1 and a_2 by their analytical expression (7), while equality $(L(A) - \overline{A}) = 0$, if A is a leaf, is a direct consequence of the definitions.

3.4 Properties

In Equation (1), $(L(A) - \overline{A})$, $(L(B) - \overline{B})$ and $(L(C) - \overline{C})$ only depend on the structure of subtrees A, B and C, respectively. Indeed, Equation (7) clearly shows that the branch length $a1$ depends on the copies in subtrees A_1, A_2 and $R = B \bigcup C$, but not on the structure of R (*i.e.* the content of B and C). The same applies with a_2. Identically, this property is valid for branch length a_{11}, which depends on the copies in subtrees A_{11}, A_{12} and $R' = A_2 \bigcup R$, but not on the structure of R', and therefore not on the structure of R. It can be established in this way that none of the branch lengths in A depends on the structure of R. Therefore, to compute $L(T)$, we independently compute the values for $(L(A) - \overline{A})$, $(L(B) - \overline{B})$ and $(L(C) - \overline{C})$, and then apply Equation (1).

For the same reasons, $(L(A_1) - \overline{A_1})$ and $(L(A_2) - \overline{A_2})$ only depend on the structure of A_1 and A_2, respectively. Therefore, to compute $(L(A) - \overline{A})$, we independently compute the values for $(L(A_1) - \overline{A_1})$ and $(L(A_2) - \overline{A_2})$, and then apply Equation (2).

4   Reconstructing Optimal Single Copy Duplication Trees under the ME Principle

The above recurrence equations enable to calculate the OLS length of given un-rooted tree topology, given a matrix of pairwise distances. In this section, we seek the duplication tree whose length is minimum, among all possible duplication trees. As we shall see, the above recurrence equations not only allow tree length estimation, but also form the basis of a dynamic programming algorithm which solves the problem at hand.

4.1 Basic Algorithm

Equation (1) consists of four independent terms $(L(A) - \overline{A})$, $(L(B) - \overline{B})$, $(L(C) - \overline{C})$ and the remaining term. As we said above, $(L(A) - \overline{A})$, $(L(B) - \overline{B})$ and

$(L(C) - \overline{C})$ only depend on the structure of subtrees A, B and C, respectively, while the remaining term consists of average distances, and therefore does not depend on the structure of A, B and C. To minimize Equation (1), we adopt a divisive strategy, which consists first in partitioning the whole set of copies into three subsets A, B and C, then in computing the structure which minimizes $(L(X) - \overline{X})$ for each of these subsets, and finally in applying Equation (1). The optimal tree is given by the optimal partitioning. Identically, to obtain the optimal structure for A, Equation (2) shows that we need to evaluate every partitioning of A into A_1 and A_2, then to compute the structure for A_1 and A_2 which minimizes $(L(X) - \overline{X})$ and finally to select the partitioning which minimizes Equation (2).

Although used in some divisive clustering methods [21], this strategy cannot be used to reconstruct optimal phylogenies when n is large, since the number of combinations of subsets is exponential. This is different with single copy duplication trees, since we only have to evaluate combinations of (two or three) adjacent intervals, and the number of combinations is polynomial.

Let S and M be two $n \times n$ matrices. $S_{p,q}$ represents the minimal value of $(L(X_{p,q}) - \overline{X_{p,q}})$ where $X_{p,q}$ is any subtree with leaves in $[p, q]$, while $M_{p,q}$ represents the value of m for which $S_{p,q}$ is minimal (see below). Let $X_{\overline{p,q}}$ represent the subset of copies that do not belong to $X_{p,q}$ (i.e. $X_{\overline{p,q}} = X_{1,p-1} \bigcup X_{q+1,n}$). Starting from an interval $[1, n]$ representing the n copies, the reconstruction algorithm for single copy duplication trees necessitates the three following steps:

(a) the first step consists in using Equation (2) to calculate $S_{p,q}$ for a growing interval $X_{p,q}$ of $[1, n]$, until $q - p = n - 3$. Computing $S_{p,q}$ requires evaluating the combination of every couple of adjacent intervals $X_{p,m}$ and $X_{m+1,q}$, with m varying from p to $q - 1$. Therefore, using Equation (2), $S_{p,q}$ is given by:

$$S_{p,q} = \min_{p \le m \le q-1} \left[\begin{array}{l} S_{p,m} + S_{m+1,q} \\ + \frac{1}{2} \Delta_{X_{p,m} X_{m+1,q}} \\ + \frac{1}{2} \left(\frac{p+q-2m-1}{q-p+1} \right) \Delta_{X_{p,m} X_{\overline{p,q}}} \\ + \frac{1}{2} \left(\frac{2m-p-q+1}{q-p+1} \right) \Delta_{X_{m+1,q} X_{\overline{p,q}}} \end{array} \right], \tag{8}$$

while $M_{p,q}$ is the value of m minimizing the above expression. Moreover, we have $S_{p,p} = 0$ for $1 \le p \le n$.

(b) the second step consists in using Equation (1) to search for the set of three adjacent intervals X_{1,m_1}, X_{m_1+1,m_2} and $X_{m_2+1,n}$ which minimizes $L(T)$.

(c) in the third step, the complete tree topology is recovered by stepping back through the optimal intervals stored in M.

This algorithm is summarized above. The number of intervals which need to be evaluated during the first step is in $O(n^2)$. Evaluating a single interval using Equation (8) necessitates the evaluation of $O(n)$ combinations of adjacent sub-intervals. Evaluating a single combination requires the average distances between the $X_{p,m}$, $X_{m+1,q}$ and $X_{\overline{p,q}}$ subsets to be computed, and necessitates $O(n^2)$ time. Therefore, the total time complexity of the first step is $O(n^5)$. As we show in the next section, the time required to evaluate a single combination

---

Algorithm 1 Single copy duplication tree reconstruction algorithm

---

$S \leftarrow n \times n$ matrix
$M \leftarrow n \times n$ matrix
for l from 1 to $n - 3$ **do**
    for i from 1 to $n - l$ **do**
        compute $S_{i,i+l}$ and $M_{i,i+l}$ using Equation (8)
    end for
end for
$L^*(T) \leftarrow \infty$
for m_1 from 1 to $n - 2$ **do**
    for m_2 from $m_1 + 1$ to $n - 1$ **do**
        compute $L(T)$ for X_{1,m_1}, X_{m_1+1,m_2}, $X_{m_2+1,n}$ using Equation (1) and S
        if $L(T) < L^*(T)$ **then**
            $L^*(T) \leftarrow L(T)$, $m_1^* \leftarrow m_1$, $m_2^* \leftarrow m2$
        end if
    end for
end for
recursively create T using the values in M, starting from M_{1,m_1^*}, $M_{m_1^*+1,m_2^*}$ and $M_{m_2^*+1,n}$

---

of adjacent sub-intervals can be lowered to $O(1)$. When using this refinement, the total time complexity of the first step is lowered to $O(n^3)$.

In the second step, we evaluate every possible combination of three adjacent intervals X_{1,m_1}, X_{m_1+1,m_2} and $X_{m_2+1,n}$. Therefore, the number of combinations that need to be tested in the second step is also in $O(n^2)$. As in the previous step, evaluating a single combination requires average distances between intervals to be computed. Therefore, the time complexity of the second step is $O(n^4)$, and can be lowered to $O(n^2)$ when using another algorithmic refinement, as described below.

Since the last step is only a depth-first tree traversal, it requires $O(n)$ time, so the total time complexity is $O(n^5)$ in the above "basic" description of our algorithm, and $O(n^3)$ using the following refinements.

4.2 $O(1)$ Computation of the Average Distances between Intervals

Consider an interval $X_{p,q}$, composed of two adjacent intervals $X_{p,m}$ and $X_{m+1,q}$. As shown above, the calculation of $(L(X_{p,q}) - \overline{X_{p,q}})$ requires three average distances $\Delta_{X_{p,m}X_{m+1,q}}$, $\Delta_{X_{p,m}X_{\overline{p,q}}}$, $\Delta_{X_{m+1,q}X_{\overline{p,q}}}$ to be computed. In this section, we describe a refinement whose basic idea is the following : since we are considering growing intervals, most of the work needed for calculating these average distances for intervals of current size l has already been done when intervals of size $(l - 1)$ and $(l - 2)$ were considered. We show below that calculating these average distances can be done in $O(1)$ time, using a memorization scheme, thus lowering the complexity of our algorithm to $O(n^3)$.

$O(1)$ Computation of $\Delta_{X_{p,m}X_{m+1,q}}$. It is easy to show that the average distance between two intervals $X_{p,m}$ and $X_{m+1,q}$ can be expressed as follows:

$$\Delta_{X_{p,m}X_{m+1,q}} = \frac{1}{(m-p+1)(q-m)} \tag{9}$$
$$\times \begin{pmatrix} (m-p+1)(q-m-1)\Delta_{X_{p,m}X_{m+1,q-1}} \\ + (m-p)(q-m)\Delta_{X_{p+1,m}X_{m+1,q}} \\ - (m-p)(q-m-1)\Delta_{X_{p+1,m}X_{m+1,q-1}} \\ + \delta_{p,q} \end{pmatrix}.$$

Let $l = q - p + 1$ be the length of the current interval. In Equation (9), both $\Delta_{X_{p,m}X_{m+1,q-1}}$ and $\Delta_{X_{p+1,m}X_{m+1,q}}$ have been calculated when we considered intervals of length $(l-1)$, while $\Delta_{X_{p+1,m}X_{m+1,q-1}}$ has been calculated when we considered intervals of length $(l-2)$. Therefore, calculating $\Delta_{X_{p,m}X_{m+1,q}}$ can be done in $O(1)$ provided the required average distances have been stored when we considered both previous interval lengths. This memorization procedure requires $O(n^2)$ space, since we only need to store average distances for intervals with lengths $l-1$ and $l-2$.

$O(1)$ Computation of $\Delta_{X_{p,m}X_{\overline{p,q}}}$ and $\Delta_{X_{m+1,q}X_{\overline{p,q}}}$. Equations (2) and (8) are identical, except that the A_1, A_2 and R notation in Equation (2) is replaced with $X_{p,m}$, $X_{m+1,q}$ and $X_{\overline{p,q}}$ in Equation (8), respectively. In this section, we use A_1, A_2 and R, for simplicity. In Equation (2), computing $(L(A) - \overline{A})$ requires both Δ_{A_1R} and Δ_{A_2R} to be known, where $A = A_1 \bigcup A_2$ and R is the set of copies which do not belong to A_1 or A_2, i.e. $R = T - (A_1 \bigcup A_2)$. It is then easy to show that $\Delta_{A_1(T-A_1)}$ is given by the following equation:

$$\Delta_{A_1(T-A_1)} = \frac{1}{n - |A_1|} \left(|A_2| \Delta_{A_1 A_2} + (n - |A_1| - |A_2|)\Delta_{A_1R} \right),$$

which yields:

$$\Delta_{A_1R} = \frac{1}{(n - |A_1| - |A_2|)} \left((n - |A_1|)\Delta_{A_1(T-A_1)} - |A_2| \Delta_{A_1 A_2} \right), \tag{10}$$

and Δ_{A_2R} is obtained by symmetry. In the previous section, we showed that $\Delta_{A_1 A_2}$ can be calculated in $O(1)$. Therefore, we only need to know the average distance between A_1 and $(T - A_1)$ to compute Δ_{A_1R} using Equation (10). The same applies to A_2. The refinement we introduce here consists in precalculating and storing $\Delta_{A_1(T-A_1)}$ for every possible interval A_1, so that Δ_{A_1R} can be calculated in $O(1)$ time.

Assume that $\Delta_{Z(T-Z)}$ has been calculated for all intervals Z with length $(l-1)$. Let now Z be an interval with length l. Let z be the leftmost (or rightmost) copy of Z, which "jumps" at the current step from $(T - (Z - \{z\})$ to Z. $Z - \{z\}$ has length $(l-1)$ and $\Delta_{(Z-\{z\})(T-(Z-\{z\}))}$ has already been computed. $\Delta_{Z(T-Z)}$ can be computed in $O(n)$ time using the following equation:

$$\Delta_{Z(T-Z)} = \frac{1}{|Z|\,(n-|Z|)} \left(\begin{array}{c} (|Z|-1)(n-|Z|+1)\Delta_{(Z-\{z\})(T-(Z-\{z\}))} \\ -\sum_{y\in(Z-\{z\})} \delta_{yz} + \sum_{y\in(T-Z)} \delta_{yz} \end{array} \right).$$

$O(n^2)$ intervals need to be considered, and each of them is evaluated in $O(n)$ time. Therefore, precomputing every $\Delta_{Z(T-Z)}$ is done in $O(n^3)$ time, and the space required to store the resulting values is in $O(n^2)$.

$O(1)$ Computation of Δ_{AB}, Δ_{AC}, Δ_{BC}. The second step of our reconstruction algorithm requires the values of Δ_{AB}, Δ_{AC}, and Δ_{BC} within Equation (1) to be calculated. It is easy to show that these values can directly be calculated in $O(1)$ from $\Delta_{A(B\bigcup C)}$, $\Delta_{B(A\bigcup C)}$ and $\Delta_{C(A\bigcup B)}$, using the following set of equations:

$$\Delta_{A(B\bigcup C)} = \Delta_{A(T-A)} = \tfrac{1}{|B|+|C|} \left(|B|\,\Delta_{AB} + |C|\,\Delta_{AC} \right),$$
$$\Delta_{B(A\bigcup C)} = \Delta_{B(T-B)} = \tfrac{1}{|A|+|C|} \left(|A|\,\Delta_{AB} + |C|\,\Delta_{BC} \right),$$
$$\Delta_{C(A\bigcup B)} = \Delta_{C(T-C)} = \tfrac{1}{|A|+|B|} \left(|A|\,\Delta_{AC} + |B|\,\Delta_{BC} \right).$$

When solving the above linear system, we obtain the following expressions for Δ_{AB}:

$$\Delta_{AB} = \frac{1}{2\,|A|\,|B|} \left(\begin{array}{c} |A|\,[|B|+|C|]\,\Delta_{A(T-A)} + |B|\,[|A|+|C|]\,\Delta_{B(T-B)} \\ -\,|C|\,[|A|+|B|]\,\Delta_{C(T-C)} \end{array} \right),$$

and Δ_{AC}, Δ_{BC} are obtained by symmetry. Since every possible $\Delta_{Z(T-Z)}$ has been computed and stored during the first step, Δ_{AB}, Δ_{AC}, and Δ_{BC} are calculated from the above expression in $O(1)$.

5   Conclusion

In this paper, we presented an exact algorithm to reconstruct single copy duplication trees from a matrix of evolutionary distances between tandemly repeated sequences, using the minimum evolution criterion. This algorithm is based on a new recurrence equation for ordinary least-squares estimation of tree length. We showed that, using a simple memorization scheme and a dynamic programming approach, computing the optimal single copy duplication tree can then be done in $O(n^3)$ time and $O(n^2)$ space. It would now be relevant to compare our algorithm with some of the classical approaches, in terms of topological accuracy. Indeed, heuristic methods such as NJ [18] or DTSCORE [22] often do well in practice. Moreover, NJ could easily be adapted to single copy duplication tree reconstruction by only agglomerating adjacent pairs of taxa. But an exact algorithm such as ours has performance guaranty and will then avoid some possible (even rare) shortcomings that would trap heuristic approaches into local minima. It follows that, in phylogenetic reconstruction as in other domains, the search for polynomiality remains of importance; on the other hand, most phylogeny problems are NP-hard. So a direction for further research would be to extend (if possible) our results to multiple duplications and to other distance criteria, such as weighted least-squares [23,24], or to demonstrate the NP-hardness of these tasks.

References

1. Ohno, S.: Evolution by gene duplication. Springer Verlag, New York (1970)
2. Smith, G.: Evolution of repeated dna sequences by unequal crossover. Science **191** (1976) 528–535
3. Fitch, W.: Phylogenies constrained by cross-over process as illustrated by human hemoglobins in a thirteen-cycle, eleven amino-acid repeat in human apolipoprotein A-I. Genetics **86** (1977) 623–644
4. Jeffreys, A., Harris, S.: Processes of gene duplication. Nature **296** (1981) 9–10
5. Elemento, O., Gascuel, O., Lefranc, M.P.: Reconstruction de l'histoire de duplication de gènes répétés en tandem. In: Actes des Journées Ouvertes Biologie Informatique Mathématiques. (2001) 9–11
6. Elemento, O., Gascuel, O., Lefranc, M.P.: Reconstructing the duplication history of tandemly repeated genes. Molecular Biological Evolution **19** (2002) 278–288
7. Benson, G., Dong, L.: Reconstructing the duplication history of a tandem repeat. In Lengauer, T., Schneider, R., Bork, P., Brutlag, D., Glasgow, J., Mewes, H.W., Zimmer, R., eds.: Proceedings of Intelligent Systems in Molecular Biology ISMB'99. (1999) 44–53
8. Tang, M., Waterman, M., Yooseph, S.: Zinc finger gene clusters and tandem gene duplication. In El-Mabrouk, N., Lengauer, T., Sankoff, D., eds.: Proceedings of RECOMB 2001. (2001) 297–304
9. Tang, M., Waterman, M., Yooseph, S.: Zinc finger gene clusters and tandem gene duplication. Journal of Computational Biology **9** (2002) 429–446
10. Jaitly, D., Kearney, P., Lin, G., Ma, B.: Methods for reconstructing the history of tandem repeats and their application to the human genome. Journal of Computer and System Sciences **65** (2002) 494–507.
11. Zhang, J., Nei, M.: Evolution of antennapedia-class homeobox genes. Genetics **142** (1996) 295–303
12. Wang, L., Gusfield, D.: Improved approximation algorithms for tree alignment. Journal of Algorithms **25** (1997) 255–273
13. Kidd, K., Sgaramella-Zonta, L.: Phylogenetic analysis: concepts and methods. American Journal of Human Genetics **23** (1971) 235–252
14. Rzhetsky, A., Nei, M.: Theoretical foundation of the minimum-evolution method of phylogenetic inference. Molecular Biological Evolution **10** (1993) 173–1095
15. Denis, F., Gascuel, O.: On the consistency of the minimum evolution principle of phylogenetic inference. Computational Molecular Biology Series, Issue IV. Discrete Applied Mathematics **127** (2003) 63–77
16. Felsenstein, J.: Cases in which parsimony or compatibility methods will be positively misleading. Systematic Zoology **27** (1978) 401–410
17. Vardi, I.: Computational Recreations in Mathematica. Addison-Wesley (1991)
18. Saitou, N., Nei, M.: The neighbor-joining method: a new method for reconstructing phylogenetic trees. Molecular Biology and Evolution **4** (1987) 406–425
19. Vach, W.: Least-squares approximation of additive trees. In Opitz, O., ed.: Conceptual and Numerical Analysis of Data, Heidelberg, Springer (1989) 230–238

20. Gascuel, O.: Concerning the NJ algorithm and its unweighted version, UNJ. In Mirkin, B., McMorris, F., Roberts, F., Rzhetsky, A., eds.: Mathematical Hierarchies and Biology. DIMACS Series in Discrete Mathematics and Theoretical Computer Science. Amer. Math. Society, Providence (1997) 149–170

21. Barthelemy, J., Guénoche, A.: Trees and proximity representations. Wiley and Sons (1991)

22. Elemento, O., Gascuel, O.: A fast and accurate distance-based algorithm to reconstruct tandem duplicatin trees. Bioinformatics **18** (2002) S92–S99 Proceedings of European Conference on Computational Biology (ECCB2002).

23. Fitch, W., Margoliash, E.: Construction of phylogenetic trees. Science **155** (1967) 279–284

24. Felsenstein, J.: An alternating least squares approach to inferring phylogenies from pairwise distances. Systematic Biology **46** (1997) 101–111

Average-Optimal Multiple Approximate String Matching

Kimmo Fredriksson[1]* and Gonzalo Navarro[2]**

[1] Department of Computer Science, University of Joensuu
kfredrik@cs.joensuu.fi
[2] Department of Computer Science, University of Chile
gnavarro@dcc.uchile.cl

Abstract. We present a new algorithm for multiple approximate string matching, based on an extension of the optimal (on average) single-pattern approximate string matching algorithm of Chang and Marr. Our algorithm inherits the optimality and is also competitive in practice. We present a second algorithm that is linear time and handles higher difference ratios. We show experimentally that our algorithms are the fastest for intermediate difference ratios, an area where the only existing algorithms permitted simultaneous search for just a few patterns. Our algorithm is also resistant to the number of patterns, being effective for hundreds of patterns. Hence we fill an important gap in approximate string matching techniques, since no effective algorithms existed to search for many patterns with an intermediate difference ratio.

1 Introduction

Approximate string matching is one of the main problems in classical string algorithms, with applications to text searching, computational biology, pattern recognition, etc. Given a text $T_{1...n}$, a pattern $P_{1...m}$, and a maximal number of differences permitted, k, we want to find all the text positions where the pattern matches the text up to k differences. The differences can be substituting, deleting or inserting a character. We call $\alpha = k/m$ the *difference ratio*, and σ the size of the alphabet Σ. For the average case analyses it is customary to assume a random text over a uniformly distributed alphabet.

A natural extension to the basic problem consists of *multipattern searching*, that is, searching for r patterns $P^1 \ldots P^r$ simultaneously in order to report all their occurrences with at most k differences. This has also several applications such as virus and intrusion detection, spelling applications, text retrieval under synonym or thesaurus expansion, several problems in computational biology, batch processing of single-pattern approximate searching, etc. Moreover, some single-pattern approximate search algorithms resort to multipattern searching of

---

* Work developed while the author was working in the Dept. of Computer Science, University of Helsinki. Supported by the Academy of Finland.
** Partially supported by Fondecyt grant 1-020831.

R. Baeza-Yates et al. (Eds.): CPM 2003, LNCS 2676, pp. 109–128, 2003.

pattern pieces. Multidimensional search problems can also be reduced to string matching. Depending on the application, r may vary from a few to thousands of patterns. The naive approach is to perform r separate searches, so the goal is to do better.

The single-pattern problem has received a lot of attention since the sixties [8]. After the first dynamic-programming-based $O(mn)$ time solution to the problem [11], many faster techniques have been proposed, both for the worst and the average case. In 1994, Chang and Marr [3] showed that the average complexity of the problem is $O((k + \log_\sigma m)n/m)$, and gave an algorithm that achieved that average-optimal cost for $\alpha < 1/3 - O(1/\sqrt{\sigma})$.

The multipattern problem has received much less attention, not because of lack of interest but because of its difficulty. There exist algorithms that search permitting only $k = 1$ difference [6], and algorithms that handle either too few patterns or too low difference ratios [2].

Hence multiple approximate string matching is a rather undeveloped area. No algorithm exists when one searches for more than a few of patterns with intermediate difference ratios. Moreover, as the number of patterns grows, the difference ratios that can be handled get reduced.

The goal of this paper is to present an algorithm that is optimal on the average and that permits searching even for thousands of patterns with low and intermediate difference ratios, thus filling an important gap in the area. We build over an average-optimal algorithm that searches for single patterns [3] and inherit its optimality, obtaining $O(n(k + \log_\sigma(rm))/m)$ average search time. We show that the algorithm is not only theoretically appealing but also good in practice thanks to several practical improvements we introduce. Since the algorithm does not work for difference ratios beyond $1/3$, we introduce a second, $O(n)$ average time variant that reaches ratios of $1/2$. The algorithms are shown to be the fastest for a wide range of values of m, r and k, for small alphabets, see Sec. 6.

2   Related Work

2.1   Multiple Approximate String Matching

The naive approach to multipattern approximate searching is to perform r separate searches, one per pattern. If we use the optimal single-pattern algorithm [3], the average search time becomes $O((k + \log_\sigma m)rn/m)$ for the naive approach. On the other hand, if we use the classical $O(mn)$ algorithm [11] the time is $O(rmn)$.

Few algorithms exist for multipattern approximate searching under the k differences model. The first one, based on hashing, was presented by Muth and Manber [6]. It permits searching with $k = 1$ differences only, but is rather tolerant to the number of patterns r, which can reach the thousands without affecting much the cost of the search. The preprocessing time is $O(rm)$ and the average search time is $O(mn(1 + rm^2/M))$, where M is the size of the hash table. This adds up $O(rm + nm(1 + rm^2/M))$, which is $O(m(r + n))$ of $M = \Omega(m^2 r)$. This is basically independent of r if n is large enough.

Baeza-Yates and Navarro [2] have presented several algorithms for this problem. One of them, *partitioning into exact search*, uses the fact that, if P is cut into $k+1$ pieces, then at least one of the pieces appears inside every occurrence with no differences. Hence the algorithm splits every pattern into $k+1$ pieces and searches for the $r(k+1)$ pieces with an exact multipattern search algorithm. The preprocessing takes $O(rm)$ time. If they used an optimal multipattern exact search algorithm like MultiBDM [4], the search time would have been $O(k \log_\sigma(rm)n/m)$ on average. For practical reasons they used another algorithm, more suitable to searching for short pieces (of length $\lfloor m/(k+1) \rfloor$), albeit with worse theoretical complexity. This technique can be applied for $\alpha < 1/\log_\sigma(rm)$, a limit that gets more and more strict as m or r increase.

They also presented other algorithms that, although can handle higher difference ratios, are linear on r, which means that they give a speedup only up to a constant number c of patterns and then just divide the search into r/c groups that are searched for separately. *Superimposition* uses a standard search technique on a set of "superimposed" patterns, which means that the i-th character of the superimposition matches the i-th character of any of the superimposed patterns. Implemented over a newer bit-parallel algorithm [7], superimposition would yield average time $O(rn/(\sigma(1 - \alpha)^2))$ for $\alpha < 1 - e\sqrt{r/\sigma}$ on patterns shorter than the number of bits in the computer word, w (typically $w = 32$ or 64). Different techniques are used to cope with longer patterns, but the times are worse. *Counting* extends a single-pattern algorithm that slides a window of length m over the text checking in linear time whether it shares at least $m - k$ characters with the pattern (regardless of the order). The multipattern version keeps several counters in a single computer word, achieving an average search time of $O(rn \log(m)/w)$ for $\alpha < e^{-m/\sigma}$.

2.2   The Algorithm of Chang and Marr

Chang and Marr [3] show that no approximate search algorithm for a single pattern can be faster than $O((k + \log_\sigma m)n/m)$ on the average. This is not hard to prove, and we give more details in Section 4.

In the same paper [3], Chang and Marr presented an algorithm achieving that optimal average time complexity. In the preprocessing phase they build a table D as follows. They choose a number ℓ in the range $1 \leq \ell \leq \lceil (m - k)/2 \rceil$, whose exact value we will consider shorty. For every string S of length ℓ (ℓ-gram), they search for S in P and store in $D[S]$ the smallest number of differences needed to match S inside P (this is a number between 0 and ℓ). Hence D requires space for σ^ℓ entries and is computed in $\sigma^\ell \ell m$ time. A numerical representation of Σ^ℓ permits constant time access to D.

The text scanning phase consists of logically dividing the text in blocks of length $b = \lceil (m - k)/2 \rceil$, which ensures that any approximate occurrence of P (which has length at least $m - k$) contains at least one whole block. Each block $T_{ib+1\ldots ib+b}$ is processed as follows. They take the first ℓ-gram of the block,

$S^1 = T_{ib+1...ib+\ell}$, and obtain $D[S^1]$. Then they take the next ℓ-gram, $S^2 = T_{ib+\ell+1...ib+2\ell}$, and obtain $D[S^2]$, and so on. If, before reaching the end of the block, they have obtained $\sum_{1 \le j \le t} D[S^j] > k$, then they can safely skip the block because no occurrence of P can contain the block, as merely matching those t ℓ-grams anywhere inside P requires more than k differences. If, on the other hand, they reach the end of the block without surpassing k total differences, the block must be checked. In order to check for $T_{ib+1...ib+b}$ they run the classical dynamic programming algorithm over $T_{ib+1-m-k+b...ib+m+k}$.

In order to keep the space requirement polynomial in m, it is required that $\ell = O(\log_\sigma m)$. On the other hand, in order to achieve the claimed complexity, it is necessary that $\ell \ge x \log_\sigma m$ for some constant x, so the space is $O(m^x)$. The optimal complexity holds as long as $\alpha < 1/3 - O(1/\sqrt{\sigma})$.

3 Our Algorithm

The basic idea of our algorithm is as follows. Given r search patterns $P^1 \ldots P^r$, we build the table D taking the minimum number of differences to match each ℓ-gram inside *any* of the patterns. The scanning phase is the same as in Section 2.2. If we surpass k differences inside a block we are sure that none of the patterns match, since there are t ℓ-grams inside the block that need more than k differences in order to be found inside any pattern. Otherwise, we check the patterns one by one over the block. Figure 1 gives the code. We present now several improvements over this basic idea.

Search $(T_{1...n},\ P^1_{1...m} \ldots P^r_{1...m},\ k)$

1. $\ell \leftarrow$ **Preprocess** ( )
2. $b \leftarrow \lceil (m-k)/2 \rceil$
3.   **For** $i \in 0 \ldots \lfloor n/b \rfloor - 1$ **Do**
4.   **VerifyBlock** ($i,\ b$)

Fig. 1. High-level description of the algorithm. The input parameters are taken as global variables in the rest of the paper, to simplify the descriptions.

3.1 Optimal Choice of ℓ-Grams

The basic single-pattern algorithm [3] uses the first consecutive ℓ-grams of the block in order to find more than k differences. This is simple, but not necessarily the best choice. Note that any set of non-overlapping ℓ-grams found inside the block whose total number of differences inside P exceeds k permits us discarding the block. Hence the question of using the best possible set is raised.

The optimization problem is as follows. Given the text block $T_{ib+1...ib+b}$ we have $b - \ell + 1$ possible ℓ-grams, namely $T_{ib+1...ib+\ell}$, $T_{ib+2...ib+\ell+1}$, ..., $T_{ib+b-\ell+1...ib+b}$. From this set we want a subset of non-overlapping ℓ-grams $S^1 \ldots S^t$ such that $\sum_{1 \leq j \leq t} D[S^j] > k$. Moreover, we want to process the set left to right and detect a good enough subset as soon as possible.

This is solved by calling M_u the maximum sum that can be obtained using ℓ-grams that start in the positions $ib+1 \ldots ib+u$. Initially we start with $M_u = 0$ for $-\ell < u \leq 0$. Then we traverse the block computing, for increasing u values,

$$M_u \quad \leftarrow \quad \max(D[T_{ib+u...ib+u+\ell-1}] + M_{u-\ell} , M_{u-1}) \tag{1}$$

where the first term accounts for the fact that we choose to use the ℓ-gram that starts at u and add to it the best previous solution that does not overlap this ℓ-gram, and the second term accounts for the fact that we do not use the ℓ-gram that starts at u.

We compute M_u for increasing u until either (*i*) $M_u > k$, in which case we abandon the block, or (*ii*) $u > b - \ell + 1$, in which case we have to verify the block. Figure 2 gives the code.

CanDiscard (i, b, D, ℓ)

1.   **For** $u \in -\ell \ldots 0$ **Do** $M_u \leftarrow 0$
2.   **For** $u \in 1 \ldots b - \ell + 1$ **Do**
3. $M_u \leftarrow \max(D[T_{ib+u...ib+u+\ell-1}] + M_{u-\ell} , M_{u-1})$
4.   **If** $M_u > k$ **Then Return** TRUE
5.   **Return** FALSE

Fig. 2. Optimization technique to choose the set of overlapping ℓ-grams that maximize the sum of differences. It returns whether the block can be discarded.

Note that the cost of choosing the best set of ℓ-grams is that, if we abandon the block after considering position x, then we work $O(x/\ell)$ with the simple method and $O(x)$ with the current one. (This assumes we can read an ℓ-gram in constant time, which is true in practice given the ℓ values used.) However, x itself may be smaller with the optimization method.

3.2   Hierarchical Verification

On the blocks that have to be verified, we could simply run the verification for every pattern, one by one. A more sophisticated choice is *hierarhical verification* (already presented in previous work [2]). We form a tree whose nodes have the form $[i, j]$ and represent the group of patterns $P^i \ldots P^j$. The root is $[1, r]$. The leaves have the form $[i, i]$. Every internal node $[i, j]$ has two children $[i, \lfloor (i+j)/2 \rfloor]$ and $[\lfloor (i + j)/2 \rfloor + 1, j]$.

The hierarchy is used as follows. For every internal node $[i, j]$ we have a table D computed using the minimum distances between ℓ-grams and patterns $P^i \ldots P^j$. This is done by computing first the leaves (that is, each pattern separately) and then computing every cell of D in the internal node as the minimum over the corresponding cell in its two children. In order to scan the text, we use the D table of the root node, which corresponds to the full set of patterns. Every time a block has to be verified with respect to a node in the hierarchy (at first, the root node), we rescan the block considering the two children of the current node. It is possible that the block can be discarded for both children, for one, or for none. We recursively repeat the process for every child that does not permit discarding the block, see Fig. 3. If we process a leaf node and still have to verify the block, then we run dynamic programming over the corresponding single pattern.

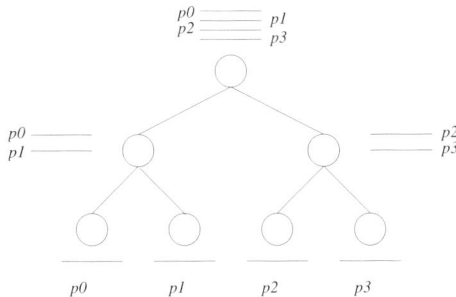

Fig. 3. Pattern hierarchy for 4 patterns.

The idea of using the hierarchy instead of plainly checking the r patterns one by one is that it is possible that the grouping of the pattern matches a block, but that none of its halves match. In this case we save verification time. The plain technique needs $O(\sigma^\ell)$ space, while hierarchical verification needs much more, $O(r\sigma^\ell)$.

Note that verification would benefit if the patterns we group together are as similar as possible, in terms of numbers of differences. A simple heuristic is to lexicographically sort the patterns before grouping them by ranges.

As a final note, we use Myers' algorithm [7] for the verification of single patterns, which makes the cost $O(m^2/w)$, where w is the number of bits in the computer word.

Figures 4 and 5 show the preprocessing and verification using hierarchical verification.

3.3 Reducing Preprocessing Time

Either if we use plain or hierarchical verification, preprocessing time is an issue. We have to search every pattern for every ℓ-gram, resulting in $O(r\ell m\sigma^\ell)$ prepro-

---

HierarchyPreprocess $(i,\ j,\ \ell)$

1.   **If** $i = j$ **Then** $D_{i,j} \leftarrow$ **PreprocessD**(P^i, ℓ)
2.   **Else**
3. $m \leftarrow \lfloor (i+j)/2 \rfloor$
4.   **HierarchyPreprocess** (i, m, ℓ)
5.   **HierarchyPreprocess** $(m + 1, j, \ell)$
6.   **For** $s \in \Sigma^\ell$ **Do**
7. $D_{i,j}[s] \leftarrow \min(D_{i,m}[s], D_{m+1,j}[s])$

Preprocess ( )

8. $\ell \leftarrow \lceil \frac{3\log_\sigma m + \log_\sigma r}{1 - c + 2c\log_\sigma c + 2(1-c)\log_\sigma(1-c)} \rceil$ // see Eq. (3)
9.   **HierarchyPreprocess**$(1, r, \ell)$

Fig. 4. Preprocessing to build the hierarchy. It is initially invoked with parameters $(1, r)$ and produces global tables $D_{i,j}$ to be used by **HierarchyVerify**. The main search table is $D_{1,r}$.

HierarchyVerify $(i,\ j,\ b,\ s)$

1.   **If** NOT **CanDiscard** $(s, b, D_{i,j}, \ell)$ **Then**
2.   **If** $i = j$ **Then** Search for P^i in $T_{sb+1-m-k+b...sb+m+k}$
3.   **Else**
4. $m \leftarrow \lfloor (i+j)/2 \rfloor$
5.   **HierarchyVerify** (i, m)
6.   **HierarchyVerify** $(m + 1, j)$

Fig. 5. Hierarchical verification. Procedure **VerifyBlock**(i, b) is then defined as **HierarchyVerify** $(1, r, b, i)$.

cessing time. In the case of hierarchical verification we pay an additional $O(r\sigma^\ell)$ time to create the D tables of the internal nodes, but this is negligible compared to the cost to compute the individual patterns.

In order to find the minimum number of differences to match an ℓ-gram S inside a pattern P, we compute the matrix $C_{i,j}$, for $0 \leq i \leq \ell$ and $0 \leq j \leq m$, as follows [11]:

$$C_{i,0} = i\ ,\quad C_{0,j}\ =\ 0$$
$$C_{i+1,j+1} = \text{if } S_{i+1} = P_{j+1} \text{ then } C_{i,j} \text{ else } 1 + \min(C_{i,j}, C_{i,j+1}, C_{i+1,j})$$

which can be computed, for example, row-wise left to right. We need only the previous row in order to compute the current row. The minimum distance is finally $\min_{0 \leq j \leq m} C_{\ell,j}$.

We present now a method to reduce the preprocessing time to $O(rm\sigma^\ell)$, which has been used before in the context of indexed approximate string matching [10]. Instead of running the ℓ-grams one by one over a pattern P, we form a trie data structure of all the ℓ-grams. For every trie node whose path from the root spells out the string S, we compute the last row of the C matrix corresponding to searching for S inside P. For this sake we use the previous matrix row, which was computed for the parent node. Hence, if we traverse the trie using a classical depth first search recursion and compute a new matrix row at each invocation, then the execution stack contains the matrix computed up to now, so we use the row computed at the invoking process to compute the row of the invoked process. Since we work $O(m)$ at every trie node and there are $O(\sigma^\ell)$ nodes, the overall process takes $O(m\sigma^\ell)$ time. It needs just space for the stack, $O(m\ell)$. By repeating this over each pattern we obtain $O(rm\sigma^\ell)$ time.

Note finally that the trie of ℓ-grams does not need to be explicitly built, as we know that we have every possible ℓ-gram and hence can use an implicit method to traverse all them without actually storing them. Only the minima over the final rows are stored into the corresponding D entries. Figure 6 shows the code.

RecPreprocessD $(P,\ i,\ \ell,\ S,\ Cold,\ D)$

1.   **If** $i = \ell$ **Then** $D[S] \leftarrow \min_{0 \le j \le m} Cold_j$
2.   **Else**
3.   **For** $s \in \Sigma$ **Do**
4. $Cnew_0 \leftarrow i$
5.   **For** $j \in 1 \ldots m$ **Do**
6.   **If** $s = P_j$ **Then** $Cnew_j \leftarrow Cold_{j-1}$
7.   **Else** $Cnew_j \leftarrow 1 + \min(Cold_{j-1}, Cold_j, Cnew_{j-1})$
8.   **RecProcessD** $(P, i+1, \ell, Ss, Cnew, D)$

PreprocessD $(P,\ \ell)$

9.   **For** $j \in 0 \ldots m$ **Do** $C_j \leftarrow 0$
10.   **RecPreprocessD** $(P, 0, \ell, \varepsilon, C, D)$
11.   **Return** D

Fig. 6. Preprocessing for a single table.

Again, we use Myers' algorithm [7] to compute the matrix rows, which makes the preprocessing time $O(rm\sigma^\ell/w)$. For this sake we need to modify the algorithm so that it takes the ℓ-gram as the text and P^i as the pattern. This means that the matrix is transposed, so the current "column" starts with zeros and at the i-th step its first cell has the value i. The necessary modifications are simple and are described, for example, in [5].

The only complication is how to obtain the value $\min_{0 \leq j \leq m} C_{\ell,j}$ from Myers' compressed representation of C as a bit vector of increments and decrements. A solution is to use bit magic, so as to store preprocessed answers that give the total increment and minimum value for every bit mask of a given length. Since C is represented using two bit vectors of m bits (one for increments and the other for decrements), we need $O(2^{2x})$ space in order to process the bit vector in $O(m/x)$ time. A reasonable choice not affecting the time complexity is $x = w/4$ for 32-bit machines or $x = w/8$ for 64-bit machines (for a table of 2^{16} entries).

3.4   Packing Counters

Our final optimization resorts to bit-parallelism, that is, to storing several values inside the same computer word (this has been also used, for example, in the counting algorithm [2]). For this sake we will denote the bitwise *and* operation as "&", the *or* as "|", and the bit complementation as "∼". Shifting i positions to the left (right) is represented as "$<< i$" ("$>> i$"), where the bits that fall are discarded and the new bits that enter are zero. We can also perform arithmetic operations over the computer words. We use exponentiation to denote bit repetition, e.g. $0^3 1 = 0001$, and write the most significant bit as the leftmost bit.

In our process of adding up differences, we start with zero differences and grow at most up to $k + \ell$ differences before abandoning the block. This means that it suffices to use $B = \lceil \log_2(k + \ell + 1) \rceil$ bits to store a counter. Instead of taking minima over several patterns, we could separately store their counters in a single computer word C of w bits ($w = 32$ or 64 in current architectures). This means that we could store $A = \lfloor w/B \rfloor = O(w/\log k)$ counters in a single machine word C.

Consequently, we should keep several difference counts in the same machine word of a D cell. We can still add up our counter and the corresponding D cell and all the counters will be added simultaneously, so the cost is exactly the same as for one single counter or pattern.

Every text block must be traversed until *all* the counters exceed k, so we need a mechanism to check for this condition over all the counters in a single operation. A solution is to initialize the counters not at zero but at $2^{B-1} - k - 1$, which ensures that the highest bit in each counter will be activated as soon as the counter reaches the value $k + 1$. However, this means that the values stored inside the counters may now reach $2^{B-1} + \ell - 1$. This will not cause overflow as long as $2^{B-1} + \ell - 1 < 2^B$, that is, $2\ell \leq 2^B$. So in fact B should be chosen such that $2^B > \max(k + \ell, 2\ell - 1)$, that is, $B = \lceil \log_2 \max(k + \ell + 1, 2\ell) \rceil$.

With this arrangement, in order to check whether all the counters have exceeded k, we simply check whether all the highest bits of all the counters are set. This is achieved using the bitwise *and* operation: Let $H = (10^{B-1})^A$ be the bit mask where all the highest bits of the counters are set. Then, all the counters have exceeded k if and only if $H \ \& \ C = H$. In this case we can abandon the block.

Note that it is still possible that our counters overflow, because we can have that some of them have exceeded $k + \ell$ while others have not. We avoid using more bits for the counters and at the same time ensure that, once a counter has its highest bit set, it will stay with this bit set. Before adding $C \leftarrow C + D[S]$, we remove all the highest bits from C, that is, we assign $O \leftarrow H \;\&\; C$, and replace the simple sum by the assignment $C \leftarrow ((C \;\&\; \sim H) + D[S]) \mid O$. Since we have selected B such that $\ell \leq 2^{B-1}$, adding $D[S]$ to a counter with its highest bit set cannot cause an overflow. Note also that highest bits that are already set are always preserved.

This technique permits us searching for $A = \lfloor w/B \rfloor$ patterns at the same time. If we have more patterns we resort to grouping. In a plain verification scenario, we can group r/A patterns in a single counter and search for the A patterns simultaneously, with the advantage of having to verify only r/A patterns instead of all the r patterns whenever a block requires verification. In a hierarchical verification scenario, the result is that our hierarchy tree has arity A instead of two, and has no root. That is, the tree has A roots that are searched for together, and each root packs r/A patterns. If one such node has to be verified, then we consider its A children nodes (that pack r/A^2 patterns each), all together, and so on. This reduces not only verification costs but also the preprocessing space, since we need less tables.

We have also to consider how this is combined with the optimization algorithm of Section 3.1, since the best choice to maximize one counter may not be the best choice to maximize another. The solution is to pack also the different values of M_u in a single computer word. The operation of Eq. (1) can be perfectly done in parallel for several counters, as long as we replace the sum by the above technique to avoid overflows. The only obstacle is the maximum, which as far as we know has never been used in a bit-parallel scenario. We do that now.

If we have to compute $\max(X, Y)$, where X and Y contain several counters properly aligned, in order to obtain the counter-wise maxima, we need an extra highest bit per counter, which is always zero. Say that counters have now $B + 1$ bits, counting this new highest bit. We precompute the bit mask $J = (10^B)^A$ (where now $A = \lfloor w/(B+1) \rfloor$) and perform the operation $F \leftarrow ((X \mid J) - Y) \;\&\; J$. The result is that, in F, each highest bit is set if and only if the counter of X is larger than that of Y. We now compute $F \leftarrow F - (F >> B)$, so that the counters where X is larger than Y have all their bits set in F, and the others have all the bits in zero. Finally, we choose the maxima as $\max(X, Y) \leftarrow (X \;\&\; F) \mid (Y \;\&\; \sim F)$.

Fig. 7 shows the bit-parallel version of the counter accummulation, and Fig. 8 shows an example of pattern hierarhy.

4   Analysis

We analyze our algorithm by following the analysis of the corresponding single pattern algorithm [3]. Two useful lemmas shown there follow (we have written them in a way more convenient for us).

Lemma 1 [3] The probability that two random ℓ-grams have a common subsequence of length $(1-c)\ell$ is at most $a\sigma^{-d\ell}/\ell$, for constants $a = (1+o(1))/(2\pi c(1-$

CanDiscard $(i,\ b,\ D,\ \ell)$

1. $B \leftarrow \lceil \log_2 \max(k + \ell + 1, 2\ell) \rceil$
2. $A \leftarrow \lfloor w/(B+1) \rfloor$
3. $H \leftarrow (010^{B-1})^A$
4. $J \leftarrow (10^B)^A$
5.   **For** $u \in -\ell \dots 0$ **Do**
6. $M_u \leftarrow (2^{B-1} - k - 1) \times (0^B 1)^A$
7.   **For** $u \in 1 \dots b - \ell + 1$ **Do**
8. $X \leftarrow M_{u-\ell}$
9. $O \leftarrow X\ \&\ H$
10. $X \leftarrow ((X\ \&\ \sim H) + D[T_{ib+u \dots ib+u+\ell-1}])\ |\ O$
11. $Y \leftarrow M_{u-1}$
12. $F \leftarrow ((X\ |\ J) - Y)\ \&\ J$
13. $F \leftarrow F - (F >> B)$
14. $M_u \leftarrow (X\ \&\ F)\ |\ (Y\ \&\ \sim F)$
15.   **If** $H\ \&\ M_u = H$ **Then Return** TRUE
16.   **Return** FALSE

Fig. 7. The bit-parallel version of **CanDiscard**. It requires that D is preprocessed by packing the values of A different patterns in the same way. Lines 1–6 can in fact be done once at preprocessing time.

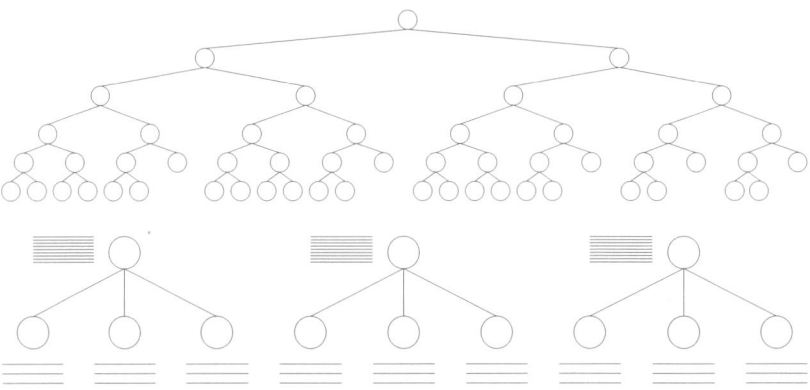

Fig. 8. Top: basic pattern hierarchy for 27 patterns. Bottom: pattern hierarchy with bit-parallel counters (27 patterns).

c)) and $d = 1 - c + 2c \log_\sigma c + 2(1 - c) \log_\sigma (1 - c)$. The probability decreases exponentially for $d > 0$, which surely holds if $c < 1 - e/\sqrt{\sigma}$.

Lemma 2 [3] If S is an ℓ-gram that matches inside a given string P (larger than ℓ) with less than $c\ell$ differences, then S has a common subsequence of length $\ell - c\ell$ with some ℓ-gram of P.

We measure the amount of work in terms of inspected characters. For a given text block, if there is a single ℓ-gram inside the block that matches inside any pattern P^i with less than $c\ell$ differences, we pessimistically assume that we verify the whole block. Otherwise, after considering $1+\lceil k/(c\ell) \rceil$ non-overlapping ℓ-grams, we abandon the block without verifying it. For the latter to be correct, it must hold $k/m = \alpha < c/(c+2)(1+O(1/m))$, since otherwise we reach the end of the block (of length $(m-k)/2$) before considering those $1+\lceil k/(c\ell) \rceil$ ℓ-grams.

Given Lemmas 1 and 2, the probability that a given ℓ-gram matches with less than $c\ell$ differences inside some P^i is at most that of having a common subsequence of length $\ell - c\ell$ with some ℓ-gram of some P^i. The probability of this is $mra\sigma^{-d\ell}/\ell$. Consequently, the probability that any ℓ-gram in the current text block matches is $m^2ra\sigma^{-d\ell}/\ell^2$, since there are m/ℓ ℓ-grams. (We assume for the analysis that we do not use the optimization of Section 3.1; this is pessimistic for every possible text block.)

Hence, with probability $m^2ra\sigma^{-d\ell}/\ell^2$ we verify the block, and otherwise we do not. In the first case we pay $O(m^2r)$ time if we use plain verification (Section 3.2, we see the case of hierarhical verification later) and dynamic programming. In the second case we pay the number of characters inspected in order to process $1 + \lceil k/(c\ell) \rceil$ ℓ-grams, that is, $\ell + k/c$. Hence the average cost is upper bounded by

$$O\left(\frac{n}{m} \left(\frac{am^4r^2}{\ell^2}\sigma^{-d\ell} + \ell + \frac{k}{c} \right) \right)$$

The first part is the cost of verifications, and we have to make it negligible compared to the second part, that is, we have to ensure that verifications are rare enough. A sufficient condition on ℓ is

$$\ell \geq \frac{4\log_\sigma m + 2\log_\sigma r}{d} = \frac{4\log_\sigma m + 2\log_\sigma r}{1 - c + 2c\log_\sigma c + 2(1-c)\log_\sigma(1-c)}$$

(in fact a slightly better, but more complicated, bound can be derived).

Note that we are free to choose any constant $2\alpha/(1-\alpha) < c < 1 - e/\sqrt{\sigma}$. If we let c approach $1 - e/\sqrt{\sigma}$, the value of ℓ goes to infinity and so does our preprocessing cost. If we let c approach $2\alpha/(1-\alpha)$, ℓ gets as small as possible but our search cost becomes $O(n)$. Having properly chosen c and ℓ, our algorithm is on average

$$O\left(\frac{n(k + \log_\sigma(rm))}{m} \right) \qquad (2)$$

character inspections. We remark that this is true as long as $2\alpha/(1-\alpha) < 1 - e/\sqrt{\sigma}$, that is, $\alpha < 1/3 - O(1/\sqrt{\sigma})$, as otherwise the whole algorithm reduces to dynamic programming.

Recall that our preprocessing cost is $O(mr\sigma^\ell/w)$. Given the value of ℓ, this is $O(m^5r^3\sigma^{O(1)}/w)$. The space with plain verification is $\sigma^\ell = m^4r^2\sigma^{O(1)}$ integers.

As a practical consideration, we have that since σ^ℓ must fit in memory, we must be able to hold $\ell \log_2 \sigma$ bits in a single computer word, so we can read a whole ℓ-gram in a single computer instruction. The number of instructions executed then becomes $O(n(1+k/\log M)/m)$, where M is the amount of memory we spend on a D table. Note that this is not true if we use the optimization method of Section 3.1, although we are not able to analyze the benefit that this method produces, on the other hand.

The fact that we perform the verification using Myers' algorithm [7] changes its cost to $O(rm^2/w)$, and this permits reducing ℓ a bit in practice, but the overall complexity does not change.

Let us now analyze the effect of hierarchical verification. This time we start with r patterns, and if the block requires verification, we run two new scans for $r/2$ patterns, and continue the process until a single pattern asks for verification. Only then we perform the dynamic programming verification. Let $p = a\sigma^{-d\ell}m^2/\ell^2$. Then the probability of verifying the root node is pr. For a non-root node, the probability that it requires verification given that the parent requires verification is $Pr(child/parent) = Pr(child \wedge parent)/P(parent) = Pr(child)/Pr(parent) = 1/2$, since if the child requires verification then the parent requires verification. Then the number of times we scan the whole block is on average

$$pr(1 + 2(1/2(1 + 2(1/2\ldots \quad = \quad pr\log_2 r$$

Hence the total character inspections for the scans that require verifications is $O(pmr\log r)$. Finally, each individual pattern is verified provided an ℓ-gram of the text block matches inside it. This accounts for $O(prm^2)$ verification cost. Hence the overall cost under hierarchical verification is

$$O\left(\frac{n}{m}\left(\frac{am^3r(m+\log r)}{\ell^2}\sigma^{-d\ell} + \ell + \frac{k}{c}\right)\right)$$

which is clearly better than the cost with plain verification. The condition on ℓ to obtain the same search time of Eq. (2) is now

$$\ell \geq \frac{\log_\sigma(m^3r(m+\log_2 r))}{d} = \frac{3\log_\sigma m + \log_\sigma r + \log_\sigma(m+\log_2 r)}{1 - c + 2c\log_\sigma c + 2(1-c)\log_\sigma(1-c)} \quad (3)$$

which is smaller and hence requires less preprocessing effort. This time the preprocessing cost is $O(m^4r^2(m + \log r)\sigma^{O(1)}/w)$, smaller than with plain verification. The space of hierarchical verification, however, is $2r\sigma^\ell = 2m^3r^2(m + \log_2 r)\sigma^{O(1)}$, which is larger than with plain verification.

Finally, let us consider the use of bit-parallel counters (Section 3.4). This time the arity of the tree is $A = \lfloor w/(1 + \lceil\log_2(k+1)\rceil)\rfloor$ and it has no root. We have r/A tables in the leaves of the hierarchical tree. The total space requirement is less than $r/(A-1)$ tables. The verification effort is now $O(pmr\log_A r)$ for scanning and re-scanning, and $O(prm^2)$ for dynamic programming. This puts a less stringent condition on ℓ:

$$\ell \geq \frac{\log_\sigma(m^3r(m+\log_A r))}{d} = \frac{3\log_\sigma m + \log_\sigma r + \log_\sigma(m+\log_A r)}{1 - c + 2c\log_\sigma c + 2(1-c)\log_\sigma(1-c)}$$

and reduces the preprocessing effort to $O(m^4 r^2 (m + \log_A r)\sigma^{O(1)}/w)$. The space requirement is $\lceil r/(A-1) \rceil \sigma^\ell = m^3 r^2 (m + \log_A r)\sigma^{O(1)}/(A-1)$. With plain verification the space requirement is still smaller, but the difference is now smaller.

To summarize, we have shown that we are able to perform, on average, $O(n(k + \log_\sigma(rm))/m)$ character inspections whenever $\alpha < 1 - e/\sqrt{\sigma}$. This requires a preprocessing time of roughly $O(m^4 r^2 (m + \log_{w/\log k} r)\sigma^{O(1)}/w)$ and an extra space of $O(m^3 r^2 (m + \log_{w/\log k} r)\sigma^{O(1)} \log(k)/w)$ by using the best techniques. The number of machine instructions for the search can be made $O(n(1 + k/\log M)/m)$ provided we use M memory for a single D table.

It has been shown that, for a single pattern, $O(n(k + \log_\sigma m)/m)$ is optimal [3]. This uses two facts. The first is that it is necessary to inspect at least $k+1$ characters in order to skip a given text window of length m, so we need at least $\Omega(kn/m)$ character inspections. The second is that the $\Omega(n \log_\sigma(m)/m)$ lower bound of Yao [14] for exact string mathcing applies to approximate searching too, as exact searching is included in the approximate search problem. When searching for r patterns, this lower bound becomes $\Omega(n \log_\sigma(rm)/m)$, as we show in the Appendix A. Hence our algorithm is optimal.

5   A Slower Algorithm for Higher Difference Ratios

A weakness of the algorithm is that it cannot cope with difference ratios beyond $1/3$. This is due in part to the use of text blocks of length $(m-k)/2$. A different alternative to fixed-position blocks is the use of a sliding window of t ℓ-grams, where $t = \lfloor (m-k+1)/\ell \rfloor - 1$. If we consider text blocks for the form $T_{i\ell+1...i\ell+t\ell}$, we are sure that every occurrence (whose minimum length is $m-k$) contains a complete block. Then, if the ℓ-grams inside the window add up more than k differences, we can move to the next block.

The main difference is that blocks overlap with each other by $t-1$ ℓ-grams, so we should be able to update our difference counter from one text block to the next in constant time. This is rather easy, although it does not permit anymore the use of the optimization technique of Section 3.1. The result is an algorithm that takes $O(n)$ time for $\alpha < 1/2 - O(1/\sqrt{\sigma})$. Figure 9 shows this algorithm.

6   Experimental Results

We have implemented the algorithms in C, compiled using gcc 3.2.1 with full optimizations. The experiments were run in 2GHz Pentium 4, with 512MB RAM, with Linux 2.4.

We ran experiments for alphabet sizes $\sigma = 4$ (DNA), $\sigma = 20$ (protein) and $\sigma = 256$ (ASCII text). The test data for DNA and protein alphabets was randomly generated. The texts were 64MB characters for DNA, and 16MB characters for protein, and the patterns were 64 characters. The texts were stored used only 2 (DNA) and 5 bits (protein) per character, which allowed $O(1)$ time access to the ℓ-grams.

Search $(T_{1...n}, \; P^1_{1...m} \cdots P^r_{1...m}, \; k)$

1. $\ell \leftarrow$ **Preprocess** ( )
2. $t \leftarrow \lfloor (m - k + 1)/\ell \rfloor - 1$
3. $b \leftarrow t\ell$
4. $M \leftarrow 0$
5.   **For** $i \in 0 \dots t - 2$ **Do** $M \leftarrow M + D[T_{i\ell+1...i\ell+\ell}]$
6.   **For** $i \in t - 1 \dots \lfloor n/\ell \rfloor\text{-}1$ **Do**
7. $M \leftarrow M + D[T_{i\ell+1...i\ell+\ell}]$
8.   **If** $M \leq k$ **Then** Verify text block
9. $M \leftarrow M - D[T_{(i-t+1)\ell+1...(i-t+1)\ell+\ell}]$

Fig. 9. High-level description of the slower algorithm. The verification of a text block can also be done hierarchically.

Table 1. Preprocessing times in seconds for various number of patterns, and for various ℓ-gram lenghts. The pattern lenghts are $m = 64$ for DNA and protein, and $m = 16$ for ASCII.

DNA	1	8	32	64	protein	1	64	256	1024	ASCII	1	64	256	1024
4	0.00	0.00	0.00	0.00	1	0.00	0.01	0.01	0.02	1	0.00	0.01	0.01	0.06
6	0.01	0.01	0.04	0.08	2	0.00	0.01	0.03	0.07	2	0.01	0.59	2.60	10.65
8	0.02	0.15	0.58	1.17	3	0.01	0.09	0.40	1.55	3	4.26			
10	0.38	3.02	12.00	24.19	4	0.04	6.09							

Table 1 gives the preprocessing times for various alphabets, number of patterns and ℓ-grams. The preprocessing timings are for the basic algorithms, without the bit-parallel counters technique, which requires slightly more time. The maximum values in practice are $\ell \leq 8$ for DNA, $\ell \leq 3$ for protein, and $\ell \leq 2$ for ASCII. The search times were measured for these maximum values.

Figs. 10, 11, and 12 give the search times for the DNA, protein, and ASCII alphabets. The abreaviations in the figures are as follows. SL: the basic sublinear time algorithm, SLO: SL with the optimal choice of ℓ-grams, L: the basic linear time filtering algorithm, SLC: SL with bit-parallel counters, SLCO: SLC with the optimal choice of ℓ-grams, LC: L with bit-parallel counters. All filters use the hierarchical verification. For comparison, Fig. 13 gives timings for the exact pattern partitioning algorithm given in [1]. This algorithm beats the new algorithms for large rm, σ, k. Fig. 14 illustrates.

Optimal choice of ℓ-grams helps only sometimes, but is usually slower due to its complexity. The linear time filtering algorithms quickly become faster than the sublinear algorithms for large rm, k. The bit-parallel counters speed-up the search for large rm. The performance of the algorithms collapse when the error ratio grows past a certain limit, and this collapse is very sharp. Before that limit, the new algorithms are very efficient.

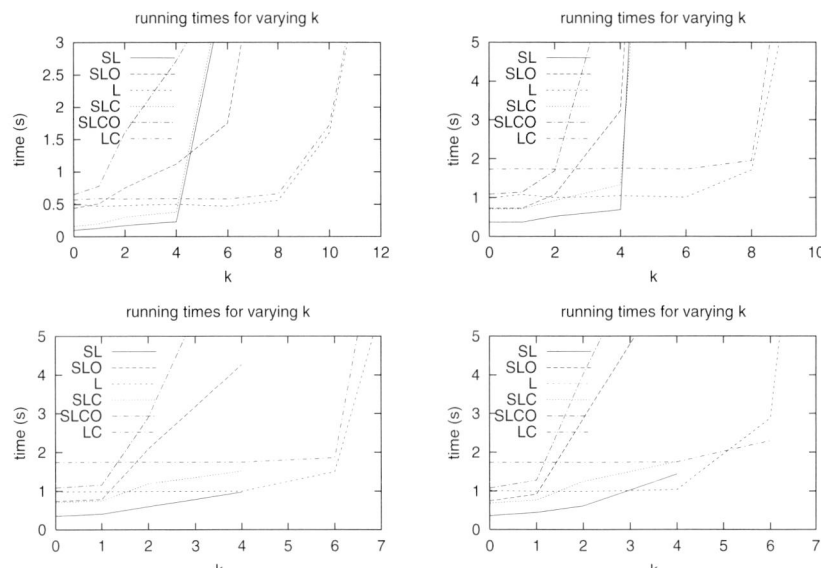

Fig. 10. Search times in seconds. Parameters are: $\sigma = 4$, $m = 64$, and $\ell = 8$. The figures are for, from left to right, top to bottom: $r = 1$, $r = 8$, $r = 32$, and $r = 64$.

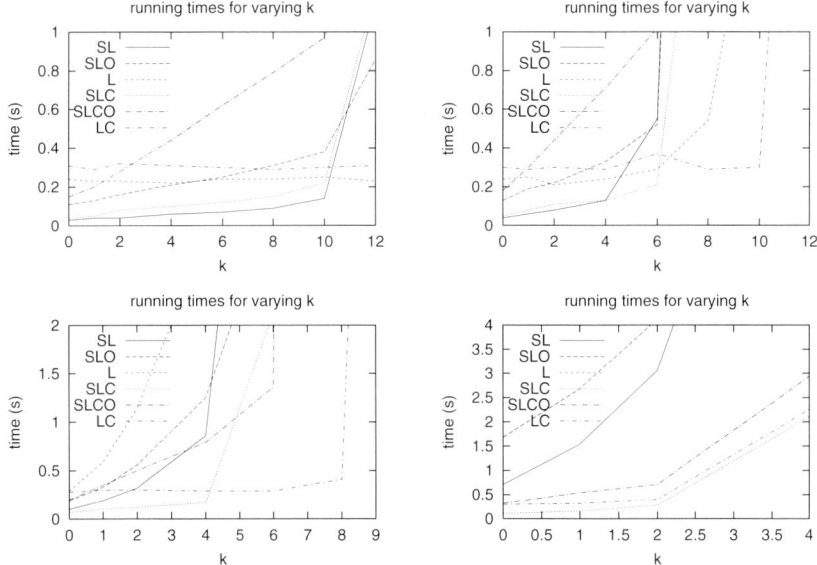

Fig. 11. Search times in seconds. Parameters are: $\sigma = 20$, $m = 64$, and $\ell = 3$. The figures are for, from left to right, top to bottom: $r = 1$, $r = 64$, $r = 256$, and $r = 1024$.

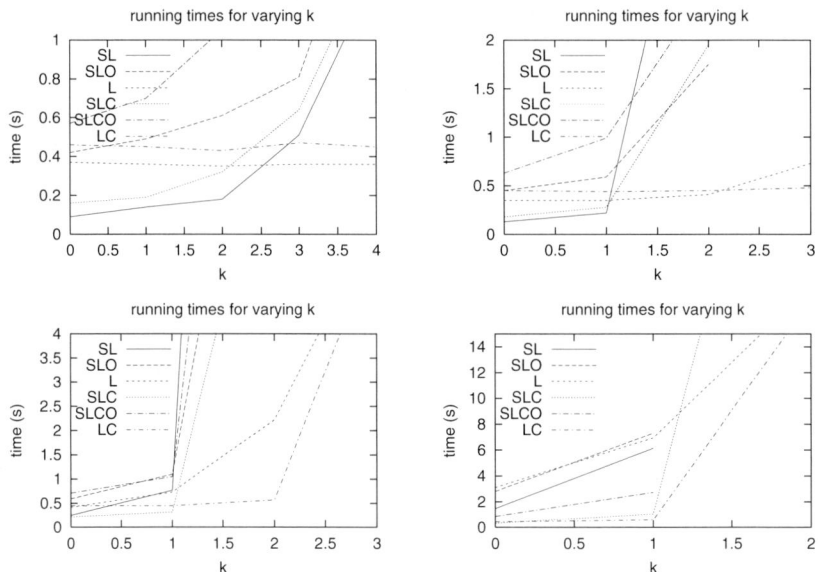

Fig. 12. Search times in seconds. Parameters are: $\sigma = 64$ (ASCII), $m = 64$, and $\ell = 3$. The figures are for, from left to right, top to bottom: $r = 1$, $r = 64$, $r = 256$, and $r = 1024$.

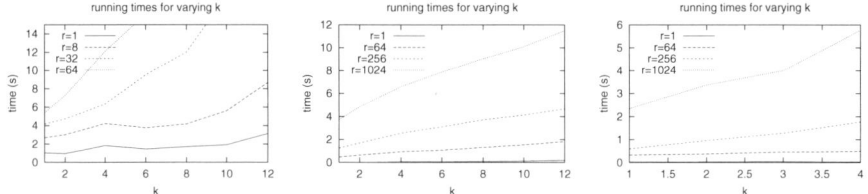

Fig. 13. Search times in seconds for exact pattern partitioning algorithm, for $\sigma = 4$ (DNA), $\sigma = 20$ (protein), and $\sigma = 64$ (ASCII) alphabets.

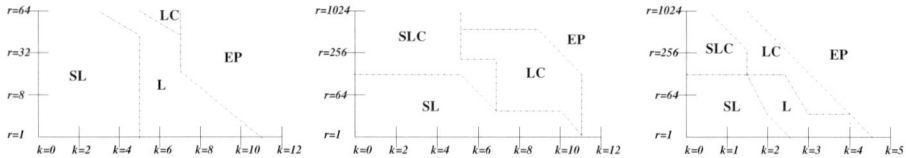

Fig. 14. Areas where each algorithm performs best.

7   Conclusions

Multiple approximate string matching is an important problem that arises in several applications, and for which the current state of the art is in a very primitive stage. Nontrivial solutions exist only for the case of very low difference ratios or very few patterns.

We have presented a new algorithm to improve this situation. Our algorithm is not only optimal on average, but also practical. A second algorithm we present is slower but handles higher difference ratios. We have shown that they perform well in handling large numbers of patterns and intermediate difference ratios. They are indeed the best alternatives for reasonably small alphabets.

The algorithms do not induce any order on the ℓ-grams, but they can appear in any order, as long as their total distance is at most k. The filtering can be still improved by requiring that the ℓ-grams from the pattern must appear in approximately same order in the text. This approach was used in [12]. The same method can be applied for multiple patterns as well.

There are several ways we plan to try in order to reduce preprocessing time and memory usage. A first one is lazy evaluations of the table cells. Instead of fully computing the D tables of size σ^ℓ for each pattern, we compute the cells only for the text ℓ-grams as they appear. If a given table cell is not yet computed, we compute it on the fly for all the patterns. This gives a preprocessing cost that is $O(rm\sigma^\ell(1 - e^{-n/\sigma^\ell}))$ on the average (using Myers' algorithm for the ℓ-grams inside the patterns, as $\lceil \ell/w \rceil = 1$). This, however, is advantageous only for very long ℓ-grams, namely $\ell + \Theta(\log \log \ell) > \log_\sigma n$.

Another possibility is to compute D only for those ℓ-grams that appear in a pattern with at most ℓ' differences, and assume that all the others appear with $\ell' + 1$ differences. This reduces the effectivity at search time but, by storing the relevant ℓ-grams in a hash table, requires $O(rm(\sigma\ell)^{\ell'})$ space and preprocessing time (either for plain or hierarchical verification), since the number of strings at distance ℓ' to an ℓ-gram is $O((\sigma\ell)^{\ell'})$ [13]. With respect to plain verification, the space is reduced for $\ell' < (\ell - \log_\sigma(rm))/(1 + \log_\sigma \ell)$, and with respect to hierarchical verification, for $\ell' < (\ell - \log_\sigma m)/(1 + \log_\sigma \ell)$. These values are reasonable.

It is also possible to improve the verification performance. A simple strategy is to sort the patterns before grouping by ranges in order to achieve some clustering in the groups. This could be handled with an algorithm designed for hierarchical clustering. This clustering could be done taking a distance defined as the number of differences necessary to convert one pattern into the other, or any other reasonable measure of similarity (Hamming distance, longest common subsequence, etc.).

Indexing consists of preprocessing the text to build a data structure (index) on it that can be used later for faster querying [9]. In general, we find that methods designed for indexed approximate string matching can be adapted to (non-indexed) multiple approximate string matching. The idea is to index the pattern set and use the text somehow as the pattern, in order to "search for the text" inside the structure of the patterns. Our present ideas are close to approximate q-gram methods, and several other techiques can be adapted too. We are currently pursuing this line.

References

1. R. Baeza-Yates and G. Navarro. Multiple approximate string matching. In F. Dehne et al., editor, *Proceedings of the 5th Annual Workshop on Algorithms and Data Structures (WADS'97)*, pages 174–184, 1997.
2. R. Baeza-Yates and G. Navarro. New and faster filters for multiple approximate string matching. *Random Structures and Algorithms (RSA)*, 20:23–49, 2002.
3. W. Chang and T. Marr. Approximate string matching and local similarity. In *Proc. 5th Combinatorial Pattern Matching (CPM'94)*, LNCS 807, pages 259–273, 1994.
4. M. Crochemore and W. Rytter. *Text Algorithms*. Oxford University Press, 1994.
5. H. Hyyrö and G. Navarro. Faster bit-parallel approximate string matching. In *Proceedings of the 13th Annual Symposium on Combinatorial Pattern Matching (CPM 2002)*, LNCS 2373, pages 203–224, 2002.
6. R. Muth and U. Manber. Approximate multiple string search. In *Proc. 7th Combinatorial Pattern Matching (CPM'96)*, LNCS 1075, pages 75–86, 1996.
7. E. W. Myers. A fast bit-vector algorithm for approximate string matching based on dynamic programming. *Journal of the ACM*, 46(3):395–415, 1999.
8. G. Navarro. A guided tour to approximate string matching. *ACM Computing Surveys*, 33(1):31–88, 2001.
9. G. Navarro, R. Baeza-Yates, E. Sutinen, and J. Tarhio. Indexing methods for approximate string matching. *IEEE Data Engineering Bulletin*, 24(4):19–27, 2001. Special issue on Managing Text Natively and in DBMSs.
10. G. Navarro, E. Sutinen, J. Tanninen, and J. Tarhio. Indexing text with approximate q-grams. In *Proc. 11th Combinatorial Pattern Matching (CPM 2000)*, LNCS 1848, pages 350–363, 2000.
11. P. Sellers. The theory and computation of evolutionary distances: pattern recognition. *Journal of Algorithms*, 1:359–373, 1980.
12. E. Sutinen and J. Tarhio. Filtration with q-samples in approximate string matching. In D. S. Hirschberg and E. W. Myers, editors, *Proceedings of the 7th Annual Symposium on Combinatorial Pattern Matching*, number 1075 in Lecture Notes in Computer Science, pages 50–63, Laguna Beach, CA, 1996. Springer-Verlag, Berlin.
13. E. Ukkonen. Finding approximate patterns in strings. *Journal of Algorithms*, 6:132–137, 1985.
14. A. C. Yao. The complexity of pattern matching for a random string. *SIAM Journal of Computing*, 8(3):368–387, 1979.

A   Lower Bound for Multipattern Matching

We extend the classical proof of Yao [14] to the case of searching for several patterns. Let us assume that we search for r random patterns of length m in a text of length n. Random patterns means that they are independently generated sequences where each character is chosen from the set Σ with uniform probability. To simplify matters, we will not assume that the patterns are necessarily different from each other, they are simply r randomly generated sequences. We prove that the lower bound of the problem on the average is $\Omega(n \log_\sigma(rm)/m)$, where $\sigma = |\Sigma|$. The bound refers to the number of character inspections made.

We use the same trick of dividing the text in blocks of length $2m - 1$, and assume that we just have to search for the presence of the patterns inside each block (which is an optimistic assumption). Since no information gathered inside one block can be used to search the other, we can regard each block in isolation. So the cost is at least $n/(2m-1)$ times the cost to search a single block. Hence what we have to prove is that we have to work $\Omega(\log_\sigma(rm))$ inside a given block.

Inside a given block $B_{1\ldots2m-1}$, each of the r patterns can match in m different positions (starting at position 1 to m). Each possible match position of each pattern will be called a *candidate* and identified by the pair (t, i), where $t \in 1 \ldots r$ is the pattern number and $i \in 1 \ldots m$ is the starting position inside the block. Hence there are rm candidates.

We have to examine enough characters to ensure that we have found every match inside the block. We will perform a sequence of accesses (block character reads) inside the block, at positions i_1, $i_2 \ldots i_k$ until the information we have gathered is enough to know that we found every pattern occurrence. Which is the same, we have to "rule out" all the rm candidates, or report those candidates that have not been ruled out after considering their m positions.

Note that each candidate has to be ruled out independently of the rest. Moreover, the only way to rule out a candidate (t, i) is to perform an access i_j such that $B_{i_j} \neq P^t_{i_j-i+1}$.

Given an access i_j to block B, the probability to rule out a candidate (t, i) with the access is at most $1 - 1/\sigma$: even assuming that the area covered by the candidate includes i_j (that is, $i \leq i_j < i + m$) and that the candidate has not been already outruled by a previous access, there is a probability of $1/\sigma$ that $B_{i_j} = P^t_{i_j-i+1}$ and hence we cannot rule out (t, i). This means that the probability that a given access does not rule out a given candidate is $\geq 1/\sigma$. Note that the way we have bounded the probability permits us considering every access independently of the others. Consequently, the probability of *not* ruling out a given candidate after k accesses is at least $1/\sigma^k$.

Since every candidate has to be ruled out independently of the others, a sequence of k accesses leaves at least rm/σ^k candidates not ruled out, on average. Each individual candidate can be directly verified by examining $\sigma/(\sigma-1)$ characters on average. Hence, our average cost is at least

$$k + \frac{rm}{\sigma^{k-1}(\sigma-1)}$$

The optimum is to keep examining characters until the average cost to directly verify the candidates equals the cost we would pay if we kept examining characters, and then switch to direct verification. This corresponds to minimizing the above formula. The optimum is

$$k^* = \log_\sigma\left(\frac{rm\sigma\ln\sigma}{\sigma-1}\right)$$

and hence the lower bound on the average cost per block is

$$\frac{1 + \ln\left(\frac{rm\sigma\ln\sigma}{\sigma-1}\right)}{\ln\sigma} = \Theta(\log_\sigma(rm))$$

which proves our claim.

Optimal Partitions of Strings: A New Class of Burrows-Wheeler Compression Algorithms[*]

Raffaele Giancarlo and Marinella Sciortino

Dipartimento di Matematica ed Applicazioni,
Università degli Studi di Palermo,
Via Archirafi 34, 90123 Palermo, Italy
{raffaele, mari}@math.unipa.it

Abstract. The Burrows-Wheeler transform [1] is one of the mainstays of lossless data compression. In most cases, its output is fed to Move to Front or other variations of symbol ranking compression. One of the main open problems [2] is to establish whether Move to Front, or more in general symbol ranking compression, is an essential part of the compression process. We settle this question positively by providing a new class of Burrows-Wheeler algorithms that use optimal partitions of strings, rather than symbol ranking, for the additional step. Our technique is a quite surprising specialization to strings of partitioning techniques devised by Buchsbaum et al. [3] for two-dimensional table compression. Following Manzini [4], we analyze two algorithms in the new class, in terms of the k-th order empirical entropy of a string and, for both algorithms, we obtain better compression guarantees than the ones reported in [4] for Burrows-Wheeler algorithms that use Move to Front.

1 Introduction

The Burrows-Wheeler transform (BWT from now on) [1] is a powerful tool for Data Compression. In simple terms, the transform rearranges the input, i.e., produces a permutation BWT(s) of an input string s, so that the string becomes easier to compress. Usually, BWT(s) is *post-processed* by applying Move to Front (MTF from now on) [5]. Finally, the output of this stage is fed to a statistical compressor such as Huffman or Arithmetic Coding (see [6]). Extensive experimental work shows that use of BWT yields methods that are competitive, if not superior, with the best known compression methods (see [2,4] and references therein). Despite its practical relevance, very few theoretic studies of BWT have been conducted, in order to get deeper insights into the process [7,4,8]. That is rather unfortunate, since compression methods based on the transform are new and likely to become the new standard in data compression. A better understanding of their

---

[*] Both authors are partially supported by Italian MURST Project of National Relevance "Bioinformatica e Ricerca Genomica". Additional support is provided to the first author by FIRB Project "Bioinformatica per la Genomica e la Proteomica" and to the second author by Italian MURST Project of National Relevance "Linguaggi Formali ed Automi: Teoria e Applicazioni".

R. Baeza-Yates et al. (Eds.): CPM 2003, LNCS 2676, pp. 129–143, 2003.
© Springer-Verlag Berlin Heidelberg 2003

nature and relationship with other methods seems to be an essential task, with potentially great rewards in practical terms.

1.1   The Burrows-Wheeler Transform and Related Open Problems

The two main research questions that BWT poses are intriguing both from the theoretical and practical point of view.

The first one asks whether MTF is an essential step for the successful use of BWT [2], i.e., whether there are alternatives to MTF and symbol ranking compression. Indeed, most of the known BWT compressors use MTF, for very good practical reasons (again see discussion in [2]). We are aware of only three papers that deal with this issue, and only at the experimental level [9,10,11]. In particular, Wirth and Moffat, based on their experiments, tend to reject the essentiality of MTF for block sorting compression. In any case, this question remains largely unexplored and completely lacks analytic results.

The second question asks for an asymptotic analysis of the compression performance of BWT compression methods. As customary in Data Compression, some optimality results have been provided by resorting to some very strong statistical assumptions, such as ergodicity, about the source emitting the string to be compressed [7,8]. Recently, Manzini [4] has made substantial progress towards a completely satisfactory settlement of the second problem, providing bounds only in terms of the empirical entropy of the input string. That is, the analysis by Manzini makes no assumption on the input string and provides bounds which hold in the worst case, i.e., for any possible string (rather than for a family of strings satisfying some statistical properties).

We need to recall the results by Manzini [4]. Given a string s and an integer k, let $H_k(s)$ denote the k-th order *empirical entropy* of s. Moreover, let $H_k^*(s)$ be the k-th order *modified empirical entropy* of s. Both functions are defined in Section 2. As pointed out in [4], the modified empirical entropy of a string is a more realistic measure to bound the worst case performance of compression methods, since its value is equal to the minimum number of bits needed to write down the length of the string (when its empirical entropy is zero).

Let BWO(s) be the output, in bits, of the original BWT algorithm, with Arithmetic Coding as the zero order compressor. Then, for any $k \geq 0$, |BWO| is bounded by

$$\approx 8|s|H_k(s) + \frac{2}{25}|s| \tag{1}$$

Notice that (1) holds *simultaneously* for all $k \geq 0$. That is, k is not a parameter known to the algorithm. Motivated by theoretic more than practical considerations, Manzini also proposed a variant of the compression algorithm that uses run length encoding. Let $\text{BWO}_{RL}(s)$ be the output, in bits, of this second algorithm. Then, for any $k \geq 0$, there exists a constant g'_k such that $|\text{BWO}_{RL}(s)|$ is bounded by

$$(5 + \epsilon)|s|H_k^*(s) + g'_k \tag{2}$$

where $\epsilon \approx 10^{-2}$ and, again, (2) holds simultaneously for all $k \geq 0$.

The analysis by Manzini gives some deep insights into the BWT transform. Indeed, application of BWT followed by MTF can be seen as a *booster* for the compression performance of any order zero compressor. That is, *preprocessing* the input string, via BWT and MTF, results in a big performance gain for a very weak compressor, as order zero compressors can be regarded. At an intuitive level, based on extensive experimental work, this fact was also pointed out by Fenwick [2].

A remarkable consequence of the investigation by Manzini is that, if there existed an ideal compressor $\mathcal{A}$, such that, for any partition $x_1, x_2, \cdots x_p$ of a string x, one would have

$$\mathcal{A}(x) \leq \Sigma_{i=1}^p |x_i| H_0(x_i) \tag{3}$$

then

$$\mathcal{A}(\text{BWT}(s)) \leq |s|H_k(s) \tag{4}$$

and (4) would hold simultaneously for all k's. That is, $\mathcal{A}(\text{BWT}(s))$ would achieve, exactly, the best possible performance, simultaneously for all k's. Analogous results would hold also for $H_k^*(s)$.

Manzini [4] conjectures that no BWT compression method can get to a bound of the form

$$|s|H_k^*(s) + g_k \tag{5}$$

$k \geq 0$ and $g_k \geq 0$ constant. Moreover, he also conjectures that the ideal algorithm $\mathcal{A}$ does not exist (indeed, if it existed, bound (4) and therefore (5) for BWT algorithms would be achievable).

Here we make progress in answering all of the questions mentioned so far. Indeed, we provide a new class of algorithms based on the BWT transform that *do not* use MTF as part of the compression process. We also provide a theoretic analysis of two of those new methods, in the worst case setting [4], showing that they are superior, both in terms of empirical and modified empirical entropy, with respect to the ones reported in [4]. To the best of our knowledge, those are the first two methods that do not use MTF and for which one can actually show better theoretic bounds with respect to the ones obtained in [4]. Moreover, we show that both conjectures are true.

Our results are based on a deep and rather surprising relationship between the BWT transform and another new compression technique, devised by Buchsbaum et al. [3], and successfully used for compression of two-dimensional tables. Before stating our results in technical terms, we give an outline of such a technique.

1.2 Table Compression via Rearrangements and Partitions

The idea of *boosting* the performance of a base compressor has also been recently introduced and successfully exploited in the realm of Table Compression [3]. Consider each record in a table to be a row in a matrix and let $\mathcal{C}$ be a base compressor, say `gzip`. A naive method of table compression is to compress, via $\mathcal{C}$, the string derived from scanning the table in row-major order. Buchsbaum et al. [3] observe experimentally that partitioning the table into contiguous intervals of columns and compressing each interval separately in this fashion can achieve significant compression improvement, i.e., it can boost the compression performance of the base compressor $\mathcal{C}$. The partition is generated by a one-time, off-line training procedure, and the resulting compression strategy is applied on-line to the table. A unified theory and extensive experiments studying table rearrangements and partitioning are given in [12]. One of the main tools is a Dynamic Program, based on the notion of Combinatorial Dependency (an analogous of Statistical Dependency), that optimally partitions the table. We refer to that dynamic Program as CD. It can be regarded as the simplest instance of a new two-dimensional transform (related to some classic optimization problems as TSP), which seems to be of a fundamentally different nature with respect to BWT (see experiments in [3]).

1.3 Method and Results

Here we are able to combine those two, apparently far and unrelated boosting techniques, to produce a new class of BWT-based compression methods, which are markedly distinct from the ones known so far. The simplest of the new methods works as follows. Fix Huffman encoding as the order zero compressor, which we denote as the base compressor. Compute BWT(s) and use CD to find the best partition of BWT(s) in terms of the base compressor. Compress BWT(s) according to that optimal partition. We denote such an algorithm as BW_{CD}. Note that it satisfies the paradigm of compression methods based on BWT transform, except that CD has replaced MTF. We also present another method in the new class, denoted $\mathrm{BW}_{CD,RL}$, which still consists of BWT and CD. However, the base compressor now is a combination of Run Length and Huffman encoding. We show that $|\mathrm{BW}_{CD}(s)|$ is bounded by

$$\approx |s|H_k(s) + |s| \tag{6}$$

Moreover, for any $k \geq 0$, there exists a constant g_k such that $|\mathrm{BW}_{CD,RL}(s)|$ is bounded by

$$2.5|s|H_k^*(s) + g_k \tag{7}$$

Both (6) and (7) hold simultaneously for all $k \geq 0$. The constant g_k in (7) is essentially the same as in (2). Both bounds compare rather favourably to the ones produced by Manzini and indeed show that MTF can be substituted by CD with substantial gains in terms of performance guarantee, in the worst case setting. Bound (7) is particularly relevant. Indeed, as mentioned by Manzini [4], a bound similar to (2) (and therefore also (7)) cannot hold for many of the best known compression algorithms. In particular, LZ77 [13], LZ78 [14] and PPMC [15]. It remains an open problem whether a similar bound can hold for DMC [16] and PPM* [17], for which no theoretical analysis (at least in terms of empirical entropy) is available.

The technique can be applied to boost the performance of any base compressor $\mathcal{C}$, satisfying requirements on its inner working stated in Section 3. We also settle the conjectures stated by Manzini which, in view of known lower bounds on coding of the integers [18,19], turn out to be easy and are given here for completeness.

The remainder of this extended abstract is organized as follows. Section 2 provides some basic definitions. Section 3 outlines the BWT transform, Combinatorial Dependency and shows how they can be combined to yield a new class of BWT algorithms. Section 4 is devoted to show the negative results related to the ideal algorithm $\mathcal{A}$. The next two sections outline two compression algorithms, which are the base compressors for the boosting. Finally, the last section gives some concluding remarks and open problems. Due to space limitations, proofs will be outlined.

2   Preliminaries

Let s be a string over the alphabet $\Sigma = \{a_1, \cdots, a_h\}$ and, for each $a_i \in \Sigma$, let n_i be the number of occurrences of a_i in s. Throughout this paper, we assume that $n_i \geq n_{i+1}$, $1 \leq i < h$. Let $P_i = n_i/|s|$, $1 \leq i \leq h$, be the empirical probability distribution for the string s. Unless otherwise stated, from now on we consider only empirical probability distributions. We need to recall some definitions and results. Additional details are in [6,4].

The *zero-th order empirical entropy* of the string s is defined as $H_0(s) = -\sum_{i=1}^{h} P_i \log(P_i)$. Let $r = |s| - n_1$. The *binary empirical entropy* is $\mathcal{H}(P_1) = -P_1 \log P_1 - (1 - P_1) \log(1 - P_1)$. Fix an integer $k \geq 0$. For any string $y \in \Sigma^k$, let y_s be the string consisting of the characters following y in s. The k-th order empirical entropy of the string s is $H_k(s) = \frac{1}{|s|} \sum_{y \in \Sigma^k} |y_s| H_0(y_s)$.

Following [4], we define the k-th order *modified empirical entropy* as $H_k^*(s) = \frac{1}{|s|} \sum_{w \in \Sigma^k} |w_s| H_0^*(w_s)$, where H_0^* is:

$$H_0^*(s) = \begin{cases} 0 & \text{if } |s| = 0 \\ (1 + \lfloor \log |s| \rfloor)/|s| & \text{if } |s| \neq 0 \text{ and } H_0(s) = 0 \\ H_0(s) & \text{otherwise.} \end{cases}$$

When $H_0(s) = 0$, i.e., s is of the form a^n, $|s| H_0^*(s)$ gives the minimum number of bits that a data compressor must use to store the length of s. We point out that such a "minimal requirement" on the information one needs to encode for strings of the given form was used by Manzini [4] to motivate the definition and use of $H_0^*(s)$ as a more realistic measure for the performance of a given compressor.

3 Boosting Compression via BWT Transform and Combinatorial Dependency

In this section we show how to combine the Burrows-Wheeler transform [1] with Combinatorial Dependency [3] in order to boost the compression performance of a base compressor $\mathcal{C}$.

3.1 The Burrows-Wheeler Transform

In 1994, Burrows and Wheeler [1] introduced a transform that turns out to be very elegant in itself and extremely useful for textual data compression. Given a string s, the transform consists of a string BWT(s) and an index I, pointing to a suitably chosen position in it. BWT(s) is a carefully chosen permutation of the characters in s. It is quite remarkable that, given BWT(s) and I, one can obtain s via an efficient algorithm. The process of computing BWT(s) and I is referred to as the *transform*, while obtaining s from BWT(s) and I is referred to as the *anti-transform*.

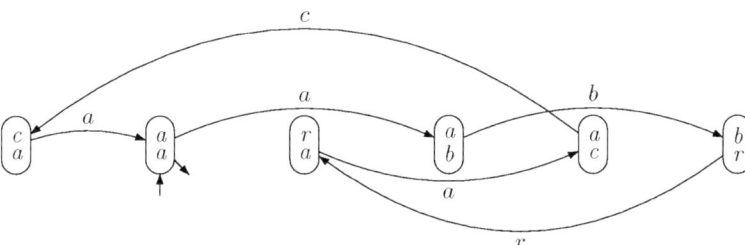

Fig. 1. The first character of each pair is on top, the second is on bottom. With reference to the matrix M (see Figure 2), the sequence of the characters on top represents the elements of L and the characters on bottom the elements of F.

We now provide a short description of the basic properties underlying the transform by means of an example, trying to highlight the generality of the method. Let $s = abraca$ and consider the graph G in Figure 1 and the matrix M in Figure 2 consisting of all cyclic shifts of s, lexicographically sorted.

G is obtained by considering s as a circular string. Each pair of contiguous characters in such a representation corresponds to a node. Each edge tells us that the second character of a pair overlaps with the first character of the pair it is pointing to. Moreover, we also mark the node containing the two characters corresponding to the "wrap around", via a pointer I. Given G, it is obvious that

$$
\begin{array}{cc}
F & L \\
\downarrow & \downarrow
\end{array}
$$

$$
\begin{array}{cccccccc}
 & 0 & a & a & b & r & a & c \\
I \to & 1 & a & b & r & a & c & a \\
 & 2 & a & c & a & a & b & r \\
 & 3 & b & r & a & c & a & a \\
 & 4 & c & a & a & b & r & a \\
 & 5 & r & a & c & a & a & b
\end{array}
$$

Fig. 2. Matrix consisting of all cyclic shifts of $s = abraca$, lexicographically sorted

we can obtain s by visiting G, starting at the node pointed to by I. It turns out that we can encode this graph via the string $\text{BWT}(s)$ and a pointer (denoted again I) to one of its characters.

In order to see this, consider the first and the last columns of M. Since M is a lexicographic sort, F consists of at most $|\Sigma|$ runs of equal symbols. Pick one such run, for instance the one consisting of a's. It is easy to show that the relative order of the a's in F must be the same in L and that such a fact is true also for other runs of identical symbols in F. Consider now the pair $(L[i], F[i])$, $0 \le i \le |s|$. Notice that it is a node in G. We can recover the edges of G as follows. By the observation on the relative order preservation, $F[0]$ must correspond to $L[1]$. That is, the first a in F must correspond to the first a in L. So, node $(L[0]F[0]) = (ca)$ must be connected to node $(L[1]F[1]) = (aa)$ with label $F[0] = a$. Such a process can be iterated. Therefore, given F and L, we can build G and recover s. It turns out that we need only L, since F is obtained by lexicographically sorting L. Finally, the algorithm computing the transform consists of lexicographically sorting all cyclic shifts of s to determine L and I, which is the row index where s appears in M. As for the anti-transform, we sort L to obtain F and use the property of relative order preservation for each run of equal characters to infer which pair follows which.

Burrows and Wheeler [1], in their seminal paper, pointed out that the power of the transform seems to rest on the fact that equal contexts in a string, i.e, substrings, ending with the same character are grouped together, resulting in a cluster of identical symbols in $\text{BWT}(s)$. That clustering makes $\text{BWT}(s)$ a better string to compress, with respect to s. For a long time, there was only experimental evidence that the transform actually makes compression "easier". Recently, Manzini has cleverly pointed out properties of the transform allowing to state the above experimental evidence in mathematical terms:

Theorem 1. *[4] Let s be a string. For each $k \ge 0$, there exists an $f \le h^k$ and a partition $\hat{s}_1, \hat{s}_2, \cdots \hat{s}_f$ of $\text{BWT}(s) = \hat{s}$ such that $H_k(s) = \frac{1}{|s|} \sum_{i=1}^{f} |\hat{s}_i| H_0(\hat{s}_i)$. An analogous result holds for $H_k^*(s)$.*

That is, the transform is so powerful because it clusters together, optimally, all equal contexts. Indeed, we can think of each $\hat{s}_i$ in the statement of Theorem 1 as a cluster of symbols. Proper identification of those clusters accounts for the

k-th order empirical entropy of s in terms of the zero-th order empirical entropy of each of the clusters. As pointed out by Manzini, a remarkable consequence of Theorem 1 is that, if the ideal compressor $\mathcal{A}$ mentioned in Section 1.1 existed, then bounds (4) and (5) would both hold.

Manzini mentions that such an ideal compressor is unlikely to exists. In fact, in Section 4, we actually show that it cannot exist, under some very natural assumptions on its bahaviour. Nevertheless, Theorem 1 suggests that in order to approximate $\mathcal{A}$ one could try to partition $\mathtt{BWT}(s)$ into pieces and then compress each piece separately. But this is exactly the idea used by Buchsbaum et al. [3] for table compression and outlined in Section 1.2 that, quite surprisingly, specializes to strings.

3.2   Combinatorial Dependency and Optimal Partitions of Strings

Let $\mathcal{C}$ be a data compressor such that, when given in input a string x, it adds a special end-of-string symbol $\#$ (not in Σ) and compresses the string $x\#$. Moreover, in a left to-right scan of the compressed string, the decoder must produce x as soon as the encoding of the end-of string symbol is met. Many data compression algorithms can be adapted to work in this way. Indeed, either $\#$ is really appended at the end of the string or the length of x is explicitly stored as a prefix of the compressed string x. For this latter approach, one can use algorithms for the variable length encoding of integers, such as the ones presented in [18].

Let $\mathcal{C}(x\#)$ be the output of such an algorithm and let $|\mathcal{C}(x\#)|$ denote its length, in bits. We also assume that the output contains a header providing information to the decoder on how to recover the string. For instance, if the algorithm is standard Huffman encoding, then the header will consist of the distinct symbols appearing in x, the tree used for the encoding and the corre-spondence between symbols and codewords. Section 5 gives details about this type of Huffman encoding. The header information, usually not accounted for in estimating the performance of a data compressor, here needs to be considered, as it will be self-evident in what follows. We now specialize the definition of Combinatorial Dependence, given in an abstract setting by Buchsbaum et al. [12], to strings:

Definition 2. Two strings x and y are *Combinatorially Dependent* if and only if $|\mathcal{C}(xy\#)| < |\mathcal{C}(x\#)| + |\mathcal{C}(y\#)|$.

The Dynamic Program that computes an optimal partition of a string s is $E[j] = \min_{0 \le k < j} E[k] + |\mathcal{C}(s[k+1, j]\#)|$.

Assume that the partition so computed is $s_1, s_2, \cdots s_f$. Then, the compression can be thought of as corresponding to the string $s_1 \# s_2 \# \cdots s_f \#$, where each piece delimited by $\#$ is compressed separately and in the order given. $E[|s|]$ gives the total number of bits so produced. Using the properties we are assuming about $\mathcal{C}$, it is simple to show that we can recover s, by properly decoding each piece of the partition.

Consider now the following data compression algorithm, based on Burrows-Wheeler Transform and Combinatorial Dependency. (A) compute $\hat{s} = \text{BWT}(s)$; (B) Optimally Partition $\hat{s}$, using $\mathcal{C}$ as the base compressor. Compress, separately, each piece of the partition. We denote such an algorithm as BWT_{OPT}.

Theorem 3. *Let $\mathcal{C}$ be a compression algorithm satisfying the assumptions stated at the beginning of this section and such that $|\mathcal{C}(x\#)| \leq \lambda|x|H_0(x) + \gamma|x| + c$, where λ, γ and c are non-negative constants. Assume that $\mathcal{C}$ is the base compressor in BWT_{OPT}. Then, given a string s, for all $k \geq 0$, the number of bits produced by BWT_{OPT} is bounded by $\lambda|s|H_k(s) + \gamma|s| + \Theta(ch^k)$. Analogous results hold for $H_0^*(s)$ and $H_k^*(s)$.*

Proof. Let us consider a partition $\hat{s}_1', \hat{s}_2', \cdots \hat{s}_{l'}'$ of $\hat{s}$. Since it may not be optimal, the number of bits produced by BWT_{OPT} is bounded by $\Sigma_{i=1}^{l'}|\mathcal{C}(s_i'\#)|$. Then, using the bound on $|\mathcal{C}(x\#)|$, we have that $|\text{BWT}_{OPT}|$ is bounded by $\Sigma_{i=1}^{l'}(\lambda|\hat{s}_i'|H_0(\hat{s}_i') + \gamma|s_i'| + c)$. Now notice that, by Theorem 1, there exists an $f \leq h^k$ and a partition $\hat{s}_1, \hat{s}_2, \cdots \hat{s}_f$ of $\hat{s}$ such that $H_k(s) = \frac{1}{|s|}\Sigma_{i=1}^{f}|\hat{s}_i|H_0(\hat{s}_i)$. The result follows. $\square$

Corollary 4. *Let HC and RHC be two compression algorithms satisfying the assumptions stated at the beginning of this section and such that $|\text{HC}(x\#)| \leq |x|(H_0(x) + 1) + \Theta(h \log h)$ and $|\text{RHC}(x\#)| \leq 2.5|x|H_0^*(x) + \Theta(h \log h)$. Given a string s, for all $k \geq 0$, the number of bits produced by BWT_{OPT} is bounded by $|s|H_k(s) + |s| + \Theta(h^k \log h)$, when the base compressor is HC and by $2.5|s|H_k^*(s) + \Theta(h^k \log h)$, when the base compressor is RHC.*

A few remarks are in order. Theorem 3 is an assessment of the fact that the Burrows-Wheeler transform combined with Combinatorial Dependency is a general boosting method for a base compressor $\mathcal{C}$. In Sections 5 and 6, we outline the compression algorithms HC and RHC that we use for boosting in Corollary 4. Any data compression algorithm exhibiting a better performance than HC or RHC and the same properties, outlined at the beginning of this Section, will yield even better compression algorithms based on BWT and Combinatorial Dependence. The time complexity of BWT_{OPT} depends critically on that of $\mathcal{C}$ and it is at least $O(n^2)$. Fortunately, if $\mathcal{C}$ has a linear time decompression algorithm, then BWT_{OPT} also admits a linear time decompression algorithm.

4   Lower Bound

In this section we address the existence of the ideal algorithm satisfying bound (3) and of compression algorithms, based on BWT transform, satisfying bound (5). We answer both questions in the negative. Those results come from the stronger finding that, independently of whether the transform is applied as a pre-processing step, the bottleneck for the performance of compression algorithms rests on the need to encode the length of the input string, when it is of the form a^n, i.e., $H_0(s) = 0$. We model such a requirement by assuming that a compressor

$\mathcal{C}$, when restricted to work on input strings a^n, $n \geq 1$, produces a codeword for n. That is, $\{\mathcal{C}(a^n)|n \geq 1\}$ is a codeword set for the integers. For technical reasons, we also assume that $|\mathcal{C}(a^n)|$ is a non-decreasing function of n. Depending on the implementation, those assumptions seem to account for "the inner working" of many of the best known compression algorithms that, either implicitly or explicitly, use codes for the integers. Indeed, the decompression algorithm must derive n from the information encoded by the compressor because, since both algorithms must work for any n, they cannot agree once and for all on its value. The lower bound now comes from Theorem 4 in [19], which we restate in our notation.

Theorem 5. *[19, Theorem 4] Let $\mathcal{C}$ be a compressor satisfying the assumptions stated at the beginning of the section. Then, there exists a countable number of strings s such that $|\mathcal{C}(s)| \geq |s|H_k^*(s) + \beta(|s|)$, where $\beta(n)$ is a diverging function of n.*

Proof. Let $s = a^n$. The compressor $\mathcal{C}$, when restricted to work on strings of that form, produces a codeword set for the integers. But, there must exist infinitely many values of n such that $|\mathcal{C}(s)| \geq \text{logsum}_2 n - (\log_2 \log_2 e) \log_2^* n$, where $\text{logsum}_2 x = \sum_{1 \leq i \leq \log_2^* x} \log_2^{(i)} x$ and $\log_2^* x = \min\{t : \log_2^{(t)} x \leq 1\}$. Then, using the definition of modified empirical entropy and the fact that the k-th order modified empirical entropy is a non-increasing function of k, we have $|\mathcal{C}(s)| \geq |s|H_k^*(s) + \beta(|s|)$, where $\beta(n)$ is a diverging function of n. □

Corollary 6. *No compression algorithm satisfying the assumptions stated at the beginning of the section can achieve bound (3) and (5), independently of whether or not the BWT is applied as a pre-processing step.*

Proof. Consider $s = a^n$ and the partition consisting of s only. Moreover, observe that $\text{bwt}(s) = s$, when s is of the form a^n. □

5   A Prefix Code Compressor

In this Section we present a prefix code compressor HC that satisfies the conditions and the performance bounds stated in Section 3, Corollary 4. Recall that $\#$ is an end-of-string marker, not in the input alphabet. Let T_1 be the Huffman tree obtained from the empirical probability distribution of s. We derive a tree T_2 from T_1 and use it to encode $s\#$. The main idea behind this modification of Huffman encoding is to take advantage of results in [20] that, when applied here, imply we can encode the end-of string symbol basically for free. That is, without any departure from the well known and classic bound on the expected codeword length of a Huffman code, stated in terms of the entropy of the information source. The transformation is as follows.

(a) Assume s consists of only one distinct symbol, i.e., $s = a^n$. T_2 consists of a root and two leaves, one assigned to a and the other to $\#$.

(b) Assume s consists of more that one distinct symbol and let b be the least frequent symbol in s. In case of ties, b is chosen arbitrarily among all least frequent symbols. Then T_2 is obtained from T_1 by making the leaf corresponding to b an internal node. One of the new leaves is assigned to b and the other to #

We now outline the encoding algorithm. We point out that the algorithm explicitly stores information needed for decoding, i.e, a proper representation of T_2 and the correspondence between symbols and codewords. Let $\mathtt{HC}(s\#)$ be the binary string produced by the algorithm. Its layout is: (a) $\lceil \log h \rceil$ bits give a count of the number c of distinct symbols of Σ appearing in $s\#$; (b) c blocks of $\lceil \log h \rceil$ bits each gives the encoding of each of the c symbols- this list is the one obtained by a preorder visit of the leaves of T_2; (c) following the convention that a right branch is encoded by 0 and a left branch by 1, we then store the binary string corresponding to the branches traversed by a preorder visit of T_2; (d) finally, the binary string y corresponding to the encoding of $s\#$ via the code represented by T_2. In what follows, given $\mathtt{HC}(s\#)$, we refer to the binary string preceding y as *the table*. Notice that the table takes at most $M = (h+1)\lceil \log h \rceil + 2h$ bits.

Due to space limitations, we omit the presentation of the decoding algorithm and limit ourselves to point out that the only non-trivial part is to recover T_2 from the binary string representing it.

Theorem 7. *Consider a string s. Let $P_1, \cdots, P_h$ be the empirical probability distribution of s (i.e., the end marker symbol is excluded) and let $H_0(s)$ be its empirical entropy. Then, $|\mathtt{HC}(s\#)| \leq |s|H_0(s) + |s| + \Theta(h \log h)$.*

Proof. $\mathtt{HC}(s\#)$ is composed of the table storing T_2 and then y. Using results in [20], one can show that $|y| \leq |s|H_0(s) + |s| + h + 1$. In addition, the table requires at most M bits. But $h + M + 1 = \Theta(h \log h)$.   □

6   A Compressor Based on Prefix and Run Length Encoding

In this section we provide an order zero compression algorithm, denoted $\mathtt{RHC}$, that combines Huffman encoding with Run Length Encoding. It satisfies the conditions and the performance bounds stated in Section 3, Corollary 4. The main idea behind the algorithm is to use knowledge about the symbol frequencies in a string so that its performance closely tracks its empirical entropy. As pointed out by Manzini [4], for low entropy strings, it is essential to use run length encoding in order to get close to their empirical entropy. Our algorithm proposes a novel combination of Huffman and Run Length encoding.

The run length encoding scheme we use depends critically on a variable length encoding of a sequence of integers. The next lemma states the abstract problem one needs to consider and a bound on the length of the binary string produced by our method, which is sketched in the proof of the lemma.

Lemma 8. *Given two positive integers t and w with $t < w$, and the integers $d_1, d_2, \ldots d_t \in [1, w]$, with $d_i < d_{i+1}$ and $1 \leq i < t$, there is an algorithm that produces a binary string encoding $d_1, d_2, \ldots, d_t$ and w. Its output size, in bits, is bounded by $2(t+1) + \frac{3}{2} \left((t+1) \log \left(\frac{w}{t} \right) + \log e \right) + 1$.*

Proof. We outline the method and its analysis. Consider $l_1 = d_1, l_2 = d_2 - d_1, \ldots, l_t = d_t - d_{t-1}, l_{t+1} = w - d_t$.

Our algorithm encodes $l_1 \# l_2 \# \ldots l_t \#$, when $l_{t+1} = 0$, and $l_1 \# l_2 \# \ldots l_{t+1} \#$, otherwise. In both cases, $\#$ is a delimiter allowing us to use a variable length encoding of each l_i. The first bit of the encoded string tells us which case we are dealing with. To fix ideas, we present the encoding for $l_1 \# l_2 \# \ldots l_{t+1} \#$. It is a two stage process.

Let **0** and **1** be two symbols different from $\#$ and not in the input alphabet. Let $B(l_i)$ denote the number $l_i + 1$ written in binary using **0** and **1** and discarding the most significant bit, $1 \leq i \leq t + 1$. For example

$$B(1) = \mathbf{0}, \ B(2) = \mathbf{1}, \ B(3) = \mathbf{00}, \ B(4) = \mathbf{01}$$

In the first stage, we compute $x' = B(l_1) \# B(l_2) \# \ldots B(l_{t+1}) \#$. The second stage encodes x' in binary as follows. Let z_0, z_1 and $z_\#$ be the number of **0**'s, **1**'s and $\#$'s in x', respectively. Assume that $z_0 \geq z_1$ (the other case is symmetric). Let x be the string obtained from x' by encoding its symbols with the code: $\mathbf{0} \rightarrow 0$, $\mathbf{1} \rightarrow 10$ and $\# \rightarrow 11$.

As for the analysis, let $p = \max_i l_i$. For $j = 1, \cdots, p$, let q_j denote the number of l_i's equal to j. Therefore, $\Sigma_{j=1}^p q_j = g$, where g is equal to t, when $l_{t+1} = 0$, and to $t + 1$, otherwise. With reference to the algorithm outlined above, we outline a proof only for the case $z_1 \leq z_0$, since the other is symmetric. We have $|x| = z_0 + 2z_1 + 2z_\# + 1 = z_0 + 2z_1 + 2g + 1$. The only nontrivial part is to bound $z_0 + z_1$ in terms of t and w. Using Jenhsen's inequality [6], one can show that $z_0 + z_1 \leq (t+1) \log \left(\frac{w}{t} \right) + \log e$.   □

For completeness, we point out that the problem posed in Lemma 8 can be solved using the many methods known for universal encoding of the integers [18]. However, those encoding schemes are asymptotically optimal and better than ours for large integers, i.e., large values of w and small t's, while we need to be efficient for all integers, i.e., all values of w and $t \leq w$.

We now describe the algorithm encoding $s\#$. It consists of four cases, based on the probability P_1 of its most frequent symbol. Intuitively, as that probability increases towards one, the modified and empirical entropy of the given string rapidly decrease and, in order to track them with some degree of accuracy, the algorithm needs to know when to use run length encoding.

Algorithm RHC

(a) Assume that $P_1 < 1/2$. Then, build and use T_2, as described in Section 5.

(b) Assume that $1/2 \le P_1 < \frac{|s|-1}{|s|}$. Here we use run length encoding. Consider s and let $\mathbf{b} = b_1 b_2 \dots b_r$ be the string resulting from s once that its most frequent symbol a has been removed. For an efficient encoding of $s\#$, we first encode $\mathbf{b}\#$ and then the positions (integers in $[1, |s|]$) where each b_i appears in s, $1 \le i \le r$. The encoding of $\mathbf{b}\#$ closely follows the scheme given in Section 5 for the encoding of $s\#$, except that we can save since we do not need to encode a. Indeed, consider the tree T_2 introduced in Section 5 to encode $s\#$ and let T_2' be the subtree not containing a (say the left subtree). Since T_2 is obtained from an Huffman tree T_1 and $1/2 \le P_1$, we have that a corresponds to a leaf at level one in T_1 and T_2. Since T_2 has at least three leaves, T_2' has at least two nodes and can be used to encode $\mathbf{b}\#$. We point out that the layout of the binary string so produced is analogous to the one described in Section 5 so that proper decoding of this part is granted. As for the positions where each b_i appears in s, $1 \le i \le r$, we use the algorithm outlined in Lemma 8 for their encoding.

(c) Assume that $\frac{|s|-1}{|s|} = P_1$. That is, s is composed of only two symbols a and b and, say, b appears only once. Let d be the position of b in s. In order to compress $s\#$, we need to give a, b as they are encoded for the input alphabet, an encoding of $|s|$ and d. For $|s|$, we encode it with a variation of the algorithm outlined in Lemma 8 . That part is followed by a field of $\lceil \log |s| \rceil$ bits, which can be used to store the standard binary encoding of d.

(d) Assume that $P_1 = 1$. All we need to do here is to encode a, as it is encoded for the input alphabet, and $|s|$. Again, we encode s with a variation of the algorithm outlined in Lemma 8.

Theorem 9. *Consider a string s. Let $P_1, \cdots, P_h$ be the empirical probability distribution of s (i.e., the end marker symbol is excluded) and let $H_0^*(s)$ be its modified empirical entropy. Then, $|\mathrm{RHC}(s\#)| \le \frac{5}{2}|s|H_0^*(s) + \Theta(h \log h)$.*

Proof. We start by noting that, depending on the various cases of Algorithm RHC, RHC($s\#$) is composed of a table, providing a representation of the alphabet symbols as they appear coded in the input string and, possibly, an encoding of a binary tree T_2 (as defined in Section 5). This part takes $\Theta(h \log h)$ bits. Now, let y be the remaining binary string output by the algorithm. We now show that $|y| \le \frac{5}{2}|s|H_0^*(s) + c$, where $c \le 2h$. That would prove the lemma. We outline only case (b), since (a) essentially follows from the results in Section 5 and (c) and (d) are simple variations of (b).

Case (b). We have $1/2 \le P_1 < \frac{|s|-1}{|s|}$. With reference to the corresponding case in Algorithm RHC, y is "composed" of two parts, one giving a prefix encoding of $\mathbf{b}\#$ and the other giving an encoding of the occurrences in s of symbols in $\mathbf{b}$. Since there are r of those symbols and we use the algorithm outlined in Lemma 8, we can use that result to bound that part of y. In order to bound the length of the encoding of **b**, let y' be the encoding of $s\#$ obtained via tree T_2 of Section 5. The two encodings we are considering are essentially the same, except that the pruning of T_2 makes the first one $|s|$ bits shorter than the second. Moreover, one can show that $|y'| \leq |s|(H_0(s)+2-\mathcal{H}(P_1)-P_1)+h+1$, $P_1 \geq \frac{1}{2}$. Putting all of those observations together, we obtain $|y| \leq 2r+\frac{r}{2}\log\left(\frac{|s|}{r}\right)+\frac{3}{2}\log\left(\frac{|s|}{r}\right)+|s|H_0(s)+c$. Using elementary calculus, one can show that $2r + \frac{r}{2}\log\left(\frac{|s|}{r}\right) + \frac{3}{2}\log\left(\frac{|s|}{r}\right) \leq 1.5|s|H_0(s)$ and the result follows. $\qquad\square$

7   Conclusions and Open Problems

We have shown how to combine the BWT transform with a partitioning technique, based on Combinatorial Dependency, to obtain a new class of compression methods based on the the BWT transform. We also exhibit two methods in the class, showing superior compression guarantees, in the worst case setting, with respect to the so far best available algorithms and analysis. Although the results we have obtained seem to be mainly of theoretic interest, they show that the BWT transform can be "followed" by a compression strategy different from MTF. That settles an open problem posed by Fenwick [2] and originally motivated by practical considerations. We also show that the bottleneck for compression algorithms to obtain "ideal performance" bounds rests on the need to encode integers. Such an observation seem to address some open problems and conjectures posed by Manzini [4].

The main open problems related to this research consist of obtaining even tighter bounds, on one hand, and of engineering some of the methods presented in this paper, on the other, in order to obtain practical implementations of the two new algorithms.

Acknowledgements. The authors are deeply indebted to G. Manzini for insightful discussions and comments and to A. L. Buchsbaum for reading an early version of this abstract.

References

1. Burrows, M., Wheeler, D.: A block sorting data compression algorithm. Technical report, DIGITAL System Research Center (1994)
2. Fenwick, P.: The Burrows-Wheeler transform for block sorting text compression. The Computer Journal **39** (1996) 731–740
3. Buchsbaum, A.L., Caldwell, D.F., Church, K.W., Fowler, G.S., Muthukrishnan, S.: Engineering the compression of massive tables: An experimental approach. In: Proc. 11th ACM-SIAM Symp. on Discrete Algorithms. (2000) 175–184
4. Manzini, G.: An analysis of the Burrows-Wheeler transform. Journal of the ACM **48** (2001) 407–430
5. Bentley, J., Sleator, D., Tarjan, R., Wei, V.: A locally adaptive data compression scheme. Comm. of ACM **29** (1986) 320–330

6. Cover, T.M., Thomas, J.A.: Elements of Information Theory. Wiley Interscience (1990)
7. Effros, M.: Universal lossless source coding with the Burrows-Wheeler transform. In: Proc. IEEE Data Compression Conference, IEEE Computer Society (1999) 178–187
8. Sadakane, K.: On optimality of variants of the block sorting compression. In: Proc. IEEE Data Compression Conference, IEEE Computer Society (1998) 570
9. Arnavut, Z., Magliveras, S.S.: Block sorting and compression. In: Proc. IEEE Data Compression Conference, IEEE Computer Society (1997) 181–190
10. Balkenhol, B., Kurtz, S.: Universal data compression based on the Burrows and Wheeler-transformation: Theory and practice. Technical Report 98-069, Sonderforshunngsbereich: Diskrete Strukturen in der Mathematik, Universität Bielefeld, Germany (1998) Available from
    http://www.mathematik.uni-bielefeld.de/sfb343/preprints.
11. Wirth, A.I., Moffat, A.: Can we do without ranks in Burrows Wheeler transform compression? In: Proc. IEEE Data Compression Conference, IEEE Computer Society (2001) 419–428
12. Buchsbaum, A.L., Giancarlo, R., Fowler, G.S.: Improving table compression with combinatorial optimization. In: Proc. 13th ACM-SIAM Symp. on Discrete Algorithms. (2002) 213–222
13. Lempel, A., Ziv, J.: A universal algorithm for sequential data compression. IEEE Trans. on Information Theory **IT-23** (1977) 337–343
14. Ziv, J., Lempel, A.: Compression of individual sequences via variable-rate coding. IEEE Trans. on Information Theory **IT-24** (1978) 530–578
15. Moffat, A.: Implementing the PPM data compression scheme. IEEE Trans. on Communication **COM-38** (1990) 1917–1921
16. Cormak, G., Horspool, R.: Data compression using dynamic markov modelling. Computer J. **30** (1987) 541–550
17. Cleary, J., Teahan, W.: Unbounded length contexts for PPM. Computer J. **40** (1997) 67–75
18. Elias, P.: Universal codeword sets and representations of the integers. IEEE Transactions on Information Theory **21** (1975) 194–203
19. Levenshtein, V.: On the redundancy and delay of decodable coding of natural numbers. (Translation from) Problems in Cybernetics, Nauka, Mscow **20** (1968) 173–179
20. Capocelli, R.M., Giancarlo, R., Taneja, I.: Bounds on the redundancy of Huffman codes. IEEE Transactions on Information Theory **32** (1986) 854–857

Haplotype Inference by Pure Parsimony

Dan Gusfield[*]

Computer Science Department, University of California, Davis, Davis CA 95616, USA
gusfield@cs.ucdavis.edu

Abstract. The next high-priority phase of human genomics will involve the development and use of a full *Haplotype Map* of the human genome [7]. A critical, perhaps dominating, problem in all such efforts is the inference of large-scale SNP-haplotypes from raw genotype SNP data. This is called the Haplotype Inference (HI) problem. Abstractly, input to the HI problem is a set of n strings over a ternary alphabet. A solution is a set of at most $2n$ strings over the binary alphabet, so that each input string can be "generated" by some pair of the binary strings in the solution. For greatest biological fidelity, a solution should be consistent with, or evaluated by, properties derived from an appropriate genetic model.

A natural model, that has been suggested repeatedly is called here the *Pure Parsimony* model, where the goal is to find a *smallest* set of binary strings that can generate the n input strings. The problem of finding such a smallest set is called the *Pure Parsimony Problem*. Unfortunately, the Pure Parsimony problem is NP-hard, and no paper has previously shown how an optimal Pure-parsimony solution can be computed efficiently for problem instances of the size of current biological interest. In this paper, we show how to formulate the Pure-parsimony problem as an integer linear program; we explain how to improve the practicality of the integer programming formulation; and we present the results of extensive experimentation we have done to show the time and memory practicality of the method, and to compare its accuracy against solutions found by the widely used general haplotyping program PHASE. We also formulate and experiment with variations of the Pure-Parsimony criteria, that allow greater practicality. The results are that the Pure Parsimony problem *can* be solved efficiently in practice for a wide range of problem instances of current interest in biology. Both the time needed for a solution, and the accuracy of the solution, depend on the level of recombination in the input strings. The speed of the solution improves with increasing recombination, but the accuracy of the solution decreases with increasing recombination.

1 Introduction

Progress on population-scale genomics has focused on the acquisition and use of SNP's and SNP-haplotypes [16,3]. Consequently, the next high-priority phase

---

[*] Research Supported by NSF grants DBI-9723346 and EIA-0220154

R. Baeza-Yates et al. (Eds.): CPM 2003, LNCS 2676, pp. 144–155, 2003.

of human genomics will involve the development of a full *Haplotype Map* of the human genome [7]. It will be used in large-scale screens of populations to associate specific SNP-haplotypes with specific complex genetic-influenced diseases. Building a Haplotype Map of the human genome has become a central NIH promoted goal [7,13]. A critical, perhaps dominating, problem in all such efforts is the computational determination of large-scale SNP-haplotypes from raw SNP data.

1.1   Introduction to SNP's, Genotypes, and Haplotypes

In diploid organisms (such as humans) there are two (not completely identical) "copies" of each chromosome (except for the sex chromosome). A description of the data from a single copy is called a *haplotype*, while a description of the conflated (mixed) data on the two copies is called a *genotype*. For many diseases, haplotype data (identifying a set of gene alleles inherited together) is a much better predictor of disease or disease susceptibility than is genotype data.

The underlying data that forms a haplotype is either the full DNA sequence in the region, or more commonly the values of *single nucleotide polymorphisms (SNP's)* in that region. A SNP is a single nucleotide site where exactly two (of four) different nucleotides occur in a large percentage of the population. The SNP-based approach is the dominant one, and worldwide efforts to characterize human molecular genetic variability has resulted in the construction of high density SNP maps (with a density of about one SNP per thousand nucleotides). As this work continues, it is essential to develop methods of deducing haplotypes from SNP data.

We focus on one chromosome with m sites (SNP's) where each site can have one of two states (alleles), denoted by 0 and 1. For each of n individuals, we would ideally like to describe the states of the m sites on each of the two chromosome copies separately, i.e., the haplotype pair for each individual. However, experimentally determining the haplotype pair is technically difficult or expensive. As a result, almost all population data consists only of the set of SNP's possessed by an individual (their genotype), rather than their haplotypes. One then uses computation to deduce haplotype information from the given genotype information. Several methods have been explored and some are intensely used for this task [2,1,5,17,6,15]. None of these methods are presently fully satisfactory, although the overal accuracy of most of these methods is very impressive.

Abstractly, for a population of n individuals, input to the haplotyping problem (HI) consists of n *genotype* vectors (or strings), each of length m, where each value in the vector is either 0,1, or 2. Each position in a vector is associated with a site of interest on the chromosome. The position in the genotype vector has a value of 0 or 1 if the associated chromosome site has that state on both copies (it is a *homozygous* site), and has a value of 2 otherwise (the chromosome site is *hetrozygous*).

Given an input set of n genotype vectors, a *solution* to the *Haplotype Inference (HI) Problem* is a set of n pairs of binary vectors (strings), one pair for each genotype. For any genotype g, the associated binary vectors v_1, v_2 must both have value 0 (or 1) at any position where g has value 0 (or 1); but for any position where g has value 2, exactly one of v_1, v_2 must have value 0, while the other has value 1. That is, v_1, v_2 must be a feasible "resolution" of g into two haplotypes that could explain how g was created. Hence, for an individual with k hetrozygous sites there are 2^{k-1} haplotype pairs that can resolve it.

For example, if the observed genotype g is 0212, then the pair of haplotypes 0110, 0011 is one feasible resolutions, out of two feasible resolutions. Of course, we want to find the resolution that actually gave rise to g, and a solution for the HI problem for the genotype data of all the n individuals. However, without additional biological insight, one cannot know which of the exponential number of solutions is the "correct one". Therefore, haplotype deduction would be impossible without the implicit or explicit use of some genetic model, either to assess the biological fidelity of any proposed solution, or to guide the algorithm in constructing a solution.

1.2   The Pure-Parsimony-Criteria

One natural approach to the HI problem that is often suggested is called here the *Pure-Parsimony* approach[1]: Find a solution to the HI problem that minimizes the total number of distinct haplotypes used.

For example, consider the set of genotypes: 02120, 22110, and 20120. There are HI solutions for this example that use six distinct haplotypes, but the solution 00100, 01110; 01110, 10110; 00100, 10110, for the three genotype vectors respectively, uses only the three distinct haplotypes 00100, 01110, and 10110.

The Pure parsimony criteria reflects the fact that in natural populations, the number of observed distinct haplotypes is vastly smaller than the number of combinatorially possible haplotypes, and this is also expected from population genetics theory. For example, in the absence of recombination and recurrent mutation, the number of distinct haplotypes observed in a population with k SNP sites is at most $k + 1$, in great contrast with the number, 2^k, of distinct haplotypes that can be formed from k binary SNPs. Even with recombination, each recombination event in the history of the sequences can only increase the number of distinct haplotypes by one. Further, general population genetic theory predicts that the number of recombinations is on the order of the number of SNP sites. So theory predicts that the number of distinct haplotypes will be relatively small. This lack of diversity in haplotypes has been observed even more strongly in the last several years in a series of papers on longer-than-expected low-diversity haplotype blocks that appear in the Human genome (for example, see [16,3]). Reported haplotype blocks have up to sixty SNP's, but are reported to contain only a small number of haplotypes (with numbers around five) in population samples of several hundred individuals. Other support for the parsimony criteria comes from the empirical results reported in [15], where the best solutions obtained by Clark's haplotyping method, were those that used

---

[1] This approach was suggested to us Earl Hubbell, who also proved that the problem of finding such solutions is NP-hard [8].

the fewest number of distinct haplotypes. There were typically around twenty solutions that used the fewest number of haplotypes, out of 10,000 solutions, and the distribution of accuracy of those twenty solutions (relative to the known correct haplotype pairs for the input) was dramatically better than for the 10,000 solutions as a whole. Moreover, the parsimony criteria is to some extent involved in existing computational methods for haplotype inference. For example, some papers have tried to explain Clark's method [2] in terms of parsimony [14], although the role of parsimony is not explicit in Clark's method. The haplotyping program PHASE [17] has been partially explained in terms of the parsimony criteria [4]. However, the indirect role of the parsimony criteria in these methods, and the complex details of the computations, makes it hard to see explicitly how the parsimony criteria influences the computation. This makes it difficult to use those methods to evaluate the effectiveness of the parsimony criteria as a genetic model.

One of the goals of the research reported here is to be able to assess the efficacy of the pure parsimony criteria, without the complicating influence of approximations and heuristics that are integral to existing haplotyping methods. However, no published paper had previously shown how an optimal Pure-Parsimony solution can be computed efficiently in a practical range of problem instances, although we announced without detail in [6] that integer programming could be used to achieve this result. Recently we learned that a similar integer programming approach was studied theoretically by Lancia et. al in [11], but without an implementation. That work differs from what is presented here in that the work in this paper is primarily empirical. We have also learned of another program was recently put on the web [18] that solves the pure parsimony problem by a branch-and-bound approach.

In this paper we detail how to compute, via integer-linear-programming, an HI solution that minimizes the number of distinct haplotypes, i.e., is guaranteed to solve the Pure-Parsimony problem. However, the worst-case running time increases exponentially with the problem size, so the empirical issue is whether this approach is practical for problem instances of current interest in population-scale genomics. We improve the practicality of the basic integer programming formulation in a way that is very effective in practice. We report on the results of extensive experimentation we have done to show the time and memory practicality of the method, and to compare its accuracy against existing HI methods that do not explicitly follow the Pure-parsimony criteria. We also formulate and experiment with two variations of the Pure-Parsimony criteria, that allow greater practicality.

Empirically, the end result is that for haplotyping problem instances of current interest, Pure-parsimony *can* be computed efficiently in most cases. However, it's accuracy is somewhat inferior to the solutions produced by the program PHASE, although this depends on the number of sites and the level of recombination.

In more detail, the practicality and accuracy of our approach depend on the level of recombination in the data (the more recombination, the more practical but less accurate is the method). We show here that the Pure-Parsimony approach is practical for genotype data of up to 30 sites and 50 individuals (which is large enough for practical use in many current haplotyping projects). Up to moderate levels of recombination, the haplotype calls are 80 to 95 percent correct, and the solutions are generally found in several seconds to minutes, except for the no-recombination case with 30 sites, where some solutions require a few hours.

While the main point of this paper is to report that Pure-Parsimony solutions can be practically obtained, the accuracy level observed is a validation of the genetic model implicit in the Pure-Parsimony objective function, for a purely randomly picked solution to the HI problem would correctly resolve only a minuscule fraction of the genotypes.

2   Pure Parsimony via Integer Linear Programming

We assume that the reader is generally familiar with linear programming: A set of linear inequalities, defined over a set of variables, must be satisfied (by assigning values to the variables), and among all feasible solutions to the inequalities, one seeks a solution that optimizes a linear objective function defined on the same variables (either maximizing or minimizing its value as specified in the formulation). Linear programming problems can be solved efficiently both in theory and practice, and commercial software exists that can efficiently solve linear programs with millions of variables and inequalities. In Integer-linear-programming, one insists that the variables take on values that are integral. In our application, the values of the variables will be further restricted to be either 0 or 1. There are no known methods to solve integer-linear programming problems efficiently in the theoretical worst-case sense, but many integer-programming formulations are efficiently solved in practice, and commercial code exists that efficiently solves huge integer-linear-programs for certain applications. The practical success of integer programming on many problems motivates this empirical study on the use of integer programming to solve the Pure-Parsimony problem.

2.1   The Conceptual TIP Formulation

We will begin by describing a *conceptual* integer-linear-programming solution to the problem of finding an HI solution that minimizes the number of distinct haplotypes used. The solution is conceptual because it would be generally impractical to use. After describing this conceptual solution, we introduce two simple observations that makes this conceptual solution practical on ranges of data of current biological interest.

Let g_i denote the i'th genotype input vector, and suppose it has h_i polymorphic sites (i.e., sites with value 2). There are 2^{h_i-1} pairs of haplotypes that could have generated g_i. In the conceptual integer-linear-programming formulation, we enumerate each one of these pairs, and create one integer-programming variable $y_{i,j}$ for each of the 2^{h_i-1} pairs. As we create these y variables, we take note of the haplotypes in the enumerated pairs. Whenever a haplotype is enumerated that has not been seen before, in any of the pairs of haplotypes enumerated for any previously examined genotype, we generate a new integer-programming variable x_k for that haplotype. Thus, no matter how many times that haplotype occurs in the haplotype pairs, over all the genotype vectors, there will only be one x variable generated for it.

Now we explain the linear-programming inequalities defined on the y and the x variables. We will describe these through the following example. For genotype $g_i = 02120$ we enumerate the two haplotype pairs 00100, 01110; and 01100, 00110, and generate the two variables $y_{i,1}$ and $y_{i,2}$ for these pairs. Assuming, that the four haplotypes above (which are all distinct) have not been seen before, we generate the four variables x_1, x_2, x_3, x_4 for them. Then, we create the inequality

$$y_{i,1} + y_{i,2} = 1$$

Recall that the variables can only be set to 0 or 1, so this inequality says that in an HI solution, we must select exactly one of the enumerated haplotype pairs as the resolution (explanation) of genotype g_i. The y variable set to 1 indicates which haplotype pair will be used in the explanation of genotype g_i.

Next, we create two inequalities for *each* variable $y_{i,j}$. These are:

$$y_{i,1} - x_1 \leq 0$$

$$y_{i,1} - x_2 \leq 0$$

$$y_{i,2} - x_3 \leq 0$$

$$y_{i,2} - x_4 \leq 0$$

To explain these inequalities, we examine the first one: $y_{i,1} - x_1 \leq 0$. It says, that if we set $y_{i,1}$ to 1, then we must also set x_1 to 1. Essentially, if we select the haplotype pair associated with variable $y_{i,1}$ to explain g_i in an HI solution, then we must use the haplotype associated with variable x_1, because that haplotype is one of the pair of haplotypes associated with variable $y_{i,1}$. The next inequality says the same thing for the haplotype associated with variable x_2.

These are the two types of inequalities that are included in the integer-programming formulation. We include such inequalities for *each* input genotype vector. Of course, in general, there will be many more inequalities generated than in the above example. If a genotype has h polymorphic sites, then there will be exactly $2^h + 1$ inequalities generated for it. This fully specifies the inequalities needed in the formulation.

For the objective function, let X denote the set of all the x variables that are generated, over all the genotypes. Recall, that there is one x variable for each distinct haplotype, no matter how many times it occurs in the enumerated pairs. Then the objective function is:

$$\text{Minimize} \sum_{x \in X} x$$

That objective function forces the x variables to be set so as to select the *minimum* possible number of distinct haplotypes. Taken together, the objective function and the inequalities, along with the restriction that the variables can only be set to 0 or 1, specifies an integer-linear-programming formulation whose solution gives an HI solution that minimizes the number of distinct haplotypes used. That is, this formulation truly solves the "Pure-Parsimony" haplotype problem. We refer to this formulation as the TIP formulation. For any specific problem instance, when the TIP formulation for that instance is solved, the set of x variables that are set to 1 in the solution specify a solution to the Pure-Parsimony problem.

3   Efficiency

The above formulation can actually be used without further modification for some problems of current biological interest. But for many problems of current interest (up to 50 individuals and 30 sites) the number of inequalities generated would make it impractical to solve the resulting integer program. For that reason, the TIP formulation is only conceptual, and additional ideas are required to make it practical.

The first idea is the following: If the haplotype pair for variable $y_{i,j}$ consists of two haplotypes that are both part of *no* other haplotype pair, then in the integer-program, there is no need to include variable $y_{i,j}$ or the two x variables for the two haplotypes in the pair associated with $y_{i,j}$. Let RTIP denote the integer programming formulation created by removing such y and x variables from TIP. If there is a genotype vector g such that all associated y variables are removed, then there is an optimal solution to the TIP formulation where we arbitrarily choose the haplotype pair for g. Otherwise, there is an optimal solution to the original TIP formulation which does not set any of the removed x or y variables to 1. Hence there is no loss in removing them, and the smaller RTIP formulation will find exactly the same optimal solution as the TIP formulation finds.

This idea is particularly effective because DNA sequences in populations have generally undergone some amount of recombination. Recombination is a process where a prefix of one string, and a suffix of another string, are concatenated to form a third string. The haplotypes in a population evolve both by mutations of single nucleotides, and by pairwise recombinations. Depending on the level of recombination that occurred in the evolution the strings, the RTIP formulation can be much smaller (fewer variables and inequalities) than the original TIP formulation. The reason is that as recombination levels rise, the haplotypes become more differentiated, and in turn the genotypes become more different from each other, so that more of the haplotypes enumerated in the TIP formulation only appear in one haplotype pair. These haplotypes are removed in the RTIP formulation. Smaller formulations allow the integer programming solution codes to run more efficiently.

The above idea reduces the size of the integer-programming formulation while preserving the optimal solution. However, if we first enumerate all the haplotypes and haplotype pairs in the TIP formulation, and then remove variables, the work involved could still make the RTIP approach impractical. For a genotype with h polymorphic sites, there are 2^h haplotypes, and so brute-force enumeration itself may be overly time consuming, even if later many variables are removed. That brute-force enumeration can be eliminated as follows: Let g_1 and g_2 be two genotype vectors, and let H_1 and H_2 be respectively the set of haplotypes that are associated with each genotype, in the TIP formulation. It is easy to identify the haplotypes in $H_1 \cap H_2$, and generate them in time proportional to $m|H_1 \cap H_2|$, where m is the length of the genotype vector. Simply scan g_1 and g_2 left to right; if a site occurs with a value of 1 in one and 0 in the other, then $H_1 \cap H_2 = \emptyset$; if a site occurs with a 2 in one vector and a 0 or 1 in the other, then set that 2 to be equal to the other value. Then if there are k remaining sites, where both g_1 and g_2 contain 2's, then there are exactly 2^k distinct haplotypes in $H_1 \cap H_2$, and we generate them by setting those k sites to 0 or 1 in all possible ways. The point is that the time for this enumeration is proportional to $m|H_1 \cap H_2|$. Moreover, each generated haplotype in $H_1 \cap H_2$ specifies a haplotype pair that will be included in RTIP, for both g_1 and g_2.

Any x variable that is included in the RTIP formulation must occur in an intersecting set for some pair of genotypes, and every pair of haplotypes that should be associated with a y variable must also be found while examining some pair of genotypes. Hence we can produce the RTIP formulation, essentially in time proportional to its size, and so if the RTIP formulation is small, we can produce it very quickly.

4 A Variation

In the Pure-Parsimony criteria we find a solution to the HI problem that minimizes the *total* number of distinct haplotypes used. However, in most data sets there are individuals whose genotypes are homozygous at every position. These are haplotypes that one knows for sure are in the population, and population genetic theory suggests that they are likely to have been used in pairs of haplotypes that create ambiguous genotypes. In fact, this idea is central in Clark's haplotyping method. So, we should preferentially try to use in the HI solution any haplotypes that appear as homozygous genotypes in the input. That leads to the *Modified-Parsimony* criteria and problem: find an HI solution that minimizes the number of distinct haplotypes used, *excluding* in the count, any haplotype that is seen as a homozygous genotype in the input. Hence this minimizes the number of *new* haplotypes generated, beyond the ones given in the input. Note that the solution to this problem can be smaller than the solution to the Pure-Parsimony problem minus the number of homozygotes in the input. Essentially, we are able to use in the solution any homozygous string from the input, at zero cost. The integer programming formulation for the Modified-Parsimony problem is created by simply removing from the objective function any x variable that is identical to a homozygous genotype in the input.

5   Boosting Accuracy

As we will detail below, for experiments with a small number of sites, or a moderate to large level of recombination, the time to generate and solve the RTIP formulation was so small, that we were able to generate multiple optimal solutions, when they existed. We are generally able to find around 100 multiple optimal solutions in under one minute. Prior work and some theoretical considerations, suggested that when multiple optimal solutions are available, a more accurate solution can be obtained by creating the "consensus" solution from the multiple optimals[15]. That is, we simply look at each genotype and use the resolution for it found in the largest number of the multiple optimal solutions. Consistently, the accuracy of the consensus solution is better than the average accuracy of the multiple optimals.

6   Experimental Results

To test the RTIP approach, we ran a widely-used program developed by R. Hudson [9,10] that uses coalescent theory to generate a simulated population of haplotypes. Haplotypes generated by the program were then randomly paired to create genotype data. The Hudson program allows one to specify a level of recombination in the simulated data through a parameter r. In this paper, we report on the results obtained from data with 50 individuals, with 10 and 30 sites, and with r set to 0, 4, 16, and 40. It is not known what setting of r corresponds to real recombination levels in nature, but the range used here is thought to be large enough to capture realistic levels of recombination.

For each of these eight different settings, we generated 15 data sets and then generated and solved the RTIP formulations (for both the Pure-Parsimony and the Modified-Parsimony criteria). We use the commercial program CPLEX from ILOG to find the optimal integer-programming solutions. We were interested in the size of the generated formulations, and the time needed to solve them, along with the quality of the solutions. We compared the solutions given by the Pure-Parsimony criteria to the solutions given by the Modified-Parsimony criteria (both with the consensus solution when practical), and to solutions obtained by running the publically-available program PHASE [17], using its default parameter settings.

We summarize the experimental results as follows: First, for ten sites, regardless of the level of recombination, the RTIP formulations were generated and solved in under one second. At low levels of recombination $r = 0$ up to $r = 16$ the quality of the Pure-Parsimony and Modified-Parsimony and PHASE solutions were essentially indistinguishable, with each method occasionally being superior or inferior to the others. The quality of the solutions depended on the level of recombination. For $r = 0$, all methods were correct for 96 to 100 percent of the haplotype calls. For example, both the Pure-Parsimony and PHASE were 100 percent correct in 6 of the 15 trials, and were 96% correct in only one of the 15 trials; they were 98% correct in all the other trials. For $r = 16$, the accuracy ranged from 72% to 96%.

As the recombination level increased, the accuracy of all the methods decreased, but the accuracy is still impressive. For 50 individuals, 10 sites and $r = 40$, the accuracy of all methods ranged from 74% to 96%.

For 30 sites, the practicality of the RTIP approach depends inversely on the level of recombination, while the accuracy of all methods depends directly on the level of recombination. For $r = 0$, the sizes of the 15 datasets were quite variable. One dataset had under 300 variables and the optimal solution was found in 0.03 seconds. Another dataset had 135,000 variables, and the optimal solution was found in 2.5 minutes. The largest dataset had nearly four million variables, and no optimal solution was obtained. A second dataset had two million variables and again no solution was obtained. Most of the datasets had under 10,000 variables and were solved in under two minutes. The accuracy of the 13 datasets that were solved, was in the 80% to 95% range. In comparison, PHASE took much more time, but was slightly more accurate in 10 out 15 datasets (excluding the two where the RTIP formulation wasn't solved), and slightly less accurate in three out of the 15 datasets.

For 30 sites and $r = 16$, the sizes of the RTIP formulations were notably smaller and all the executions ran in seconds. The accuracy was in the 72% to 92% range, but were consistently inferior (but only by small amounts) to the solutions obtained by PHASE.

For 30 sites and $r = 40$, the RTIP formulations again ran in seconds, and the accuracy was in the 50% to 80% range, but PHASE had notably superior accuracy in 13 of the 15 datasets, with accuracy between 68% and 92%.

As expected, the RTIP formulations were generally significantly smaller than the TIP formulation. Two typical instances are: for 50 individuals and 30 sites and $r = 4$, the TIP formulation had 28,580 haplotype variables, while the RTIP formulation had only 5,418; for 50 individuals and 40 sites and $r = 16$, the TIP formulation had 548,352 haplotype variables, while the RTIP formulation had only 129,814.

7   A Hybrid Approach

In this paper we have concentrated on datasets with at most 30 sites and 50 individuals. But we have also explored datasets with up to 150 individuals and 100 sites. Datasets of that size are too large for the Pure-Parsimony approach via integer programming. However, we also have an integer programming formulation that builds on Clark's well-known method.

A. Clark [2] suggested a (non-deterministic) algorithm to solve the HI problem. This algorithm is widely used, and is the basis for several large-scale haplotype studies [1,5,12]. The algorithm starts with a set of unresolved genotypes and a set of haplotypes that are known to be in the population. The key idea of Clark's algorithm is to repeatedly use the following "resolution rule" until no more unresolved genotypes remain, or no more can be resolved: If an unresolved genotype g can be resolved by a haplotype pair h, h', where h is a haplotype that is already known, or an already deduced haplotype, and h' is its "conjugate" (which is forced, given h and g), then declare g resolved with the pair h, h'. If h' has not previously been deduced (or known), then add it to the set of available haplotypes that can be used in additional deductions.

In [6] we showed how to maximize, via integer programming, the number of resolutions that Clark's method achieves. By combining that formulation with the ideas in the RTIP formulation, we now can minimize the number of distinct haplotypes used, within the group of solutions that maximizes the number of resolutions. The integer programming formulations produced are much smaller than those produced by RTIP. With this method, we can find solutions in seconds to minutes even for datasets with 100 sites and 150 individuals. For datasets that can be solved by Pure-Parsimony, this hybrid approach is a only a bit less accurate than the Pure-Parsimony solution, so its ability to solve large datasets quickly makes it of interest for further study.

8   Conclusions

The parsimony objective function has been frequently mentioned in the haplotyping literature, and is a part of the conceptual underpinning of some existing haplotyping methods. However, because those methods are complex, it is difficult to use their performance to access the effectiveness of the Pure-Parsimony objective function. We show here that for genotype data of the size of current interest, the Pure-Parsimony problem can be efficiently solved via integer-linear-programming. However, except in the case when the number of sites is small (around 10 sites), these methods, while impressively accurate in an absolute sense, generally produce solutions that are less accurate than those produced by the PHASE program.

While the main point of this paper is to report that Pure-Parsimony solutions *can* be practically obtained, so that the efficacy of the criteria can be accessed, the level of accuracy observed is a validation of the genetic model implicit in the Pure-Parsimony objective function. Purely randomly picked solutions to the HI problem would correctly resolve only a minuscule fraction of the genotypes, in contrast to the observed 80 to 90 percent accuracy of the Pure-parsimony-based method.

References

1. A. Clark, K. Weiss, and D. Nickerson et. al. Haplotype structure and population genetic inferences from nucleotide-sequence variation in human lipoprotein lipase. *Am. J. Human Genetics*, 63:595–612, 1998.
2. A. Clark. Inference of haplotypes from PCR-amplified samples of diploid populations. *Mol. Biol. Evol*, 7:111–122, 1990.

3. M. Daly, J. Rioux, S. Schaffner, T. Hudson, and E. Lander. High-resolution haplotype structure in the human genome. *Nature Genetics*, 29:229–232, 2001.
4. P. Donnelly. Comments made in a lecture given at the DIMACS conference on Computational Methods for SNPs and Haplotype Inference, November 2002.
5. M. Fullerton, A. Clark, Charles Sing, and et. al. Apolipoprotein E variation at the sequence haplotype level: implications for the origin and maintenance of a major human polymorphism. *Am. J. of Human Genetics*, pages 881–900, 2000.
6. D. Gusfield. Inference of haplotypes from samples of diploid populations: complexity and algorithms. *Journal of computational biology*, 8(3), 2001.
7. L. Helmuth. Genome research: Map of the human genome 3.0. *Science*, 293(5530):583–585, 2001.
8. E. Hubbel. Personal Communication, August 2000.
9. R. Hudson. Gene genealogies and the coalescent process. *Oxford Survey of Evolutionary Biology*, 7:1–44, 1990.
10. R. Hudson. Generating samples under the Wright-Fisher neutral model of genetic variation. *Bioinformatics*, 18(2):337–338, 2002.
11. G. Lancia, C. Pinotti, and R. Rizzi. Haplotyping populations: Complexity and approximations, technical report DIT-02-082. Technical report, University of Trento, 2002.
12. D. Nickerson, S. Taylor, K. Weiss, and A. Clark et. al. DNA sequence diversity in a 9.7-kb region of the human lipoprotein lipase gene. *Nature Genetics*, 19:233–240, 1998.
13. NIH. Report on variation and the haplotype map:
http://www.nhgri.nih.gov/About_NHGRI/Der/variat.htm.
14. T. Niu, Z. Qin, X. Xu, and J.S. Liu. Bayesian haplotype inference for multiple linked single-nucleotide polymorphisms. *Am. J. Hum. Genet*, 70:157–169, 2002.
15. S. Orzack, D. Gusfield, and V. Stanton. The absolute and relative accuracy of haplotype inferral methods and a consensus approach to haplotype inferral. Abstract Nr 115 in Am. Society of Human Genetics, Supplement 2001.
16. N. Patil and D. R. Cox et al. Blocks of limited haplotype diversity revealed by high-resolution scanning of human chromosome 21. *Science*, 294:1719–1723, 2001.
17. M. Stephens, N. Smith, and P. Donnelly. A new statistical method for haplotype reconstruction from population data. *Am. J. Human Genetics*, 68:978–989, 2001.
18. L. Wang et al. http://www.cs.cityu.edu.hk/~lwang/hapar/.

A Simpler 1.5-Approximation Algorithm for Sorting by Transpositions

Tzvika Hartman

Dept. of Computer Science and Applied Mathematics, Weizmann Institute of Science,
Rehovot 76100, Israel, tzvi@cs.weizmann.ac.il
http://www.cs.weizmann.ac.il/~tzvi

Abstract. An important problem in genome rearrangements is sorting permutations by transpositions. Its complexity is still open, and two rather complicated 1.5-approximation algorithms for sorting linear permutations are known (Bafna and Pevzner, 96 and Christie, 98). In this paper, we observe that the problem of sorting circular permutations by transpositions is equivalent to the problem of sorting linear permutations by transpositions. Hence, all algorithms for sorting linear permutations by transpositions can be used to sort circular permutations. Our main result is a new 1.5-approximation algorithm, which is considerably simpler than the previous ones, and achieves running time which is equal to the best known. Moreover, the analysis of the algorithm is significantly less involved, and provides a good starting point for studying related open problems.

1 Introduction

When trying to determine evolutionary distance between two organisms using genomic data, one wishes to reconstruct the sequence of evolutionary events that have occurred, transforming one genome into the other. One of the most promising ways to trace the evolutionary events is to compare the order of appearance of identical (or orthologous) genes in two different genomes. In the 1980's, evidence was found that different species have essentially the same set of genes, but their order may differ between species [16,12]. This suggests that global rearrangement events (such as reversals and transpositions of genome segments) can be used to trace the evolutionary path between genomes. Such rare events may provide more accurate and robust clues to the evolution than local mutations (i.e. insertions, deletions, and substitutions of nucleotides).

In the last decade, a large body of work was devoted to genome rearrangement problems. Genomes are represented by permutations, where each element stands for a gene. Circular genomes (such as bacterial and mitochondrial genomes) are represented by circular permutations. The basic task is, given two permutations, to find a shortest sequence of rearrangement operations that transforms one permutation into the other. Assuming that one of the permutations is the identity permutation, the problem is to find the shortest way of sorting a permutation using a given rearrangement operation (or set of operations). For more background on genome rearrangements refer to [19,17,18,20].

R. Baeza-Yates et al. (Eds.): CPM 2003, LNCS 2676, pp. 156–169, 2003.

The problem of sorting permutations by reversals has been studied extensively. It was shown to be NP-hard [6], and several approximation algorithms have been suggested [2,7,5]. On the other hand, for signed permutations (every element of the permutation has a sign, + or -, which represents the direction of the gene), a polynomial algorithm for sorting by reversals was first given by Hannenhalli and Pevzner [11]. Subsequent work improved the running time of the algorithm, and simplified the underlying theory [13,4,1]. The problem of sorting signed circular permutations by reversals was shown to be equivalent to the linear case [15].

There has been less progress in studying the problem of sorting by transpositions. A transposition is a rearrangement operation, in which a segment is cut out of the permutation, and pasted in a different location. The complexity of sorting by transpositions is still open. It was first studied by Bafna and Pevzner [3], who devised a rather complicated 1.5-approximation algorithm, which runs in quadratic time. Christie [8] gave a somewhat simpler $O(n^4)$ algorithm with the same approximation ratio. Eriksson et al. [9] provided an algorithm that sorts any given permutation on n elements by at most $2n/3$ transpositions, but has no approximation guarantee. The problem of sorting by both reversals and transpositions was addressed in [21,10,14].

In this paper we study the problem of sorting permutations by transpositions. First, we prove that the problem of sorting circular permutations by transpositions is equivalent to the problem of sorting linear permutations by transpositions. Hence, all algorithms for sorting linear permutations by transpositions can be used to sort circular permutations. Then, we derive our main result: A new quadratic 1.5-approximation algorithm, which is considerably simpler than the previous ones [3,8]. Thus, the algorithm achieves running time which is equal to the best known [3], with the advantage of being much simpler. Moreover, the analysis of the algorithm is significantly less involved, and provides a good starting point for studying related open problems.

The paper is organized as follows. In Section 2 we first prove the equivalence between the problem of sorting linear and circular permutations by transpositions. Then, we review some classical genome rearrangement results, and show that every permutation can be transformed into a so-called simple permutation. Our main result, a new and simple 1.5-approximation algorithm for sorting permutations by transpositions, is introduced in Section 3. We conclude with a short discussion and some open problems (Section 4).

2   Preliminaries

2.1   Linear and Circular Permutations

Let $\pi = [\pi_1 \ \ldots \ \pi_n]$ be a permutation on n elements. Define a *segment* A in π as a consecutive sequence of elements $\pi_i, \ldots, \pi_k$ $(k \geq i)$. Two segments $A = \pi_i, \ldots, \pi_k$ and $B = \pi_j, \ldots, \pi_l$ are *contiguous* if $j = k+1$ or $i = l+1$. A *transposition* τ on π is the exchange of two disjoint contiguous segments (Figure 1a). If the segments are

$A = \pi_i, \ldots, \pi_{j-1}$ and $B = \pi_j, \ldots, \pi_{k-1}$, then by performing τ on π, the resulting permutation, denoted $\tau \cdot \pi$, is $[\pi_1 \ \ldots \ \pi_{i-1} \ \pi_j \ \ldots \ \pi_{k-1} \ \pi_i \ \ldots \ \pi_{j-1} \ \pi_k \ \ldots \ \pi_n]$ (note that the end segments can be empty if $i = 1$ or $k - 1 = n$). We shall say that τ *cuts* π *before* positions i, j and k. We say that τ *involves* index l if $i \leq l < k$, i.e., if l belongs to one of the two exchanged segments.

In circular permutations, one can define analogously a transposition as the exchange of two contiguous segments. Note that here the indices are cyclic, so the disjointness of the exchanged segments is a meaningful requirement. The transposition partitions a circular permutation into three segments, as opposed to at most four in a linear permutation (see Figure 1). Since there are only two cyclic orders on three segments, and each two of the three segments are contiguous, the transposition can be represented by exchanging any two of them. Note that the number of possible transpositions on a linear n-permutation is $\binom{n+1}{3}$, since there are $n + 1$ possible cut points of segments. In contrast, in a circular n-permutation there are only $\binom{n}{3}$ possibilities.

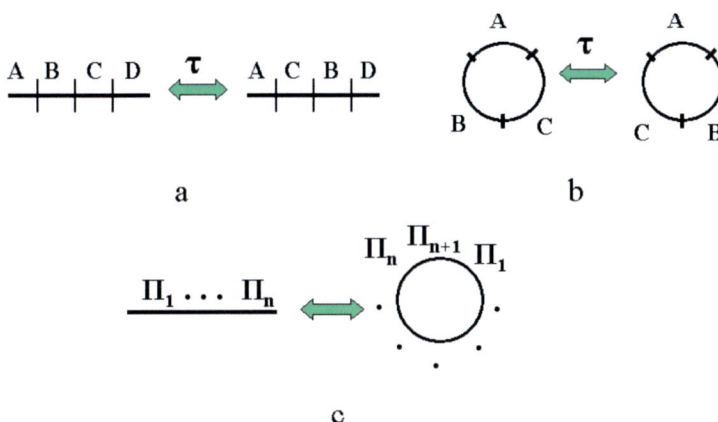

Fig. 1. (a) A transposition τ, which is applied on a linear permutation, and exchanges segments B and C. (b) A transposition τ, which is applied on a circular permutation. τ can be viewed as exchanging A and B, or B and C, or A and C. (c) A one-to-one transformation between linear and circular permutations. In the circular permutation, a new element, π_{n+1}, is introduced.

The problem of finding a shortest sequence of transpositions, which transforms a (linear or circular) permutation into the identity permutation, is called *sorting by transpositions*. The *transposition distance* of a permutation π, denoted by $d(\pi)$, is the length of the shortest sorting sequence.

Theorem 1 *The problem of sorting linear permutations by transpositions is linearly equivalent to the problem of sorting circular permutations by transpositions.*

Proof. Given a linear n-permutation, circularize it by adding an $n+1'$st element $\pi_{n+1} = x$, and closing the circle (see Figure 1c). Call the new circular permutation π^c. By the discussion above, any transposition on π^c can be represented by the two segments that do not include x. Hence, there is an optimal sequence of transpositions that sorts π^c, and none of them involves x. The same sequence can be viewed as a sequence of transpositions on the linear permutation π, by ignoring x. This implies that $d(\pi) \leq d(\pi^c)$. On the other hand, any sequence of transpositions on π is also a sequence of transpositions on π^c, so $d(\pi^c) \leq d(\pi)$. Hence, $d(\pi) = d(\pi^c)$. Moreover, an optimal sequence for π^c provides an optimal sequence for π.

For the other direction, starting with a circular permutation, we can linearize it by removing an arbitrary element, which plays a role of x above (see Figure 1c). The same arguments imply that an optimal solution for the linear permutation translates to an optimal solution for the circular one.

In the rest of the paper, we will discuss only circular permutations. As implied by Theorem 1, all the results on circular permutations hold also for linear ones. We prefer to work with circular permutations since it simplifies the analysis.

2.2   The Breakpoint Graph

We transform a permutation π on n elements into a permutation $f(\pi)$ on $2n$ elements, by replacing each element i by two elements $2i-1, 2i$. On the doubled permutation $f(\pi)$, we allow only transpositions that cut before odd positions. This ensures that no transposition cuts between $2i-1$ and $2i$, and therefore every transposition on π can be mimicked by a transposition on $f(\pi)$. We call such transposition *legal*. We now define the circular breakpoint graph, which is the circular version of the breakpoint graph [2]. Throughout, in both indices and elements, we identify $2n+1$ and 1.

Definition 1 *Let $\pi = (\pi_1 \ \ldots \ \pi_n)$ be a circular permutation, and $f(\pi) = \pi' = (\pi'_1 \ \ldots \ \pi'_{2n})$. The breakpoint graph $G(\pi)$ is an edge-colored graph on $2n$ vertices $\{1, 2, \ldots, 2n\}$. For every $1 \leq i \leq n$, π'_{2i} is joined to π'_{2i+1} by a black edge (denoted by b_i), and $2i$ is joined to $2i+1$ by a gray edge.*

Note that unlike previous studies of transpositions [3,8], we chose to double the number of vertices and work with an undirected graph, as done in the signed case [2]. It is convenient to draw the breakpoint graph on a circle, such that black edges are on the circumference and gray edges are chords (see Figure 2). We shall use this representation throughout the paper.

Since the degree of each vertex is exactly 2, the graph uniquely decomposes into cycles. Denote the number of cycles in $G(\pi)$ by $c(\pi)$. The length of a cycle is the number of black edges it contains. A k-*cycle* is a cycle of length k, and it is *odd* if k is odd. The number of odd cycles is denoted by $c_{odd}(\pi)$. Define $\Delta c(\pi, \tau) = c(\tau \cdot \pi) - c(\pi)$, and $\Delta c_{odd}(\pi, \tau) = c_{odd}(\tau \cdot \pi) - c_{odd}(\pi)$.

Bafna and Pevzner proved the following useful lemma (This - and other results we quote - was proved for linear permutations, but holds also for circular ones):

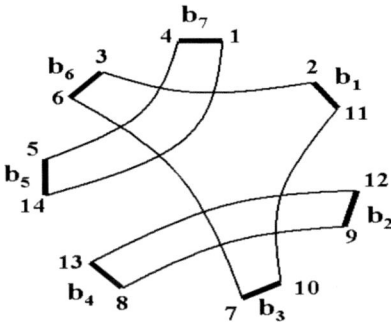

Fig. 2. The circular breakpoint graph of the permutation $\pi = (1\ 6\ 5\ 4\ 7\ 3\ 2)$, for which $f(\pi) = (1\ 2\ 11\ 12\ 9\ 10\ 7\ 8\ 13\ 14\ 5\ 6\ 3\ 4)$. Black edges are represented as thick lines on the circumference, and gray edges are chords.

Lemma 2 (Bafna and Pevzner [3]) *For all permutations π and transpositions τ, it holds that $\Delta c(\pi, \tau) \in \{-2, 0, 2\}$, and $\Delta c_{odd}(\pi, \tau) \in \{-2, 0, 2\}$.*

Let $n(\pi)$ denote the number of black edges in $G(\pi)$. The maximum number of cycles is obtained iff π is the identity permutation. In that case, there are $n(\pi)$ cycles, and all of them are odd (in particular, they are all of length 1) . Starting with π with c_{odd} odd cycles, Lemma 2 implies the following lower bound on $d(\pi)$:

Theorem 3 (Bafna and Pevzner [3]) *For all permutations π, $d(\pi) \geq (n(\pi) - c_{odd}(\pi))/2$.*

By definition, every legal transposition must cut three black edges. The transposition that cuts black edges b_i, b_j and b_k is said to *act on* these edges. A transposition τ is a *k-transposition* if $\Delta c_{odd}(\pi, \tau) = k$. A cycle is called *oriented* if there is a 2-transposition that acts on three of its black edges; otherwise, it is *unoriented*.

Observation 4 *There are only two possible configurations of 3-cycles that can be obtained by legal transpositions.*

The two possibilities are shown in Figure 3. It is easy to verify that the left 3-cycle is unoriented, and the right one is oriented.

Given a cyclic sequence of elements $i_1, \ldots, i_k$, an *arc* is an interval in the cyclic order, i.e., a set of contiguous elements in the sequence. The pair (i_j, i_l) ($j \neq l$) defines two disjoint arcs: $i_j, \ldots, i_{l-1}$ and $i_l, \ldots, i_{j-1}$. Similarly, a triple defines a partition of the cycle into three disjoint arcs. We say that two pairs of black edges (a, b) and (c, d) are *intersecting* if a and b belong to different arcs of the pair (c, d). A pair of black edges intersects with cycle C, if it intersects with a pair of black edges that belong to C. Cycles C and D intersect if there is a pair of black edges in C that intersect with D (see Figure 4a). Triples of black

Fig. 3. The only two possible configurations of 3-cycles. The left one is unoriented, and the right one is oriented.

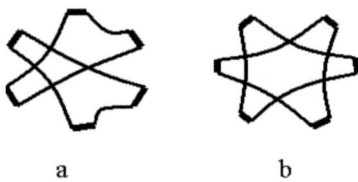

a   b

Fig. 4. (a) A pair of intersecting 3-cycles. (b) A pair of interleaving 3-cycles.

edges are *interleaving* if each of the edges of one triple belongs to a different arc of the second triple. Two 3-cycles are interleaving if their edges interleave (see Figure 4b).

Throughout the paper, we use the term permutation also when referring to the breakpoint graph of the permutation (as will be clear from the context). For example, when we say that π contains an oriented cycle, we mean that $G(\pi)$ contains an oriented cycle.

2.3   Transformation into Equivalent Simple Permutations

A k-cycle in the breakpoint graph is called *short* if $k \leq 3$; otherwise, it is called *long*. A breakpoint graph is called *simple* if it contains only short cycles. A permutation π is called *simple* if $G(\pi)$ is simple. Following [11,14], we show how to transform an arbitrary permutation into a simple one, while maintaining the lower bound of Theorem 3.

Let $b = (v_b, w_b)$ be a black edge and $g = (v_g, w_g)$ be a gray edge belonging to the same cycle $C = (\ldots, v_b, w_b, \ldots, w_g, v_g, \ldots)$ in $G(\pi)$. A (g, b)-*split* of $G(\pi)$ is a sequence of operations on $G(\pi)$, resulting in a new graph $\hat{G}(\pi)$ with one more cycle, as follows:

- Removing edges b and g.
- Adding two new vertices v and w.
- Adding two new black edges (v_b, v) and (w, w_b).
- Adding two new gray edges (w_g, w) and (v, v_g).

Figure 5 shows a (g, b)-split transforming a cycle C in $G(\pi)$ into two cycles C_1 and C_2 in $\hat{G}(\pi)$. Note that the order of the nodes of each edge along the cycle

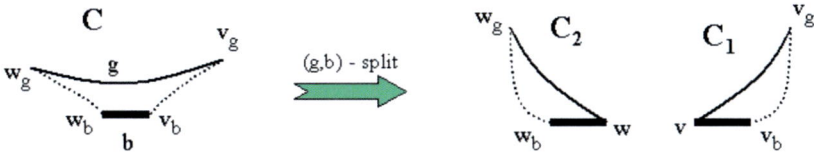

Fig. 5. A (g, b)-split. A dashed line indicates a path.

is important, as other orders may not split the cycle. Hannenhalli and Pevzner
[11] show that for every (g, b)-split on a permutation π of n elements, there is a
permutation $\hat{\pi}$ of $n + 1$ elements, which is obtained by inserting an element into
π, such that $\hat{G}(\pi) = G(\hat{\pi})$. Thus, a (g, b)-split can be viewed as a transformation
from π to $\hat{\pi}$. A (g, b)-split is called *safe* if $n(\pi) - c_{odd}(\pi) = n(\hat{\pi}) - c_{odd}(\hat{\pi})$, i.e.,
if it maintains the lower bound of Theorem 3.

Lemma 5 (Lin and Xue [14]) *Every permutation can be transformed into a
simple one by safe splits.*

Proof. Let π be a permutation that contains a long cycle C. Let b_1 be a black
edge in C. Denote by b_2 and b_3 the black edges that are connected to b_1 via a gray
edge. Let g be the gray edge that is connected to b_2 but not to b_1. Then a (g, b_3)-
split breaks C into a 3-cycle and a $(k - 2)$-cycle in $\hat{\pi}$. Clearly, $n(\hat{\pi}) = n(\pi) + 1$,
and $c_{odd}(\hat{\pi}) = c_{odd}(\pi) + 1$, so the split is safe. This process can be repeated until
a simple permutation is eventually obtained.

We say that permutation π is *equivalent* to permutation $\hat{\pi}$ if $n(\pi) - c_{odd}(\pi) = n(\hat{\pi}) - c_{odd}(\hat{\pi})$.

Lemma 6 (Hannenhalli and Pevzner [11]) *Let $\hat{\pi}$ be a simple permutation that
is equivalent to π, then every sorting of $\hat{\pi}$ mimics a sorting of π with the same
number of operations.*

In the following, we show how to sort a simple permutation by transpositions. We prove that the number of transpositions is within a factor of 1.5
from the lower bound of Theorem 3. Thus, we obtain a 1.5-approximation algorithm for sorting simple permutations. The above discussion implies that this
algorithm translates into a 1.5-approximation algorithm for an arbitrary permutation: Transform the permutation into an equivalent simple permutation
(Lemma 5), sort it, and then mimic the sorting on the original permutation
(Lemma 6).

3   The Algorithm

In this section we provide a 1.5-approximation algorithm for sorting permutations by transpositions. We first develop an algorithm for simple permutations,

and then use the results of Section 2.3 to prove the general case. Recall that the breakpoint graph of a simple permutation contains only 1-, 2- and 3-cycles. Our goal is to obtain a graph with only 1-cycles, which is the breakpoint graph of the identity permutation. Thus, the sorting can be viewed as a process of transforming the 2- and 3-cycles into 1-cycles.

First we deal with the case that the permutation contains a 2-cycle:

Lemma 7 (Christie [8]) *If π is a permutation that contains a 2-cycle, then there exists a 2-transposition on π.*

By definition, an oriented 3-cycle can be eliminated by a 2-transposition that acts on its black edges. Suppose from now on that all 2-cycles were eliminated by applying Lemma 7, and all oriented 3-cycles were eliminated. The only remaining problem is how to handle unoriented 3-cycles. This is the case we analyze henceforth.

A *(0,2,2)-sequence* is a sequence of three transpositions, of which the first is a 0-transposition, and the next two are 2-transpositions. Note that a $(0, 2, 2)$-sequence increases the number of odd cycles by 4 out of 6 that are the maximum possible in 3 steps, and thus a series of $(0, 2, 2)$-sequences preserves a 1.5 approximation ratio. We shall show below that such a sequence is always possible.

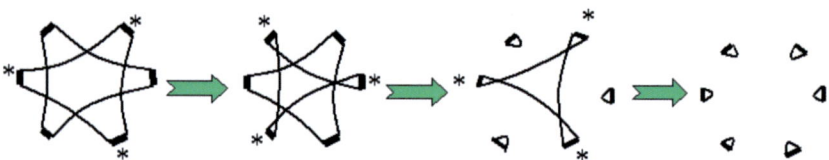

Fig. 6. A $(0, 2, 2)$-sequence of transpositions for two interleaving unoriented 3-cycles. At each step the transposition acts on the three black edges that are marked by a star.

Lemma 8 *Let π be a permutation that contains two interleaving unoriented 3-cycles. Then, there exists a $(0, 2, 2)$-sequence of transpositions on π.*

Proof. The $(0, 2, 2)$-sequence is described in Figure 6.

Lemma 9 *Let C and D be two intersecting unoriented 3-cycles that are not interleaving. Then, there exists a transposition which transforms C and D into a 1-cycle and an oriented 5-cycle.*

Proof. Let c_1, c_2 and c_3 be the three black edges of C. Assume, without loss of generality, that (c_1, c_2) intersects with D. We shall in fact prove a stronger statement, namely, for any choice of a black edge $d \in D$ such that (d, c_3) intersects

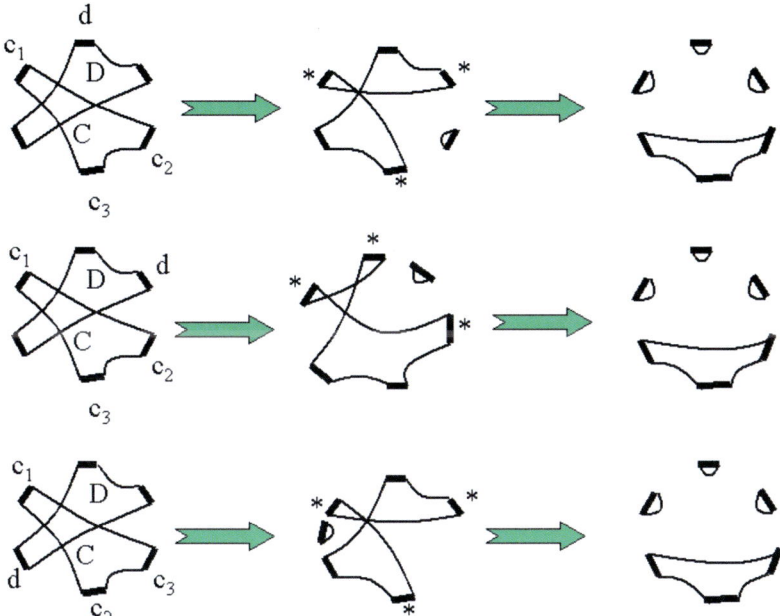

Fig. 7. The three possible cases of two intersecting unoriented 3-cycles that are not interleaving. In each case, the transposition that acts on edges c_1, c_2, and d, transforms C and D into a 1-cycle and an oriented 5-cycle.

with (c_1, c_2), the transposition on c_1, c_2 and d satisfies the lemma. Depending on the number of black edges of D in the arc of (c_1, c_2) that avoids c_3, there are three possible cases to consider for d, which are shown in Figure 7. In each case, the first transposition, which acts on c_1, c_2 and d, transforms 3-cycles C and D into a 1-cycle and a 5-cycle. Then, in order to show that the 5-cycle is oriented, a 2-transposition which acts on three edges of its edges is shown.

The following lemma claims that every pair of black edges in an unoriented cycle must intersect with some other cycle.

Lemma 10 (Bafna and Pevzner [3]) *Let C be an unoriented cycle in $G(\pi)$, and let b_i and b_j be black edges of C. Then, (b_i, b_j) intersects with some cycle in $G(\pi)$.*

We say that cycle E is *shattered* by cycles C and D if every pair of edges in E intersects with a pair of edges in C or with a pair of edges in D.

Corollary 11 *Let π be a simple permutation that contains no 2-cycles, no oriented 3-cycles, and no pair of interleaving 3-cycles. Then, there exists a triple of 3-cycles C, D and E in $G(\pi)$, such that E is shattered by C and D.*

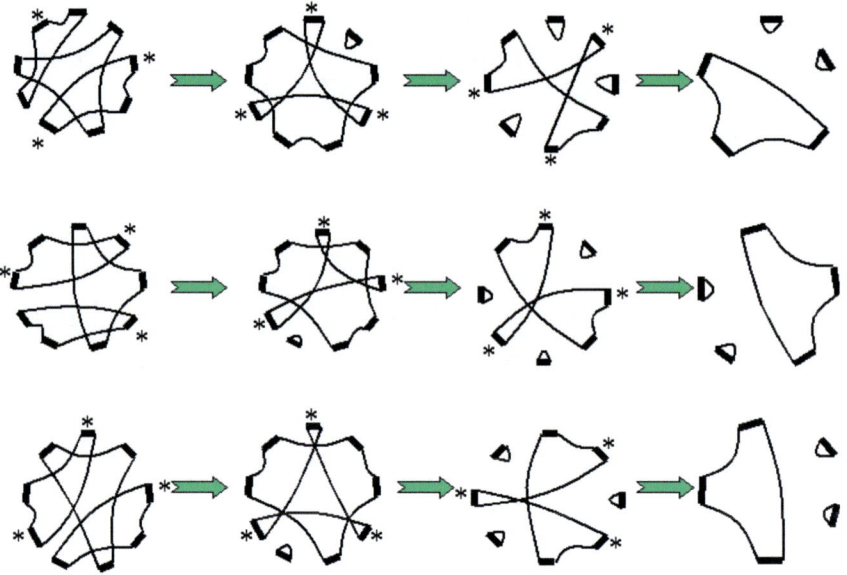

Fig. 8. The three possible cases of three intersecting unoriented 3-cycles, such that no pair is interleaving and two out of the three are non-intersecting. In each case, a $(0, 2, 2)$-sequence of transpositions is shown. For simplicity, every 1-cycle is shown only when it is formed and not in subsequent graphs (since it is not affected by transpositions in later steps).

Proof. Let E be an unoriented 3-cycle in $G(\pi)$, which consists of the black edges b_1, b_2 and b_3. By Lemma 10, there is a cycle C that intersects with the pair (b_1, b_2). This means that cycle C intersects also, without loss of generality, with the pair (b_2, b_3). Since E and C are not interleaving, there is another cycle D, that intersects with the pair (b_1, b_3) (again by Lemma 10). Hence, E is shattered by C and D.

Lemma 12 *Let π be a permutation that contains three unoriented 3-cycles C, D and E, such that E is shattered by C and D. Then, there exists a $(0, 2, 2)$-sequence of transpositions on π.*

Proof. If two out of the three cycles are interleaving, the (0,2,2)-sequence follows from Lemma 8. Otherwise, there are two general cases:

1. Two out of the three cycles are non-intersecting. In this case, there are three possible configurations of the cycles, which are shown in Figure 8. For every sub-case, a $(0, 2, 2)$-sequence is shown.

2. The three cycles are mutually intersecting. The general case is illustrated in Figure 9. Since cycles C and D are unoriented, the condition of the proof of Lemma 9 is fulfilled. Thus, we can apply a 0-transposition that acts on edges c_1, c_2, and d, and obtain a new oriented cycle F. Now we apply a 2-transposition on E (which has also become oriented). Cycle F remains oriented, since the latter transposition does not change its structure. Thus, another 2-transposition is possible on the edges of F, which completes the $(0, 2, 2)$-sequence.

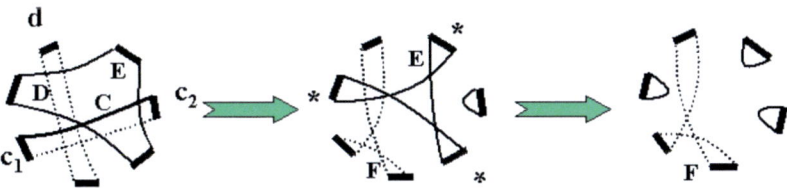

Fig. 9. Three mutually intersecting unoriented cycles. A dashed line represents a path of length 1 or 3. Note that edges c_1 and c_2 are connected by a single gray edge.

Our algorithm is described in Figure 10. Note that none of the steps 2 - 5 create a long cycle, and thus the permutation remains simple throughout the algorithm. Note also that in steps 4 and 5 we do not create 2-cycles or oriented 3-cycles, and hence, there is no need to iterate over steps 2 and 3. Corollary 11 guarantees us that for every simple permutation, at least one of the steps 2 - 5 may be applied, unless the permutation is sorted (and contains only 1-cycles).

Algorithm *Sort* (π)

1. Transform permutation π into an equivalent simple permutation $\hat{\pi}$ (Lemma 5).
2. While $G = G(\hat{\pi})$ contains a 2-cycle, apply a 2-transposition (Lemma 7).
3. While G contains an oriented 3-cycle, apply a 2-transposition on it.
4. If two interleaving 3-cycles are found in G, apply a $(0, 2, 2)$-sequence (Lemma 8).
5. If a shattered unoriented 3-cycle is found in G, apply a $(0, 2, 2)$-sequence (Lemma 12).
6. If G contains a cycle of length greater than 1, go to step 4.
7. Mimic the sorting of π using the sorting of $\hat{\pi}$ (Lemma 6).

Fig. 10. 1.5-approximation algorithm for sorting by transpositions.

Performing only steps 2 - 6 is an algorithm in its own, and is denoted by *Algorithm SortSimple*. The following theorem claims that algorithm SortSimple is a 1.5-approximation algorithm for sorting simple permutations:

Theorem 13 *Algorithm* SortSimple *is a 1.5-approximation algorithm for simple permutations, and it runs in time* $O(n^2)$.

Proof. The sequence of transpositions that is generated by the algorithm contains only 2-transpositions and $(0, 2, 2)$-sequences of transpositions. Therefore, every sequence of three transpositions increases the number of odd cycles by at least 4 out of 6 possible in 3 steps (as implied from the lower bound of Theorem 3). Hence, the approximation ratio is 1.5.

We now analyze the running time of the algorithm. The number of transpositions performed is $O(n)$, and so is the time to perform a transposition. Thus, it suffices to show that at each step, we can decide in linear time which transposition to apply. Steps 2 and 3 are trivial. Checking whether two pairs of black edges intersect can be done in constant time. Let A be an unoriented 3-cycle. For each pair of black edges (b_i, b_j) in A, scan all 3-cycles (in linear time) and check if they intersect with (b_i, b_j). If there is a 3-cycle B that intersects with all three pairs of edges of A, then A interleaves with B, and we can apply step 4. Otherwise, there are two cycles B and C such that A is shattered by B and C, and step 5 can be applied.

Now we are ready to prove the main result of the paper:

Theorem 14 *Algorithm* Sort *is a 1.5-approximation algorithm for general permutations, and it runs in time* $O(n^2)$.

Proof. By Theorem 13, we are guaranteed that $alg(\hat{\pi}) \leq 1.5 \cdot d(\hat{\pi})$, where $alg(\hat{\pi})$ is the number of transpositions used by Algorithm SortSimple to sort $\hat{\pi}$. Thus, by Theorem 3,

$$alg(\hat{\pi}) \leq 1.5 d(\hat{\pi}) \leq 1.5 \left(\frac{n(\hat{\pi}) - c_{odd}(\hat{\pi})}{2} \right) = 1.5 \left(\frac{n(\pi) - c_{odd}(\pi)}{2} \right) \leq 1.5 \cdot d(\pi)$$

Using Lemma 6, we can sort π by $alg(\hat{\pi})$ transpositions, which implies an approximation ratio of 1.5.

Since steps 1 and 7 can be done in linear time, Theorem 13 implies that the running time of Algorithm Sort is $O(n^2)$.

4   Discussion and Open Problems

In this paper we studied the problem of sorting permutations by transpositions, gave a simple quadratic 1.5-approximation algorithm for the problem, and simplified the underlying theory. We believe that this is an important step towards solving some related open problems. The main open problem is to determine the complexity of sorting by transpositions. Devising algorithms with better approximation ratio and/or faster running time is also desirable. Another direction, which is more biologically relevant, is to consider algorithms for sorting permutations by a set of rearrangement operations (such as reversals, transpositions and translocations).

Acknowledgements. I am very grateful to my advisor, Ron Shamir, for many invaluable and helpful discussions and for reading the manuscript carefully. I would like to thank Roded Sharan for fruitful discussions, and Vineet Bafna for help in understanding the complexity of the Bafna-Pevzner Algorithm. This work was supported in part by the Israel Science Foundation (grant 309/02).

References

1. D.A. Bader, B. M.E. Moret, and M. Yan. A linear-time algorithm for computing inversion distance between signed permutations with an experimental study. *Journal of Computational Biology*, 8(5):483–491, 2001.
2. V. Bafna and P. A. Pevzner. Genome rearragements and sorting by reversals. *SIAM Journal on Computing*, 25(2):272–289, 1996.
3. V. Bafna and P. A. Pevzner. Sorting by transpositions. *SIAM Journal on Discrete Mathematics*, 11(2):224–240, May 1998.
4. A. Bergeron. A very elementary presentation of the Hannenhalli-Pevzner theory. In *Proc. 12th Annual Symposium on Combinaotrial Pattern Matching (CPM '01)*, 2001.
5. P. Berman, S. Hannanhalli, and M. Karpinski. 1.375-approximation algorithm for sorting by reversals. In *Proc. of 10th Eurpean Symposium on Algorith,s (ESA'02)*, pages 200–210. Springer, 2002. LNCS 2461.
6. A. Caprara. Sorting permutations by reversals and Eulerian cycle decompositions. *SIAM Journal on Discrete Mathematics*, 12(1):91–110, February 1999.
7. D. A. Christie. A 3/2-approximation algorithm for sorting by reversals. In *Proc. ninth annual ACM-SIAM Symp. on Discrete Algorithms (SODA 98)*, pages 244–252. ACM Press, 1998.
8. D. A. Christie. *Genome Rearrangement Problems*. PhD thesis, University of Glasgow, 1999.
9. H. Eriksson, K. Eriksson, J. Karlander, L. Svensson, and J. Wastlund. Sorting a bridge hand. *Discrete Mathematics*, 241(1-3):289–300, 2001.
10. Q. P. Gu, S. Peng, and H. Sudborough. A 2-approximation algorithm for genome rearrangements by reversals and transpositions. *Theoretical Computer Science*, 210(2):327–339, 1999.
11. S. Hannenhalli and P. Pevzner. Transforming cabbage into turnip: Polynomial algorithm for sorting signed permutations by reversals. *Journal of the ACM*, 46:1–27, 1999. (Preliminary version in Proceedings of the Twenty-Seventh Annual ACM Symposium on Theory of Computing 1995 (STOC 95), pages 178–189).
12. S. B. Hoot and J. D. Palmer. Structural rearrangements, including parallel inversions, within the chloroplast genome of Anemone and related genera. *J. Molecular Evoolution*, 38:274–281, 1994.
13. H. Kaplan, R. Shamir, and R. E. Tarjan. Faster and simpler algorithm for sorting signed permutations by reversals. *SIAM Journal of Computing*, 29(3):880–892, 2000. (Preliminary version in Proceedings of the eighth annual ACM-SIAM Symposium on Discrete Algorithms 1997 (SODA 97), ACM Press, pages 344–351).
14. G. H. Lin and G. Xue. Signed genome rearrangements by reversals and transpositions: Models and approximations. *Theoretical Computer Science*, 259:513–531, 2001.
15. J. Meidanis, M. E. Walter, and Z. Dias. Reversal distance of signed circular chromosomes. manuscript, 2000.

16. J. D. Palmer and L. A. Herbon. Tricircular mitochondrial genomes of Brassica and Raphanus: reversal of repeat configurations by inversion. *Nucleic Acids Research*, 14:9755–9764, 1986.

17. P. A. Pevzner. *Computational Molecular Biology: An Algorithmic Approach*. MIT Press, 2000.

18. D. Sankoff and N. El-Mabrouk. Genome rearrangement. In T.Jiang, T. Smith, Y. Xu, and M. Q. Zhang, editors, *Current Topics in Computational Molecular Biology*. MIT Press, 2002.

19. J. Setubal and J. Meidanis. *Introduction to Computational Biology*. PWS Publishing Co., 1997.

20. R. Shamir. Algorithms in molecular biology: Lecture notes, 2002. Available at http://www.math.tau.ac.il/~rshamir/algmb/01/algmb01.html.

21. M. E. Walter, Z. Dias, and J. Meidanis. Reversal and transposition distance of linear chromosomes. In *String Processing and Information Retrieval: A South American Symposium (SPIRE 98)*, 1998.

Efficient Data Structures and a New Randomized Approach for Sorting Signed Permutations by Reversals

Haim Kaplan and Elad Verbin

School of Computer Science, Tel Aviv University, Tel Aviv, Israel.
{haimk, eladv}@cs.tau.ac.il

Abstract. The problem of sorting signed permutations by reversals (SBR) is a fundamental problem in computational molecular biology. The goal is, given a signed permutation, to find a shortest sequence of reversals that transforms it into the positive identity permutation, where a reversal is the operation of taking a segment of the permutation, reversing it, and flipping the signs of its elements.

In this paper we describe a randomized algorithm for SBR. The algorithm tries to sort the permutation by performing a random walk on the oriented Caylay-like graph of signed permutations under its oriented reversals, until it gets "stuck". We show that if we get stuck at the identity permutation, then we have found a shortest sequence. Empirical testing shows that this process indeed succeeds with high probability on a random permutation.

To implement our algorithm we describe an efficient data structure to maintain a permutation under reversals and draw random oriented reversals in sub-linear time per operation. With this data structure we can implement the random walk in time $O(n^{3/2}\sqrt{\log n})$, thus obtaining an algorithm for SBR that almost always runs in subquadratic time. The data structures we present may also be of independent interest for developing other algorithms for SBR, and for other problems.

1 Introduction

A permutation over $1, \ldots, n$, where each element has a sign, $+$ or $-$, is called a *signed permutation*. The problem of *sorting signed permutations by reversals (SBR)* is defined as follows. Given a signed permutation, find a shortest sequence of reversals that transforms it to the positive identity permutation, $id = (+1, \ldots, +n)$, where a reversal is the operation of taking a segment of the permutation, reversing it, and flipping the signs of its elements. The reversal $\rho(i, j)$ transforms $\pi - (\pi_1, \ldots, \pi_n)$ into $\pi' = (\pi_1, \ldots, \pi_{i-1}, -\pi_j, -\pi_{j-1}, \ldots, -\pi_i, \pi_{j+1}, \ldots, \pi_n)$. The length of such a shortest sequence of reversals is denoted $d(\pi)$, and called the *reversal distance* of π from id.

R. Baeza-Yates et al. (Eds.): CPM 2003, LNCS 2676, pp. 170–185, 2003.

This problem of sorting signed permutations by reversals is of great interest in computational biology, because it allows one to find a possible evolutionary path that transformed one species into another. For more details on this application see the book by Pevzner [12].

The unsigned variant of the problem, in which we deal with unsigned permutations, is NP-hard [5]. Somewhat surprisingly, Hannenhalli and Pevzner [8], in 1995, showed that the problem of sorting a *signed* permutation by reversals is in fact polynomial. They proved a duality theorem that equates the reversal distance with the sum of three combinatorial parameters associated with the permutation. Based on this theorem, they described an algorithm that sorts signed permutations by reversals in $O(n^4)$ time. More recently, Kaplan, Shamir, and Tarjan [10] simplified the underlying combinatorial structure and described an algorithm that finds a shortest sorting sequence in $O(n^2)$ time. Last year Bergeron [3] simplified the underlying combinatorial structure even further and described a somewhat different algorithm that runs in $O(n^3)$ time (which can be reduced to $O(n^2)$ on a bit-vector machine). Unfortunately, Bergeron has not been able to use her simplified analysis of the underlying structure to beat the $O(n^2)$ algorithm of Kaplan et. al., which is still the asymptotically fastest algorithm to date. We further note that if one is interested only in calculating $d(\pi)$, and does not require an actual sequence, then this can be calculated in linear time [1].

The structure of the rest of this paper is as follows. Section 2 gives some definitions. In Section 3 we describe our randomized algorithm. This algorithm repeatedly executes a random-walk process, which we call *SBR-RandWalk*. Each run of SBR-RandWalk either finds a minimum sorting sequence or fails. Section 4 gives the results of empirical tests that we performed with SBR-RandWalk. These results indicate that the average over all permutations of the expected number of times we need to run SBR-RandWalk to get a minimum sorting sequence is a small constant (namely, 1.6). We consider these empirical results to be quite strong, in the sense that the success rate seems to be independent of the size of the permutation. However, we could not find any mathematical proof of this observed behavior and nalyzing SBR-RandWalk theoretically is an open problem. In Section 5 we describe a data structure for representing the permutation so that we can draw a random oriented reversal and apply it in $O(\sqrt{n \log n})$ time. Thus obtaining an implementation of SBR-RandWalk that runs in time $O(n^{3/2}\sqrt{\log n})$. Combining this with the empirical study we conjecture that the expected running time of our algorithm on a random permutation is $O(n^{3/2}\sqrt{\log n})$. The data structure we present may be interesting in its own right, because if we could expand it to maintain some more information about the permutation then we would get a provably worst-case sub-quadratic algorithm for SBR. We address this, as well as other open questions, in Section 6.

We also believe that our data structure presented in Section 5 is of independent interest, since it seems to be quite versatile. For example, it might be useful to speed up an approximation algorithm for Sorting by Transpositions of Hartman [9].

2   Preliminaries

We begin with some definitions and background that will be used throughout the paper. Note that we will always augment our permutations by defining $\pi_0 = +0$, $\pi_{n+1} = n + 1$. This is common practice in SBR papers, and allows us to avoid some special cases.

Suppose we want to sort a signed permutation, π. Now, to sort the permutation we must begin by locating a reversal that transforms π into π', where $d(\pi') = d(\pi) - 1$ – this is called a *safe* reversal . The underlying combinatorial theory (discovered by [2] and refined in [8] and [10]) distinguishes a class of reversals called *oriented reversals*. An oriented reversal is a reversal that makes consecutive elements in the permutation adjacent with the same sign[1]. It is easily seen that:

Proposition 1 *A permutation has an oriented reversal iff it has a negative element.*

The theory guarantees that:

Proposition 2 *If a permutation has an oriented reversal then it also has an oriented reversal which is safe.*

All existing algorithms for solving SBR invest at least linear time in searching for a safe oriented reversal, "performing it" (i.e. updating the data structures so that they represent π' rather than π), and repeating the process. Since the distance of a permutation is at most $n + 2$ (and is almost always $\Theta(n)$) this means that in order to get a sub-quadratic algorithm we must take a different approach (Ozery-Flato and Shamir [11] give a more comprehensive treatment of this question, and prove that all existing algorithms have permutations on which they spend $\Theta(n^2)$ time). Our approach will be to choose a random oriented reversal, and "hope" it is safe – see section 3.

There is one extra complication, though: For many permutations the process of repeatedly picking a safe oriented reversal and applying it will indeed generate a shortest sorting sequence. However, this is not true for all permutations. There are permutations containing structures called *unoriented components* (which we will not define here, see [10]), that this process will *always* fail to sort. For permutations that contain unoriented components *all* sequences of safe oriented reversals will end with a permutation that has no oriented reversals (i.e. a positive permutation), but is not the identity permutation. The theory suggests algorithmic ways of handling these special permutations: One can detect in linear time whether a permutation contains unoriented components or not. Furthermore, in case π contains unoriented components one can find in linear time a "good" sequence of reversals that "clears these components", i.e. a sequence of t reversals that transforms π into π' that has no unoriented components and $d(\pi') = d(\pi) - t$. It is also important to note that, according to

---

[1] By adjacent we mean either $i, i+1$ or $-(i+1), -i$, in the augmented permutation

the theory, an oriented reversal is safe iff by applying it we do not create new unoriented components. So, to sum it all up, if we start with a permutation with no unoriented components and perform an unsafe oriented reversal we end up with a permutation that contains an unoriented component. Furthermore, subsequent oriented reversals cannot clear this unoriented component, and so once we perform an unsafe oriented reversal, the process of repeatedly picking a safe oriented reversal and performing it will not succeed in fully sorting the permutation.

The conclusion from the background given in the last paragraph is the following observation, which is central to our algorithm:

Observation 3 *If we start from a permutation that contains no unoriented components, and repeatedly perform oriented reversals, then if we got the identity permutation in the end of the process, we are sure, in retrospect, that all reversals we have performed along the way were safe, and thus the sequence sorts the permutation. This is because if we had performed an unsafe reversal along the way then we would have been stuck with a permutation that is unsortable using oriented reversals.*

Given an augmented signed permutation, $\pi = (+0, \pi_1, \pi_2, \ldots, \pi_n, n+1)$, we consider *pairs* (π_i, π_j) so that $i < j$ and π_i, π_j are consecutive integers (that is, $|\pi_i| - |\pi_j| = \pm 1$). Such a pair is called an *oriented pair* if its elements are of opposite signs, and *unoriented* otherwise. Conventionally, 0 is considered to be positive, and not negative. Note that there are exactly $n+1$ pairs in π, and that there are no oriented pairs iff the permutation is positive (i.e. all of its elements are positive).

The reversal $\rho = \rho(i, j)$, which changes the permutation $\pi = (\pi_0, \pi_1, \ldots, \pi_{n+1})$ into $\rho \cdot \pi = (\pi_0, \ldots, \pi_{i-1}, -\pi_j, -\pi_{j-1}, \ldots, -\pi_i, \pi_{j+1}, \ldots, \pi_{n+1})$ is oriented if either $\pi_i + \pi_{j+1} = +1$ or $\pi_{i-1} + \pi_j = -1$ (that is, if it makes a consecutive pair of elements adjacent and identically signed). Alternatively, oriented reversals can be derived from oriented pairs: if (π_i, π_j) is an oriented pair, then the reversal

$$\begin{cases} \rho(i, j-1) & \text{if } \pi_i + \pi_j = +1 \\ \rho(i+1, j) & \text{if } \pi_i + \pi_j = -1 \end{cases}$$

is an oriented reversal; We can get all oriented reversals this way. Note that we get a specific oriented reversal from either one or two different oriented pairs.

3   Exploiting Randomization to Sort by Reversals

The algorithm we define in this section is motivated by empirical studies on random permutations showing that, typically, a very large portion of oriented reversals are safe, and so we may be able to wave the complex task of finding a safe oriented reversal, instead just picking a random oriented reversal and hoping it is safe. This argument is supported by the fact that an unsafe oriented reversal creates an unoriented component, but easy calculations show that the fraction of the permutations of length n that contain an unoriented component is very small, that is $O(\frac{1}{n^2})$. More results that support our claims were presented in a recent paper by Bergeron et. al. [4].

We next describe our algorithm. Given a signed permutation, π, we first clear the unoriented components (if there are any) using the method of [10]. To sort the "cleared" permutation we iterate the following "random walk"-like process, called SBR-RandWalk. This random walk repeatedly picks a random oriented reversal and applies it (without bothering to check if it is safe or not). SBR-RandWalk repeats this process until it gets a permutation with no oriented reversals (i.e. a positive permutation). The theory above, and Observation 3 in particular, implies that this walk generates a shortest sorting sequence iff it ends with the identity permutation. If SBR-RandWalk fails (i.e. we ended with a positive permutation that is not the identity permutation) we run SBR-RandWalk again and again, each time on the original permutation. If it fails $f(n)$ times, where $f(n)$ is some slowly growing function of n, we resort to running one of the known algorithms (e.g. [10] or [3]) to sort the permutation. The function $f(n)$ can be any function that tends to ∞ as $n \to \infty$, and only affects the worst-case running time of the algorithm. For concreteness we use $f(n) = \log n$.

Alternatively we can run a variant of this procedure which instead of selecting a random oriented reversal selects a random oriented pair and performs the oriented reversal which corresponds to it. Computationally the two procedures are equivalent – if we can select a random oriented pair then we can also select a random oriented reversal in roughly the same time as follows: We pick a random oriented pair p that defines an oriented reversal ρ and check whether there is another oriented pair which defines ρ. If only one pair defines ρ, we return ρ. If two pairs define ρ, we flip a coin; If the coin comes out heads, we return ρ, and if tails, we pick at random a new oriented pair and repeat this procedure. Clearly the average number of oriented pairs we need to draw to get a random oriented reversal is no more than two. Likewise, given a method for choosing a random oriented reversal we can select a random oriented pair in approximately the same time.

It is easy to see that after at most $n + 1$ reversals SBR-RandWalk reaches a positive permutation. However, a naive implementation of SBR-RandWalk would spend linear time per reversal (This implementation traverses the permutation, finds all oriented reversals, and picks one at random.), for a total running time of $\Theta(n^2)$.

Another straightforward implementation picks a random oriented pair by drawing a random index $i \in [0, n]$ and checking if the pair whose elements are $\pm i$ and $\pm(i+1)$ is oriented. If it is, we select it; Otherwise, we draw another index and try again. The latter method may be more efficient for some permutations, but there are permutations for which it still takes $\Theta(n)$ expected time to draw an oriented reversal, in $\Theta(n)$ iterations, leading to $\Theta(n^2)$ expected running time on a worst case permutation [2].

---

[2] One such example is a permutation of Ozery-Flato and Shamir that will be discussed shortly.

Fortunately, we found a data structure that maintains a signed permutation under reversals and allows us to draw a random oriented reversal and perform it all in sub-linear time. In section 5.2 we describe this data structure, which allows us to draw a random oriented reversal and perform it in time $O(n^{1/2}\sqrt{\log n})$. Using this representation SBR-RandWalk can be implemented to run in $O(n^{3/2}\sqrt{\log n})$ time in the worst case.

Once we see that a single run of SBR-RandWalk can be implemented in sub-quadratic time, the next question to ask is whether our suggested algorithm that repeats calling SBR-RandWalk up to $f(n)$ times runs in sub-quadratic expected time on the worst-case permutation. Unfortunately this is not the case. There are permutations for which with very high probability SBR-RandWalk fails a superpolynomial (in n) number of times. One such example is the permutation $\pi = (2, 4, 6, \ldots, 2k - 2, -1, -3, -5, \ldots, -(2k - 3), 2k - 1)$ of size $2k-1$ of Ozery-Flato and Shamir [11]. No matter which reversals we pick to sort it, throughout the sorting sequence we alternate between a permutation in which all oriented reversals are safe (and symmetric) and a permutation with exactly two oriented reversals, only one of which is safe. Therefore, the chances of succeeding in one run of SBR-RandWalk are extremely low, namely $\frac{1}{2^{k-1}}$.[3] The probability that we succeed in one out of $f(n)$ trials is $O(\frac{f(2k-1)}{2^{k-1}}) = O(\frac{\log(2k-1)}{2^{k-1}})$ which still goes rapidly to zero with k. So for this permutation our algorithm almost surely resorts to run one of the standard algorithms for SBR, both of which take $\Omega(n^2)$ on most permutations (including this one, see [11]). To conclude, the worst-case complexity of our algorithm is the same as the complexity of the fallback algorithm we choose to run.

However, it could be that the permutations for which SBR-RandWalk fails $f(n)$ times are rare and on a random permutation the running time is indeed $o(n^2)$. We address this question empirically in the next section.

4   Empirical Results and Conjectures about SBR-RandWalk

As previously stated, we expect that iterating SBR-RandWalk would work well on most permutations. This is because typically a large fraction of the oriented reversals are safe. To support this intuition we measured empirically the performance of our algorithm on random permutations. Our results are summarized in Table 1 and in Figure 1. For a random permutation without unoriented components of size ranging from 10 to 10000 we estimated the probability that a single run of SBR-RandWalk succeeds. We also estimated the average number of

---

[3] Surprisingly, this is not much better than an easily proven lower bound for the probability of success of SBR-RandWalk on any permutation with no unoriented components: That probability is guaranteed to be at least $\frac{1}{(n+1)!}$, since the number of oriented pairs starts from at most $n + 1$ and decreases by at least 1 every time a safe reversal is performed, and every permutation without unoriented components has at least one safe reversal.

runs of SBR-RandWalk that our algorithm does until it gets a sorting sequence. Specifically, we drew many permutations while throwing away those that had unoriented components (such permutations are rare and have appeared only for smaller values of n). Then we ran SBR-RandWalk repeatedly until success for each of the permutations. Column 3 in Table 1 lists the percentage of permutations that were solved in our first run of SBR-RandWalk, and column 4 lists the average over the selected permutations of the number of tries of SBR-RandWalk we needed until one succeeded. Figure 1 shows the distribution of the number of trials until success for our tests of permutations of size 500. Results for other values of n look almost identical and exhibit the same decay. Note that Fig. 1 shows that the decay of the number of trials to success is exponential. This hints that an overwhelming majority of permutations have a very good chance of being sorted by SBR-RandWalk.

Table 1. The performance of SBR-RandWalk on a random permutation.

n	number of permutations tested	percentage of success at first try	average number of tries until success
10	300000000	66.20%	1.642
20	100000000	63.86%	1.607
50	30000000	63.01%	1.594
100	100000000	62.82%	1.594
200	3000000	62.76%	1.594
500	600000	62.69%	1.596
1000	200000	62.62%	1.596
2000	50000	63.35%	1.586
5000	6000	63.7%	1.58
10000	2000	63%	1.59

The conclusions from our experiments are as follows:

- If we select a random permutation that has no unoriented component uniformly of all such permutations of size n and run Procedure SBR-RandWalk on it once then it has about 0.63 chance of success, regardless of n. Also, for a specific permutation the chance of success of SBR-RandWalk is well-behaved, and is for almost all permutations between 0.55 and 0.75, also regardless of n.
- If we select a permutation as above and run SBR-RandWalk repeatedly until we get a sorting sequence we will need roughly 1.6 trials on average, also regardless of n.
- We also found that running the variant of SBR-RandWalk that chooses a random oriented pair instead of a random oriented reversal we obtained roughly the same results. We also obtained similar results when we ran on a random permutation having $n + 1$ breakpoints (which are, in some sense, the only permutations of "true" size n)

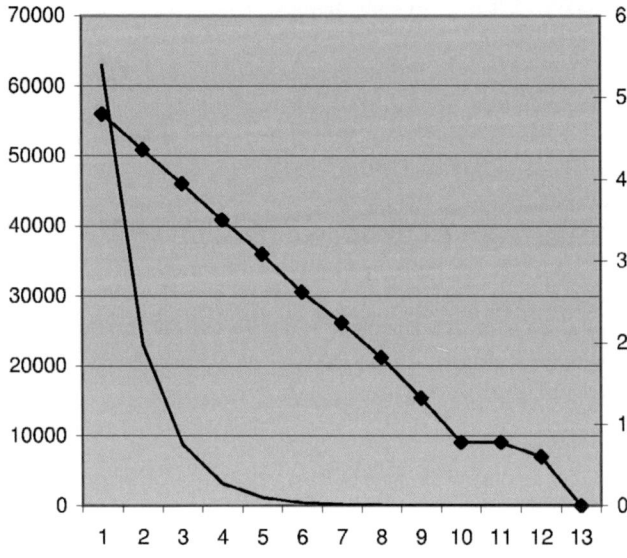

Fig. 1. Distribution of the number of SBR-RandWalk runs needed to sort a permutation of size $n = 500$. The x-axis is the number of trials. The thin line is the number of permutations that are sorted with exactly that number of trials (with the scale on the left showing the values). The thick line presents the same data, but with a logarithmic scale in the y-axis, so it can be observed that the decay is very close to exponential decay. Its scale is on the right.

We leave open the question of finding analytical support to our experimental findings. We conjecture that the following theorem holds.

Conjecture 4 *The average number of runs of SBR-RandWalk needed to find a sorting sequence of a uniformly-chosen signed permutation with no unoriented components of size n is $O(1)$.*

It would immediately follow from this conjecture that the average running time of our algorithm on a random permutation is $O(n \cdot \sqrt{n \log n})$. Recall that at most $O(1/n^2)$ fraction of all permutations contain unoriented components. Therefore the average complexity of our algorithm on a randomly chosen permutation and on a randomly chosen permutation with no unoriented components differ by at most a constant factor if our fallback procedure when SBR-RandWalk fails $f(n)$ times runs in $O(n^3)$ time.

5 Data Structures for Handling Permutations

Existing algorithms for SBR maintain the current permutation via two integers arrays, one containing π and the other containing π^{-1} (that is, the inverse array, which for each element gives its location in π, disregarding its current sign).

With this representation we can locate elements of the permutation in $O(1)$ time, but it would take time proportional to $j - i$ to perform a reversal $\rho(i,j)$, i.e. to update the arrays so that they represent the new permutation. Thus on the worst case (i.e. for long reversals), it takes $\Theta(n)$ time to update the arrays. It follows that any algorithm for SBR that looks for one reversal at a time and runs in sub-quadratic time must manipulate the permutation via a better representation that allows to perform a reversal in $o(n)$ time. For a discussion of the drawbacks of current algorithms that solve SBR, see [11].

Another way to hold a permutation would be to represent π and π^{-1} by threading the elements in two doubly linked lists, one representing π and the other representing π^{-1}. With this representation we can perform a reversal in $O(1)$ time. It is harder however to locate various elements of the permutation. In particular, it is not clear how using this representation one would draw a random oriented reversal in sub-linear time.

It follows that we cannot implement SBR-RandWalk to run in sub-linear time with either of these two obvious representations. In this section we present alternative representations that do allow faster implementation of SBR-RandWalk. Any such representation should support two basic operations: 1) Draw a random oriented reversal (or say that none exist), 2) perform a reversal ρ on π, and they should take sub-linear time each.

To get started we show two simple representations that allow three operations: 1) query: locating π_i given i, 2) inverse query: locating π_i^{-1} given i, 3) performing a reversal, all in sub-linear time. This allows us to demonstrate the basic techniques in a simpler setup. Later we show how to use these techniques to construct more complicated representations that also allow to draw a random oriented reversal in sub-linear time.

5.1   Maintaining the Permutation

The two simple representations we describe here were used by Chrobak et. al. [6] and further investigated by Fredman et. al. [7]. In both cases they were used to efficiently implement a common local improvement heuristics for the travelling salesman problem. These data structures are quite practical and perform well in practice.

Our first representation is based on balanced binary search trees. Using this representation we can perform queries, inverse queries and reversals all in $O(\log n)$ time. Our second representation is based on a partition of the permutation into blocks of size $\Theta(\sqrt{n})$. Using this representation we can perform queries in $O(\log n)$ time, inverse queries in $O(1)$ time and reversals in $O(\sqrt{n})$ time. Our later data structures that allow to draw a random oriented reversal borrow ideas from both these two simple structures.

Theorem 5. *There exists a data structure that maintains a (signed) permutation, and allows querying for π_i, π_i^{-1}, and performing a reversal $\rho(i,j)$, all in logarithmic time. The data structure is of size linear in n, and, given a permutation, initialization can be performed in linear time.*

Proof. This data structure is built upon any balanced binary search tree data structure that is re-balanced via rotations and supports split and concatenate operations, such as splay trees [14], red-black trees, 2-3 trees, and AVL trees. Fredman et. al. used splay trees which offer good constants and for which the implementations of split and concatenate are particularly elegant, and so our description, too, will use splay trees (and our running times will therefore be amortized. We can avoid amortization by using some other type search tree).

In our representation we hold a tree with n nodes containing the elements of the permutation, such that an in-order traversal of the tree gives us the permutation. We complicate matters by introducing a *"reversed"* flag for each node, which, if turned on, indicates that the subtree rooted at that node should be read in reversed, that is – from right to left, and the signs of its elements should be flipped. Further reversed flags down the tree can once again alter the order of the implied permutation. The invariant we keep is that an in-order traversal of the tree, modified by the reversed flags, always gives us the permutation.

We note that we can clear the reversed flag in an internal node by exchanging its children, flipping the state of the reversed flag in each of them, and flipping the sign of the element at the node. The reversed flag of a leaf can be cleared if we flip the sign of the element contained in that leaf. One can view this process as pushing down the reversed flags. Using this ability to clear a reversed flag we can implement a rotation operation in a splay tree so that it correctly maintains the permutation. This is done by pushing down the reversed flag of the nodes involved in the rotation before we actually carry it out. Clearly a rotation on nodes with reversed flags turned off keeps the order of the permutation.

We implement queries by maintaining subtree counts in the nodes. If we want the value of π_i we just look for the ith node in the (altered) in-order traversal and return the element in it, with the correct sign according to the reversed flags of the nodes of the path that led us to it. Given a pointer to the node x containing i we can find π_i^{-1} by counting how many nodes are to the left of the path from x to the root, again taking into account the reversed flags. The value of π_i^{-1} is larger by one than this count. We get the pointer to node x by keeping an additional array of pointers mapping an element i to the node x containing it. This array is static if we keep each element inside the same node at all times. When performing a query we splay the corresponding node. The cost of the query is proportional to the cost of the splay. This proves the logarithmic running time.

To execute a reversal $\rho(i, j)$ we split the tree at the node corresponding to π_j to a tree T_1 containing all items with indices at most j, and a tree T_2 containing all items with indices larger than j. Then we split T_1 at π_i to a tree T_3 containing all items with indices smaller than i and to a tree T_4 containing all items with indices at least i. Finally we flip the reversed flag of the root of T_4, concatenate T_3 to T_4 and the resulting tree to T_2. It is easy to see that the resulting tree represents the permutation after we performed the reversal. The $O(\log n)$ time bound follows from the logarithmic time bounds on split and concatenate operations of splay trees.   □

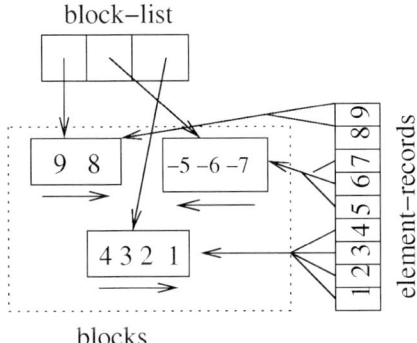

Fig. 2. A representation of the permutation $\pi = (9, 8, 7, 6, 5, 4, 3, 2, 1)$

Theorem 6. *There exists a data structure that maintains a (signed) permutation, and allows querying for π_i in logarithmic time, querying for π_i^{-1} in constant time, and performing a reversal $\rho(i, j)$ in time $O(\sqrt{n})$. The data structure is of size linear in n, and, given a permutation, initialization can be performed in linear time.*

Proof. The second data structure is in many ways a two-level version of the previous structure. The structure is based on partitioning the permutation into blocks, each of size $\Theta(\sqrt{n})$. Each block is a contiguous fragment of the permutation, and is accompanied by a reversed flag, which, if *on*, indicates that the block should be read in reversed direction and the elements in it have opposite signs. That way, we can reverse a block simply by flipping its *reversed* flag. We maintain the blocks in a list, where each block points to a list of the items it contains (and also stores its size) and each item points to the block containing it. See Figure 2. We also maintain the following invariant.

Invariant 7 *At the end of each operation the number of elements in each block is between $\frac{1}{2}\sqrt{n}$ and $2\sqrt{n}$.*

Note that this implies that there are always $\Theta(\sqrt{n})$ blocks, each of size $\Theta(\sqrt{n})$. It is easy to see that one can initialize the data structure in linear time.

We perform a reversal by the following three steps.

1. The extreme elements of the reversal are found, as well as the blocks containing them. We split each of these two blocks that also contains elements that do not belong to the reversal into two blocks. One of the resulting blocks contains only elements that belong to the reversal, and the other contains only elements that do not belong to the reversal. Note that this may violate Invariant 7. We reinsert the blocks to the block-list so that the blocks that contain the elements belonging to the reversal are consecutively placed.
2. Now the reversal consists of a subsequence of complete blocks. We reverse the order of those blocks in the block-list, and flip their reversed flags.

3. Finally, we reinstate Invariant 7 – we look at the four blocks that may now be of size smaller than $\frac{\sqrt{n}}{2}$, and we merge each of them to their neighbors until they are all above $\frac{\sqrt{n}}{2}$. Now some blocks may be larger than $2\sqrt{n}$ (but all are still smaller than $3\sqrt{n}$). We split the large blocks to smaller ones that satisfy the invariant.

It is easy to see that using our representation we can perform these three steps in $O(\sqrt{n})$ time. Figure 3 shows an example of this process.

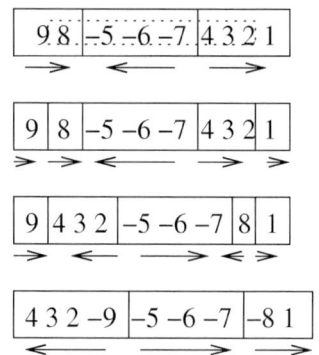

Fig. 3. Performing the reversal $\rho(2,8)$ on $\pi = (9,8,7,6,5,4,3,2,1)$ results in $\pi' = (9,-2,-3,-4,-5,-6,-7,-8,1)$. Note that we allow blocks' sizes to be between 2 and 6. The arrows represent the reversed flags.

It is easy to see how we can use this data structure to answer queries and inverse queries in time $O(\sqrt{n})$. However by maintaining some additional information we can answer queries even faster. To speed up queries we maintain the list of blocks in a search tree where the key of each block is the number of elements in blocks to its left (these are the "running totals" of the block sizes). Within each block we maintain the items in an array. With this representation we can locate the block we are looking for by performing a search on the tree, and then go directly to the required element within the block.

In order to implement an inverse query in time $O(1)$ we maintain with each element its index within its block and maintain with each block the number of elements in blocks preceding it (this is the key of the block in the search tree mentioned above). With this representation we can get the location of an element by adding up its index to the key of its block.   □

5.2   Adding the Ability to Draw a Random Oriented Pair

Theorem 8. *There exists a data structure for maintaining a signed permutation, that supports queries for π_i and π_i^{-1} in logarithmic time, allows to perform* a reversal in $O(\sqrt{n \log n})$ time, and allows to select an oriented reversal uniformly at random in logarithmic time. The data structure also allows to find the number of oriented reversals in π in constant time. The data structure is of size linear in n, and, given a permutation, we can initialize it in linear time.

Proof. Our data structure will be based on the structure in Theorem 6. Like it, the new data structure divides the permutation into blocks. We maintain the old structure but add to each block a tree whose purpose is to store the pairs related to that block, and maintain their orientations. Each of these trees will be similar to a single instance of the structure of Theorem 5 — it will be a balanced binary search tree based on rotations, such as a Splay Tree. Each node in the tree will hold a pair and a boolean variable that indicates whether the pair is oriented, as well as a reversed flag whose interpretation will be described shortly.

The tree at a block contains, for each element x in the block, the pair whose elements are $\pm x$ and $\pm(x + 1)$.[4] An in-order read of the tree will order the pairs according to the location of their other end in the permutation (that is, according to increasing order of $\pi^{-1}(\pm(x + 1))$). With each pair we also store its orientation. As was already stated, each node in the tree will have a reversed flag. If the reversed flag of a node v is *on* it means that the subtree T_v rooted at v should be read in reversed order, and the orientations of the pairs in T_v is opposite to what is indicated by the corresponding variable. As before, further reversed flags down the tree can again alter the orientations and the order. We also associate an additional meaning to reversed flag of the *block* itself – now if a block's reversed flag is *on* it also means that the orientations of the pairs associated with the block's elements (that is, the orientations of the pairs in the block's tree) are once again opposite to what is indicated by the tree corresponding to the block.

As in Section 5.1, we facilitate fast query time by maintaining subtree counts of the oriented pairs in the nodes (these can be updated while doing the rotations, as before) and adding "running totals" of oriented pairs to the search tree of the block-list, where we used to keep the running totals of the block sizes. Using these it is easy to return the number of oriented reversals in constant time (it is listed in the "grand total" which is the last of the running totals), and to uniformly draw a random oriented pair in time $O(\log n)$ (by randomly selecting the index of the oriented pair we are looking for, searching on the running totals to find the tree it is in, and then looking for the right index in the tree).

To execute a reversal $\rho(i, j)$ we operate on the blocks like we did in Theorem 6. When blocks are split and concatenated we rebuild the trees associated with them "from scratch". This takes time linear in the size of the trees involved, that is $\Theta(\sqrt{n})$. This leaves us with the problem of updating the trees (both those associated with blocks that are in the segment defining the reversal and those associated with blocks that are out of the reversal) so that the orientations and the order of pairs inside the trees will be correct (when we assume that the reversal spans an integral number of blocks).

---

[4] Except, of course, for the element $x = n + 1$ which does not have a pair associated with it.

To complete the execution of $\rho(i, j)$ we go over all blocks. For each block we split the corresponding tree into three trees, much like we did in Theorem 5. The middle tree will contain all pairs whose "remote" element is part of the reversal (i.e. all pairs for which $\pi^{-1}(\pm(x+1)) \in [i, j]$), and the other two trees will contain the pairs whose "remote" element is out of the reversal. We then flip the reversed flag of the root of the middle tree, and concatenate the trees in their original order. After this operation, the order of the pairs is correct for all trees. Regarding the orientations, we leave to the reader to verify that they are correct for trees in blocks that do not belong in the reversal, while in the blocks that do belong in the reversal, the orientations are exactly opposite of their correct setting. This happens because the orientation is flipped exactly for those pairs for which one of the elements can be found inside the reversal and the other is outside of it. However, note that we expanded the role of the blocks' reversed flags to indicate that the orientations of the block's pairs are flipped once again, and this can be seen to fix the orientations of pairs in those blocks that *are* in the reversal so that they are now at the correct state as well. Thus, once we have done this splitting, flipping, and concatenating, all invariants are satisfied.

In order to fine-tune the asymptotic cost of performing a reversal in this structure we keep the sizes of the blocks between $\frac{\sqrt{n \log n}}{2}$ and $2 \cdot \sqrt{n \log n}$. This makes the entire procedure of performing an oriented reversal run in $O(\sqrt{n \log n})$ time. It takes $O(\sqrt{n \log n})$ time to perform the operations on the part of the data structure that was inherited from Theorem 5.2 and additional $O(\sqrt{n \log n})$ time to update $\Theta(\sqrt{\frac{n}{\log n}})$ trees as outlined above, in logarithmic time per tree.   □

6   Conclusion and Further Questions

We described a randomized algorithm for SBR that gives good results on random permutations. To implement this algorithm efficiently we described a data structure to maintain a permutation under reversals such that it is possible to draw a random oriented reversal, and to perform other queries in sublinear time.

One problem that we left open is proving Conjecture 4. This would: 1) support theoretically the good empirical performance of our algorithm, and 2) indicate that there are many minimum sorting sequences of reversals, so the biologist may want to add constraints in order to get a meaningful sequence. Research in this direction was initiated by Pinter and Skiena in [13] and by Bergeron et. al. in [4].

Another interesting problem is expanding the data structures of Section 5.2 to maintain some information about the permutation in order to get a worst-case (or almost worst-case) sub-quadratic algorithm. We can achieve this, for example, by adding to the data structure the ability to check whether the permutation it represents contains an unoriented component in sub-linear time. That would mean that we can perform a walk as before, but checking at each step retroactively whether the oriented reversal we performed was safe or not, and if it was not safe we can backtrack and try to find another one. This random walk would give an algorithm that takes $o(n^2)$ expected time for every permutation, provided we can prove that at least a constant fraction of oriented reversals is always safe (a phenomenon that we have observed empirically). Another (maybe harder) way to augment the data structure could be to find a way to maintain with the pairs whether the reversals they define are safe or not, such that reversals on this data structure can still be done in sub-linear time. Then you could solve SBR by performing only safe reversals until the permutation is sorted.

Finally, we think there may be further applications for our data structure in the area of sequence analysis, and in other areas. It also seems plausible that our data structure can be improved, maybe even so that both performing a reversal and getting a random oriented reversal could be done in logarithmic time. This would immediately give us an improvement in the running time of SBR-RandomWalk. This is, in some sense, a geometric data structure (if you think of the pairs as points in N^2), so techniques from computational geometry (such as those for orthogonal range searching) may be useful.

Acknowledgments. We thank Michal Ozery-Flato, Tzvika Hartman, and Ron Shamir for many helpful discussions.

References

1. D. A. Bader, B. M. E. Moret, and M. Yan, *A linear-time algorithm for computing inversion distance between signed Permutations with an experimental study*, Workshop on Algorithms and Data Structures, 2001, pp. 365–376.
2. V. Bafna and P. A. Pevzner, *Genome rearragements und sorting by reversals*, SIAM Journal on Computing **25** (1996), no. 2, 272–289.
3. A. Bergeron, *A very elementary presentation of the hannenhalli-pevzner theory*, CPM (2001), 106–117.
4. A. Bergeron, C. Chauve, T. Hartman, and K. Saint-Onge, *On the properties of sequences of reversals that sort a signed Permutation*, proceedings of JOBIM, June 2002, pp. 99–108.
5. A. Caprara, *Sorting by reversals is difficult*, Proceedings of the First International Conference on Computational Molecular Biology (RECOMB), ACM Press, 1997, pp. 75–83.
6. M. Chrobak, T. Szymacha, and A. Krawczyk, *A data structure useful for finding hamiltonian cycles*, Theoretical Computer Science **71** (1990), no. 3, 419–424.
7. M. L. Fredman, D. S. Johnson, L. A. McGeoch, and G. Ostheimer, *Data structures for traveling salesmen*, Journal of Algorithms **18** (1995), no. 3, 432–479.
8. S. Hannenhalli and P. A. Pevzner, *Transforming cabbage into turnip (polynomial algorithm for sorting signed Permutations by reversals)*, Proceedings of the Twenty-Seventh Annual ACM Symposium on Theory of Computing (Las Vegas, Nevada), 29 May–1 June 1995, pp. 178–189.
9. T. Hartman, *A simpler 1.5-approximation algorithm for sorting by transpositions*, CPM '03, in these proceedings.
10. H. Kaplan, R. Shamir, and R. E. Tarjan, *A faster and simpler algorithm for sorting signed permutations by reversals*, SIAM J. Comput. **29** (1999), no. 3, 880–892.

11. M. Ozery-Flato and R. Shamir, *Two notes on genome rearrangements*, Journal of Bioinformatics and Computational Biology (to appear).
12. P. A. Pevzner, *Computational molecular biology: An algorithmic approach*, The MIT Press, Cambridge, MA, 2000.
13. R. Y. Pinter and S. Skiena, *Genomic sorting with length-weighted reversals*, manuscript.
14. D.D. Sleator and R.E. Tarjan, *Self-adjusting binary search trees*, J. Assoc. Comput. Mach. **32** (1985), 652–686.

Linear-Time Construction of Suffix Arrays
(Extended Abstract)

Dong Kyue Kim[1][*], Jeong Seop Sim[2], Heejin Park[3], and Kunsoo Park[3][**]

[1] School of Electrical and Computer Engineering, Pusan National University
dkkim1@pusan.ac.kr
[2] Electronics and Telecommunications Research Institute, Daejeon 305-350, Korea
simjs@etri.re.kr
[3] School of Computer Science and Engineering, Seoul National University
{hjpark,kpark}@theory.snu.ac.kr

Abstract. The time complexity of suffix tree construction has been shown to be equivalent to that of sorting: $O(n)$ for a constant-size alphabet or an integer alphabet and $O(n \log n)$ for a general alphabet. However, previous algorithms for constructing suffix arrays have the time complexity of $O(n \log n)$ even for a constant-size alphabet.

In this paper we present a linear-time algorithm to construct suffix arrays for integer alphabets, which do not use suffix trees as intermediate data structures during its construction. Since the case of a constant-size alphabet can be subsumed in that of an integer alphabet, our result implies that the time complexity of directly constructing suffix arrays matches that of constructing suffix trees.

1 Introduction

The suffix tree due to McCreight [19] is a compacted trie of all the suffixes of a string T. It was designed as a simplified version of Weiner's position tree [26]. The suffix array due to Manber and Myers [18] and independently due to Gonnet et al. [11] is basically a sorted list of all the suffixes of a string T. There are also some other index data structures such as suffix automata [3].

When we consider the complexity of index data structures, there are three types of alphabets from which string T of length n is drawn: (i) a constant-size alphabet, (ii) an integer alphabet where symbols are integers in the range $[0, n^c]$ for a constant c, and (iii) a general alphabet in which the only operations on string T are symbol comparisons.

The time complexity of suffix tree construction has been shown to be equivalent to that of sorting [7]. For a general alphabet, suffix tree construction has time bound of $\Theta(n \log n)$. And suffix trees can be constructed in linear time for a constant-size alphabet due to McCreight [19] and Ukkonen [24] or for an integer alphabet due to Farach-Colton, Ferragina, and Muthukrishnan [6,7].

---

[*] Supported by KOSEF grant R01-2002-000-00589-0.
[**] Supported by BK21 Project and IMT2000 Project AB02.

R. Baeza-Yates et al. (Eds.): CPM 2003, LNCS 2676, pp. 186–199, 2003.
© Springer-Verlag Berlin Heidelberg 2003

Despite simplicity of suffix arrays among index data structures, the construction time of suffix arrays has been larger than that of suffix trees. Two known algorithms for constructing suffix arrays by Manber and Myers [18] and Gusfield [12] have the time complexity of $O(n \log n)$ even for a constant-size alphabet. Of course, suffix arrays can be constructed by way of suffix trees in linear time, but it has been an open problem whether suffix arrays can be constructed in $o(n \log n)$ time without using suffix trees.

In this paper we solve the open problem in the affirmative and present a linear-time algorithm to construct suffix arrays for integer alphabets. Since the case of a constant-size alphabet can be subsumed in that of an integer alphabet, we will consider only the case of an integer alphabet in describing our result.

We take the recent divide-and-conquer approach for our algorithm [6,7,8,15, 22], i.e., (i) construct recursively a suffix array SA_o for the set of odd positions, (ii) construct a suffix array SA_e for the set of even positions from SA_o, and (iii) merge SA_o and SA_e into the final suffix array SA_T. The hardest part of this approach is the merging step and our main contribution is a new merging algorithm.

Our new merging algorithm is quite different from Farach-Colton et al.'s [6, 7] that is designed for suffix trees. Whereas Farach-Colton et al.'s uses a coupled depth-first search in the merging, ours uses equivalence relations defined on factors of T [5,12] (and thus it is more like a breadth-first search). Also, Farach-Colton et al.'s algorithm goes back and forth between suffix trees and suffix arrays during its construction, while ours uses only suffix arrays during its construction.

Recently, Kärkkäinen and Sanders [16] and Ko and Aluru [17] also proposed simple linear-time construction algorithms for suffix arrays. Burkhardt and Kärkkäinen [4] gave another construction algorithm that takes $O(n \log n)$ time using only $O(n/\sqrt{\log n})$ extra space.

Space reduction of a suffix array is an important issue [20,9,13,21] because the amount of text data is continually increasing. Grossi and Vitter [13] proposed the *compressed* suffix array of $O(n \log |\Sigma|)$-bits size and Sadakane [21] improved it by adding the *lcp* information. Since their compressions also exploit the divide-and-conquer approach mentioned above, we can directly build the compressed suffix array from a given string in linear time by applying our proposed algorithm to their compressions.

2   Preliminaries

2.1   Definitions and Notations

We first give some definitions and notations that will be used in our algorithms. Consider a string T of length n over an alphabet Σ. Let $T[i]$ denote the ith symbol of string T and $T[i, j]$ the substring starting at position i and ending at position j in T. We assume that $T[n]$ is a special symbol $\#$ which is lexicographically smaller than any other symbol in Σ. S_i, $1 \leq i \leq n$, denotes the suffix of

T that starts at position i. The prefix of length k of a string α is denoted by $\text{pref}_k(\alpha)$. We denote by $\text{lcp}(\alpha, \beta)$ the longest common prefix of two strings α and β and by $\text{lcp}_i(\alpha, \beta)$ the longest common prefix of $\text{pref}_i(\alpha)$ and $\text{pref}_i(\beta)$. When string α is lexicographically smaller than string β, we denote it by $\alpha \prec \beta$.

We define the suffix array $SA_T = (A_T, L_T)$ of string T as a pair of arrays A_T and L_T [7].

- The *sort array* A_T is the lexicographically ordered list of all suffixes of T. That is, $A_T[i] = j$ if S_j is lexicographically the ith suffix among all suffixes $S_1, S_2, \ldots, S_n$ of T. The number i will be called the *index* of suffix S_j, denoted by $\text{index}(j) = i$.
- The *lcp array* L_T stores the length of the longest common prefix of adjacent suffixes in A_T, i.e., $L_T[i] = |\text{lcp}(S_{A_T[i]}, S_{A_T[i+1]})|$ for $1 \leq i < n$. We set $L_T[0] = L_T[n] = -1$.

We define odd and even arrays of a string T. The *odd array* $SA_o = (A_o, L_o)$ is the suffix array of all suffixes beginning at odd positions in T. That is, the sort array A_o of SA_o is the lexicographically ordered list of all suffixes beginning at odd positions of T, and the lcp array L_o has the length of the longest common prefix of adjacent odd suffixes in A_T. Similarly, the *even array* $SA_e = (A_e, L_e)$ is the suffix array of all suffixes beginning at even positions in T.

For a subarray $A[x, y]$ of sort array A, we define $P_A(x, y)$ as the longest common prefix of the suffixes $S_{A[x]}, S_{A[x+1]}, \ldots, S_{A[y]}$. If $x = y$, $P_A(x, x)$ is defined as the suffix $S_{A[x]}$ itself. Lemma 1 gives some properties of P_A in a subarray of sort array A.

Lemma 1. *Let $A[x, y]$ be a subarray of sort array A for $x < y$ and L be a corresponding lcp array.*

(a) $P_A(x, y) = \text{lcp}(S_{A[x]}, S_{A[y]})$.
(b) $|P_A(x, y)|$ *is equal to the minimum value in $L[x, y - 1]$.*

In order to find $|P_A(x, y)|$ efficiently, we define the following problem.

Definition 1. *Given an array A of size n whose elements are integers in the range $[0, n - 1]$ and two indices a and b $(1 \leq a < b \leq n)$ in array A, the range-minimum query $\text{MIN}(A, a, b)$ is to find the smallest index $a \leq j \leq b$ such that $A[j] = \min_{a \leq i \leq b} A[i]$.*

This MIN query can be answered in constant time after linear-time pre-processing of array A. The first solution [10] for the range minima problem constructed the cartesian tree [25] of array A and answered nearest common ancestor computations [14,23] on the tree. In the Appendix, we described another solution for the problem, which is a modification of Berkman and Vishkin's simple solution. We remark that this method uses only arrays without making any kinds of trees. Similar results were given by Bender and Farach-Colton [1] and Sadakane [21]. By a MIN query, we get the following theorem.

string	1	2	3	4	5	6	7	8	9	10	11	12	13	14	15	16	17	18	19	20
	a	a	a	a	b	b	b	b	a	a	a	b	b	b	a	a	b	b	b	#

The suffix array

index	1	2	3	4	5	6	7	8	9	10	11	12	13	14	15	16	17	18	19	20
sort array	20	1	9	2	15	10	3	16	11	4	19	8	14	18	7	13	17	6	12	5
lcp array	0	3	6	2	5	5	1	4	4	0	1	3	1	2	4	2	3	5	3	0

Equivalence Classes

E_1	#	a	a	a	a	a	a	a	a	a	b	b	b	b	b	b	b	b	b	b
E_2	a	a	a	a	a	a	b	b	b	#	a	a	b	b	b	b	b	b	b	
E_3	a	a	a	b	b	b	b	b	b	a	a	#	a	a	b	b	b	b		
E_4	a	b	b	b	b	b	b	b	b	a	b	a	a	#	a	a	b			
E_5	b	b	b	b	b	#	a	b	a	b	a	a								
E_6	b	b	#	a	b	a	b	a	a											
	a	b		A[5,7]																

Fig. 1. Equivalence classes and subarrays of a sort array.

Theorem 1. *Given a suffix array (A, L) and $x < y$, $|P_A(x, y)|$ (i.e., the smallest index $x \le j < y$ such that $L_T[j] = \min_{x \le i < y} L_T[i]$) can be computed in constant time.*

An advantage of suffix trees is that *suffix links* are defined on suffix trees. When $\text{lcp}(S_i, S_j) = a\alpha$ for $a \in \Sigma$ and $\alpha \in \Sigma^*$, $\text{lcp}(S_{i+1}, S_{j+1}) = \alpha$. Suffix links enable us to find α from S_i and S_j. In suffix arrays this can be done by finding $\text{lcp}(S_{i+1}, S_{j+1})$ using a MIN query. This method will be used in Section 3.4 with the following lemma.

Lemma 2. *Let i and j ($i < j$) be two positions in string T. If $T[i]$ and $T[j]$ match, $|\text{lcp}(S_i, S_j)| = |\text{lcp}(S_{i+1}, S_{j+1})| + 1$; otherwise, $|\text{lcp}(S_i, S_j)| = 0$.*

2.2   Equivalence Classes

In this section, we will define equivalence relation E_l on sort arrays such as A_T, A_o, and A_e, and explain the relationship between equivalence classes of E_l on a sort array and subarrays of the sort array.

Let A be a sort array of size m and L be the corresponding lcp array. Equivalence relation E_l ($l \ge 0$) on A is:

$$E_l = \{(i, j) \mid \text{pref}_l(S_{A[i]}) = \text{pref}_l(S_{A[j]})\}.$$

That is, two suffixes $S_{A[i]}$ and $S_{A[j]}$ have a common prefix of length l if and only if i and j are in the same equivalence class of E_l on A.

We describe the relationship between equivalence classes of E_l on A and subarrays of A. Since the integers in A are sorted in the lexicographical order of the suffixes they represent, we get the following fact from the definition of E_l.

Fact 1 *Subarray $A[p, q]$, $1 \leq p \leq q \leq m$, is an equivalence class of E_l, $0 \leq l \leq n$, on A if and only if $L[p-1] < l$, $L[q] < l$, and $L[i] \geq l$ for all $p \leq i < q$.*

We now describe how an equivalence class of E_l on A is partitioned into equivalence classes of E_{l+1}. Let $A[p, q]$ be an equivalence class of E_l. By Fact 1, $L[i] \geq l$ for all $p \leq i < q$. Let $p \leq i_1 < i_2 < \cdots < i_r < q$ denote all the indices such that $L[i_1] = L[i_2] = \cdots = L[i_r] = l$. Since $L[i] \geq l+1$ for $i \notin \{i_1, i_2, \ldots, i_r\}$ and $p \leq i < q$, $A[p, i_1]$, $A[i_1+1, i_2]$, $\ldots$, $A[i_r+1, q]$ are equivalence classes of E_{l+1}. We can find $i_1, i_2, \ldots, i_r$ in $O(r)$ time by Theorem 1 and we get the following lemma.

Lemma 3. *An equivalence class of E_l can be partitioned into equivalence classes of E_{l+1} in $O(r)$ time, where r is the number of the partitioned equivalence classes of E_{l+1}.*

An equivalence class of E_l can be an equivalence class of E_k for $k \neq l$. Consider $A[5, 7]$ in Fig. 1 where $L[i] \geq 5$ for $5 \leq i < 7$, $L[4] = 2$, and $L[7] = 1$. Then, $A[5, 7]$ is an equivalence class of E_3, E_4, and E_5. In general, we have the following fact.

Fact 2 *A subarray $A[p, q]$ is an equivalence class of E_i for $a \leq i \leq b$ if and only if $\max\{L[p-1], L[q]\} = a - 1$ and $b = |P_A(p, q)| (= \min_{p \leq i < q} L[i])$.*

The integers a and b are called the *start stage* and the *end stage* of the equivalence class $A[p, q]$.

3   Linear-Time Construction

We present a linear-time algorithm for constructing suffix arrays for integer alphabets. Our construction algorithm follows the divide-and-conquer approach used in [6,7,8,15,22], and it consists of the following three steps.

1. Construct the odd array SA_o recursively. Preprocess L_o for range-minimum queries.
2. Construct the even tree SA_e from SA_o. Preprocess L_e for range-minimum queries.
3. Merge SA_o and SA_e to get the final suffix array SA_T.

The first two steps are essentially the same as those in [6,7,8] and our main contribution is a new merging algorithm in step 3.

3.1   Constructing Odd Array

Construction of the odd array SA_o is based on recursion and it takes linear time besides recursion.

1. Encode the given string T into a string of a half size: We make pairs $(T[2i-1], T[2i])$ for every $1 \leq i \leq n/2$. Radix-sort all the pairs in linear time, and map the pairs into integers in the range $[1, n/2]$. If we convert the pairs in T into corresponding integers, we get a new string of a half size, which is denoted by T'.

2. Recursively construct suffix array $SA_{T'}$ of T'.
3. Compute SA_o from $SA_{T'}$: We get A_o by $A_o[i] = 2A_{T'}[i] - 1$ for all i. Since two symbols in T are encoded into one symbol in T', we get L_o from $L_{T'}$ as follows. If the first different symbols of T' in adjacent suffixes $S_{A_o[i]}$ and $S_{A_o[i+1]}$ have the same first symbol of T, then $L_o[i] = 2L_{T'}[i] + 1$; otherwise, $L_o[i] = 2L_{T'}[i]$.

3.2   Constructing Even Array

The even array SA_e is constructed from SA_o in linear time. We first compute the sort array A_e and then compute the lcp array L_e as follows.

1. Make the sort array A_e: An even suffix is one symbol followed by an odd suffix. We make tuples for even suffixes: the first element of a tuple is $T[2i]$ and the second element is suffix S_{2i+1}. First, the tuples are sorted by the second elements (this result is given in A_o). Then we stably sort the tuples by the first elements and we get A_e.
2. Compute the lcp array L_e: Consider two even suffixes S_{2i} and S_{2j}. By Lemma 2, if $T[2i]$ and $T[2j]$ match, $|\mathtt{lcp}(S_{2i}, S_{2j})| = |\mathtt{lcp}(S_{2i+1}, S_{2j+1})| + 1$; otherwise, $\mathtt{lcp}(S_{2i}, S_{2j}) = 0$. We can get $|\mathtt{lcp}(S_{2i+1}, S_{2j+1})|$ from the odd array SA_o in constant time as follows. Let $x = \mathrm{index}(2i + 1)$ and $y = \mathrm{index}(2j + 1)$ in SA_o. By Lemma 1, $|\mathtt{lcp}(S_{2i+1}, S_{2j+1})| = |\mathrm{P}_{A_o}(x, y)|$, which is computed by a $\mathrm{MIN}(L_o, x, y)$ query.

3.3   Merging Odd and Even Arrays

We will show how to obtain suffix array $SA_T = (A_T, L_T)$ from SA_o and SA_e in $O(n)$ time, where n is the length of T. The main task in merging is to compute the sort array A_T. The lcp array L_T is computed as a by-product during the merging. The *target entry* of an entry $A_o[i]$ (resp. $A_e[i]$) is the entry of A_T that stores the integer in $A_o[i]$ (resp. $A_e[i]$) after we merge A_o and A_e. To merge A_o and A_e, we first compute the target entries of entries in A_o and A_e and then store all the integers in A_o and A_e into A_T. Hence, the problem of merging is reduced to the problem of computing target entries of entries in A_o and A_e.

We first introduce some notions on equivalence classes of E_i on A_o and A_e. For brevity, we define notions only on equivalence classes on A_o. (They are defined on A_e similarly.) An equivalence class $A_o[w, x]$ of E_i is *i-coupled* with an equivalence class $A_e[y, z]$ of E_i if and only if all the suffixes represented by the integers in $A_o[w, x]$ and $A_e[y, z]$ have the common prefix of length i, i.e., $\mathtt{pref}_i(\mathrm{P}_{A_o}(w, x)) = \mathtt{pref}_i(\mathrm{P}_{A_e}(y, z))$. The integers in $A_o[w, x]$ and $A_e[y, z]$ (that are *i-coupled* with each other) form an equivalence class of E_i on A_T after we merge A_o and A_e because each odd suffix represented by an integer in $A_o[w, x]$ and each even suffix represented by an integer in $A_e[y, z]$ have the common prefix $\mathtt{pref}_i(\mathrm{P}_{A_o}(w, x)) = \mathtt{pref}_i(\mathrm{P}_{A_e}(y, z))$ and the other odd or even suffixes do not have $\mathtt{pref}_i(\mathrm{P}_{A_o}(w, x))$ as their prefixes.

Lemma 4. *The integers in $A_o[w, x]$ and $A_e[y, z]$ that are i-coupled with each other form an equivalence class $A_T[w + y - 1, x + z]$.*

An equivalence class $A_o[w, x]$ of E_i is i-*uncoupled* if it is not i-coupled with any equivalence class of E_i on A_e. If an equivalence class $A_o[w, x]$ of E_i is i-uncoupled, no suffix represented by an integer in A_e has $\mathtt{pref}_i(\mathrm{P}_{A_o}(w, x))$ as its prefix and thus the integers in an i-uncoupled equivalence class $A_o[w, x]$ form an equivalence class of E_i on A_T, which is $A_T[a + w, a + x]$ for some a, after we merge A_o and A_e.

We now explain the notion of a *coupled pair*, which is central in our merging algorithm. Consider an equivalence class $A_o[w, x]$ whose start stage is l_o and end stage is k_o and an equivalence class $A_e[y, z]$ whose start stage is l_e and end stage is k_e such that $l = \max\{l_o, l_e\} \leq k = \min\{k_o, k_e\}$ and $A_o[w, x]$ and $A_e[y, z]$ are l-coupled with each other. We call $C = \langle A_o[w, x], A_e[y, z] \rangle$ a *coupled pair*. Since $A_o[w, x]$ and $A_e[y, z]$ is l-coupled with each other, the integers in $A_o[w, x]$ and $A_e[y, z]$ form an equivalence class $A_T[w + y - 1, x + z]$ after we merge A_o and A_e. We define the start stage and the end stage of coupled pair C as the start stage and the end stage of equivalence class $A_T[w + y - 1, x + z]$. Since l is the smallest integer such that $A_o[w, x]$ is l-coupled with $A_e[y, z]$, l is the start stage of $A_T[w + y - 1, x + z]$ and thus l is the start stage of C. Now we are interested in the end stage of C. Since one of $A_o[w, x]$ and $A_e[y, z]$ will be partitioned into several classes of E_{k+1}, $A_T[w + y - 1, x + z]$ cannot be an equivalence class of E_{k+1}. In the sense that the end stage of C cannot be larger than k, the value k is called the *limit stage* of C. The actual end stage of C is the value of $|\mathtt{lcp}(\mathrm{P}_{A_o}(w, x), \mathrm{P}_{A_e}(y, z))|$, and it is in the range of $[l, k]$. In our algorithm, we maintain coupled pairs in multiple queues $Q[k]$ for $0 \leq k < n$. Each queue $Q[k]$ contains coupled pairs whose limit stage is k.

We now describe the outline of computing the target entries of entries in A_o and A_e. We will compute target entries only for *uncoupled* equivalence classes on A_o and A_e. Since all equivalence classes of E_i on A_o and A_e will eventually be uncoupled as we increase i, we can find target entries of all entries in A_o and A_e in this way.

Our merging algorithm consists of at most n stages, and it maintains the following invariants.

Invariant: At the end of stage $s \geq 0$, the equivalence classes that constitute coupled pairs whose start stages are at most s and limit stages are at least s are stored in $Q[i]$ for $s \leq i \leq n - 1$. For every i-uncoupled equivalence class for $0 \leq i \leq s$ that does not constitute such a coupled pair, the target entries for the equivalence class have been computed.

We will call an equivalence class for which target entries have been computed a *processed* equivalence class.

We describe the outline of stages. At stage s, we do the following for each coupled pair $C = \langle A_o[w, x], A_e[y, z] \rangle$ stored in $Q[s - 1]$. We first compute the end stage of C by solving the following coupled-pair lcp problem. In the next section, we show how to solve the coupled-pair lcp problem in $O(1)$ time.

Definition 2 (The coupled-pair lcp problem). *Given a coupled pair $C = \langle A_o[w, x], A_e[y, z] \rangle$ whose limit stage is $s - 1$, compute the end stage of C. Furthermore, if the end stage of C is less than $s - 1$, determine whether $\mathrm{P}_{A_o}(w, x) \prec \mathrm{P}_{A_e}(y, z)$ or $\mathrm{P}_{A_o}(w, x) \succ \mathrm{P}_{A_e}(y, z)$.*

After solving the coupled-pair lcp problem for C, we have two cases depending on whether or not the end stage of C is $s - 1$.

- If the end stage of C is $s - 1$, $A_o[w, x]$ is $(s - 1)$-coupled with $A_e[y, z]$. We first partition $A_o[w, x]$ and $A_e[y, z]$ into equivalence classes of E_s. Every partitioned equivalence class will be either s-coupled or s-uncoupled. The s-coupled equivalence classes constitute coupled pairs whose start stages are at most s and limit stages are at least s, and thus we store each coupled pair in $Q[k]$ for $s \leq k \leq n - 1$, where k is the limit stage of the coupled pair. For the s-uncoupled equivalence classes, we find the target entries for them.
- If the end stage of C is smaller than $s - 1$, $A_o[w, x]$ and $A_e[y, z]$ are $(s - 1)$-uncoupled. We find the target entries for $A_o[w, x]$ and $A_e[y, z]$.

It is not difficult to see that the invariant is satisfied after stage s.

In our merging algorithm, we will use four arrays $\mathtt{ptr}_o$, $\mathtt{ptr}_e$, $\mathtt{fin}_o$, and $\mathtt{fin}_e$. Since $\mathtt{ptr}_e$ and $\mathtt{fin}_e$ are similar to $\mathtt{ptr}_o$ and $\mathtt{fin}_o$, we explain $\mathtt{ptr}_o$ and $\mathtt{fin}_o$ only. At the end of stage s, the values stored in $\mathtt{ptr}_o$ and $\mathtt{fin}_o$ are as follows.

- $\mathtt{fin}_o$ stores target entries for A_o, i.e., $\mathtt{fin}_o[i]$ for $1 \leq i \leq n_o$ is defined if $A_o[i]$ is an entry of a processed equivalence class and it stores the index of the target entry of $A_o[i]$.

- $\mathtt{ptr}_o[i]$ for $1 \leq i \leq n_o$ is defined if $A_o[i]$ is either the last entry of a coupled equivalence class or an entry of a processed equivalence class.

  - If $A_o[i]$ is the last entry of an equivalence class $A_o[a, b]$ (i.e., $i = b$) coupled with $A_e[c, d]$ (i.e., $\langle A_o[a, b], A_e[c, d] \rangle$ is stored in $Q[k]$ for some $s \leq k \leq n - 1$), $\mathtt{ptr}_o[b]$ stores d.
  - If $A_o[i]$ is an entry of a processed equivalence class $A_o[a, b]$:
    - If $A_o[i]$ is not the last entry of $A_o[a, b]$ (i.e., $a \leq i < b$), $\mathtt{ptr}_o[i]$ stores b.
    - Otherwise, $\mathtt{ptr}_o[b]$ stores β such that $A_e[\beta]$ is the last entry of a partitioned equivalence class $A_o[\alpha, \beta]$ and that β satisfies $|\mathtt{lcp}(S_{A_o[b]}, S_{A_e[\beta]})| \geq |\mathtt{lcp}(S_{A_o[b]}, S_{A_e[\delta]})|$ for any other $1 \leq \delta \leq n_e$. In addition, $|\mathtt{lcp}(S_{A_o[b]}, S_{A_e[\beta]})|$ is stored in $L_T[\mathtt{fin}_o[b]]$ if $\mathtt{fin}_o[b] < \mathtt{fin}_e[\beta]$ and $L_T[\mathtt{fin}_e[\beta]]$ otherwise.

We describe stages in detail. Initially, we are given a coupled pair $\langle A_o[1, n_o], A_e[1, n_e] \rangle$ whose start stage and limit stage is 0. In stage 0, we store $\langle A_o[1, n_o], A_e[1, n_e] \rangle$ into $Q[0]$ and initialize $\mathtt{ptr}_o[n_o] = n_e$, $\mathtt{ptr}_e[n_e] = n_o$, $L_T[0] = L_T[n] = -1$. In stage s, $1 \leq s \leq n$, we do nothing if $Q[s - 1]$ is empty. Otherwise, for every coupled pair $C = \langle A_o[w, x], A_e[y, z] \rangle$ stored in $Q[s - 1]$, we compute the end stage of C by solving the coupled-pair lcp problem. We have two cases depending on whether or not the end stage of C is $s - 1$.

Case 1: If the end stage of C is $s-1$, $A_o[w, x]$ is $(s-1)$-coupled with $A_e[y, z]$. We first partition $A_o[w, x]$ and $A_e[y, z]$ into equivalence classes of E_s. Let C_o and C_e denote the set of equivalence classes into which $A_o[w, x]$ and $A_e[y, z]$ are partitioned respectively. We denote equivalence classes in C_o by $A_o[w_i, x_i]$, $1 \leq i \leq r_1$, such that $\mathsf{P}_{A_o}(w_j, x_j) \prec \mathsf{P}_{A_o}(w_k, x_k)$ if $j < k$ and equivalence classes in C_e by $A_e[y_i, z_i]$, $1 \leq i \leq r_2$, such that $\mathsf{P}_{A_e}(y_j, z_j) \prec \mathsf{P}_{A_e}(y_k, z_k)$ if $j < k$. Partitioning $A_o[w, x]$ and $A_e[y, z]$ into equivalence classes of E_s takes $O(r_1 + r_2)$ time by Lemma 3.

Each equivalence class in C_o (resp. C_e) is either s-coupled or s-uncoupled. We find every coupled pair $\langle A_o[w_i, x_i], A_e[y_j, z_j] \rangle$, store it into $Q[\min\{|\mathsf{P}_{A_o}(w_i, x_i)|, |\mathsf{P}_{A_e}(y_j, z_j)|\}]$, set $\mathtt{ptr}_o[x_i] = z_j$, $\mathtt{ptr}_e[z_j] = x_i$, and compute $L_T[x_i + z_j]$ appropriately. For each s-uncoupled equivalence class $A_o[w_i, x_i]$, we find target entries for $A_o[w_i, x_i]$, store them in $\mathtt{fin}_o[\alpha]$ for $w_i \leq \alpha \leq x_i$, and compute $\mathtt{ptr}_o[k]$ for $w_i \leq k \leq x_i$ and $L_T[\mathtt{fin}_o[x_i]]$. We perform a similar operation for each s-uncoupled equivalence class $A_e[y_j, z_j]$. The following procedure shows the operations in detail. (We assume $a_{r_1+1} = b_{r_2+1} = \$$ where $\$ \succ a$ for any $a \in \Sigma$, $w_{r_1+1} = x_{r_1} + 1$, $x_{r_1+1} = x_{r_1}$, $y_{r_2+1} = z_{r_2} + 1$, and $z_{r_2+1} = z_{r_2}$.)

Procedure MERGE(C_o, C_e)
```
 1:    i ← 1 and j ← 1
 2:    while i ≤ r₁ or j ≤ r₂ do
 3:        aᵢ ← the sth symbol of P_{A_o}(wᵢ, xᵢ)
 4:        bⱼ ← the sth symbol of P_{A_e}(yⱼ, zⱼ)
 5:        if aᵢ = bⱼ then              // A_o[wᵢ, xᵢ] and A_e[yⱼ, zⱼ] are s-coupled.
 6:            k ← min{ |P_{A_o}(wᵢ, xᵢ)|, |P_{A_e}(yⱼ, zⱼ)| }
 7:            store ⟨A_o[wᵢ, xᵢ], A_e[yⱼ, zⱼ]⟩ into Q[k]
 8:            if i + j < r₁ + r₂ then L_T[xᵢ + zⱼ] ← s − 1 fi
 9:            if i < r₁ then ptr_o[xᵢ] ← zⱼ fi
10:            if j < r₂ then ptr_e[zⱼ] ← xᵢ fi
11:            i ← i + 1 and j ← j + 1
12:        else if aᵢ ≺ bⱼ then                      // A_o[wᵢ, xᵢ] is s-uncoupled.
13:            if i + j < r₁ + r₂ then L_T[xᵢ + yⱼ − 1] ← s − 1 fi
14:            fin_o[k] ← k + yⱼ − 1 for wᵢ ≤ k ≤ xᵢ
15:            ptr_o[k] ← xⱼ for wᵢ ≤ k < xᵢ
16:            if i < r₁ then ptr_o[xᵢ] ← zⱼ
17:            i ← i + 1
18:        else                              // A_e[yⱼ, zⱼ] is s-uncoupled.
19:            if i + j < r₁ + r₂ then L_T[wᵢ + zⱼ − 1] ← s − 1 fi
20:            fin_e[k] ← k + wᵢ − 1 for yⱼ ≤ k ≤ zⱼ
21:            ptr_e[k] ← zⱼ for yⱼ ≤ k < zⱼ
22:            if j < r₂ then ptr_e[zⱼ] ← xᵢ fi
23:            j ← j + 1
24:        fi
25: od
end
```

For each equivalence class $A_o[w_i, x_i]$, we show that $\mathtt{fin}_o[\alpha]$ and $\mathtt{ptr}_o[\alpha]$ for $w_i \leq \alpha \leq x_i$ store correct values. (Similarly for $A_e[y_j, z_j]$.) We only show that $\mathtt{ptr}_o[x_i]$ stores a correct value when $A_o[w_i, x_i]$ is s-uncoupled (so processed) because setting other values is trivial. From the description of procedure $\mathrm{MERGE}(C_o, C_e)$, $\mathtt{ptr}_o[x_i]$ is z_j for some $1 \leq j \leq r_2$.

Claim. z_j satisfies $|\mathtt{lcp}(S_{A_o[x_i]}, S_{A_e[z_j]})| \geq |\mathtt{lcp}(S_{A_o[x_i]}, S_{A_e[\alpha]})|$ for $1 \leq \alpha \leq n_e$ and $|\mathtt{lcp}(S_{A_o[x_i]}, S_{A_e[z_j]})|$ is stored in $L_T[\mathtt{fin}_o[x_i]]$ if $\mathtt{fin}_o[x_i] < \mathtt{fin}_e[z_j]$ and in $L_T[\mathtt{fin}_e[z_j]]$ otherwise.

Proof of Claim: Since $A_o[w, x]$ and $A_o[y, z]$ is $(s-1)$-coupled and $A_o[w_i, x_i]$ is s-uncoupled, $|\mathtt{lcp}(S_{A_o[x_i]}, S_{A_e[z_j]})| = s - 1$. Since $A_o[w_i, x_i]$ is s-uncoupled, $|\mathtt{lcp}(S_{A_o[x_i]}, S_{A_e[\alpha]})| \leq s - 1$ for $1 \leq \alpha \leq n_e$. Hence, z_j satisfies $|\mathtt{lcp}(S_{A_o[x_i]}, S_{A_e[z_j]})| \geq |\mathtt{lcp}(S_{A_o[x_i]}, S_{A_e[\alpha]})|$ for $1 \leq \alpha \leq n_e$. If $\mathtt{fin}_o[x_i] < \mathtt{fin}_e[z_j]$, $\mathtt{fin}_o[x_i] < x + z$ and thus $L_T[\mathtt{fin}_o[x_i]]$ is set to $s - 1$, which is $|\mathtt{lcp}(S_{A_o[x_i]}, S_{A_e[z_j]})|$. Otherwise, $\mathtt{fin}_e[z_j] < x + z$ and thus $L_T[\mathtt{fin}_e[z_j]]$ is set to $s - 1$.

Case 2: If the end stage of C is smaller than $s - 1$, $A_o[w, x]$ and $A_e[y, z]$ are $(s-1)$-uncoupled. Assume without loss of generality that $\mathrm{P}_{A_o}(w, x) \prec \mathrm{P}_{A_e}(y, z)$. We first find the target entries for $A_o[w, x]$ and $A_e[y, z]$. We set $\mathtt{fin}_o[i] = i + y - 1$ for $w \leq i \leq x$ and $\mathtt{fin}_e[i] = i + x$ for $y \leq i \leq z$. We also set $\mathtt{ptr}_o[i] = x$ for $w \leq i < x$, $\mathtt{ptr}_e[i] = z$ for $y \leq i < z$, and $L_T[x + y - 1] = |\mathtt{lcp}(\mathrm{P}_{A_o}(w, x), \mathrm{P}_{A_e}(y, z))|$. We already set $\mathtt{ptr}_o[x]$ as z and $\mathtt{ptr}_e[z]$ as x and set $L_T[x + z]$ appropriately when we were storing C into $Q[s - 1]$ and the values stored in $\mathtt{ptr}_o[x]$, $\mathtt{ptr}_o[z]$, and $L_T[x + z]$ are still effective.

Consider the time complexity of the merging algorithm. Procedure MERGE (except $\mathtt{fin}$ and $\mathtt{ptr}$) takes time proportional to the total number of partitioned equivalence classes in A_o and A_e. Since there are at most n_o partitioned equivalence classes in A_o and at most n_e classes in A_e, MERGE takes $O(n)$ time. Since each entry of $\mathtt{fin}$ and $\mathtt{ptr}$ is set only once throughout stages, it takes $O(n)$ time overall. The rest of the merging algorithm takes time proportional to the total number of coupled pairs inserted into $Q[k]$. Since a couple pair corresponds to an equivalence class on A_T, the total number of coupled pairs is at most $n - 1$. Therefore, the time complexity of merging is $O(n)$.

3.4   The Coupled-Pair lcp Problem

Recall the coupled-pair lcp problem: Given a coupled pair $C = \langle A_o[w, x], A_e[y, z] \rangle$ whose limit stage is $s - 1$, compute the end stage of C. And if the end stage of C is less than $s - 1$, determine whether $\mathrm{P}_{A_o}(w, x) \prec \mathrm{P}_{A_e}(y, z)$ or $\mathrm{P}_{A_o}(w, x) \succ \mathrm{P}_{A_e}(y, z)$. The problem is easy to solve when s is 1 or 2. When $s = 1$, $|\mathrm{P}_{A_o}(w, x)|$ and $|\mathrm{P}_{A_e}(y, z)|$ are 0 and thus the end stage of C is 0. When $s = 2$, the end stage of C is 1. From now on, we describe how to compute the end stage of C when $s \geq 3$. Assume without loss of generality that the end stage of $A_o[w, x]$ is $s - 1$.

We first show that when $s \geq 3$, the problem of computing the end stage of C (i.e., $|\mathtt{lcp}(\mathrm{P}_{A_o}(w, x), \mathrm{P}_{A_e}(y, z))|$) is reduced to the problem of computing the longest common prefix of two other suffixes.

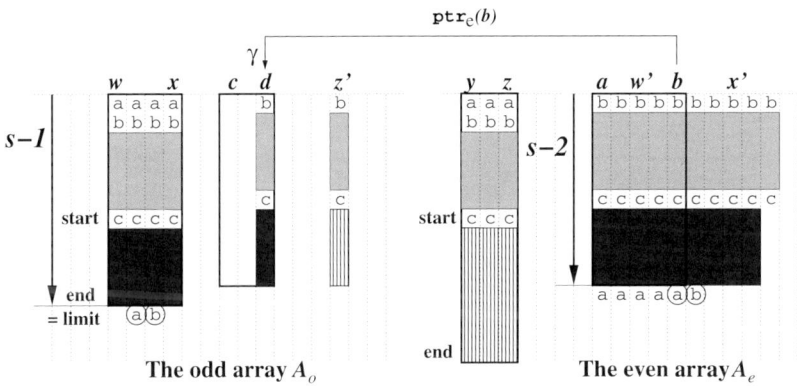

Fig. 2. Finding γ at stage s.

$$
\begin{aligned}
|\text{lcp}(\text{P}_{A_o}(w,x), \text{P}_{A_e}(y,z))| &= |\text{lcp}_{s-1}(\text{P}_{A_o}(w,x), \text{P}_{A_e}(y,z))| \\
&= |\text{lcp}_{s-1}(S_{A_o[w]}, S_{A_e[z]})| \\
&= |\text{lcp}_{s-2}(S_{A_o[w]+1}, S_{A_e[z]+1})| + 1
\end{aligned}
$$

The first equality holds because the end stage of $A_o[w,x]$ is $s-1$. The second equality holds because $\text{pref}_{s-1}(\text{P}_{A_o}(w,x)) = \text{pref}_{s-1}(S_{A_o[w]})$ and $\text{pref}_{s-1}(\text{P}_{A_e}(y,z)) = \text{pref}_{s-1}(S_{A_e[z]})$. The third equality holds because the start stage of the coupled pair is at least 1 which implies that the first characters of $S_{A_o[w]}$ and $S_{A_e[z]}$ are the same. From now on, let $w' = \text{index}_e(A_o[w] + 1)$ and $z' = \text{index}_o(A_e[z] + 1)$ for brevity.

We show how to compute $t = |\text{lcp}_{s-2}(S_{A_e[w']}, S_{A_o[z']})|$ in $O(1)$ time. We first define an index γ of A_o as follows.

Definition 3. *Let γ be an index of array A_o such that $|\text{lcp}_{s-2}(S_{A_e[w']}, S_{A_o[\gamma]})| \geq |\text{lcp}_{s-2}(S_{A_e[w']}, S_{A_o[\delta]})|$ for any other index δ of A_o.*

By definition of γ, t is the minimum of $t_1 = |\text{lcp}_{s-2}(S_{A_e[w']}, S_{A_o[\gamma]})|$ and $t_2 = |\text{lcp}_{s-2}(S_{A_o[\gamma]}, S_{A_o[z']})|$. To compute t, we first find γ and compute t_1. Let $A_e[a,b]$ be the partitioned equivalence class including $A_e[w']$ after stage $s-1$. We will show $\gamma = \text{ptr}_e[b]$. There are two cases whether or not $A_e[a,b]$ constitutes a coupled pair stored in $Q[k]$ just after stage $s-1$.

If $A_e[a,b]$ constitutes a coupled pair stored in $Q[k]$ for $s-1 \leq k < n$, let $A_o[c,d]$ be the equivalence class coupled with $A_e[a,b]$. See Fig. 2.

Lemma 5. *The start stages of $A_e[a,b]$ and $\langle A_o[c,d], A_e[a,b] \rangle$ are both $s-1$.*

We show that γ is $\text{ptr}_e[b] = d$ and t_1 is $s-2$. Since the start stage of C' is $s-1$ and $a \leq w' \leq b$, $|\text{lcp}(S_{A_e[w']}, S_{A_o[d]})| \geq s-1$ and thus $|\text{lcp}_{s-2}(S_{A_e[w']}, S_{A_o[d]})| = s-2$. Since $|\text{lcp}_{s-2}(S_{A_e[w']}, S_{A_o[d]})|$ is at most $s-2$, γ in definition 3 is d and $t_1 = |\text{lcp}_{s-2}(S_{A_e[w']}, S_{A_o[\gamma]})| = s-2$. We have only to show how to find γ $(= d)$

in $O(1)$ time. Since $A_e[w']$ and $A_e[x']$ are in the same equivalence class of E_{s-2} and $A_e[x']$ is not in $A_e[a, b]$ whose start stage is $s - 1$, we can compute b from w' and x' in $O(1)$ time by a $\mathrm{MIN}(L_e, w', x')$ query. Once b is computed, we get d from $\mathtt{ptr}_e[b]$.

If $A_e[a, b]$ is processed after stage $s - 1$, $A_e[a, b]$ is an i-uncoupled equivalence class for some $0 \leq i \leq s - 1$ by the invariant. Since $A_e[a, b]$ is i-uncoupled, $\mathtt{pref}_i(S_{A_e[b]}) = \mathtt{pref}_i(S_{A_e[j]})$ and $\mathtt{pref}_i(S_{A_e[j]}) \neq \mathtt{pref}_i(S_{A_o[k]})$ for $a \leq j \leq b$ and $1 \leq k \leq n_o$ and thus $|\mathtt{lcp}(S_{A_e[w']}, S_{A_o[k]})| = |\mathtt{lcp}(S_{A_e[b]}, S_{A_o[k]})|$ for all $1 \leq k \leq n_o$. Hence, γ in definition 3 is $\mathtt{ptr}_e[b]$ by definition of $\mathtt{ptr}_e$. We can compute γ in $O(1)$ time because $\gamma = \mathtt{ptr}_e[b]$ and $b = \mathtt{ptr}_e[w']$ if $w' \neq b$ by definition of $\mathtt{ptr}_e$. We can also compute $|\mathtt{lcp}_{s-2}(S_{A_e[b]}, S_{A_o[\gamma]})|$ in $O(1)$ time by definition of $\mathtt{ptr}_e$.

Finally, $t_2 = |\mathtt{lcp}_{s-2}(S_{A_o[\gamma]}, S_{A_o[z']})|$ is the minimum of $s-2$ and $|\mathtt{lcp}(S_{A_o[\gamma]}, S_{A_o[z']})|$, where $|\mathtt{lcp}(S_{A_o[\gamma]}, S_{A_o[z']})|$ can be obtained in $O(1)$ time by the query $\mathrm{MIN}(L_o, \gamma, z' - 1)$ or $\mathrm{MIN}(L_o, z', \gamma - 1)$.

References

1. M. Bender and M. Farach-Colton, The LCA Problem Revisited, *In Proceedings of LATIN 2000*, LNCS 1776, 88–94, 2000.
2. O. Berkman and U. Vishkin, Recursive star-tree parallel data structure, *SIAM J. Comput.* 22 (1993), 221–242.
3. A. Blumer, J. Blumer, D. Haussler, A. Ehrenfeucht, M. T. Chen and J. Seiferas, The smallest automaton recognizing the subwords of a text, *Theoret. Comput. Sci.* 40 (1985), 31–55.
4. S. Burkhardt and J. Kärkkäinen, Fast lightweight suffix array construction and checking, *Accepted to Symp. Combinatorial Pattern Matching* (2003).
5. M. Crochemore, An optimal algorithm for computing the repetitions in a word, *Inform. Processing Letters* 12 (1981), 244–250.
6. M. Farach, Optimal suffix tree construction with large alphabets, *IEEE Symp. Found. Computer Science* (1997), 137–143.
7. M. Farach-Colton, P. Ferragina and S. Muthukrishnan, On the sorting-complexity of suffix tree construction, *J. Assoc. Comput. Mach.* 47 (2000), 987–1011.
8. M. Farach and S. Muthukrishnan, Optimal logarithmic time randomized suffix tree construction, *Int. Colloq. Automata Languages and Programming* (1996), 550-561.
9. P. Ferragina and G. Manzini, Opportunistic data structures with applications, *IEEE Symp. Found. Computer Science* (2001), 390–398.
10. H.N. Gabow, J.L. Bentley, and R.E. Tarjan, Scaling and Related Techniques for Geometry Problems, *ACM Symp. Theory of Computing* (1984), 135–143.
11. G. Gonnet, R. Baeza-Yates, and T. Snider, New indices for text: Pat trees and pat arrays. In W. B. Frakes and R. A. Baeza-Yates, editors, Information Retrieval: Data Structures & Algorithms, *Prentice Hall* (1992), 66–82.
12. D. Gusfield, An "Increment-by-one" approach to suffix arrays and trees, *manuscript* 1990.
13. R. Grossi and J.S. Vitter, Compressed suffix arrays and suffix trees with applications to text indexing and string matching, *ACM Symp. Theory of Computing* (2000), 397–406.

14. D. Harel and R.E. Tarjan. Fast algorithms for finding nearest common ancestors, *SIAM J. Comput.* 13 (1984), 338–355.
15. R. Hariharan, Optimal parallel suffix tree construction, *J. Comput. Syst. Sci.* 55 (1997), 44–69.
16. J. Kärkkäinen and P. Sanders, Simpler linear work suffix array construction, *Accepted to Int. Colloq. Automata Languages and Programming* (2003).
17. P. Ko and S. Aluru, Space-efficient linear time construction of suffix arrays, *Accepted to Symp. Combinatorial Pattern Matching* (2003).
18. U. Manber and G. Myers, Suffix arrays: A new method for on-line string searches, *SIAM J. Comput.* 22 (1993), 935–938.
19. E.M. McCreight, A space-economical suffix tree construction algorithm, *J. Assoc. Comput. Mach.* 23 (1976), 262–272.
20. J. I. Munro, V. Raman and S. Srinivasa Rao Space Efficient Suffix Trees, FST & TCS 18, in Lecture Notes in Computer Science, (Springer-Verlag), Dec. 1998.
21. K. Sadakane, Succinct representation of lcp information and improvement in the compressed suffix arrays, *ACM-SIAM Symp. on Discrete Algorithms* (2002), 225–232.
22. S.C. Sahinalp and U. Vishkin, Symmetry breaking for suffix tree construction, *IEEE Symp. Found. Computer Science* (1994), 300–309.
23. B. Schieber and U. Vishkin, On finding lowest common ancestors: simplification and parallelization, *SIAM J. Comput.* 17, (1988), 1253–1262.
24. E. Ukkonen, On-line construction of suffix trees, *Algorithmica* 14 (1995), 249–260.
25. J. Vuillemin, A unifying look at data structures, *Comm. ACM* Vol. 24, (1980), 229–239.
26. P. Weiner, Linear pattern matching algorithms, *Proc. 14th IEEE Symp. Switching and Automata Theory* (1973), 1–11.

Appendix: Range Minima Problem

We define the *range-minima problem* as follows:

> Given an array $A = (a_1, a_2, \ldots, a_n)$ of integers $0 \leq a_i \leq n - 1$, preprocess A so that any query $\text{MIN}(A, i, j)$, $1 \leq i < j \leq n$, requesting the index of the leftmost minimum element in $(a_i, \ldots, a_j)$, can be answered in constant time.

We first describe two preprocessing algorithms for the range-minima problem: algorithm E takes exponential time and algorithm L takes $O(n \log n)$ time. Then, we present a linear-time preprocessing algorithm using both algorithms E and L. Finally, we describe how to answer a range-minimum query in constant time. Our algorithm is a modification of Berkman and Vishkin's solution for the range minima problem [2].

Algorithm E: Since the elements of A are integers in the range $[0, n-1]$, the number of possible input arrays of size n is n^n. If we regard an array of size n as a string $S \in \Sigma^n$ over an integer alphabet $\Sigma = \{0, 1, \ldots, n-1\}$, it is mapped to an integer k ($1 \leq k \leq n^n$) such that S is lexicographically the kth string among the n^n possible strings. We make a table $T_n(k, i, j)$ that stores the answer to query $\text{MIN}(A, i, j)$, where A is mapped to k. The size of table T_n is $O(n^{n+2})$ and it takes $O(n^{n+2})$ time to make T_n.

Algorithm L: We now describe an $O(n \log n)$-time algorithm. We define the prefix and suffix minima as follows. The *prefix minima* of A are $(c_1, c_2, \ldots, c_n)$ such that $c_i = \min\{a_1, \ldots, a_i\}$ for $1 \leq i \leq n$. Simiarly, the *suffix minima* of A are $(d_1, d_2, \ldots, d_n)$ such that $d_j = \min\{a_j, \ldots, a_n\}$ for $1 \leq j \leq n$. The prefix minima and suffix minima of A can be computed in linear time. The preprocessing of algorithm L constructs a complete binary tree T whose leaves are the elements of the input array A. Let A_u be the list of the leaves of the subtree rooted at node u. Each internal node u of T maintains the prefix minima and suffix minima of A_u. It takes $O(n \log n)$ time to construct T. Since T is a complete binary tree, it can be easily implemented by arrays.

Suppose that we are now given a range-minima query $\text{MIN}(A, i, j)$. Find the lowest common ancestor u of two leaves a_i and a_j in T. Let v and w be the left and right children of u, respectively. Then, $[i, j]$ is the union of a suffix of A_v and a prefix of A_w. The answer to the query is the minimum of the following two elements: the minimum of the suffix of A_v and the minimum of the prefix of A_w. These operations take constant time using T.

We now describe a linear-time preprocessing algorithm for the range-minima problem.

- Let $m = \log \log n$. Partition the input array A into n/m blocks A_i of size m. We map each block A_i into an array B_i whose elements are the rankings in the sorted list of A_i (i.e., the elements of B_i are integers in the range $[0, m-1]$). We can sort n/m blocks A_i at the same time using n buckets in $O(n)$ time. Apply algorithm E to all possible arrays of size m. Since $m^{m+2} = O(n)$, we can make table T_m in $O(n)$ time using $O(n)$ space.
- Partition A into $n/\log n$ blocks A_i of size $\log n$ and find the minimum in each block. Apply algorithm L to an array of these $n/\log n$ minima. Also, we do the following for each block A_i. Partition A_i into subblocks of size $\log \log n$, and find the minimum in each subblock. Apply algorithm L to these $\log n/\log \log n$ minima. The total time and space are $O(n)$.

When a query $\text{MIN}(A, i, j)$ is given, the range $[i, j]$ can be divided into at most five subranges, and the minimum in each subrange can be found in constant time by the preprocessing above. The answer to the query is the minimum of these five minima [2].

Space Efficient Linear Time Construction of Suffix Arrays*

Pang Ko and Srinivas Aluru**

[1] Laurence H. Baker Center for Bioinformatics and Biological Statistics
[2] Iowa State University
Ames, IA 50011
{kopang, aluru}@iastate.edu

Abstract. We present a linear time algorithm to sort all the suffixes of a string over a large alphabet of integers. The sorted order of suffixes of a string is also called suffix array, a data structure introduced by Manber and Myers that has numerous applications in pattern matching, string processing, and computational biology. Though the suffix tree of a string can be constructed in linear time and the sorted order of suffixes derived from it, a direct algorithm for suffix sorting is of great interest due to the space requirements of suffix trees. Our result improves upon the best known direct algorithm for suffix sorting, which takes $O(n \log n)$ time. We also show how to construct suffix trees in linear time from our suffix sorting result. Apart from being simple and applicable for alphabets not necessarily of fixed size, this method of constructing suffix trees is more space efficient.

1 Introduction

Suffix trees and suffix arrays are important fundamental data structures useful in many applications in string processing and computational biology. The suffix tree of a string is a compacted trie of all the suffixes of the string. The suffix tree of a string of length n over an alphabet Σ can be constructed in $O(n \log |\Sigma|)$ time and $O(n)$ space, or in $O(n)$ time and $O(n|\Sigma|)$ space [McC76, Ukk95,Wei73]. These algorithms are suitable for small, fixed size alphabets. Subsequently, Farach [FM96] presented an $O(n)$ time and space algorithm for the more general case of constructing suffix trees over integer alphabets. For numerous applications of suffix trees in string processing and computational biology, see [Gus97].

The suffix array of a string is the lexicographically sorted list of all its suffixes. In 1993, Manber and Myers introduced the suffix array data structure [MM93] as a space-efficient substitute for suffix trees. As a lexicographic-order traversal of a suffix tree can be used to produce the sorted list of suffixes, suffix arrays can be constructed in linear time and space using suffix trees. However, this defeats

---

* Research supported by IBM Faulty Award and NSF under ACI-0203782.
** Dept. of Electrical and Computer Engineering

R. Baeza-Yates et al. (Eds.): CPM 2003, LNCS 2676, pp. 200–210, 2003.

the whole purpose if the goal is to avoid suffix trees. Hence, Manber and Myers presented direct construction algorithms that run in $O(n \log n)$ worst-case time and $O(n)$ expected time, respectively. Since then, the study of algorithms for constructing suffix arrays and for using suffix arrays in computational biology applications has attracted considerable attention.

The suffix array is often used in conjunction with another array, called *lcp* array, containing the lengths of the longest common prefixes between every pair of consecutive suffixes in sorted order. Manber and Myers also presented algorithms for constructing *lcp* array in $O(n \log n)$ worst-case time and $O(n)$ expected time, respectively [MM93]. More recently, Kasai *et al.* [KLA+01] presented a linear time algorithm for constructing the *lcp* array directly from the suffix array. While the classic problem of finding a pattern P in a string T of length n can be solved in $O(|P|)$ time for fixed size Σ using a suffix tree of T, Manber and Myers' pattern matching algorithm takes $O(|P| + \log n)$ time, without any restriction on Σ. Recently, Abouelhoda *et al.* [AOK02] have improved this to $O(|P|)$ time using additional linear time preprocessing, thus making the suffix array based algorithm superior. In fact, many problems involving top-down or bottom-up traversal of suffix trees can now be solved with the same asymptotic run-time bounds using suffix arrays [AKO02,AOK02]. Such problems include many queries used in computational biology applications including finding exact matches, maximal repeats, tandem repeats, maximal unique matches and finding all shortest unique substrings. For example, the whole genome alignment tool MUMmer [DKF+99] uses the computation of maximal unique matches.

While considerable advances are made in designing optimal algorithms for queries using suffix arrays and for computing auxiliary information that is required along with suffix arrays, the complexity of direct construction algorithms for suffix arrays remained $O(n \log n)$ so far. Several alternative algorithms for suffix array construction have been developed, each improving the previous best algorithm by an additional constant factor [IT99,LS99]. We close this gap by presenting a direct linear time algorithm for constructing suffix arrays over integer alphabets. Contemporaneous to our result, Kärkkänen *et al.* [KS03] and Kim *et al.* [KSPP03] also discovered suffix array construction algorithms with linear time complexity.

It is well known that the suffix tree of a string can be constructed from the sorted order of its suffixes and the *lcp* array [FM96]. Because the *lcp* array can be inferred from the suffix array in linear time [KLA+01], our algorithm can also be used to construct suffix trees in linear time for large integer alphabets, and of course, for the special case of fixed size alphabets. Our algorithm is simpler and more space efficient than Farach's linear time algorithm for constructing suffix trees for integer alphabets. In fact, it is simpler than linear time suffix tree construction algorithms for fixed size alphabets [McC76,Ukk95,Wei73]. A noteworthy feature of our algorithm is that it does not construct or use suffix links, resulting in additional space advantage. To the best of our knowledge, this is the first suffix tree construction algorithm that achieves linear run-time without exploiting the use of suffix links.

T	M	I	S	S	I	S	S	I	P	P	I	$
Type	L	S	L	L	S	L	L	S	L	L	L	L/S
Pos	1	2	3	4	5	6	7	8	9	10	11	12

Fig. 1. The string "MISSISSIPPI$" and the types of its suffixes.

The remainder of this paper is organized as follows: In Section 2, we present our linear time suffix sorting algorithm. An implementation strategy that further improves the run-time in practice is presented in Section 3. Section 4 concludes the paper.

2   Suffix Sorting Algorithm

Consider a string $T = t_1 t_2 \ldots t_n$ over the alphabet $\Sigma = \{1 \ldots n\}$. Without loss of generality, assume the last character of T occurs nowhere else in T, and is the lexicographically smallest character. We denote this character by '$'. Let $T_i = t_i t_{i+1} \ldots t_n$ denote the suffix of T starting with t_i. To store the suffix T_i, we only store the starting position number i. For strings α and β, we use $\alpha \prec \beta$ to denote that α is lexicographically smaller than β. Throughout this paper the term *sorted order* refers to lexicographically ascending order.

We classify the suffixes into two types: Suffix T_i is of type S if $T_i \prec T_{i+1}$, and is of type L if $T_{i+1} \prec T_i$. The last suffix T_n does not have a next suffix, and is classified as both type S and type L.

Lemma 1. *All suffixes of T can be classified as either type S or type L in $O(n)$ time.*

Proof. Consider a suffix T_i $(i < n)$.

  Case 1: If $t_i \neq t_{i+1}$, we only need to compare t_i and t_{i+1} to determine if T_i is of type S or type L.
  Case 2: If $t_i = t_{i+1}$, find the smallest $j > i$ such that $t_j \neq t_i$.
    if $t_j > t_i$, then suffixes $T_i, T_{i+1}, \ldots, T_{j-1}$ are of type S.
    if $t_j < t_i$, then suffixes $T_i, T_{i+1}, \ldots, T_{j-1}$ are of type L.

Thus, all suffixes can be classified using a left to right scan of T in $O(n)$ time.
□

The type of each suffix of the string MISSISSIPPI$ is shown in Figure 1. An important property of type S and type L suffixes is, if a type S suffix and a type L suffix both begin with the same character, the type S suffix is always lexicographically greater than the type L suffix. The formal proof is presented below.

Lemma 2. *A type S suffix is lexicographically greater than a type L suffix that begins with the same first character.*

Proof. We prove this by contradiction. Suppose a type S suffix T_i and a type L suffix T_j be two suffixes that start with the same character c, such that $T_i \prec T_j$. We can write $T_i = c\alpha c_1 \beta$ and $T_j = c\alpha c_2 \gamma$, where $c_1 \neq c_2$ and α, β, and γ are (possibly empty) strings.

   Case 1: α contains a character other than c. Let c_3 be the leftmost character in α that is different from c. Because T_i is a type S suffix, it follows that $c_3 > c$. Similarly, for T_j to be a type L suffix, $c_3 < c$, a contradiction.

   Case 2: α does not contain any character other than c. In this case, we have the following:

$$T_i \text{ of type } S \Rightarrow c_1 \geq c$$
$$T_j \text{ of type } L \Rightarrow c_2 \leq c$$
$$c_2 \leq c \text{ and } c \leq c_1 \Rightarrow c_2 \leq c_1$$

But $T_i \prec T_j \Rightarrow c_1 < c_2$, a contradiction. $\square$

Corollary 1. *In the suffix array of T, among all suffixes that start with the same character, the type S suffixes appear after the type L suffixes.*

Proof. Follows directly from Lemma 2.

   Let A be an array containing all suffixes of T, not necessarily in sorted order. Create an array Rev such that $R[i] = k$ if $A[k] = i$, i.e., $R[i]$ indicates the position where suffix T_i is stored in A. We need to keep Rev up-to-date, thus any change made to A is also reflected in Rev. Let B be an array of all suffixes of type S, sorted in lexicographic order. Using B, we can compute the lexicographically sorted order of all suffixes of T as follows:

1. Bucket all suffixes of T according to their first character in array A. Each bucket consists of all suffixes that start with the same character. This step takes $O(n)$ time.
2. Scan B from right to left. For each suffix encountered in the scan, move the suffix to the current end of its bucket in A, and advance the current end by one position to the left. More specifically, the move of a suffix in array A to a new position should be taken as swapping the suffix with the suffix currently occupying that position. After the scan of B is completed, by Corollary 1, all type S suffixes are in their correct positions in A. The time taken is $O(|B|)$, which is bounded by $O(n)$.
3. Scan A from left to right. For each entry $A[i]$, if $T_{A[i]-1}$ is a type L suffix, move it to the current front of its bucket in A, and advance the front of the bucket by one. This takes $O(n)$ time. At the end of this step, A contains all suffixes of T in sorted order.

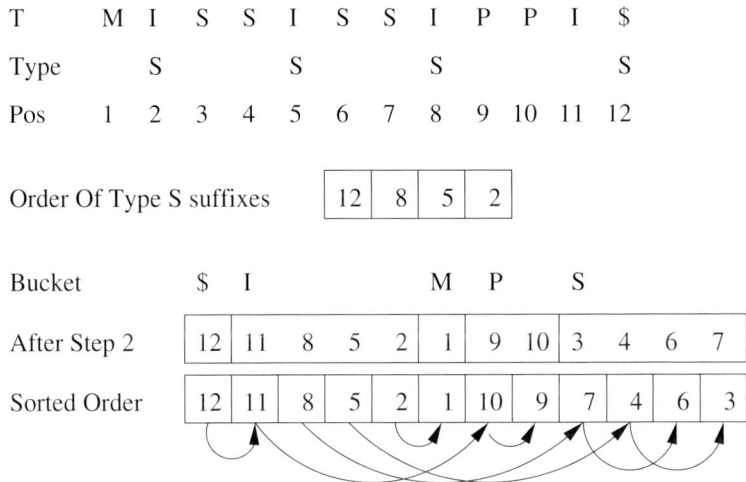

Fig. 2. Illustration of how to obtain the sorted order of all suffixes, from the sorted order of type S suffixes of the string MISSISSIPPI$.

In Figure 2, the suffix pointed by the arrow is moved to the current front of its bucket when the scan reaches the suffix at the origin of the arrow. The following lemma proves the correctness of the procedure in Step 3.

Lemma 3. *In step 3, when the scan reaches $A[i]$, then suffix $T_{A[i]}$ is already in its sorted position in A.*

Proof. By induction on i. To begin with, the smallest suffix in T must be of type S and hence in its correct position $A[1]$. By inductive hypothesis, assume that $A[1], A[2], \ldots, A[i]$ are the first i suffixes in sorted order. We now show that when the scan reaches $A[i+1]$, then the suffix in it, i.e., $T_{A[i+1]}$ is already in its sorted position. Suppose not. Then there exists a suffix referenced by $A[k]$ $(k > i+1)$ that should be in $A[i+1]$ in sorted order, i.e., $T_{A[k]} \prec T_{A[i+1]}$. As all type S suffixes are already in correct positions, both $T_{A[k]}$ and $T_{A[i+1]}$ must be of type L. Because A is bucketed by the first character of the suffixes prior to step 3, and a suffix is never moved out of its bucket, $T_{A[k]}$ and $T_{A[i+1]}$ must begin with the same character, say c. Let $T_{A[i+1]} = c\alpha$ and $T_{A[k]} = c\beta$. Since $T_{A[k]}$ is type L, $\beta \prec T_{A[k]}$. From $T_{A[k]} \prec T_{A[i+1]}$, $\beta \prec \alpha$. Since $\beta \prec T_{A[k]}$, and the correct sorted position of $T_{A[k]}$ is $A[i+1]$, β must occur in $A[1] \ldots A[i]$. Because $\beta \prec \alpha$, $T_{A[k]}$ should have been moved to the current front of its bucket before $T_{A[i+1]}$. Thus, $T_{A[k]}$ can not occur to the right of $T_{A[i+1]}$, a contradiction. □

So far, we showed that if all type S suffixes are sorted, then the sorted position of all suffixes of T can be determined in $O(n)$ time. In a similar manner, the sorted position of all suffixes of T can also be determined from the sorted order of all suffixes of type L. To do this, we bucket all suffixes of T based on their first characters into an array A. We then scan the sorted order of type L suffixes from left to right and determine their correct positions in A by moving them to the current front of their respective buckets. We then scan A from right to left and when $A[i]$ is encountered, if $T_{A[i]-1}$ is of type S, it will be moved to the current end of its bucket.

Once the suffixes of T are classified into type S and type L, we choose to sort those type of suffixes which are fewer in number. Without loss of generality, assume that type S suffixes are fewer. We now show how to recursively sort these suffixes.

Define position i of T to be a type S position if the suffix T_i is of type S, and similarly to be a type L position if the suffix T_i is of type L. The substring $t_i \ldots t_j$ is called a type S substring if both i and j are type S positions, and every position between i and j is a type L position.

Our goal is to sort all the type S suffixes in T. To do this we first sort all the type S substrings. The sorting generates buckets where all the substrings in a bucket are identical. The buckets are numbered using consecutive integers starting from 1. We then generate a new string T' as follows: Scan T from left to right and for each type S position in T, write the bucket number of the type S substring starting from that position. This string of bucket numbers forms T'. Observe that each type S suffix in T naturally corresponds to a suffix in the new string T'. In Lemma 4, we prove that sorting all type S suffixes of T is equivalent to sorting all suffixes of T'. We sort T' recursively.

We first show how to sort all the type S substrings in $O(n)$ time. Consider the array A, consisting of all suffixes of T bucketed according to their first characters. For each suffix T_i, define its $S\text{-}distance$ to be the distance from its starting position i to the nearest type S position to its left (excluding position i). If no type S position exists to the left, the $S\text{-}distance$ is defined to be 0. Thus, for each suffix starting on or before the first type S position in T, its $S\text{-}distance$ is 0. The type S substrings are sorted as follows (illustrated in Figure 3):

1. For each suffix in A, determine its $S\text{-}distance$. This is done by scanning T from left to right, keeping track of the distance from the current position to the nearest type S position to the left. While at position i, the $S\text{-}distance$ of T_i is known and this distance is recorded in array $Dist$. The $S\text{-}distance$ of T_i is stored in $Dist[i]$. Hence, the $S\text{-}distances$ for all suffixes can be recorded in linear time.

2. Let m be the largest $S\text{-}distance$. Create m lists such that list j $(1 \leq j \leq m)$ contains all the suffixes with an $S\text{-}distance$ of j, listed in the order in which they appear in array A. This can be done by scanning A from left to right in linear time, referring to $Dist[A[i]]$ to put $T_{A[i]}$ in the correct list.

3. We now sort the type S substrings using the lists created above. The sorting is done by repeated bucketing using one character at a time. To begin with, the bucketing based on first character is determined by the order in which type S suffixes appear in array A. Suppose the type S substrings are bucketed according to their first $j - 1$ characters. To extend this to j characters, we scan list j. For each suffix T_i encountered, move the type S substring starting

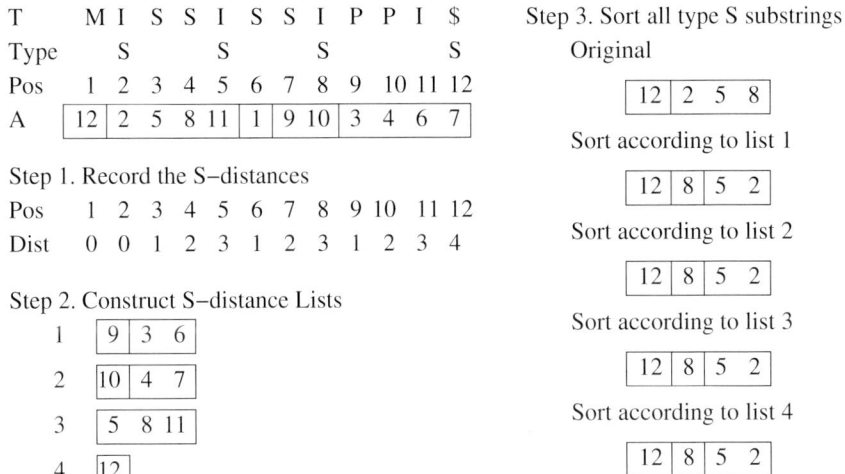

Fig. 3. Illustration of the sorting of type S substrings of the string MISSISSIPPI$.

at t_{i-j} to the current front of its bucket. Because the total size of all the lists is $O(n)$, the sorting of type S substrings only takes $O(n)$ time.

The sorting of type S substrings using the above algorithm respects lexicographic ordering of type S substrings, with the following important exception: If a type S substring is the prefix of another type S substring, the bucket number assigned to the shorter substring will be larger than the bucket number assigned to the larger substring. This anomaly is designed on purpose, and is exploited later in Lemma 4.

As mentioned before, we now construct a new string T' corresponding to all type S substrings in T. Each type S substring is replaced by its bucket number and T' is the sequence of bucket numbers in the order in which the type S substrings appear in T. Because every type S suffix in T starts with a type S substring, there is a natural one-to-one correspondence between type S suffixes of T and all suffixes of T'. Let T_i be a suffix of T and $T'_{i'}$ be its corresponding suffix in T'. Note that $T'_{i'}$ can be obtained from T_i by replacing every type S substring in T_i with its corresponding bucket number. Similarly, T_i can be obtained from $T'_{i'}$ by replacing each bucket number with the corresponding substring and removing the duplicate instance of the common character shared by two consecutive type S substrings. This is because the last character of a type S substring is also the first character of the next type S substring along T.

Lemma 4. Let T_i and T_j be two suffixes of T and let $T'_{i'}$ and $T'_{j'}$ be the corresponding suffixes of T'. Then, $T_i \prec T_j \Leftrightarrow T'_{i'} \prec T'_{j'}$.

Proof. We first show that $T'_{i'} \prec T'_{j'} \Rightarrow T_i \prec T_j$. The prefixes of T_i and T_j corresponding to the longest common prefix of $T'_{i'}$ and $T'_{j'}$ must be identical. This is because if two bucket numbers are the same, then the corresponding substrings must be the same. Consider the leftmost position in which $T'_{i'}$ and $T'_{j'}$ differ. Such a position exists and the characters (bucket numbers) of $T'_{i'}$ and $T'_{j'}$ in that position determine which of $T'_{i'}$ and $T'_{j'}$ is lexicographically smaller. Let k be the bucket number in $T'_{i'}$ and l be the bucket number in $T'_{j'}$ at that position. Since $T'_{i'} \prec T'_{j'}$, it is clear that $k < l$. Let α be the substring corresponding to k and β be the substring corresponding to l. Note that α and β can be of different lengths, but α cannot be a proper prefix of β. This is because the bucket number corresponding to the prefix must be larger, but we know that $k < l$.

Case 1: β is not a prefix of α. In this case, $k < l \Rightarrow \alpha \prec \beta$, which implies $T_i \prec T_j$.

Case 2: β is a proper prefix of α. Let the last character of β be c. The corresponding position in T is a type S position. The position of the corresponding c in α must be a type L position.

Since the two suffixes that begin at these positions start with the same character, by Corollary 1, the type L suffix must be lexicographically smaller then the type S suffix. Thus, $T_i \prec T_j$.

From the one-to-one correspondence between the suffixes of T' and the type S suffixes of T, it also follows that $T_i \prec T_j \Rightarrow T'_{i'} \prec T'_{j'}$.   □

Corollary 2. *The sorted order of the suffixes of T' determines the sorted order of the type S suffixes of T.*

Proof. Let $T'_{i'_1}, T'_{i'_2}, T'_{i'_3}, \ldots$ be the sorted order of suffixes of T'. Let $T_{i_1}, T_{i_2}, T_{i_3}, \ldots$ be the sequence obtained by replacing each suffix $T'_{i'_k}$ with the corresponding type S suffix T_{i_k}. Then, $T_{i_1}, T_{i_2}, T_{i_3}, \ldots$ is the sorted order of type S suffixes of T. The proof follows directly from Lemma 4.   □

Hence, the problem of sorting the type S suffixes of T reduces to the problem of sorting all suffixes of T'. Note that the characters of T' are consecutive integers starting from 1. Hence our suffix sorting algorithm can be recursively applied to T'.

If the string T has fewer type L suffixes than type S suffixes, the type L suffixes are sorted using a similar procedure – Call the substring $t_i, \ldots, t_j$ a type L substring if both i and j are type L positions, and every position between i and j is a type S position. Now sort all the type L substrings and construct the corresponding string T' obtained by replacing each type L substring with its bucket number. Sorting T' gives the sorted order of type L suffixes.

Thus, the problem of sorting the suffixes of a string T of length n can be reduced to the problem of sorting the suffixes of a string T' of size at most $\lceil \frac{n}{2} \rceil$, and $O(n)$ additional work. This leads to the recurrence

$$T(n) = T\left(\left\lceil \frac{n}{2} \right\rceil\right) + O(n)$$

Theorem 1. *The suffixes of a string of length n can be lexicographically sorted in $O(n)$ time and space.*

We now consider the space required for the execution of our suffix array construction algorithm. By applying several implementation strategies, some of which are similar to those presented by Manber and Myers [MM93], it is possible to derive an implementation of our algorithm that uses only 3 integer arrays of size n and 3 boolean arrays (2 of size n and one of size $\lceil \frac{n}{2} \rceil$). Assuming each integer representation takes 4 bytes of space, the space requirement of our algorithm is $12n$ bytes plus $2.5n$ bits. This compares favorably with the best space-efficient implementations of linear time suffix tree construction algorithms, which still require $20n$ bytes [AOK02]. Hence, direct linear time construction of suffix arrays using our algorithm is more space-efficient.

In case the alphabet size is constant, it is possible to further reduce the space requirement. Note that the maximum space utilization by our algorithm occurs in the first iteration. As the size of the string reduces at least by half in each iteration, so does the space required by the algorithm. We take advantage of this fact by designing a more space-efficient implementation for the first iteration, which is applicable only for constant sized alphabets. The main underlying idea is to eliminate the construction of the lists used in sorting type S substrings. This reduces the space required to only $8n$ bytes plus $0.5n$ bits for the first iteration. Note that this idea cannot be used in subsequent iterations because the string T' to be worked on in the second iteration will still be based on integer alphabet. So we resort to the traditional implementation for this and all subsequent iterations. As a result, the space requirement for the complete execution of the algorithm can be reduced to $8n$ bytes plus $1.25n$ bits. This is competitive with Manber and Myers' $O(n \log n)$ time algorithm for suffix array construction [MM93], which requires only $8n$ bytes. In many practical applications, the size of the alphabet is a small constant. For instance, computational biology applications deal with DNA and protein sequences, which have alphabet sizes of 4 and 20, respectively.

3   Reducing the Size of T'

In this section, we present an implementation strategy to further reduce the size of T'. Consider the result of sorting all type S substrings of T. Note that a type S substring is a prefix of the corresponding type S suffix. Thus, sorting type S substrings is equivalent to bucketing type S suffixes based on their respective type S substring prefixes. The bucketing conforms to the lexicographic ordering of type S suffixes. The purpose of forming T' and sorting its suffixes is to determine the sorted order of type S suffixes that fall into the same bucket. If a bucket contains only one type S substring, the position of the corresponding type S suffix in the sorted order is already known.

Let $T' = b_1 b_2 \ldots b_m$. Consider a maximal substring $b_i \ldots b_j$ $(j < m)$ such that each b_k $(i \leq k \leq j)$ contains only one type S substring. We can shorten T' by replacing each such maximal substring $b_i \ldots b_j$ with its first character b_i.

Since $j < m$ the bucket number corresponding to '$' is never dropped, and this is needed for subsequent iterations. It is easy to directly compute the shortened version of T', instead of first computing T' and then shortening it. Shortening T' will have the effect of eliminating some of the suffixes of T', and also modifying each suffix that contains a substring that is shortened. We already noted that the final positions of the eliminated suffixes are already known. It remains to be shown that the sorted order of other suffixes is not affected by the shortening.

Consider any two suffixes $T'_k = b_k \ldots b_m$ and $T'_l = b_l \ldots b_m$, such that at least one of the suffixes contains a substring that is shortened. Let $j \geq 0$ be the smallest integer such that either b_{k+j} or b_{l+j} (or both) is the beginning of a shortened substring. The first character of a shortened substring corresponds to a bucket containing only one type S substring. Hence, the bucket number occurs nowhere else in T'. Therefore $b_{k+j} \neq b_{l+j}$, and the sorted order of $b_k \ldots b_m$ and $b_l \ldots b_m$ is determined by the sorted order of $b_k \ldots b_{k+j}$ and $b_l \ldots b_{l+j}$. In other words, the comparison of any two suffixes never extends beyond the first character of a shortened substring.

4   Conclusions

In this paper we present a linear time algorithm for sorting the suffixes of a string over integer alphabet, or equivalently, for constructing the suffix array of the string. Our algorithm can also be used to construct suffix trees in linear time. Apart from being the first direct algorithm for constructing suffix arrays in linear time, the simplicity and space advantages of our algorithm are likely to make it useful in suffix tree construction as well.

References

[AKO02]   M. I. Abouelhoda, S. Kurtz, and E. Ohlebusch. The enhanced suffix array and its applications to genome analysis. In *2nd Workshop on Algorithms in Bioinformatics*, pages 449–63, 2002.

[AOK02]   M. I. Abouelhoda, E. Ohlebusch, and S. Kurtz. Optimal exact string matching based on suffix arrays. In *International Symposium on String Processing and Information Retrieval*, pages 31–43. IEEE, 2002.

[DKF+99]   A. L. Delcher, S. Kasif, R. D. Fleischmann, J. Peterson, O. White, and S. L. Salzberg. Alignment of whole genomes. *Nucleic Acids Research*, 27:2369–76, 1999.

[FM96]   M. Farach and S. Muthukrishnan. Optimal logarithmic time randomized suffix tree construction. In *Proc. of 23rd International Colloquium on Automata Languages and Programming*, 1996.

[Gus97]   D. Gusfield. *Algorithms on Strings Trees and Sequences*. Cambridge University Press, New York, New York, 1997.

[IT99]   H. Itoh and H. Tanaka. An efficient method for in memory construction of suffix array. In *International Symposium on String Processing and Information Retrieval*, pages 81–88. IEEE, 1999.

[KLA⁺01]   T. Kasai, G. Lee, H. Arimura, S. Arikawa, and K. Park. Linear-time longest-common-prefix computation in suffix arrays and its applications. In *12th Annual Symposium, Combinatorial Pattern Matching*, pages 181–92, 2001.

[KS03]   J. Kärkkänen and P. Sanders. Simpler linear work suffix array construction. In *International Colloquium on Automata, Languages and Programming*, page to appear, 2003.

[KSPP03]   D. K. Kim, J. S. Sim, H. Park, and K. Park. Linear-time construction of suffix arrays. In *14th Annual Symposium, Combinatorial Pattern Matching*, 2003.

[LS99]   N. J. Larsson and K. Sadakane. Faster suffix sorting. Technical Report LU-CS-TR:99-214, LUNDFD6/(NFCS-3140)/1–20/(1999), Department of Computer Science, Lund University, Sweden, May 1999.

[McC76]   E. M. McCreight. A space-economical suffix tree construction algorithm. *Journal of the ACM*, 23:262–72, 1976.

[MM93]   U. Manber and G. Myers. Suffix arrays: a new method for on-line search. *SIAM Journal on Computing*, 22:935–48, 1993.

[Ukk95]   E. Ukkonen. On-line construction of suffix-trees. *Algorithmica*, 14:249–60, 1995.

[Wei73]   P. Weiner. Linear pattern matching algorithms. In *14th Symposium on Switching and Automata Theory*, pages 1–11. IEEE, 1973.

Tuning String Matching for Huge Pattern Sets

Jari Kytöjoki, Leena Salmela, and Jorma Tarhio[*]

Department of Computer Science and Engineering
Helsinki University of Technology
P.O. Box 5400, FIN-02015 HUT, Finland

Abstract. We present three algorithms for exact string matching of multiple patterns. Our algorithms are filtering methods, which apply q-grams and bit parallelism. We ran extensive experiments with them and compared them with various versions of earlier algorithms, e.g. different trie implementations of the Aho-Corasick algorithm. Our algorithms showed to be substantially faster than earlier solutions for sets of 1,000–100,000 patterns. The gain is due to the improved filtering efficiency caused by q-grams.

1 Introduction

We consider exact string matching of multiple patterns. Many good solutions have been presented for this problem, e.g. Aho-Corasick [1], Commentz-Walter [5,14], and Rabin-Karp [11,12] with their variations. However, most of the earlier algorithms have been designed for pattern sets of moderate size, i.e. a few dozens, and they do not unfortunately scale very well to larger pattern sets. In this paper we concentrate on practical methods that can efficiently handle several thousand patterns even in a small main memory (e.g. in a handheld device). Such algorithms are needed in intrusion detection [8], in content scanning, and in specific data mining problems [9]. The focus is on finding the occurrences of rare patterns or on checking that unwanted patterns do not occur at all.

The text $T = t_1 t_2 \cdots t_n$ is a string of n characters in an alphabet of size c. There are r patterns $P_1, \ldots, P_r$ of length m in the same alphabet. If the lengths of the patterns are not equal, we select a substring from each pattern according to the length of the shortest pattern. We consider cases where m varies between 4 and 32 and r between 100 and 100,000 mostly for $c=256$. All exact occurrences of the patterns should be reported.

As our main contribution we will present three algorithms HG, SOG, and BG based on the Boyer-Moore-Horspool [10], shift-or [3], and BNDM [13] algorithms, respectively. Our algorithms are filtering algorithms, which operate in three phases. The patterns are first preprocessed. The second phase reports candidates for matches, which are verified in the third phase. A common feature of our algorithms is matching of q-grams instead of single characters. We search for

---

[*] Corresponding author: `jorma.tarhio@hut.fi`.

R. Baeza-Yates et al. (Eds.): CPM 2003, LNCS 2676, pp. 211–224, 2003.

occurrences of a single generalized pattern of q-grams such that the pattern includes all the original patterns. In addition, SOG and BG apply bit parallelism. Related methods for a single pattern have been suggested by Fredriksson [7].

It is well known (see e.g. [2,4]) that the use of q-grams can increase the average length of shift in the algorithms of Boyer-Moore type. This can also be applied to matching of multiple patterns [15]. We use q-grams in a different way in order to improve filtration efficiency by changing the alphabet.

In order to show the applicability of our algorithms, we ran extensive tests and compared them with various implementations of earlier algorithms. We used a random text, which ensures the rareness of matches in our setting. Our algorithms showed to be very fast in practice. For example, HG is 15 times faster than the well-known Aho-Corasick algorithm in the case of random patterns for $r=10,000$, $m=8$, and $c=256$. In addition, the filtering phase of our algorithms does not require much memory: 64 kB is enough in the specified case. The filtering efficiency of our algorithms will continue beyond 100,000 patterns if more memory is used.

2   Earlier Solutions

The classical Aho-Corasick algorithm [1] has been widely used for multiple pattern matching. Although it works rather well for small pattern sets, it is not suitable for huge pattern sets because of intolerable memory requirements. And the algorithm gets slower when the number of patterns increases.

2.1   Rabin-Karp Approach

A well-known solution [9,12,17] to cope with large pattern sets with less memory is to combine the Rabin-Karp algorithm [11] with binary search. During preprocessing, hash values for all patterns are calculated and stored in an ordered table. Matching can then be done by calculating the hash value for each m-character string of the text and searching the ordered table for this hash value using binary search. If a matching hash value is found, the corresponding pattern is compared with the text.

We implemented this method for $m = 8$, 16, and 32. The hash values for patterns of eight characters are calculated as follows. First a 32-bit integer is formed of the first four bytes of the pattern and another from the last four bytes of the pattern. These are then xor'ed together resulting in the following hash function where ˆ denotes the xor-operation:

$$Hash(x_1 \ldots x_8) = x_1 x_2 x_3 x_4 \char`^ x_5 x_6 x_7 x_8$$

The hash values for $m = 16$ and 32 are calculated in a similar fashion:

$$Hash16(x_1 \ldots x_{16}) = (x_1 x_2 x_3 x_4 \char`^ x_5 x_6 x_7 x_8) \char`^ (x_9 x_{10} x_{11} x_{12} \char`^ x_{13} x_{14} x_{15} x_{16})$$

$$Hash32(x_1 \ldots x_{32}) = ((x_1 x_2 x_3 x_4 \char`^ x_5 x_6 x_7 x_8) \char`^ \ldots \char`^ (x_{25} x_{26} x_{27} x_{28} \char`^ x_{29} x_{30} x_{31} x_{32}))$$

Muth and Manber [12] use two-level hashing to improve the performance of the Rabin-Karp method. The second hash is calculated from the first one by xor'ing together the lower 16 bits and the upper 16 bits. At preprocessing time, a bitmap of 2^{16} bits is constructed. The i'th bit is zero, if no pattern has i as its second hash value, and one, if there is at least one pattern with i as its second hash value. When matching, one can quickly check from the bit table, when the first hash value does not need further inspection, and thus avoiding the time consuming binary search in many cases. In the following, we use the shorthand RKBT for the Rabin-Karp algorithm combined with binary search and two-level hashing.

2.2   Set Horspool

The Commentz-Walter algorithm [5] for multiple patterns has been derived from the Boyer-Moore algorithm [4]. A simpler variant of this algorithm is called set Horspool [14]. (The same algorithm is called set-wise Boyer-Moore in [8].) This algorithm is based on the Boyer-Moore-Horspool algorithm [10] for single patterns. In the Boyer-Moore-Horspool algorithm, the bad character function $B(a)$ is defined as the distance from the end of the pattern $p_1 p_2 \cdots p_m$ to the last occurrence of the character a: $B(a) = min\{h \mid p_{m-h} = a\}$. This function can be generalized for multiple patterns. The bad character function for the set of patterns is defined as the minimum of the bad character functions of individual patterns.

The reversed patterns are stored in a trie. The initial endpoint is the length of the shortest pattern. The text is compared from right to left with the trie until no matching entry is found for a character in the text. Then the bad character function is applied to the endpoint character and the pattern trie is shifted accordingly.

The Wu-Manber algorithm [15] is a variation of the set Horspool algorithm. It uses a hash table of the last q-grams of patterns. The famous agrep tool [16] includes an implementation of the Wu-Manber algorithm.

3   Multi-pattern Horspool with q-Grams

The Boyer-Moore-Horspool algorithm [10] can be applied to multiple patterns also in another way. We call the resulting filtering algorithm HG (short for Horspool with q-Grams). Given patterns of m characters, we construct a bit table for each of the m pattern positions as follows. The first table keeps track of characters which appear in the first position in any pattern, the second table keeps track of characters which appear in the first or second position in any pattern and so on. Figure 1a shows the six tables corresponding to the pattern 'qwerty'.

These tables can then be used for pattern matching as follows. First the m'th character is compared with the m'th table. If the character does not appear in this table, the character cannot appear in positions $1 \dots m$ in any pattern and

```
1-gram tables:
1.  2.  3.  4.  5.  6.
q   q   q   q   q   q
    w   w   w   w   w
        e   e   e   e
            r   r   r
                t   t
                    y
```

(a)

```
HGMatcher(T, n)
  i = 0;
  while(i < n-6)
    j = 6;
    while (1)
      if (not 1GramTable[j][T[i+j]])
        i = i+j;
        break
      else if (j = 0)
        Verify-match(i);
        i = i+1;
        break
      else
        j = j-1
```

(b)

Fig. 1. The HG algorithm: (a) the data structures for the pattern 'qwerty' and (b) the pseudo-code for m=6.

a shift of m characters can be made. If the character is found in this table, the $m-1$'th character is compared to the $m-1$'th table. A shift of $m-1$ characters can be made if the character does not appear in this table and therefore not in any pattern in positions $1, \ldots, m - 1$. This process is continued until the algorithm has advanced to the first table and found a match candidate there. The pseudo-code for m=6 is shown in Figure 1b. Given this procedure, it is clear that all matches are found. However, also false matches can occur. E.g. 'qqqqqq' is a false candidate in our example. The candidates are verified by using the RKBT method described in Section 2.1.

As the number of patterns grows, the filtering efficiency of the above scheme decreases until almost all the text positions are candidates because there only c different characters. A substantial improvement in the filtering efficiency can be achieved by using q-grams, $q \geq 2$, instead of single characters since there are c^q different q-grams. For an alphabet with 256 characters and for $q = 2$ this means that the alphabet size grows from 256 to 65,536. When using 2-grams, a pattern of m characters is transformed into a sequence of $m-1$ 2-grams. Thus the pattern 'qwerty' would yield the 2-gram string 'qw-we-er-rt-ty'. The HG algorithm can be applied to these 2-grams just as it was applied to single characters. With even larger pattern sets, 3-grams could be used instead of 2-grams. Because this would require quite a lot of memory, we implemented a 3-gram version of the algorithm with a hashing scheme. Before a 3-gram is used to address the tables, each character is hashed to a 7-bit value. This diminishes the number of different 3-grams from 2^{24} to 2^{21}.

4   Multi-pattern Shift-Or with q-Grams

The shift-or algorithm [3] can be extended to a filtering algorithm for multiple patterns in a straightforward way. Rather than matching the text against exact patterns, the set of patterns is transformed to a single general pattern containing classes of characters. For example if we have three patterns, 'abcd', 'pony', and 'abnh', the characters {a, p} are accepted in the first position, characters {b, o} in the second position, characters {c, n} in the third position and characters {d, h, y} in the fourth position. This approach has been used for extended string matching (see e.g. [14]). Given this scheme, it is clear that all actual occurrences of the patterns in the text are candidates. However, there are also false candidates. In our example 'aocy' would also match. Therefore, each candidate must be verified.

When the number of patterns grows, this approach is no longer adequate. As in the case of HG, the filtering capability of this approach can be considerably improved by using q-grams instead of single characters. Then the pattern is a string of $m - q + 1$ q-gram classes. We call our modification SOG (short for Shift-Or with q-Grams). Again, the RKBT method is used for verification.

The improved efficiency of this approach is achieved at the cost of space. If the alphabet size is 256, storing the 2-gram bit vectors requires 2^{16} bytes for $m=8$ while the single character vectors only take 2^8 bytes. We implemented SOG for 2-grams and 3-grams as in the case of HG.

Baeza-Yates and Gonnet [3] present a way to extend the shift-or algorithm for multiple patterns for small values of r. Patterns $P_1 = p_1^1 \cdots p_m^1, \ldots, P_r = p_1^r \cdots p_m^r$ are concatenated into a single pattern:

$$P = p_1^1 p_1^2 \ldots p_1^r p_2^1 p_2^2 \ldots p_2^r \ldots p_m^1 p_m^2 \ldots p_m^r.$$

The patterns can then be searched in the same way as a single pattern except that the shift of the state vector will be for r bits and a match is found, if any of the r bits corresponding to the highest positions is 0. This method can also be applied in a different way to make the SOG algorithm faster for short patterns. The pattern set is divided into four or two subsets based on the first 2-gram. Each subset is then treated like a single pattern in the extension method of Baeza-Yates and Gonnet.

5   Multi-pattern BNDM with q-Grams

Our third filtering algorithm is based on the backward nondeterministic DAWG matching (BNDM) algorithm by Navarro and Raffinot [13]. The BNDM algorithm itself has been developed from the backward DAWG matching (BDM) algorithm [6].

In the BDM algorithm [6], the pattern is preprocessed by forming a DAWG (directed acyclic word graph) of the reversed pattern. The text is processed in windows of size m where m is the length of the pattern. The window is searched for the longest prefix of the pattern from right to left with the DAWG. When this search ends, we have either found a match (i.e. the longest prefix is of length m) or the longest prefix. If a match was not found, we can shift the start position of the window to the start position of the longest prefix.

The BNDM algorithm [13] is a bit-parallel simulation of the BDM algorithm. It uses a nondeterministic automaton instead of the deterministic one in the BDM algorithm. For each character x, a bit vector $B[x]$ is initialized. The i'th bit is 1 in this vector if x appears in the reversed pattern in position i. Otherwise the i'th bit is 0. The state vector D is initialized to 1^m. The same kind of right to left scan in a window of size m is performed as in the BDM algorithm. The state vector is updated in a similar fashion as in the shift-and algorithm. If the m'th bit is 1 after this update operation, we have found a prefix starting at position j. If j is the first position in the window, a match has been found.

The BNDM algorithm can be extended to multiple patterns in the same way as we did with the shift-or algorithm. We call this modification BG (short for Bndm with q-Grams). The matching is done with a general pattern containing classes of characters. The bit vectors are initialized so that the i'th bit is 1 if the corresponding character appears in any of the reversed patterns in position i. As with HG and SOG, all match candidates reported by this algorithm must be verified. Just like in SOG, 2- and 3-grams can be used to improve the efficiency of the filtering. Also the division to subsets, presented for the SOG algorithm, can be used with the BG algorithm. This scheme works in the same way as with SOG algorithm except that the subsets are formed based on the last 2-gram of the patterns.

6   Analysis

Let us consider the time complexities of the SOG and BG algorithms without division to subsets. The algorithms can be divided in three phases: preprocessing, scanning, and checking. Let us assume that $m \leq w$ holds, where w is the word length of the computer. When considering the average case complexity, we assume the standard random string model, where each character of the text and the pattern is selected uniformly and independently.

In the best case, no match candidates are found and then checking needs no time. In the worst case there are $h = n - m + 1$ candidates, and then the checking time $O(nh \log r) = O(nm \log r)$ dominates. Here $O(\log r)$ comes from binary search and $O(m)$ from pairwise inspection.

The preprocessing phase of the both algorithms is similar and it works in $O(rm)$. In addition, the initialization of the descriptor bit vectors needs $O(c^q)$.

In SOG the scanning phase is linear. The expected number of candidates C_1 depends on r, c, and m:

$$C_1 = h \cdot (1 - (1 - 1/c)^r)^m.$$

This number can be reduced by utilizing q-grams. With q-grams, we estimate this expression by

$$C_q = h \cdot (1 - (1 - 1/c^q)^r)^{m-q+1}.$$

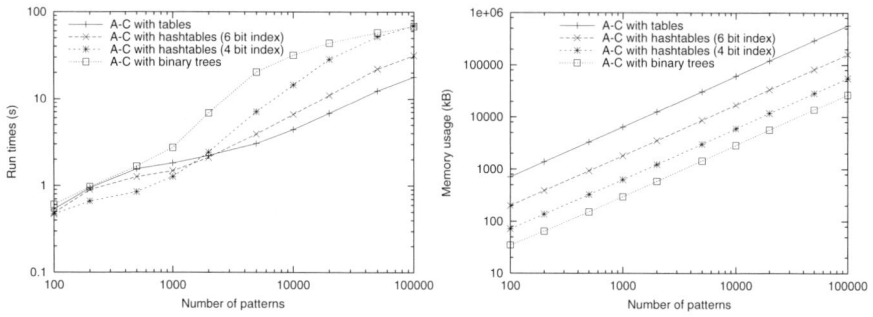

Fig. 2. Performance of different trie implementations of the Aho-Corasick algorithm.

Note that even C_2 is not accurate, because consecutive overlapping 2-grams are not independent. However, the difference from the exact value is insignificant for huge sets of patterns.

Let us then consider BG. The worst case complexity of the basic BNDM is $O(nm)$. We did not want to apply any linear modification, because the checking phase of BG is not linear, and the linear versions of BNDM are slower in practice [13]. The average searching time of the BNDM algorithm is $O(n \log_{c'} m/m)$, where c' is the size of the alphabet for the original BNDM. In our approach we need to replace c' by $1/d$ where $d = 1 - (1 - 1/c)^r$ is the probability that a single position of a generalized pattern matches. Clearly $\log_{1/d} m < m$ holds for suitable values of c, r, and m, and BG is then sublinear on the average, i.e. it does not inspect every text character. Switching to q-grams, $q \geq 2$, guarantees the sublinearity for smaller values of c and larger values of r.

7   Experiments with Earlier Algorithms

We ran tests on several algorithms. We used a 32 MB randomly created text in the alphabet of 256 characters. Also the patterns were randomly generated in the same alphabet. Note that in our case random data is in a sense harder than real data. For example, 'zg' is rarer in an English text than in a random text.

If not otherwise stated, $m=8$ and $c=256$ hold. The times are averages over 10 runs using the same text and patterns. Both the text and the patterns reside in the main memory in the beginning of each test in order to exclude reading times. The tests were made on a computer with a 1.8 GHz Pentium 4 processor, 1 GB of memory, and 256 kB on-chip cache. The computer was running Linux 2.4.18. The algorithms were written in C and compiled with the gcc compiler.

Aho-Corasick. We used a code based on the case-sensitive implementation by Fisk and Varghese [8] to test the Aho-Corasick algorithm [1]. We tested three alternative implementations of the goto-function: table, hash table, and binary tree. The hash table version was tested with table sizes 16 and 64 (resulting in 4- and 6-bit indexes), see Figure 2.

Although the speed of the Aho-Corasick algorithm is constant for small pattern sets, the situation is different for large sets even in an alphabet of moderate size. The run time graph of Figure 2 shows a steady increase. Given the memory graph of Figure 2, the hierarchical memory could explain this behavior. For pattern set sizes between 100 and 2,000, the hash table version of the goto-function is preferable. When there are more than 2,000 patterns, the table version is the fastest but its memory requirement does not make it very attractive.

RKBT. The Rabin-Karp approach was tested both with and without two-level hashing. The use of the second hash table of 2^{16} bits significantly improves the performance of the algorithm when the number of patterns is less than 100,000. When there are more patterns, a larger hash table should be considered, because this hash table tends to be full of 1's and the gain of two-level hashing disappears.

Set Horspool. We used the code of Fisk and Varghese [8] to test the set Horspool algorithm. The same variations as for the Aho-Corasick algorithm were made. The results on memory usage were similar to those of the Aho-Corasick algorithm because the trie structure is very similar. Also the test results on run times resemble those of the Aho-Corasick algorithm especially with very large pattern sets. This is probably due to the memory usage. Differences with less than 1,000 patterns were not significant between modifications.

Agrep. We also tested the agrep tool [16]. Since agrep is row-oriented, some characters, like newline, were left out of the alphabet. In the agrep tool, lines are limited to 1024 characters so we chopped the text to lines each containing 1024 characters. The run times measured do not contain the time used to preprocess the patterns.

In the experiments of Navarro and Raffinot [14] agrep was the fastest algorithm for 1,000 patterns for $m=8$. This holds true also for our experiments (excluding the new algorithms). The agrep tool is the fastest up to 2,000 patterns, the RKBT method is the fastest between 2,000 and 20,000 patterns and the set Horspool algorithm is the fastest with more than 20,000 patterns although its memory usage is excessive.

Figure 3 shows a comparison of the four earlier algorithms mentioned above. The times include verification but exclude preprocessing.

8   Experiments with New Algorithms

The test setting is the same as in the previous section. The new algorithms are all filtering algorithms, which use the RKBT method for verification of candidates. Each run time contains the verification time (if not otherwise specified) and excludes the preprocessing time.

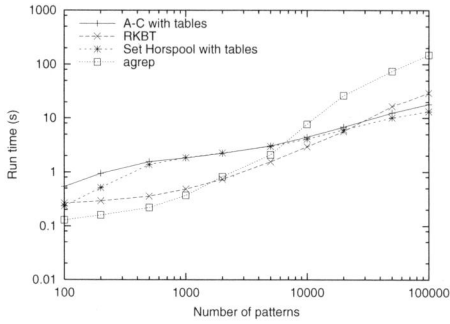

Fig. 3. Run time comparison of the earlier algorithms.

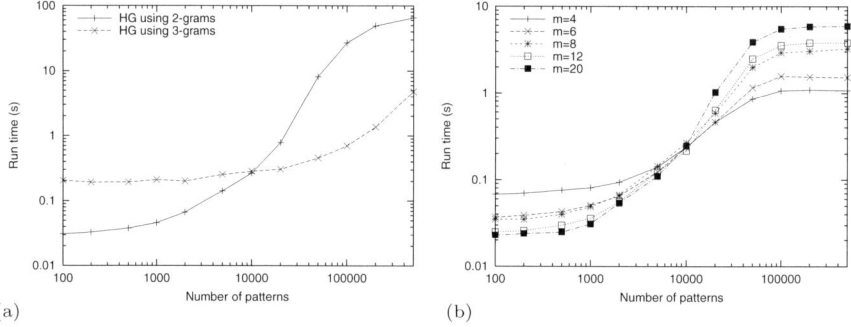

(a)   (b)

Fig. 4. The HG algorithm: (a) comparison of 2-gram and 3-gram versions and (b) run times of the 2-gram version for different pattern lenghts.

8.1   HG

The HG algorithm was tested both with the 2-gram and 3-gram versions for $m=8$, see Figure 4a. The 3-gram version is faster when the pattern set size is greater than 10,000. This is due to the better filtering efficiency of the 3-gram approach. However, when there are less than 10,000 patterns, the 2-gram version is much faster because of the hashing overhead and memory requirement of the 3-gram approach.

We tested the HG algorithm also with several pattern lengths. The verification of candidates was not carried out in this case since we implemented the RKBT method only for $m = 8$, 16, and 32. If the verification would be done, the performance of the algorithm would worsen for those set sizes that produce spurious hits. Most of the candidates reported by the HG algorithm are false matches because the probability of finding a real match is very low.

Figure 4b shows the results of these tests for the 2-gram version of the algorithm. With 50,000 patterns, the number of matches reported by the HG algorithm is roughly the same regardless of the pattern length. For $c=256$ there are $2^{16} = 65,536$ different 2-grams. So, when there are more than 50,000 pat-

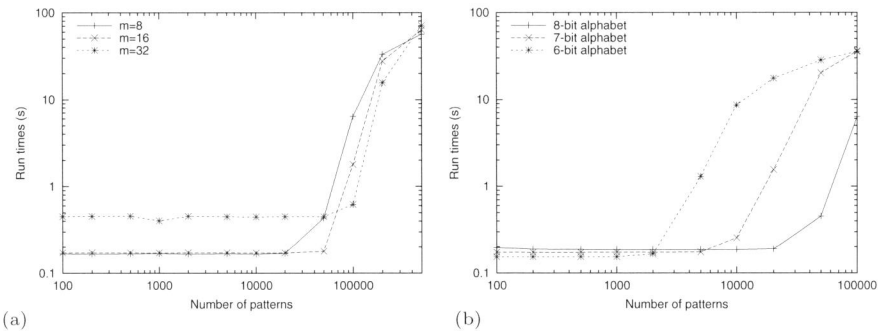

Fig. 5. The SOG algorithm: (a) the effect of pattern length and (b) the effect of alphabet size.

terns, nearly all text positions will match. Figure 4b shows that, when there are less than 10,000 patterns, HG is faster for longer patterns, because they allow longer shifts. When the number of false matches grows, the algorithm is faster for shorter patterns, because most positions match anyway and the overhead with shorter patterns is smaller.

8.2   SOG

We tested the SOG algorithm with several pattern lengths and alphabet sizes. The 3-gram variation and the division of patterns to subsets were also tried.

The tests with pattern length were made for $m = 8$, 16, and 32, see Figure 5a. The performance of the SOG algorithm degrades fast when the number of patterns reaches 100,000. This is the same effect that was found with the HG algorithm; Almost all text positions match because there are only 65,536 different 2-grams. When the pattern set size is less than 20,000, the run time of the algorithm is constant because no false matches are found.

Figure 5a also shows that the algorithm is significantly slower for $m=32$ than for $m=8$ and 16. This is likely due to the data cache. The structures of the SOG algorithm take 64 kB memory for $m=8$, 128 kB for $m=16$, and 256 kB for $m=32$. Given the cache size of 256 kB, it is clear that the structures for $m=32$ cannot be held in the cache all the time because also the text to be searched has to be there.

The behavior of SOG with alphabet sizes 64, 128, and 256 is shown in Figure 5b. Given the alphabet size 64, there are 4,096 different 2-grams, and so the performance of the SOG algorithm was expected to degrade after 4,000 patterns. Using the same reasoning, the performance of the SOG algorithm using the 7-bit alphabet was expected to degrade after 16,000 patterns and the 8-bit alphabet version after 65,000 patterns. The graphs of Figure 5b follow nicely this prediction.

The 3-gram version of the SOG algorithm was tested for $m=8$. Figure 6a shows a comparison of the 2-gram and 3-gram versions. With less than 500,000

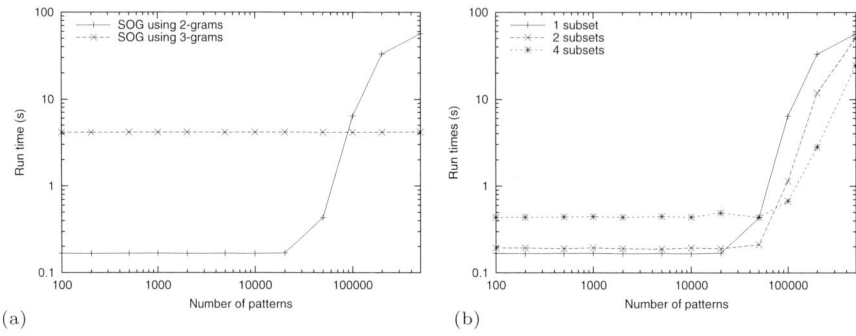

(a)   (b)

Fig. 6. The SOG algorithm: (a) comparison of the 2- and 3-gram versions and (b) the effect of one, two and four subsets.

patterns the run time of the 3-gram SOG algorithm is constant and there are only a few false matches because given our hashing scheme there are about $2 \cdot 10^6$ different 3-grams. The 3-gram version is, however, much slower than the 2-gram version due to the hashing overhead and the greater memory requirement which causes cache misses.

The use of subsets with the SOG algorithm was tested for $m=8$. We tried versions with one, two and four subsets, see Figure 6b. The versions using one or two subsets are almost as fast up to 20,000 patterns. After that the version using two subsets is slightly faster. The version using four subsets is significantly slower than the other two versions with small pattern set sizes. The problem here is that the table needed to store the 32-bit vectors is as large as the data cache. In computers with larger caches this version would likely perform as well as the other two. Given r patterns, using four subsets should result in roughly as many false matches as using one subset with $r/4$ patterns because in the version with four subsets only one subset can match at a given position. The results of the tests show that there are a little more matches than that. This is due to the more homogeneous sets produced by the division of patterns.

8.3   BG

We tested the performance of the BG algorithm for $m = 8$, 16 and 32. The algorithm is faster for $m=16$ than for $m=8$. In the case of $m=32$, the algorithm suffers from the large table which cannot be kept in the cache all the time. However, the filtering efficiency improves slightly with longer patterns.

The 3-gram version of the BG algorithm was also tested. The result was similar to that of SOG. With less than 50,000 patterns, the 2-gram approach is clearly faster but after that the 3-gram version performs faster. The 3-gram version is slower mainly because of its memory usage. The hashing scheme used also slows it down.

Table 1. Run times of the algorithms when r varies for $m=8$ and $c=256$.

	100	200	500	1,000	2,000	5,000	10,000	20,000	50,000	100,000
Aho-Corasick	0.538	0.944	1.559	1.824	2.221	3.055	4.433	6.804	12.427	17.951
RKBT	0.265	0.293	0.358	0.483	0.735	1.551	2.942	5.660	16.423	29.567
Set Horspool	0.235	0.513	1.375	1.848	2.252	2.990	4.083	6.068	10.154	13.225
agrep	0.130	0.160	0.220	0.370	0.820	2.090	7.670	26.480	74.370	148.690
SOG	0.167	0.166	0.167	0.168	0.166	0.167	0.166	0.169	0.435	6.357
HG	0.031	0.033	0.038	0.046	0.067	0.142	0.266	0.784	8.182	26.884
BG	0.027	0.029	0.035	0.041	0.056	0.106	0.151	0.206	0.649	7.389

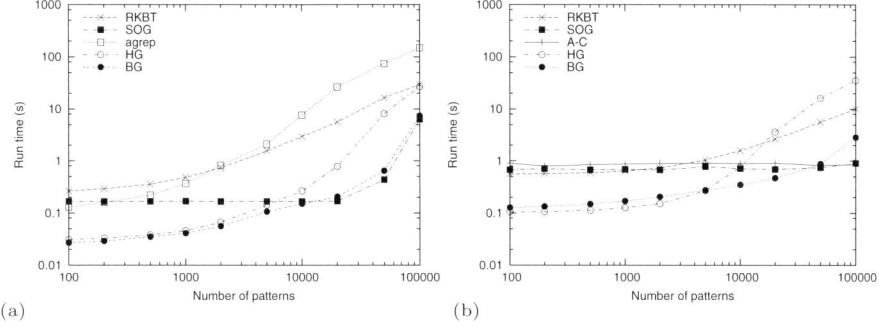

Fig. 7. Run time comparison of the algorithms for (a) random data ($m=8$, $c=256$) and (b) dna data ($m=32$).

The use of subsets with the BG algorithm was tested for $m=8$ with one, two and four subsets. The results of these tests were very similar to the ones of the SOG algorithm.

8.4   Comparison of the Algorithms

A run-time comparison of the algorithms is shown in Figures 3 and 7a based on Table 1. These times include verification but exclude preprocessing.

The memory usage and the preprocessing times of the algorithms are shown in Table 2. These are results from tests with patterns of eight characters, where HG, SOG, and BG use 2-grams.

Figure 7a shows that our algorithms are considerably faster than the algorithms presented earlier. The HG and BG algorithms are the fastest, when there are less than 5,000 patterns. Between 5,000 and 100,000 patterns the SOG and BG algorithms are the fastest. The BG algorithm has the best overall efficiency. With larger patterns sets, the use of subsets with these algorithms would be advantageous. Our algorithms scale to even larger pattern sets by using larger q-grams if there is enough memory available.

Table 2. Memory usage and preprocessing times of the algorithms for $r = 100$ and 100,000.

Algorithm	Memory (kB)		Preproc. (s)	
	100	100,000	100	100,000
RKBT	13	1,184	0.02	0.20
HG	69	1,240	0.03	0.23
SOG	77	1,248	0.03	0.21
BG	77	1,248	0.03	0.21
A-C (with tables)	799	663,174	0.54	5.10
Set Horspool (with tables)	793	656,338	0.19	1.68

Table 2 shows that the preprocessing phase of our algorithms is fast. Table 2 also shows that the memory usage of our algorithms is fairly small. In fact, the memory usage of our filtering techniques is constant. Because our algorithms use RKBT as a subroutine, their numbers cover also all the structures of RKBT including the second hash table. The space increase in Table 2 is due to the need to store the patterns for the verification phase. The space for the patterns could be reduced by using clever hash values. For example for $m=8$, we could store only four characters of each pattern and use a 32-bit hash value such that the other four characters can be obtained from these characters and the hash value.

We also run a preliminary test on dna data. Our text was the genome of fruit fly (20 MB). We used random patterns of 32 characters for $q=8$. The results are shown in Figure 7b. This test was made on a 1.0 GHz computer with 256 MB of memory and 512 kB cache. The algorithms HG and BG worked very well for sets of less than 10,000 patterns.

9   Concluding Remarks

We have presented efficient solutions for multiple string matching based on filtering with q-grams and bit-parallelism. We showed that on random test data, our algorithms perform faster and use a smaller amount of memory than the earlier ones. The preprocessing phase of our algorithms is fast. We tuned the algorithms to handle efficiently up to 100,000 patterns of eight characters. Our algorithms suit especially well to the searching of huge static sets of rare patterns.

Our approach seems to be sensitive to cache effects. We need to test the algorithms in several computers of different types in order to get additional information on their behavior.

We utilized overlapping q-grams. We tested our algorithms also with consecutive non-overlapping q-grams, but this modification brought clearly worse results. We used mainly the alphabet $c=256$. In the near future we will try small alphabets and compare our algorithm with the SBOM algorithm of Navarro and Raffinot [14]. We will also consider approximate matching (see e.g. [12]).

References

1. A. Aho, M. Corasick: Efficient string matching: An aid to bibliographic search. *Communications of the ACM* 18, 6 (1975), 333–340.
2. R. Baeza-Yates. Improved string searching. *Software – Practice and Experience*, 19, 3 (1989), 257–271.
3. R. Baeza-Yates, G. Gonnet: A new approach to text searching. *Communications of ACM* 35, 10 (1992), 74–82.
4. R. Boyer, S. Moore: A fast string searching algorithm. *Communications of the ACM* 20 (1977), 762–772.
5. B. Commentz-Walter: A string matching algorithm fast on the average. *Proc. 6th International Colloquium on Automata, Languages and Programming, Lecture Notes on Computer Science* 71, 1979, 118–132.
6. M. Crochemore, W. Rytter: *Text algorithms*. Oxford University Press, 1994.
7. K. Fredriksson: Fast string matching with super-alphabet. *Proc. SPIRE '02, String Processing and Information Retrieval, Lecture Notes in Computer Science* 2476, 2002, 44–57.
8. M. Fisk, G. Varghese: Fast content-based packet handling for intrusion detection. UCSD Technical Report CS2001-0670, 2001.
9. B. Gum, R. Lipton: Cheaper by the dozen: batched algorithms. *Proc. First SIAM International Conference on Data Mining*, 2001
10. N. Horspool: Practical fast searching in strings. *Software – Practice and Experience* 10 (1980), 501–506.
11. R. Karp, M. Rabin: Efficient randomized pattern-matching algorithms. *IBM Journal of Research and Development* 31 (1987), 249–260.
12. R. Muth, U. Manber: Approximate multiple string search. *Proc. CPM '96, Combinatorial Pattern Matching, Lecture Notes in Computer Science* 1075, 1996, 75–86.
13. G. Navarro, M. Raffinot: Fast and flexible string matching by combining bit-parallelism and suffix automata. *ACM Journal of Experimental Algorithms* 5, 4 (2000), 1–36.
14. G. Navarro, M. Raffinot: *Flexible pattern matching in strings*. Cambridge University Press, 2002.
15. S. Wu, U. Manber: A fast algorithm for multi-pattern searching. Report TR-94-17, Department of Computer Science, University of Arizona, 1994.
16. S. Wu, U. Manber: Agrep – A fast approximate pattern-matching tool. *Proc. Usenix Winter 1992 Technical Conference*, 1992, 153–162.
17. R. Zhu, T. Takaoka: A technique for two-dimensional pattern matching. *Communications of the ACM* 32 (1989), 1110–1120.

Sparse LCS Common Substring Alignment

Gad M. Landau[1⋆], Baruch Schieber[2⋆⋆], and Michal Ziv-Ukelson[1,2⋆⋆⋆]

[1] Department of Computer Science, Haifa University, Haifa 31905, Israel
{landau, michal}@cs.haifa.ac.il,
[2] IBM T.J. Watson Research Center, P.O. Box 218 Yorktown Heights, NY 10598
sbar@watson.ibm.com

Abstract. The "Common Substring Alignment" problem is defined as follows. The input consists of a set of strings $S_1, S_2 \ldots S_c$, with a common substring appearing at least once in each of them, and a target string T. The goal is to compute similarity of all strings S_i with T, without computing the part of the common substring over and over again.

In this paper we consider the Common Substring Alignment problem for the LCS (Longest Common Subsequence) similarity metric. Our algorithm gains its efficiency by exploiting the *sparsity* inherent to the LCS problem. Let Y be the common substring, n be the size of the compared sequences, L_y be the length of the LCS of T and Y, denoted $|LCS[T, Y]|$, and L be $\max\{|LCS[T, S_i]|\}$. Our algorithm consists of an $O(nL_y)$ time encoding stage that is executed once per common substring, and an $O(L)$ time alignment stage that is executed once for each appearance of the common substring in each source string. The additional running time depends only on the length of the parts of the strings that are not in any common substring.

1 Introduction

The problem of comparing two sequences A and B to determine their similarity is one of the fundamental problems in pattern matching. One of the basic forms of the problem is to determine the *longest common subsequence (LCS)* of A and B. The LCS string comparison metric measures the subsequence of maximal length common to both sequences [1]. Longest Common Subsequences have many

⋆ Department of Computer Science, Haifa University, Haifa 31905, Israel, phone: (972-4) 824-0103, FAX: (972-4) 824-9331; Department of Computer and Information Science, Polytechnic University, Six MetroTech Center, Brooklyn, NY 11201-3840; landau@poly.edu; partially supported by NSF grant CCR-0104307, by the Israel Science Foundation grant 282/01, by the FIRST Foundation of the Israel Academy of Science and Humanities, and by IBM Faculty Partnership Award.
⋆⋆ IBM T.J. Watson Research Center, P.O. Box 218 Yorktown Heights, NY 10598; Tel: (914) 945-1169; Fax: (914) 945-3434; sbar@watson.ibm.com;
⋆⋆⋆ Department of Computer Science, Haifa University, Haifa 31905, Israel; On Education Leave from the IBM T.J. Watson Research Center; michal@cs.haifa.ac.il; partially supported by the Israel Science Foundation grant 282/01, and by the FIRST Foundation of the Israel Academy of Science and Humanities.

R. Baeza-Yates et al. (Eds.): CPM 2003, LNCS 2676, pp. 225–236, 2003.
© Springer-Verlag Berlin Heidelberg 2003

applications, including sequence comparison in molecular biology as well as the widely used *diff* file comparison program. The LCS problem can be solved in $O(mn)$ time, where m and n are the lengths of strings A and B, using dynamic programming [8]. The dynamic programming creates an $m \times n$ "DP Table" that contains in its (i,j) entry the LCS of the prefix of A of size i and the prefix of B of size j.

More efficient LCS algorithms, which are based on the observation that the LCS solution space is highly redundant, try to limit the computation only to those entries of the DP Table which convey essential information, and exploit in various ways the *sparsity* inherent to the LCS problem. Sparsity allows us to relate algorithmic performances to parameters other than the lengths of the input strings. Most LCS algorithms that exploit sparsity have their natural predecessors in either Hirshberg [8] or Hunt-Szymanski [9].

All Sparse LCS algorithms are preceded by an $O(n \log |\Sigma|)$ preprocessing [1]. The Hirshberg algorithm uses $L = |LCS[A, B]|$ as a parameter, and achieves an $O(nL)$ complexity. The Hunt-Szymanski algorithm utilizes as parameter the number of matches between A and B, denoted r, and achieves an $O(r \log n)$ complexity. Apostolico and Guerra [2] achieve an $O(L \cdot m \cdot \min(\log |\Sigma|, \log m, \log(2n/m))$ algorithm, where $m \leq n$, and another $O(m \log n + d \log(nm/d))$ algorithm, where $d \leq r$ is the number of dominant matches (as defined by Hirschberg [8]). This algorithm can also be implemented in time $O(d \log \log \min(d, nm/d))$ [6]. Note that in the worst case both d and r are $\Omega(n^2)$, while L is always bounded by n.

The *Common Substring Alignment Problem* is defined in [12] as follows: The input consists of a set of one or more strings $S_1, S_2 \ldots S_c$ and a target string T. It is assumed that a common substring Y appears in all strings S_i at least once. Namely, each S_i can be decomposed in at least one way to $S_i = B_i Y F_i$. (See Figure 1.) The goal is to compute the similarity of all strings S_i with T, without computing the part of Y over and over again. It is assumed that the locations of the common subsequence Y in each source sequence S_i are known. However, the part of the target T with which Y aligns, may vary according to each B_i and F_i combination.

$$
\begin{array}{llllll}
T & = & \text{"BCBADBDCD"} & Y & = & \text{"BCBD"} \\
S_1 & = & \text{"BC } BCBD \text{ C"} & B_1 & = & \text{"BC"} & F_1 & = & \text{"C"} \\
S_2 & = & \text{"E } BCBD \text{ D} BCBD \text{ A"} & B_{2a} & = & \text{"E"} & F_{2a} & = & \text{"D} BCBD \text{ A"} \\
& & & B_{2b} & = & \text{"E} BCBD \text{ D"} & F_{2b} & = & \text{"A"}
\end{array}
$$

Fig. 1. An example of two different source strings S_1, S_2 sharing a common substring Y, and a target T.

More generally, the common substring Y could be shared by different source strings competing over similarity with a common target, or could appear repeatedly in the same source string. Also, in a given application, we could of course be dealing with more than one repeated or shared sub-component. (See Figure 1.)

Common Substring Alignment algorithms are usually composed of a *prepro-cessing* stage that depends on data availability, an *encoding* stage and an *align-ment* stage. During the encoding stage, a data structure is constructed which encodes the comparison of Y with T. Then, during the alignment stage, for each comparison of a source S_i with T, the pre-compiled data structure is used to speed up the part of aligning each appearance of the common substring Y.

In most of the applications for which Common Substring Alignment is in-tended, the source sequence database is prepared *off-line*, while the target can be viewed as an "unknown" sequence which is received *online*. The source strings can be pre-processed *off-line* and parsed into their optimal common substring representation. Therefore, we know well beforehand where, in each S_i, Y begins and ends. However, the comparison of Y and T can not be computed until the target is received. Therefore, the encoding stage, as well as the alignment stage, are both *online* stages.

Even though both stages are *online*, they do not bear an equal weight on the time complexity of the algorithm. The efficiency gain is based on the fact that the encoding stage is executed only once per target, and then the encoding results are used, during the alignment stage, to speed up the alignment of each appearance of the common substring in any of the source strings.

To simplify notation we assume from now on that all compared strings are of size n. Our results can be extended easily to the case in which the strings are of different length. We use L to denote $\max\{|LCS[T, S_i]|\}$, and L_y to denote $|LCS[T, Y]|$. (Note that $L_y \leq |Y|$, $L_y \leq L$, and $L \leq n$.)

Results. In this paper we address the following challenge: can a more efficient common substring alignment algorithm, which exploits the *sparsity* inherent to the LCS metric, be designed for the LCS metric. We show how to exploit sparsity, by replacing the traditional matrix which is used to encode the comparison of Y and T, with a smaller matrix. We show that this smaller matrix can be computed using Myers' Consecutive Suffix Alignments algorithm [14]. We also prove that this smaller matrix preserves the Total Monotonicity condition, thus enabling efficient adaptation of a matrix searching algorithm of Aggarwal et al. [3].

Our algorithm consists of an $O(nL_y)$ encoding stage, and an $O(L)$ alignment stage. When the problem is sparse ($L_y << |Y|$, $L << n$), our time bounds are better than those of previous algorithms. Even when the data is dense, our solution for the problem is no worse than the best-known algorithms.

The first Common Substring Alignment algorithm for the LCS metric was given in [11]. It presents an $O(n^2 + n|Y|)$ encoding stage, and an $O(n)$ alignment stage. In [12] a Common Substring Alignment algorithm for the LCS, Edit Dis-tance and more extended metrics is given. This algorithm consists of an $O(n|Y|)$ encoding stage, and an $O(n)$ alignment stage.

The remainder of this paper is organized as follows. Section 2 contains Com-mon Substring Alignment preliminaries. The new algorithm is described in sec-tion 3. Section 4 contains an analysis and assertion of some of the properties of

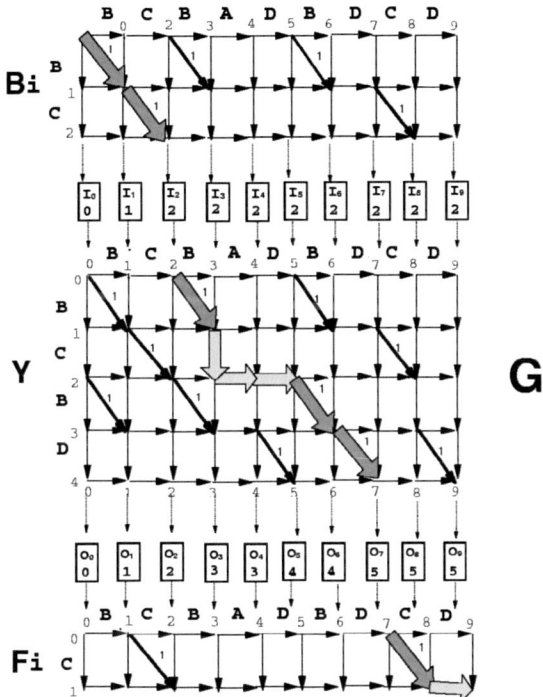

Fig. 2. The LCS Dynamic Programming Graph for the comparison of $S_1 = "BCBCBDC"$ with $T = "BCBADBCDC"$. The highlighted path in the graph corresponds to the common subsequence $"BCBBDC"$.

the new data representation which allow for the efficiency gain. Conclusions and open problems are given in Section 5.

2   Common Substring Alignment Preliminaries

In the literature the DP Table used for computing the alignment is also viewed as a directed acyclic graph (DAG), called the Dynamic Programming (DP) Graph [7]. The DP Graph for S and T, contains $(|S|+1)(|T|+1)$ vertices, each labeled with a distinct pair $(x, w)(0 \leq x \leq |S|, 0 \leq w \leq |T|)$. The vertices are organized in a matrix of $(|S|+1)$ rows and $(|T|+1)$ columns. (See Figure 2.)

When using the LCS metric, the DP Graph contains a directed edge with weight zero from each vertex (x, w) to each of the vertices $(x, w+1)$, $(x+1, w)$. It also contains a diagonal edge with weight one from vertex (x, w) to vertex $(x+1, w+1)$ if $S[x+1] = T[w+1]$. Maximal weight paths in the Dynamic Programming graph represent optimal alignments of S and T. The Dynamic

Programming algorithm will set the value of vertex (i, j) in the graph to the total weight of the highest scoring path which originates in vertex $[0, 0]$ of the graph and ends in vertex $[i, j]$.

The Dynamic Programming Graph used for computing the similarity between a source string $S_i = B_i Y F_i$ and a target string T can be viewed as a concatenation of 3 sub-graphs, where the first graph represents the similarity between B_i and T, the second graph represents the similarity between Y and T, and the third graph represents the similarity between F_i and T.

In this partitioned solution, the weights of the vertices in the last row of the first graph serve as input to initialize the weights of the vertices in the first row of the second graph. The weights of the last row of the second graph can be used to initialize the first row of the third graph.

The motivation for breaking the solution into 3 sub-graphs is that the second sub-graph, which represents the comparison of Y with T, is identical in each of its appearances in all DP graphs comparing any of the strings S_i with T. More specifically, both the structure and the weights of the edges of all DP sub-graphs comparing Y with T are identical, but the weights to be assigned to the vertices during the LCS computation may vary according to the prefix B_i which is specific to the source string. Therefore, an initial investment in the learning of this graph as an encoding stage, and in its representation in a more informative data structure, may pay off later on.

For a string A of length n and for $1 \leq u \leq z \leq n$, let A_u^z denote the substring of A from index u (inclusive) up to index z (inclusive). We define the following notations with respect to the Dynamic Programming Graph used for computing the LCS of $S_i = B_i Y F_i$ and T. Let G denote the second sub-graph which represents the comparison of Y with T. Let I denote the first row of G. Namely, $I[j] = |LCS[T_1^j, B_i]|$, for $j = 1, \ldots, n$. Let O denote the last row of G. Namely, $O[j] = |LCS[T_1^j, B_i Y]|$, for $j = 1, \ldots, n$.

Our solution will focus on the work necessary for an appearance of a given common substring: Given a target string T, a source substring Y and an input row I, compute the output row O.

As described above the online work for each common substring consists of two stages.

1. *Encoding Stage*: Study the structure of G and represent it in an efficient way.
2. *Alignment Stage*: Given I and the encoding of G, constructed in the previous stage, compute O.

Observe that, due to the monotonicity and unit-step properties of LCS, both I and O are monotone staircases with at most $L + 1$ unit steps. This enables the following definitions of the vectors PI and PO.

Definition 1. *For $k = 0, \ldots, L$, the entry $PI[k]$ contains the smallest index in the input row I whose value is k, if such exists.*

Definition 2. *For $k = 0, \ldots, L$, the entry $PO[k]$ contains the smallest index in the output row O whose value is k, if such exists.*

We claim that only the positions $PI[r]$ for $r = 0, \ldots, L$ are interesting as intermediate points in an optimal, leftmost k-path from vertex $(0,0)$ to $PO[k]$. To see this consider some $k \in [0..L]$ and assume that $i_1 = PI[k]$ is defined. Let i_3 be $PI[k+1]$ if it is defined or $n+1$ otherwise. Note that for any index i_2, where $i_1 < i_2 < i_3$, $I[i_1] + |LCS[T^j_{i_1+1}, Y]| \geq I[i_2] + |LCS[T^j_{i_2+1}, Y]|$, for any $j \in [i_2..n]$. This is true since $I[i_2] = I[i_1] = k$ and $|LCS[T^j_{i_1+1}, Y]| \geq |LCS[T^j_{i_2+1}, Y]|$ for any $i_1 < i_2 < i_3$. (The second inequality follows since the concatenation of prefix $T^{i_2}_{i_1+1}$ to the string $T^j_{i_2+1}$ can either increase the size of its common subsequence with Y or leave it unchanged.)

3   The Algorithm

The objective of the algorithm is to compute $PO[k]$, for $k = 1, \ldots, L$ given the vector PI. (Note that $PO[0] = 0$.) Recall that $PO[k]$ is the smallest index of an entry in O with a value of k. In terms of optimal paths in the alignment graph, this means that $PO[k]$ is the index of the leftmost vertex in the output border of G to end a path of weight k, which originates in vertex $(0,0)$ of the alignment graph. Note that such a path could enter G through any one of its input border vertices whose value is $\leq k$. However, in section 2 we have shown that the PI indices are sufficient for representing all the potential entry points of I, and therefore only the values $PI[0 \ldots k]$ will be considered as relevant representative entry point indices for the sought k-path.

Now, consider any path of total weight k, which originates in vertex $(0,0)$ of the alignment graph, enters G through a given input border vertex $PI[r]$, and ends in some vertex j on the output border of G. This path could be decomposed into two parts: the sub-path connecting vertex $(0,0)$ with the selected input border entry $PI[r]$, followed by the sub-path from the selected input border entry to vertex j on the output border. The weight of the sub-path from vertex $(0,0)$ to the selected input border entry is the value $I[PI[r]] = r$. The weight of the sub-path from vertex $PI[r]$ on the input border to vertex j on the output border is $|LCS[T^j_{PI[r]+1}, Y]|$. By definition, the sum of the weights of these two sub-paths, whose concatenation gives the total k-path, must be k.

Now recall that among all such potential k-paths, which could enter G through any of the $PI[r]$, $r = 0 \ldots k$ input border vertices, we are actually interested in the ones which are optimal in the sense that they end in the leftmost possible output border vertex. Therefore, the value of $PO[k]$, for $1 \leq k \leq L$, is computed as follows.

$$PO[k] = \min_{r=0}^{k} \{j \mid r + |LCS[T^j_{PI[r]+1}, Y]| = k\} \qquad (1)$$

We show how to obtain PO in two stages.

1. The encoding stage which is executed only once for each common substring Y, in which we compute $\min\{j \mid |LCS[T^j_{i+1}, Y]| = k\}$ for each pair (i, k), $i = 0 \ldots n$ and $k = 1 \ldots L_y$.

2. The alignment stage in which we compute PO, using Equation 1, given PI and the values computed in the encoding stage.

3.1   The Encoding Stage

In the encoding stage we compute the $n \times L_y$ table S, where

$$S[i, k] = \min\{j \mid |LCS[T_{i+1}^j, Y]| = k\},$$

if such an index exists. In other words, $S[i, k]$ contains the smallest index of a vertex in the last row of the DP Graph defined by the LCS of T and Y that can be reached from vertex i in the first row of this graph via a path of weight k. We observe that Myers' Consecutive Alignments algorithm [14] can be used to construct the table S.

Complexity. Given two strings Y and T over a constant alphabet Myers [14] constructs the table S for the comparison of Y versus T in $O(nL_y)$ time and space.

3.2   The Alignment Stage

During each execution of the alignment stage the objective is to compute PO from PI, using the table S computed in the encoding stage. Recall that $PO[0] = 0$ and for $k > 0$, $PO[k] = \min_{r=0}^{k}\{S[PI[r], k - r]\}$.

Note that when the alignment stage is executed the L values of the PI for this specific alignment stage are known, and S has already been computed. It follows that the representation of the competing indices in the output row O can be reduced to the $(L+1) \times L$ matrix $LEFT$ in which $LEFT[r, k] = S[PI[r], k - r]$, for $1 \le k \le L$ and $0 \le r \le k$.

Note that in the DP Graph of S_i and T the entry $LEFT[r, k]$ is the smallest index of a vertex in row O that is an endpoint of a path of weight k that starts in vertex $(0, 0)$, and goes through vertex $PI[r]$ in row I.

Clearly, for $k \in [1..L]$, $PO[k]$ is the minimum of the k-th column in the above $LEFT$ matrix. In Section 4 we prove that $LEFT$ is convex totally monotone. Hence, a recursive algorithm by Aggarwal et al. [3], nicknamed $SMAWK$ in the literature, can be used to compute the column minima of $LEFT$.

Complexity. Given S and PI, computing an element of $LEFT$ requires $O(1)$ time and space. The $SMAWK$ algorithm computes the column minima of the $(L+1) \times L$ totally monotone matrix $LEFT$ in $O(L)$ time and space, by querying only $O(L)$ entries of the array. Hence, PO is computed during the alignment stage in $O(L)$ time and space.

4   *LEFT* as a Totally Monotone Rectangular Matrix

Definition 3. *A matrix* $M[0 \ldots m, 0 \ldots n]$ *is* **totally monotone** *if either condition 1 or 2 below holds for all* $a, b = 0 \ldots m$; $c, d = 0 \ldots n$:

1. **convex condition:** $M[a, c] \geq M[b, c] \Longrightarrow M[a, d] \geq M[b, d]$ *for all* $a < b$
   and $c < d$.
2. **concave condition:** $M[a, c] \leq M[b, c] \Longrightarrow M[a, d] \leq M[b, d]$ *for all* $a < b$
   and $c < d$.

In this section we prove that $LEFT$ can be safely transformed into a full, rectangular, convex totally monotone matrix, as needed for the implementation of $SMAWK$ algorithm in the alignment stage. We start by noting that, originally, $LEFT$ is not a full, rectangular matrix, since some of its entries are undefined. Recall that $LEFT[r, k]$ is defined as $S[PI[r], k - r]$. Since I is a monotone staircase, $PI[r]$ is defined for $0 \leq r \leq L_I$, where L_I denotes $|LCS[T, B_i]|$, and undefined for $r > L_I$. We consider only the first L_I rows of $LEFT$ and thus we may assume that $PI[r]$ is always defined. It follows that an entry $LEFT[r, k]$ is undefined whenever the entry $S[PI[r], k - r]$ is undefined.

Definition 4. *Let* k_r, *for* $0 \leq r \leq L_I$, *denote the greatest column index of an entry in row* r *of* $LEFT$ *whose value is defined.*

Lemma 1. *Each row in* $LEFT$ *consists of a (possibly empty) span of undefined entries, called the undefined prefix, followed by a span of defined entries, and by a (possibly empty) span of undefined entries, called the undefined suffix. For a given row* r *of* $LEFT$, *the span of defined entries starts in entry* $LEFT[r, r]$ *and ends in some entry* $LEFT[r, k_r]$, *such that* $r \leq k_r \leq L$.

Proof. Suppose that for $c < e$, both $LEFT[r, c]$ and $LEFT[r, e]$ are defined. From the definition of the table S it follows that all the entries $S[PI[r], d - r]$, for $c \leq d \leq e$ are also defined and thus also $LEFT[r, d]$. This means that the defined entries in each row of $LEFT$ form a consecutive interval. Following the definition of the S and $LEFT$ tables, $S[PI[r], 0]$ is always defined and therefore $LEFT[r, r] = S[PI[r], r - r]$ is always defined. $S[PI[r], -1]$, on the other hand, is never defined and therefore $LEFT[r, r - 1]$ is never defined. Thus we conclude that the span of consecutive defined entries in row r of $LEFT$ begins in $LEFT[r, r]$ and ends in $LEFT[r, k_r]$.   □

Our next goal is to prove that the defined entries of $LEFT$ follow the convex total monotonicity property. Later, we show how to complement the undefined entries of LEFT, without changing its column minima, and still maintain this property.

Lemma 2. *For any* a, b, c, *such that* $a < b$ *and all four entries:* $LEFT[a, c]$, $LEFT[b, c]$, $LEFT[a, c + 1]$, *and* $LEFT[b, c + 1]$ *are defined, if* $LEFT[a, c] \geq LEFT[b, c]$, *then* $LEFT[a, c + 1] \geq LEFT[b, c + 1]$.

Proof. The proof is based on a crossing paths contradiction [[4], [10], [13], [15]].

We consider two paths in the DP Graph of $B_i Y$ and T. Let path A_{c+1} denote an optimal path (of weight c+1) connecting vertex $(0, 0)$ of the graph with vertex $LEFT[a, c + 1]$ of O and going through vertex $PI[a]$ of I. Note

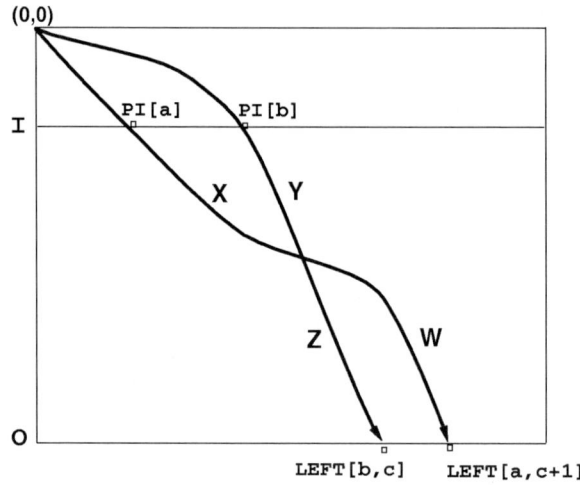

Fig. 3. Optimal paths which must cross.

that by our definition such a path always exists. Similarly, let path B_c denote an optimal path connecting vertex $(0,0)$ of the graph with vertex $LEFT[b,c]$ of O and going through vertex $PI[b]$ of I. Figure 3 shows paths B_c and A_{c+1}. Note that, since $LEFT[b,c] \leq LEFT[a,c] < LEFT[a,c+1]$, the two paths B_c and A_{c+1} must intersect at some column of the DP Graph before or at column $LEFT[b,c]$. Let X and Y be the prefixes of A_{c+1} and B_c up to the intersection point, and let W and Z be their suffixes from this intersection point.

There are two cases to be addressed, depending on the outcome of the comparison of the weights of X and Y, denoted $|X|$ and $|Y|$.

Case 1. $|X| \leq |Y|$. Then $|Y|+|W| \geq c+1$. By monotonicity of LCS this implies that $LEFT[b,c+1] \leq LEFT[a,c+1]$, in agreement with the right side of the Total Monotonicity condition.

Case 2. $|X| > |Y|$. Then $|X| + |Z| > c$. By monotonicity of LCS this implies that $LEFT[a,c] < LEFT[b,c]$, in contradiction to the assumption of the proof. $\square$

Note that the two assertions of Lemmas 1 and 2 lead, by definition, to the following conclusion.

Conclusion 1. *The defined entries of LEFT follow the convex total monotonicity property.*

We now turn to handle the undefined entries of $LEFT$. By Lemma 1 we know that the span of defined entries in a given row r starts in $LEFT[r,r]$ and ends in $LEFT[r,k_r]$. This implies, by definition, that the undefined prefix of row r consists of its first $r-1$ entries, and thus the undefined prefixes form a lower left triangle in $LEFT$.

The next Lemma assists in defining the shape formed by the undefined suffixes of the rows of $LEFT$.

Lemma 3. *For any two rows a, b of LEFT, where $a < b$, if $k_a > k_b$, then $LEFT[a, \ell] \leq LEFT[b, \ell]$, for all the defined elements in rows a and b of LEFT.*

Proof. Since $k_a > k_b$, it follows that $LEFT[a, k_b]$, $LEFT[b, k_b]$ and $LEFT[a, k_b + 1]$ are defined, yet $LEFT[b, k_b + 1]$ is undefined. Note that for every index $b \leq j \leq k_b$ both $LEFT[b, j]$ and $LEFT[a, j]$ are defined. Suppose there was some index $b \leq j \leq k_b$ such that $LEFT[a, j] > LEFT[b, j]$.

We consider two paths in the DP Graph of $B_i Y$ and T. Let path A_{k_b+1} denote an optimal path (of weight $k_b + 1$) connecting vertex $(0, 0)$ of the graph with vertex $LEFT[a, k_b + 1]$ of O and going through vertex $PI[a]$ of I. Note that by our definition such a path always exists. Similarly, let path B_j denote an optimal path (of weight j) connecting vertex $(0, 0)$ of the graph with vertex $LEFT[b, j]$ of O and going through vertex $PI[b]$ of I.

Similarly to the proof of Lemma 2, one can show that these two paths cross and that since $LEFT[a, j] > LEFT[b, j]$ there must be a path of weight at least $k_b + 1$ that originates in vertex $(0, 0)$, ends in O and goes through vertex $PI[b]$ of I, in contradiction to the assumption that $LEFT[b, k_b + 1]$ is undefined. We conclude that, for all the defined elements in rows a and b of $LEFT$, $LEFT[a, j] \leq LEFT[b, j]$.   □

Lemma 3 implies that, for any row r of $LEFT$, if there exists another row r' in $LEFT$, such that $r' < r$ and $k_r < k_{r'}$, then row r can be skipped in the search for column minima. After removing all such rows we are guaranteed that the undefined suffixes form an upper right triangle in $LEFT$.

The undefined entries of $LEFT$ can be complemented in constant time each, similarly to the solution described in [5], as follows.

1 Lower Left Triangle: These entries can be complemented by setting the value of any $LEFT[r, j]$ in the missing lower-left triangle to $(n + r + 1)$.
2 Upper Right Triangle: The value of any undefined entry $LEFT[r, j]$ in this triangle can be set to ∞.

In the next lemma we prove that $LEFT$ can be safely converted into a full, rectangular, convex totally monotone matrix.

Lemma 4. *The LEFT matrix can be transformed into a full, rectangular matrix that is totally monotone, without changing its column minima.*

Proof. We have shown (Conclusion 1) that the defined entries of $LEFT$ follow the convex total monotonicity property. It remains to show that complementing the undefined entries of $LEFT$, as described above, preserves its column minima, and still maintains this property.

1 Lower Left Triangle: The greatest possible index value in $LEFT$ is n. Since r is always greater than or equal to zero, the complemented values in the lower left triangle are lower-bounded by $(n + 1)$ and no new column minima are introduced. Also, for any complemented entry $LEFT[b, c]$ in the lower left triangle, $LEFT[a, c] < LEFT[b, c]$ for all $a < b$, and therefore the convex total monotonicity condition holds.

2 **Upper Right Triangle:** All scores in $LEFT$ are finite. Therefore, no new column minima are introduced by the re-defined entries.

Due to the upper right corner triangular shape of this ∞-patched area, for any two rows a, b, where $a < b$, if $LEFT[b, c] = \infty$, then surely $LEFT[a, c] = \infty$. Let d be the smallest index such that $LEFT[a, d] = \infty$. It follows that $LEFT[a, e] \geq LEFT[b, e]$ for all $e \geq d$, and the convex total monotonicity property is preserved. $\qquad\square$

Time Complexity. We show that the overhead associated with the marking of the undefined elements in $LEFT$, as well as the removal of redundant rows in $LEFT$ in order to obtain the crisp undefined upper right triangle, does not slow down the time complexity of the suggested Common Substring Alignment algorithm.

1. **Encoding Stage.** During the encoding stage, $LCS[T_{r+1}^n, Y]$, for $r = 0, \ldots, n-1$ is computed and stored. This information will be used later, during the alignment stage, to mark the beginning of the undefined suffix in each row. $LCS[T_{r+1}^n, Y]$ corresponds to the greatest column index of a defined entry in row r of S. Therefore, all n values of $LCS[T_{r+1}^n, Y]$, for $r = 0, \ldots, n-1$, can be queried from the constructed S table without changing the original $O(nL_y)$ complexity of the encoding stage algorithm.

2. **Alignment Stage.**
   a) **Marking the Defined and Undefined Intervals in Each Row of** $LEFT$. Given a row index r and the value $LCS[T_{r+1}^n, Y]$ which was computed in the encoding stage, the undefined prefix and suffix of row r of $LEFT$ can each be identified and marked in a constant time. Since there are $O(L)$ rows in $LEFT$, this work amounts to an additional $O(L)$ time.
   b) **Removing the Redundant Rows to Create an Upper Right Triangle.** Since PI is available in the beginning of the alignment stage, the $O(L)$ representative rows of $LEFT$ can be scanned in increasing order as a first step in the alignment stage, prior to the activation of $SMAWK$, and the rows which become redundant identified and removed.

   Altogether, this additional $O(L)$ work does not change the original $O(L)$ time complexity of the alignment stage.

5   Conclusions and Open Problems

The Sparse LCS Common Substring Alignment algorithm described in this paper consists of an $O(nL_y)$ time encoding stage and an $O(L)$ time alignment stage. It is intended for those applications where the source strings contain shared and repeated substrings. Note that we just leverage on the appearance of common substrings in the source strings. It is an open problem whether a more efficient algorithm exists when the target string contains encoded repetitions as well as the source strings.

Another remaining open challenge is to try to extend the solutions presented in this paper to more general metrics, such as Edit Distance.

References

1. A. Apostolico, String editing and longest common subsequences. In G. Rozenberg and A. Salomaa, editors, *Handbook of Formal Languages*, Vol. 2, 361–398, Berlin, 1997. Springer Verlag.
2. Apostolico A., and C. Guerra, The longest common subsequence problem revisited. *Algorithmica*, **2**, 315–336 (1987).
3. Aggarwal, A., M. Klawe, S. Moran, P. Shor, and R. Wilber, Geometric Applications of a Matrix-Searching Algorithm, *Algorithmica*, **2**, 195–208 (1987).
4. Benson, G., A space efficient algorithm for finding the best nonoverlapping alignment score, *Theoretical Computer Science*, **145**, 357–369 (1995).
5. Crochemore, M., G.M. Landau, and M. Ziv-Ukelson, A Sub-quadratic Sequence Alignment Algorithm for Unrestricted Cost Matrices, *Proc. Symposium On Discrete Algorithms*, 679–688 (2002).
6. Eppstein, D., Z. Galil, R. Giancarlo, and G.F. Italiano, Sparse Dynamic Programming I: Linear Cost Functions, *JACM*, **39**, 546–567 (1992).
7. Gusfield, D., Algorithms on Strings, Trees, and Sequences. *Cambridge University Press*, (1997).
8. Hirshberg, D.S., "Algorithms for the longest common subsequence problem", *JACM*, **24**(4), 664–675 (1977).
9. Hunt, J. W. and T. G. Szymanski. "A fast algorithm for computing longest common subsequences." *Communications of the ACM*, **20**, 350–353 (1977).
10. Kannan, S. K., and E. W. Myers, An Algorithm For Locating Non-Overlapping Regions of Maximum Alignment Score, *SIAM J. Comput.*, **25**(3), 648–662 (1996).
11. Landau, G.M., and M. Ziv-Ukelson, On the Shared Substring Alignment Problem, *Proc. 11th Annual ACM-SIAM Symposium on Discrete Algorithms*, 804–814 (2000).
12. Landau, G.M., and M. Ziv-Ukelson, On the Common Substring Alignment Problem, *Journal of Algorithms*, **41**(2), 338–359 (2001)
13. Monge, G., Déblai et Remblai, *Mémoires de l'Academie des Sciences*, Paris (1781).
14. Myers, E. W., "Incremental Alignment Algorithms and their Applications," *Tech. Rep. 86-22, Dept. of Computer Science, U. of Arizona.* 1986.
15. Schmidt, J.P., All Highest Scoring Paths In Weighted Grid Graphs and Their Application To Finding All Approximate Repeats In Strings, *SIAM J. Comput*, **27**(4), 972–992 (1998).

On Minimizing Pattern Splitting in Multi-track String Matching

Kjell Lemström and Veli Mäkinen

Department of Computer Science, University of Helsinki
P.O. Box 26 (Teollisuuskatu 23), FIN-00014 Helsinki, Finland
{klemstro,vmakinen}@cs.helsinki.fi

Abstract. Given a pattern string $P = p_1 p_2 \cdots p_m$ and K parallel text strings $\mathbb{T} = \{T^k = t_1^k \cdots t_n^k \mid 1 \leq k \leq K\}$ over an integer alphabet Σ, our task is to find the smallest integer $\kappa > 0$ such that P can be split into κ pieces $P = P^1 \ldots P^\kappa$, where each P^i has an occurrence in some text track T^{k_i} and these partial occurrences retain the order. We study some variations of this minimum splitting problem, such as splittings with limited gaps and transposition invariance, and show how to use sparse dynamic programming to solve the variations efficiently. In particular, we show that the minimum splitting problem can be interpreted as a shortest path problem on line segments.

1   Introduction

In multi-track string matching [4,11] the task is to find occurrences of a pattern across parallel strings. Given a pattern string $P = p_1 \cdots p_m$ and K parallel text strings $\mathbb{T} = \{T^k = t_1^k \cdots t_n^k \mid 1 \leq k \leq K\}$ over an integer alphabet Σ, there is such an occurrence at j iff $p_1 = t_{j-m+1}^{k_1}, \cdots, p_m = t_j^{k_m}$, where $k_i \in [1, K]$ for $1 \leq j \leq m$.

This basic problem is a generalization of exact string matching for multi-track texts. An efficient solution is achievable by casting the problem into one of subset matching; the recent result of Cole and Hariharan [3] gives an $O(Kn \log^2 Kn)$ time solution. One can also deploy bit-parallelism. It is easy to implement the naïve algorithm to work in time $O((Kn + mn)\lceil |\Sigma|/w \rceil)$ by generalizing the solution of Iliopoulos and Kurokawa [9] for the case $|\Sigma| \leq w$, w denoting the size of machine words in bits.

In a variation of the problem the occurrences may be transposed, i.e., in the formulation above there is an occurrence if $p_1 = t_{j-m+1}^{k_1}+c, \cdots, p_m = t_j^{k_m}+c$ for some constant c. Lemström and Tarhio [11] give an efficient bit-parallel filtering algorithm for solving the transposition invariant multi-track string matching problem. After an $O(nK\lceil |\Sigma|/w \rceil)$ preprocessing phase, the algorithm runs in time $O(n\lceil m/w \rceil + m + d)$ and space $O(n\lceil |\Sigma|/w \rceil + e)$ where d and e denote factors dependent on the size of the alphabet. The found candidates can then be confirmed, e.g., by a checking algorithm working in time $O(mKn)$ and space $O(m\lceil |\Sigma|/w \rceil)$ [11].

R. Baeza-Yates et al. (Eds.): CPM 2003, LNCS 2676, pp. 237–253, 2003.

In some cases it is useful to allow *gaps* between the matching elements, i.e., $p_i = t_j^{k_i}, p_{i+1} = t_{j+1+\ell_i}^{k_{i+1}}$ for $1 \le i \le m - 1$ with some $0 \le \ell_i \le n - j - 1$. The gaps ℓ_i can be controlled by an integer $0 \le \alpha \le \infty$ by requiring $\ell_i \le \alpha$ for all i. Limited gaps are considered, for instance, in [6,9] while unlimited gaps ($\alpha = \infty$) in [13,14].

In this paper, we are interested in finding the smallest integer $\kappa > 0$ such that P may be split into κ pieces $P = P^1 \ldots P^\kappa$ where each P^i has an occurrence in some text track T^{k_i} and the occurrences of the consecutive pieces must retain the order. We control the gaps between the occurrences of P^{i-1} and P^i by using the integer α.

We study this *minimum splitting problem* with its variations. Our solutions use sparse dynamic programming based on the match set $M = \{(i, j, k) \mid p_i = t_j^k\}$. The problem of unlimited gaps is solved in time $O(|M| + m + Kn)$ and space $O(|\Sigma| + |M| + m + n)$. Applying this algorithm to the transposition invariant case results in time complexity $O(mKn)$. Moreover, the result generalizes to the case of α-limited gaps. In the case where the matching symbol pairs of M form relatively long diagonal runs, called *line segments*, we obtain a more efficient algorithm solving the problem. Having constructed the set of maximal line segments $\hat{S}$ in time $O(|\Sigma|^2 + |\Sigma|m + m^2 + Kn + \hat{S} \log K)$, the minimum splitting problem is solved in time $O(|\hat{S}| \log n)$. Modifications of this algorithm find the α-limited gap occurrences in time $O(|R| \log n)$, where $|R| \le \min(|\hat{S}|^2, |M|)$. A more efficient algorithm, running in time $O(\kappa Kn)$, is given for the case $\alpha = 0$.

The problem has a natural application in content-based music retrieval: the pattern is a short excerpt of a (monophonic) query melody that may be given, e.g. by humming. Then the melody is searched for in a multi-track string representing a music database of polyphonic (multi instrument) music. However, it is musically more pertinent not to allow the pattern to be totally distributed across the tracks, but to search for occurrences of the pattern that make as few jumps as possible between different instruments. The continuity of a "hidden melody" is considered by limiting the lengths of the gaps.

Consider, for instance, the excerpt given in Fig. 1. The excerpt can be represented by a pitch string $P = efghc$, or in a transposition invariant way, by using the pitch intervals between the notes: $P' = +1, +2, +4, +1$. Fig. 2 is an excerpt from a polyphonic, four-track vocal music by Selim Palmgren. The four tracks (denoted TI, TII, BI, and BII) are divided into two systems, both containing two tracks. The two tracks within a system can be discriminated by observing the direction of the note beams; the upper track contains notes with upper beams, the lower tracks with lower beams. The excerpt in Fig. 1 has two 0-limited ($\alpha = 0$), transposed occurrences of $\kappa = 2$ in Fig. 2. The first starts at the second note in the lowest, BII track, and is distributed across two tracks (BII and TII). The second occurrence starts at the eighth note and is distributed across tracks BII and TI. It is noteworthy, that the latter occurrence captures the transition of the musical melody from the lowest tone to the uppermost tone (the perceived melody resides in the 10 first notes of track BII, and in the 5 last notes of TI).

Fig. 1. A monophonic query melody.

Fig. 2. An excerpt of a four-track vocal composition. The query melody in Fig. 1 has two transposed occurrences such that $\kappa = 2$ and $\alpha = 0$.

2   Definitions

Let Σ denote a fixed integer *alphabet*, i.e. $\Sigma \subset \mathbb{Z}$. The size of Σ is defined as $|\Sigma| = \max(\Sigma) - \min(\Sigma) + 1$. A *string* A over an integer alphabet Σ is a sequence $A = a_1 a_2 \cdots a_m$, where each *character* $a_i \in \Sigma$. The length of A is denoted by $|A| = m$. The string of length 0 is denoted by ϵ.

A sequence $a_i \cdots a_j = A_{i \ldots j}$ is a *substring* of A, where $1 \le i \le j \le m$. Substring $A_{1 \ldots j}$ is called a *prefix* and substring $A_{j \ldots m}$ is called a *suffix* of A.

A *multi-track text* $\mathbb{T}$ is a set of equal length strings. The cardinality of $\mathbb{T}$ is denoted by $\|\mathbb{T}\| = K$ and the length of each string in $\mathbb{T}$ is denoted by $|\mathbb{T}| = n$. A multi-track text $\mathbb{T}$ can be written as $\{T^k = t_1^k \cdots t_n^k \mid 1 \le k \le K\}$, where T^k denotes the string at track k.

Problem 1 *Given a pattern string P and a multi-track text $\mathbb{T} = \{T^k \mid 1 \le k \le K\}$, the* minimum splitting problem *is to find the smallest integer κ such that P can be split into κ pieces $P = P^1 P^2 \cdots P^\kappa$, where each P^i occurs in some text track T^{k_i} and the consecutive pieces must retain the order, i.e., $P^i = T^{k_i}_{j_i - |P^i| + 1 \ldots j_i}$, and $j_{i-1} < j_i - |P^i| + 1 \le n - |P^\kappa| + 1$ ($j_0 = 0$) for $k_i \in [1, K]$.*

3   Solution Based on Sparse Dynamic Programming

We next describe a sparse dynamic programming algorithm for solving the
minimum splitting problem. Let P, $\mathbb{T}$ be an instance of the problem. Let
$M = M(P, \mathbb{T})$ be the set of matching character triples between P and each
text in $\mathbb{T}$, i.e.

$$M = \{(i, j, k) \mid p_i = t_j^k, 1 \le i \le m, 1 \le j \le n, 1 \le k \le K\}. \tag{1}$$

Our algorithm fills (sparsely) an $m \times n \times K$ array (d_{ijk}) such that d_{ijk} stores the
minimum splitting κ needed between $p_1 \cdots p_i$ and $t_{j'}^{k_1} \cdots t_j^{k_\kappa}$, where $1 \le j' < j$,
$k_i \in [1, K]$, and $k_\kappa = k$. Note that only cells $(i, j, k) \in M$ need to be filled. This
sparse matrix is visualized in Fig. 3.

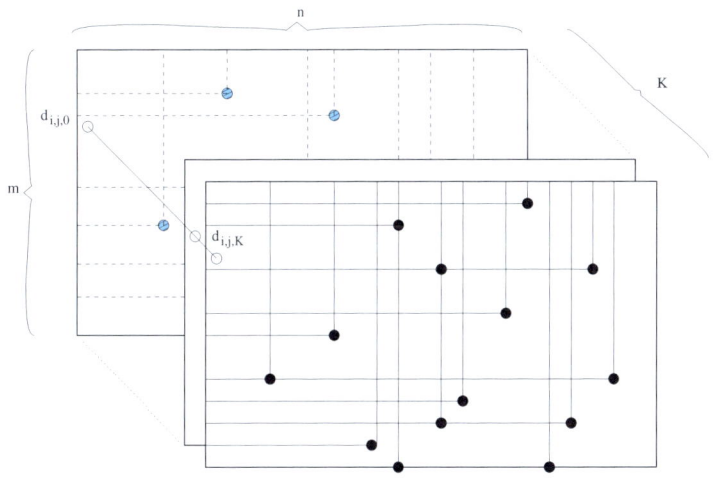

Fig. 3. The sparse matrix M.

3.1   Basic Algorithm

The recurrence for values $d_{i,j,k}$ is easily derivable from the definition. For
$(i, j, k) \in M$, it is

$$d_{i,j,k} = \min \begin{cases} \text{if } (i-1, j-1, k) \in M \text{ then } d_{i-1,j-1,k} \text{ else } \infty \\ d_{i-1,j',k'} + 1 \quad \text{where } (i-1, j', k') \in M, j' < j \ (*) \end{cases}, \tag{2}$$

with the boundary condition that $d_{1,j,k} = 0$ for each $(1, j, k) \in M$. Value $\kappa =$
$\min\{d_{m,j,k} \mid (m, j, k) \in M\}$ gives the solution to the minimum splitting problem.

An $O(mn)$ time dynamic programming algorithm is immediate from recur-
rence (2). We next show how to obtain $O(|M|)$ time. First, we need M con-
structed in some proper evaluation order: if $d_{i',j',k'}$ is needed for computing
$d_{i,j,k}$, then (i', j', k') must precede (i, j, k) in M. We use the following order.

Definition 2 *The* reverse column-by-column order *of* $M \subset \{(i,j,k)\}$ *is given by sorting* M *first by* i *in ascending order, then by* j *in descending order, and then by* k *in ascending order. That is,* (i',j',k') *precedes* (i,j,k) *iff* $i' < i$ *or* $(i' = i$ *and* $(j' > j$ *or* $(j' = j'$ *and* $k' < k)))$.

Let us, for now, assume that we are given M sorted in reverse column-by-column order into an array $L(1 \ldots |M|)$, where $L(p) = (i,j,k,d)$, $1 \le p \le |M|$. The value d in $L(p)$ will be used to store value $d_{i,j,k}$. We denote by $L(p).\aleph$ the element $\aleph$ of $L(p)$. Moreover, let us assume that given a cell $L(p) = (i,j,k,d)$ we can access an existing cell $diag(p) = L(p') = (i-1,j-1,k,d')$ in constant time. With such representation of M, the following algorithm solves the minimum splitting problem.

Algorithm 1.
(1) $C(0) \leftarrow \infty$
(2)   **for** $p \leftarrow 1$ **to** $|M|$ **do begin**
(3) $L(p).d \leftarrow \min(diag(p).d, C(L(p).i - 1) + 1)$
(4) $C(L(p).i) \leftarrow \min(C(L(p).i), L(p).d)$ **end**

Having run Algorithm 1, value $\kappa = \min\{L(p).d \mid L(p).i = m\}$ gives the solution.

To see that Algorithm 1 evaluates recurrence (2), consider the following. Array $C(0 \ldots m)$ is for keeping the row minima, corresponding to line 2 of recurrence (2). The reverse column-by-column evaluation guarantees that the row minimum $C(L(p).i - 1)$ is indeed taken over the correct values once computing $L(p).d$. Line 1 of recurrence (2) is taken into account in $diag(p).d$ of row (3): If $diag(p)$ does not exist, either $L(p).i = 1$ (first row) in which case we interpret $diag(p).d = 0$, or $L(p).i \neq 1$ and we interpret $diag(p).d = \infty$. This corresponds to the boundary condition of recurrence (2).

Algorithm 1 runs in time $O(|M|)$.

3.2   Constructing Set M

Let us now show how to achieve an $O(|\Sigma| + m + Kn + |M|)$ total running time by constructing an array $L(1 \ldots |M|)$ with pointers $diag(p)$ to represent M as required.

Lemma 3 *Given a pattern string P of length m and a multi-track text $\mathbb{T}$ of cardinality K and of length n, the array $L(1 \ldots |M|)$ giving the reverse column-by-column order of match set $M = M(P, \mathbb{T})$ can be constructed in time $O(|\Sigma| + m + Kn + |M|)$. Also the pointers $diag(p) = L(p') = (i-1, j-1, k, d')$ from each $L(p) = (i,j,k,d)$ can be assigned within the same time bound.*

Proof. First, we construct for each character $\sigma \in \Sigma$ the list of its consecutive occurrences in $\mathbb{T}$. This is done for all characters at once by traversing through

$\mathbb{T}$ in an order $t_1^1, t_1^2, \ldots, t_1^K, t_2^1, \ldots, t_n^K$. At t_j^k we insert (j, k) into the end of a list $Occ(t_j^k)$. Each insertion takes constant time, since we can use an array $Occ(1 \ldots |\Sigma|)$ to store lists. Thus, this phase is accomplished in $O(|\Sigma| + Kn)$ time.

Second, we traverse through P, and at p_i we get the ith *row* of M by copying list $Occ(p_i)$ and by adding i into each cell (j, k) to produce cell (i, j, k). At the end, we have produced M, but it is in row-by-row order. However, it is easy to produce the reverse column-by-column order; traverse through M and add each cell (i, j, k) to the beginning of a list $Col(j)$. Concatenating lists $Col(1)Col(2) \cdots Col(n)$ gives M in reverse column-by-column order. The second phase requires $O(m + |M| + n)$ time.

Finally, we assign the pointers $diag(p)$: We sort M in *diagonal-by-diagonal order*, first by values $j - i$, then by j, and then by k. This takes $O(|M|)$ time (implementation is analogous to the case considered above). Once M is ordered, the pointers are easy to assign. Consider first that $K = 1$. Then the cells to be linked are consecutive, and pointers can be assigned in constant time. When $K > 1$, one can still do this in constant time per cell by merging the sets $F(i, j) = \{(i - 1, j - 1, k) \mid (i - 1, j - 1, k) \in M\}$ and $S(i, j) = \{(i, j, k) \mid (i, j, k) \in M\}$ by value k. When $F(i, j)$ and $S(i, j)$ are merged, the cells to be linked become consecutive. Each merging can be done in linear time in the size of sets $F(i, j)$ and $S(i, j)$, since both sets can be obtained from M already in correct order. Each cell of M belongs to (at most) one $F(i, j)$ and to (at most) one $S(i, j)$, which gives the $O(|M|)$ time for assigning all $diag(p)$ pointers.

Summing all the running times, we get the claimed bound.   □

3.3   Limiting Gaps with α

Let us now consider the case of α-limited gaps. We will show that the time bound $O(|\Sigma| + m + Kn + |M|)$ of the unlimited case, can be achieved in this case, as well.

When gaps are limited, the condition (*) in recurrence (2) becomes $j - \alpha \leq j' < j$. A straightforward generalization of Algorithm 1 would result in time complexity $O(|\Sigma| + m + Kn + |M|\alpha)$. We can do better by applying a queue data structure Q supporting the following operations:

- $v \leftarrow Q.Min()$: Return the minimum value v of elements in the queue;
- $Q.Add(j, v)$: Add new element j as the first element with value v;
- $j' \leftarrow Q.KeyOfLast()$: Return the key of the last element; and
- $Q.Remove()$: Remove the last element.

The modification for Algorithm 1 is evident: Instead of maintaining each row minimum in $C(i)$, use a queue $Q(i)$ for each row i. The algorithm changes into the following.

Algorithm 2.
(1)   **for** $p \leftarrow 1$ **to** $|M|$ **do begin**
(2) $L(p).d \leftarrow \min(diag(p).d, Q(L(p).i - 1).Min() + 1)$
(3) $Q(L(p).i).Add(L(p).j, L(p).d)$
(4)   **while** $L(p).j - \alpha > Q(i).KeyOfLast()$ **do** $Q(i).Remove()$ **end**

Analogously to Algorithm 1, we interpret $diag(p).d = \infty$ if $diag(p)$ does not exist, except when $L(p).i = 1$ we interpret $diag(p).d = 0$.

If the queues are implemented as min-dequeues [7], all the above mentioned operations can be supported in constant time.[1] Thus, the algorithm runs in time $O(|\Sigma| + m + Kn + |M|)$.

3.4   Transposition Invariance

The $O(|M|)$ algorithms are useful for the transposition invariant case, as well. Let M^c be the set of matches for transposition c, i.e. $M^c = M^c(P, \mathbb{T}) = \{(i, j, k) \mid p_i + c = t_j^k\}$. Then $\sum_{c \in \mathbb{C}} |M^c| = mKn$, where $\mathbb{C} = \{t_j^k - p_i\}$ [12]. Note that $\mathbb{C}$ is the set of relevant transpositions, since for other $c \in \mathbb{R} \setminus \mathbb{C}$, the corresponding set M^c is empty. The sets $\{M^c \mid c \in \mathbb{C}\}$ can be constructed in $O(|\Sigma| + mKn)$ time on integer alphabets [12]. By repeating the previous $O(|M|)$ algorithms for each $c \in \mathbb{C}$, we get total running time $O(|\Sigma| + mKn + \sum_{c \in \mathbb{C}} |M^c|) = O(|\Sigma| + mKn)$.

The results of Sect. 3 are summarized below.

Theorem 4 *Given a pattern string P of length m and a multi-track text $\mathbb{T}$ of cardinality K and of length n, there is an algorithm solving the minimum splitting problem (with or without the α–restriction on gaps) in $O(|\Sigma| + m + Kn + |M|)$ time on integer alphabet, where $M = \{(i, j, k) \mid p_i = t_j^k\}$. The transposition invariant case can be solved in $O(|\Sigma| + mKn)$ time.*

4   Interpretation as a Shortest Path Problem on Line Segments

In this section, we interpret the minimum splitting problem as a geometric path problem on line segments and define the minimum jump distance. For simplicity, let us assume that we have only one text; the minor modifications for the case of multiple texts are given in Sect. 4.4.

The set of possible matches M is written as $M = M(P, T) = \{(i, j) \mid p_i = t_j\}$ during this section. Consider the $m \times n$ matrix, where pairs in M are sparse set of points. Diagonally consecutive points $(i, j), (i+1, j+1), \ldots, (i+\ell-1, j+\ell-1) \in M$ form a line segment S in this grid. Let us denote by $\hat{S} = \hat{S}(P, T)$ the set of all *maximal* line segments of M. A line segment $S = (i, j) \cdots (i+\ell-1, j+\ell-1)$

---

[1] Min-deques were also used by Crochemore et al. [6] for a similar problem.

is maximal if $(i-1, j-1), (i+\ell, j+\ell) \notin M$. Let us denote by $start(S) = (i,j)$ and $end(S) = (i+\ell-1, j+\ell-1)$ the two end points of line segment $S \in \hat{S}$. Our problem now equals the following geometric path problem: Find a minimum cost path from row 1 to row m such that only diagonal steps are allowed when inside a line segment, otherwise only horizontal steps are allowed. To be precise, we must also take the discrete nature of the problem into account; one can jump (using horizontal steps) from segment S' to segment S only if there is $p \geq 2$ such that $(i-1, j-p) \in S'$ for some $(i,j) \in S$. The cost of a path is the number of line segments in the path -1. *Minimum jump distance* equals the minimum cost path from row 1 to row m. This interpretation as a shortest path problem is visualized in Fig. 4.

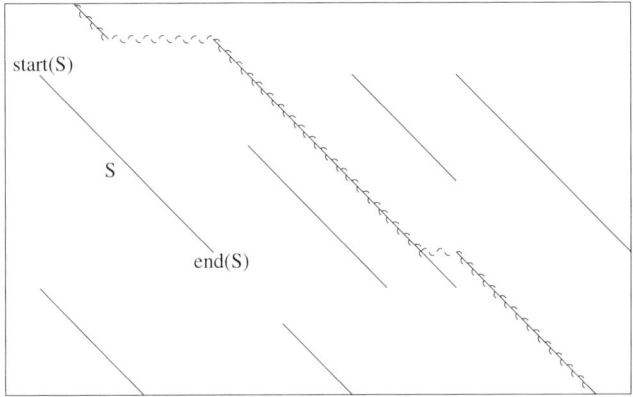

Fig. 4. Example of line segments. An optimal path from top to the bottom is visualized.

The following subsections considers, whether: (i) the set of line segments $\hat{S}$ can be constructed in a time proportional to $|\hat{S}|$, and (ii) the minimum jump distance can be computed in a similar time.

4.1 Constructing Set $\hat{S}$

Next we give an algorithm that, after $O(m^2 + |\Sigma|m + |\Sigma|^2)$ time preprocessing, constructs set $\hat{S}$ in time $O(n + |\hat{S}|)$. We use and extend some results of Ukkonen and Wood [15].

Let $Prefix(A, B)$ denote the length of the longest common prefix of A and B. Let $Maxprefix(j)$ be $\max\{Prefix(P_{i...m}, T_{j...n}) \mid 1 \leq i \leq m\}$, and $H(j)$ some index i of P that gives the maximum. Let $Jump(i,j)$ denote $Prefix(P_{i...m}, T_{j...n})$. We have the following connection.

Lemma 5 ([15]) $Jump(i,j) = \min(Maxprefix(j), Prefix(P_{i...m}, P_{H(j)...m}))$.

Ukkonen and Wood show that $Maxprefix(j)$ and $H(j)$ are computable, for all j, in $O(m^2 + |\Sigma| + n)$ time by using an Aho-Corasick automaton [1] for all suffices of the pattern (or more precisely its minimized version, the suffix automaton [5]). At preprocessing, one can construct an array $Pl(1\ldots m, 1\ldots m)$ in $O(m^2)$ time such that $Pl(i', i) = Prefix(P_{i'\ldots m}, P_{i\ldots m})$ [10,8]. From Lemma 5 it follows that, after $O(m^2 + |\Sigma| + n)$ time preprocessing, $Jump(i,j)$ can be supported in constant time for all pairs i, j.

Therefore, if we manage to call $Jump(i,j)$ only at points $(i,j) = start(S)$ for $S \in \hat{S}$, we will be able to construct $\hat{S}$ in $O(|\hat{S}|)$ time. Note that a point (i,j) equals $start(S)$ for some $S \in \hat{S}$ iff $p_{i-1} \neq t_{j-1}$ and $p_i = t_j$. We construct a data structure $\mathcal{L}$ that supports an operation $\mathcal{L}.List(x, y)$ giving the list of positions $\{i\}$ in P, where $p_{i-1} \neq x$ and $p_i = y$. Assuming that such data structure exists, the following algorithm constructs set $\hat{S}$ (represented as pairs of end points) in time $O(n + |\hat{S}|)$.

Algorithm 3.
(1)   /* construct $\mathcal{L}$ */
(2) $\hat{S} \leftarrow \emptyset$
(3)   **for** $j \leftarrow 1$ **to** n **do begin**
(4) $L \leftarrow \mathcal{L}.List(t_{j-1}, t_j)$ /* $t_0 = \epsilon$ */
(5)   **for** $k \leftarrow 1$ **to** $|L|$ **do begin**
(6) $\ell \leftarrow Jump(L(k), j)$
(7) $\hat{S} \leftarrow \hat{S} + ((L(k), j), (L(k) + \ell - 1, j + \ell - 1))$ **end end**

Let us now explain how to construct $\mathcal{L}$. We build a search tree of depth two representing all choices to read two characters. Let xy be two characters leading to leaf l. We associate a list of pattern positions $\{i\}$ to l such that $p_{i-1} \neq x$ and $p_i = y$. Then $\mathcal{L}.List(x, y)$ is found in linear time in the size of the output by traversing the path xy (to l) and printing the list associated with l. Now, the size of this structure is $O(|\Sigma|^2 + |\Sigma|m)$, since there are $O(|\Sigma|^2)$ leaves and each position i of P is stored in exactly $|\Sigma| - 1$ lists (except position 1 which is stored in $|\Sigma|$ lists, since we assume $p_0 \neq x$ for all x). The naïve algorithm to construct the structure (which inserts each i from $i = 1$ to $i = m$ into $\mathcal{L}$) runs in time proportional to the size of the structure.

Lemma 6 *Given a pattern string P and a text T of lengths m and n, respectively, Algorithm 3 constructs the set of maximal line segments $\hat{S}(P,T)$ in $O(m^2 + |\Sigma|m + |\Sigma|^2 + n + |\hat{S}|)$ time. Set $\hat{S}$ can be obtained in diagonal-by-diagonal order by a small adjustment of the algorithm.*

Proof. The result should be clear from the above discussion, expect for the diagonal-by-diagonal order. At line (7) of Algorithm 3, we can add the new line segment into the end of a list $D(j - L(k))$. Finally, we can catenate all the lists $D(-m), \ldots, D(n)$ to get $\hat{S}$ in diagonal-by-diagonal order. $\square$

4.2   Computing Minimum Jump Distance

Let $d((i,j))$ give the minimum jump distance from row 1 to (i,j). To get a bound close to $O(|\hat{S}|)$, we need an algorithm doing computation only at end points of line segments. To this end, we have the following lemma.

Lemma 7 *The minimum jump distance $d(end(S))$ equals to the minimum jump distance $d(start(S))$, for each $S \in \hat{S}$.*

Proof. If this is not true, then there must be a point $(i'',j'') \neq start(S)$ in S and a point $(i''-1,j''-p'')$ in some $S'' \in \hat{S}$ (where $p'' \geq 2$) such that $d((i''-1,j''-p'')) < d(start(S)) - 1$.

On the other hand, there must be a point $(i'-1,j'-p')$ in some $S' \in \hat{S}$ such that $start(S) = (i',j')$, $p' \geq p''$, and an optimal path from row 1 to point $(i''-1,j''-p'')$ traverses through $(i'-1,j'-p')$. From which follows that $d((i'-1,j'-p')) \leq d((i''-1,j''-p''))$.

Thus, $d((i',j')) \leq d((i'-1,j'-p')) + 1 \leq d((i''-1,j''-p'')) + 1$, which contradicts the counter-argument, and the lemma follows. $\qquad\square$

As a consequence of Lemma 7, we also know that $d((i'',j'')) = d(start(S))$ for each $(i'',j'') \in S$. Thus, it is enough to consider only end points of line segments, which are traversed e.g. in row-by-row order. Let $d(S)$ denote the minimum jump distance on line segment S. A data structure $\mathcal{T}$ stores the diagonal number $j-i$ of each line segment S together with $d(S)$. Structure $\mathcal{T}$ is updated only when some line segment starts or ends. At any point, $\mathcal{T}$ contains only those line segments that intersect row $i-1$.

Consider now that we are computing value $d((i,j))$ where $(i,j) = start(S)$ for some $S \in \hat{S}$. We need to query $\mathcal{T}.Min([-\infty, j-i))$, which gives us the value $\min\{d((i',j')) \mid S' \in \hat{S}, start(S') = (i',j'), S' \cap ([i-1] \times [0,n]) \neq \emptyset, j'-i' < j-i\}$. Then $d((i,j)) = \mathcal{T}.Min([-\infty, j-i)) + 1$. These queries can be supported by a binary search tree: use $j-i$ as the sort key, store values $d(S)$ in leaves, and maintain subtree minima in each internal node. A range query for the minimum value in a range $[l, r]$ can then be easily supported in $O(\log n)$ time. The update operations can be supported, too: When inserting a line segment, the tree can be rebalanced and the subtree minima need only be updated in the path from the new leaf to the root. Same applies for deleting a line segment. Each query takes $O(\log n)$ time, since there can be only $O(n)$ active line segments at a time.

The order of queries on $\mathcal{T}$ with respect to segment insertions and deletions is important; we must delay deletions at row $i-1$ until all queries at row i are done, and insertions at row i until all queries in that row are done. As an answer to question (ii), we conclude that the minimum jump distance can be computed in time $O(|\hat{S}| \log n)$, once the end points of the line segments are first given in row-by-row order. By Lemma 6, we get $\hat{S}$ in diagonal-by-diagonal order. It is easy to convert $\hat{S}$ into row-by-row order in $O(\hat{S})$ time using similar techniques as in the proof of Lemma 3.

To clarify the above steps, we now give the algorithm in more detail. Let us first fix the operations described above on $\mathcal{T}$:

- $v \leftarrow \mathcal{T}.Min(I)$: Return the minimum value v of elements having keys in range I;
- $\mathcal{T}.Insert(e, k, v)$: Insert new element e with key k and value v; and
- $\mathcal{T}.Delete(k)$: Remove the element with key k.

Let us then partition the end points of line segments in $\hat{S}$ by row number and the type of end point; lists $B(i)$ and $E(i)$ contain all start and end points of line segments starting/ending at row i, respectively. Each list $B(i)$, $E(i)$, $1 \ldots m$, is in increasing order of column numbers j. Each element s of $B(i)$ (or $E(i)$) contains values (i, j, seg), where seg is a pointer to the line segment S whose end point (i, j) is. Each line segment S is associated with value $S.d$ that is used for storing the minimum jump distance from row 1 to S. The algorithm is given below.

Algorithm 4.
(1) **for each** $s \in B(1)$ **do begin** /* initialize first row to zero */
(2) $s.seg.d \leftarrow 0$; $\mathcal{T}.Insert(s, s.j - s.i, s.seg.d)$; **end**
(3) **for** $i \leftarrow 2$ to m **do begin** /* compute row-by-row */
(4)   **for each** $s \in B(i)$ **do** /* compute values at row i */
(5) $s.seg.d \leftarrow \mathcal{T}.Min([-\infty, s.j - s.i)) + 1$
(6)   **for each** $s \in E(i-1)$ **do** /* remove those ending at row $i-1$ */
(7) $\mathcal{T}.Delete(s.j - s.i)$
(8)   **for each** $s \in B(i)$ **do** /* add new line segments starting at row i */
(9) $\mathcal{T}.Insert(s, s.j - s.i, s.seg.d)$; **end**

Having executed Algorithm 4, the minimum jump distance d from row 1 to row m is $d = \mathcal{T}.Min([-\infty, \infty])$.

Summing up the preprocessing time for constructing $\hat{S}$ and the computation time of Algorithm 4, we get the following result.

Theorem 8 *Given a pattern string P and text T of lengths m and n, respectively, the minimum splitting problem can be solved in time $O(|\Sigma|^2 + |\Sigma|m + m^2 + n + |\hat{S}| \log n)$, where $\hat{S}$ is the set of maximal line segments formed by the diagonally consecutive pairs of $M = \{(i, j) \mid p_i = t_j\}$.*

4.3   Limiting Gaps with α

Note that Lemma 7 does not hold if the lengths of the gaps are limited by a constant α. To see this, consider situation where there are no line segments within α distance from $start(S)$. In this case $d(start(S)) = \infty$. Nevertheless, there can be line segments within α distance from some internal points of S, in which case $d(end(S)) < d(start(S))$ may hold.

We give a weaker lemma stating that there is a sparse set of points R, $R \subseteq M$, such that similar computation as with $\hat{S}$ is possible. The size of R is bounded by $\min(|\hat{S}|^2, |M|)$. The intuition of the lemma is, that at each line segment S it is enough to consider only points s, such that, there is an optimal path that traverses through a start point of a line segment S' and that leads to s. Moreover, the part of the path from S' to S is of a certain, fixed form.

To state the lemma, we need to introduce some concepts. The feasible paths in the sparse set M can be represented in $\hat{S}$ as follows; a *feasible path* in $\hat{S}$ is a sequence of line segments visited by the path, and a list of entry and exit points on them. Let us now define the fixed form of path mentioned above. An *α-straight path* Π from $start(S')$ to S ($S', S \in \hat{S}$) is a sequence $\Pi = \pi_1 \cdots \pi_p$ of line segments of $\hat{S}$ such that $\pi_1 = S'$, $\pi_p = S$. Path Π must satisfy the following conditions:

(i) the gaps between consecutive line segments of Π are limited by α, i.e. $j_q - j_{q-1} - 1 \le \alpha$, where $1 < q \le p$ and $start(\pi_q) = (i_q, j_q)$;

(ii) only a single point in each line segment is visited, i.e. point $(i_1 + q - 1, j_q + q + i_1 - i_q - 1)$ must be inside line segment π_q, where $start(\pi_q) = (i_q, j_q)$, $1 \le q \le p$.

Definition 1. *An α-greedy path is an α-straight path Π whose line segments are picked greedily maximizing the length of gaps under constraints (i-ii).*

An α-greedy path is visualized in Fig. 5.

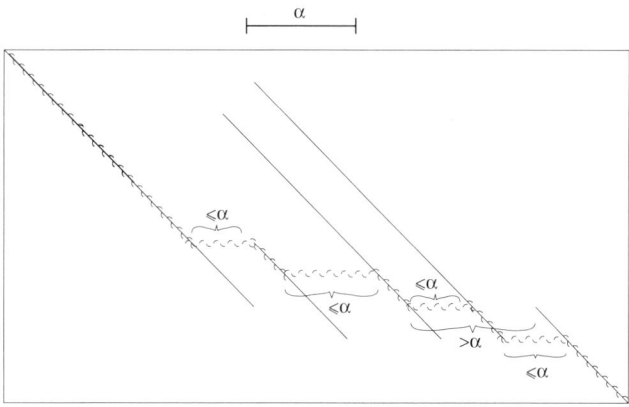

Fig. 5. An optimal path from top to the bottom. The path contains an α-greedy part from the start of the second line segment to the point where it enters the last line segment.

Lemma 9 *Let $s = (i, j)$ be a point on line segment $S \in \hat{S}$. Then there is an optimal path from row 1 to point s such that the last part of the path is an α-greedy path starting from point $start(S')$ of some line segment $S' \in \hat{S}$ to $(i - c, j - c) \in S$, $c \ge 0$.*

Proof. We will first prove that there is an optimal path that has an α-straight path as a suffix, and then we show that each α-straight path can be replaced by an α-greedy path.

Let S' be the last line segment whose start point $start(S')$ is on an optimal path from row 1 to point s. Such line segment always exists. Let us consider the last part of this path from $start(S')$ to s. We can straighten this path by removing one diagonal movement from each line segment it visits. Notice that the path is still feasible since the gaps between consecutive line segments do not change. We can continue straightening without changing the cost of the path until we meet the first start point of some line segment S'' or until the path is optimally straight, i.e. it is an α-straight path. In the latter case we are done. In the first case, we can start the same process from S'' and then by induction we finally find some line segment S''' which has an α-straight path from $start(S''')$ to point $(i - c, j - c) \in S$. It might happen that line segment S is the one that limits the straightening process at some point. This is a special case, since then we have a zero length α-straight path from $start(S)$ to $start(S)$.

Now, we still need to show that we can safely replace an α-straight path with an α-greedy path. Let $X = x_1 x_2 \cdots$ be the sequence of line segments visited by an α-straight path and $Y = y_1 y_2 \cdots$ the sequence visited by an α-greedy path such that both sequences start and end with the same line segments. If we replace x_1 with y_1, x_2 can still be reached from y_1, or x_2 is a predecessor of y_1 in the diagonal order. In the first case, we can continue on replacing x_2 with y_2. In the second case, we can omit x_2 since x_3 can be reached from y_1, or x_3 is a predecessor of y_1. Iterating this replacement for the whole X we get an equal or cheaper cost path.   □

An algorithm computing the minimum jump distance under the α constraint follows from Lemma 9. In addition to start and end points of line segments, some computation on the *intersection set* R, that contains points at the intersection of $\hat{S}$ and the α-greedy paths, is needed. Set R can be computed on-the-fly by extending Algorithm 4 of previous subsection.

We now describe the required modifications to Algorithm 4. Recall data structure $\mathcal{T}$ and lists $B(i)$, $E(i)$ from Section 4.2. We need two additional operations on $\mathcal{T}$:

- $\mathcal{T}.Update(e, k, v)$: Update the value of element having key k to v (or $Insert(e, k, v)$ if key k is not found).
- $L = \mathcal{T}.List(I)$: Return the list of elements having keys in range I.

Update-operation can be supported in time $O(\log n)$ just like *Insert*. *List*-operation can be supported in time $O(\log n + occ)$, where occ is the number of elements returned. The latter operation is used for listing all line segments in α distance from the current start point (to update values in the α-greedy paths). We cannot, however, use this operation as such at each start point, since then we could report the same intersection points multiple times. To obtain total running time $O(\min(|S|^2, |M|) \log n)$, we must ensure that we report each intersection point only once. This is done by maintaining for each row i a list $I(i-1)$

of distinct ranges, where we need to call the *List*-operation. After computing the values at row $i-1$, and collected list $I(i-1)$, we query each range of $I(i-1)$ from $\mathcal{T}$. We get an increasing list A of line segments. We can merge this list with $B(i)$ (in linear time in their sizes) to get the list F of line segments whose values need to be updated at row i. To proceed, we need to remove from A those line segments that do not continue any α-greedy path (to ensure $|S|^2$) bound). This is done by a simultaneous scan over A and over the points that formed $I(i-1)$. We get list G of line segments that continue α-greedy paths. By merging G and $B(i)$ we get a list U of line segments that continue or start new α-greedy paths. From U, we get list $I(i)$ of ranges that need to be updated at row $i+1$. We have now described in an informal level one step of the algorithm.

The detailed algorithm is given below. We need two instances of $\mathcal{T}$ since we need to be able to query two rows, $i-1$ and i, at the same time. We only give the update operations for $\mathcal{T}$ corresponding to row $i-1$. Structure $\mathcal{T}'$ for row i can be maintained similarly; at any point it contains all line segments that intersect row i, but do not start at that row. We assume that all non-initialized values are set to ∞.

Algorithm 5. /* Update operations for $\mathcal{T}'$ are not given */
(1)   **for each** $s \in B(1)$ **do begin** /* initialize first row to zero */
(2) $s.seg.d \leftarrow 0$; $\mathcal{T}.Insert(s, s.j - s.i, s.seg.d)$; **end**
(3) $U \leftarrow B(1)$
(4)   **for** $i \leftarrow 2$ to m **do begin** /* compute row-by-row */
(5) $I(i-1) \leftarrow GetRanges(U, \alpha)$; $A \leftarrow \emptyset$
(6)   **for each** range r in $I(i-1)$ **do** $A \leftarrow Concatenate(A, \mathcal{T}'.List(r))$
(7) $F \leftarrow Merge(A, B(i))$ /* line segments to be updated */
(8)   **for each** $s \in F$ **do** /* compute values at row i */
(9) $s.seg.d \leftarrow \mathcal{T}.Min([s.j - s.i - \alpha, s.j - s.i]) + 1$
(10)   **for each** $s \in E(i-1)$ **do** /* remove those ending at row $i-1$ */
(11) $\mathcal{T}.Delete(s.j - s.i)$
(12)   **for each** $s \in F$ **do** /* update values at row i */
(13) $\mathcal{T}.Update(s, s.j - s.i, s.seg.d)$
(14) $G \leftarrow Reduce(A, U)$ /* remove those not in α-greedy paths */
(15) $U \leftarrow Merge(G, B(i))$; /* get the set of active line segments */ **end**

Since the implementations of *Concatenate*() and *Merge*() are straight-forward, we only describe how to implement *GetRanges*() and *Reduce*(). In *GetRanges*(U, α), we scan over U and report each range once it is ready; Let $[l, r]$ be the current unfinished range and $s \in U$ the next segment to be processed. If $s.j - s.i \leq r$ then $[l, r]$ becomes $[l, s_j - s_i + \alpha]$. Otherwise, $[l, r]$ is reported and a new range $[s_j - s_i + 1, s_j - s_i + \alpha]$ becomes the new unfinished range. In *Reduce*(A, U), the logics is the following; For each $s' \in U$ there is at most one $s \in A$ such that s can be reached from s' with a jump $\leq \alpha$ and the jump is maximal among all segments that can be reached from s' with jump $\leq \alpha$ (thus, s

continues an α-greedy path). With a simultaneous scan over U and A we can find *all* the segments of A that are *not maximal* for *any* $s' \in U$. This is, because once a maximal segment s is found for some $s' \in U$, we can continue with the next segment in U. Then by scanning A starting from s, we can remove all segments in between until we found the new maximal segment. Hence, both *GetRanges*() and *Reduce*() can be implemented in linear time in the total lengths of the input lists. Same applies for *Merge*(). Each call to function *Concatenate*() takes constant time.

The correctness of Algorithm 5 follows from the fact that the minimum jump distance to each line segment is updated when the line segment starts, and at the intersection points of α-greedy paths.

Due to Lemma 9, these are the only points where updating is needed. The running time $O(\min(|S|^2, |M|)\log n)$ follows from two facts: (1) Each of the $|S|$ α-greedy paths can intersect at most $|S| - 1$ line segments. Algorithm 5 (due to lines 14-15) only does computation on the intersections of α-greedy paths and line segments (those intersection points that are removed in line 14 are end points of some α-greedy paths, so the computation on them in lines 5-9 is not only allowed but necessary). (2) Not depending on how many α-greedy paths intersect a point in M, each such point is accessed constant number of times (due to lines 5-7). The $\log n$ factor is from queries on $\mathcal{T}$ and $\mathcal{T}'$.

The following theorem summarizes the result.

Theorem 10 *Given a pattern string P and text string T of lengths m and n, respectively, the minimum splitting problem with α-limited gaps can be solved in time $O(|\Sigma|^2 + |\Sigma|m + m^2 + n + |R|\log n)$, $|R| \leq \min(|\hat{S}|^2, |M|)$, where $\hat{S}$ is the set of line segments formed by the diagonally consecutive pairs of $M = \{(i, j) \mid p_i = t_j\}$.*

4.4   Handling Multiple Texts

The algorithms above work with minor modifications also when $K > 1$; the preprocessing now takes $O(|\Sigma|^2 + |\Sigma|m + m^2 + Kn + |\hat{S}|\log K)$ time. The extra $\log K$ factor comes from the fact that we need to merge the sets of line segments, which are constructed for each text T^k separately. That is, we have to merge K lists that are already in row-by-row order. This takes $O(|\hat{S}|\log K)$ time. The algorithm itself takes $O(|\hat{S}|\log n)$ time or $O(|R|\log n)$ time if gaps are restricted with α. Here $\hat{S}$ is the set of (possibly parallel) line segments corresponding to the diagonally consecutive points in the (i, j) plane at fixed k in the sparse matrix $M = \{(i, j, k) \mid p_i = t_j^k\}$. Set R is the intersection set containing points at the intersection of $\hat{S}$ and the α-greedy paths.

Basically, the only modification is that the query range is changed from $j' - i' < j - i$ to $j' - i' \leq j - i$ in the algorithms (to allow jumps between parallel line segments). Also, the data structure $\mathcal{T}$ must be modified so that the key associated with a line segment is not simply $j - i$, but a combination of $j - i$ and k; the value of k can be ignored in range queries, but it must be used for separating parallel line segments.

5 Allowing No Gaps

We finally consider the special case where gaps are not allowed, i.e. $\alpha = 0$. The following fact is easy to see.

Fact 11 *Let there be a splitting of the pattern into κ pieces, starting at position j of the multi-track text $\mathbb{T}$, without gaps between the consecutive pattern piece occurrences. Then there is an equally good occurrence that can be found by the following greedy algorithm: Select the text T^k whose jth suffix has the longest common prefix with the pattern. Let this common prefix have length l. Iterate the same algorithm from position $j + l$ with pattern suffix P_{l+1}, until a splitting to κ pieces is found.*

The above greedy algorithm is given more formally below.

Algorithm 6.
(1) **for** $j = 1$ to $n - m + 1$ **do begin**
(2) $(k, i, l) \leftarrow (1, 1, 0)$
(3)   **while** $k \neq 0$ **do begin**
(4) $(l', k) \leftarrow JumpK(i + l, j + l); l \leftarrow l + l'$
(5)   **if** $l == m$ /* An occurrence found */ **end end**

Function $JumpK(i, j)$, in Algorithm 6, returns pair (l, k) such that the jth suffix of text track T^k has the longest common prefix (of length l) with the pattern suffix $P_{i...m}$. If no common prefixes are found, it returns $(0, 0)$.

We can again use the result of Ukkonen and Wood [15] (see Sect. 4.1), and support $JumpK(i, j)$ in $O(K)$ time (by computing values $Maxprefix(j)$ for each text separately). Given a threshold κ, we can limit the while-loop at lines (3–5) of Algorithm 6 so that at most $\kappa + 1$ calls to function $JumpK(i, j)$ are made at each position j. This gives an $O(m^2 + |\Sigma| + \kappa K n)$ time algorithm ($O(m^2 + |\Sigma| + K n)$ comes from the preprocessing for $JumpK(i, j)$ queries).

Theorem 12 *Given a pattern string $P = p_1 p_2 \cdots p_m$ and a multi-track text $\mathbb{T} = \{T^k = t_1^k \cdots t_n^k \mid 1 \leq k \leq K\}$, the minimum splitting problem with $\alpha = 0$ can be solved in $O(m^2 + |\Sigma| + \kappa K n)$ time, where κ is a given threshold.*

6 Conclusions

We have studied the minimum splitting problem of string patterns in multi-track texts, and introduced efficient solutions to certain variations of the problem. Our solutions based on sparse dynamic programming (given in Sect. 3), have a direct application in music retrieval. The algorithm summarized in Theorem 4 has been implemented and included in our C-BRAHMS music retrieval engine [2].

Interpreting the problem as a shortest path problem on line segments seems to give even more efficient algorithms. For random strings the line segments are rather short, and the algorithms based on line segments are not practical. In music, where repetitions are common, the line segment based algorithms are expected to perform somewhat better. Nevertheless, we expect these algorithms to be more theoretically than practically interesting.

However, a simple modification to the problem statement makes the line segment based algorithms practically interesting. Suppose that we are given a limit γ to the minimum length of the pieces the pattern can be split into. Our preprocessing can be modified to construct a set $\hat{S}^\gamma$ of maximal line segments of length at least γ in $O(m^2 \log m + Kn \log m + \gamma Kn + |\hat{S}^\gamma| \log K)$ time. This already is practical, since set $\hat{S}^{\gamma=2}$ is expected to be much smaller than $\hat{S}^{\gamma=1}$. We leave for further work to improve upon the γKn term.

References

1. A.V. Aho and M.J. Corasick. Efficient string matching. *Communications of the ACM*, 18(6):333–340, 1975.
2. C-BRAHMS. http://www.cs.helsinki.fi/group/cbrahms/demoengine/.
3. R. Cole and R. Hariharan. Verifying candidate matches in sparse and wildcard matching. In *Proc. Symposium on Theory of Computing (STOC'2002)*, pages 592–601. ACM Press, 2002.
4. T. Crawford, C.S. Iliopoulos, and R. Raman. String matching techniques for musical similarity and melodic recognition. *Computing in Musicology*, 11:71–100, 1998.
5. M. Crochemore. String matching with constraints. In *MFCS*, pages 44–58, 1988.
6. M. Crochemore, C.S. Iliopoulos, C. Makris, W. Rytter, A. Tsakalidis, and K. Tsichlas. Approximate string matching with gaps. *Nordic Journal of Computing*, 9(1):54–65, 2002.
7. H. Gajewska and R. Tarjan. Deques with heap order. *Information Processing Letters*, 12(4):197–200, 1986.
8. Z. Galil and K. Park. An improved algorithm for approximate string matching. *SIAM J. Comput.*, 19:989–999, 1990.
9. C.S. Iliopoulos and M. Kurokawa. String matching with gaps for musical melodic recognition. In *Proc. Prague Stringology Conference 2002 (PSC'2002)*, pages 55–64, Prague, 2002.
10. G.M. Landau and U. Vishkin. Fast string matching with k differences. *Journal of Computers and Systems*, 37:63–78, 1988.
11. K. Lemström and J. Tarhio. Transposition invariant pattern matching for multi-track strings. 2003. (submitted).
12. V. Mäkinen, G. Navarro, and E. Ukkonen. Algorithms for transposition invariant string matching. In *Proceedings of the 20th International Symposium on Theoretical Aspects of Computer Science (STACS'2003)*, volume 2607 of *Springer-Verlag LNCS*, pages 191–202, 2003.
13. D. Meredith, G.A. Wiggins, and K. Lemström. Pattern induction and matching in polyphonic music and other multi-dimensional datasets. In *the 5th World Multi-Conference on Systemics, Cybernetics and Informatics (SCI'2001)*, volume X, pages 61–66, Orlando, FLO, July 2001.
14. E. Ukkonen, K. Lemström, and V. Mäkinen. Sweepline the music! In *Computer Science in Perspective*, volume 2598 of *Springer-Verlag LNCS*, pages 330–342, 2003.
15. E. Ukkonen and D. Wood. Fast approximate string matching with suffix automata. *Algorithmica*, 10:353–364, 1993.

Alignment between Two Multiple Alignments*

Bin Ma, Zhuozhi Wang, and Kaizhong Zhang

Dept. of Computer Science, University of Western Ontario,
London, Ont. N6A 5B7, Canada
{bma,zzwang,kzhang}@csd.uwo.ca

Abstract. Alignment of two multiple alignments arises naturally when constructing approximate multiple sequence alignments progressively. In this paper, we consider the problem of alignment of two multiple alignments with SP-score and linear gap costs.
When there is no gap opening cost, this problem can be solved using the well-known dynamic programming algorithm for two sequences by viewing each column in the multiple alignments as an element. However if there are gap opening costs (sometimes referred as affine gap costs) then the problem becomes non-trivial. Gotoh [4] suggested a procedure for this problem and stated that "the total arithmetic operations used is close to (quadratic) in typical cases". Kececioglu and Zhang [7] gave heuristic algorithms based on optimistic and pessimistic gap counts and conjectured that this problem is NP-complete. In this paper we prove that this problem is indeed NP-complete and therefore settle this open problem. We then propose another heuristic algorithm for this problem.

1 Introduction

Sequence alignment algorithms are important in bioinformatics research. Alignment of a family of protein or nucleotide sequences is a fundamental step in determining the function-structure relationship. Optimal alignment between two sequences can be solved by the well-known dynamic programming solution [9]. Optimal multiple sequence alignment is a more difficult problem. It has been proved that this problem is NP-hard [11,2,6]. Linear approximation algorithms have been developed in view of this situation [5,11]. Many different methods have been developed to obtain good multiple sequence alignments [1,8].

Alignment of two multiple alignments arises naturally in many iterative or progressive approximate multiple sequence alignment methods [8]. Since an alignment can be considered as a sequence where each element is a column of characters, it may seem that the problem of aligning two alignments should be no harder than aligning two sequences. This is indeed true when gap opening cost is not considered. This problem becomes challenging when the objective function for the resulting multiple alignment is the commonly used sum-of-pairs measure with gaps having the general affine costs.

---

* Research supported partially by the Natural Sciences and Engineering Research Council of Canada under Grant OGP0046373 and RGP0238748 and a Sharcnet research fellowship.

R. Baeza-Yates et al. (Eds.): CPM 2003, LNCS 2676, pp. 254–265, 2003.

Gotoh [4] presented several algorithms for this problem. One of the algorithm can produce optimal alignment between two alignments when gap costs are rigorously considered. Although the worst case time complexity for this algorithm is not given, it is suggested that for this algorithm "the total arithmetic operations used is close to (quadratic) in typical cases". Kececioglu and Zhang [7] presented heuristic algorithms based on optimistic and pessimistic gap counts where when the local information of the last two columns of the alignments is not enough to determine whether there is a gap opening, a gap opening cost is either always applied (pessimistic) or always not applied (optimistic). They further conjectured that this problem is NP-complete.

In this paper we first prove that this problem is NP-complete and therefore settle this open problem. This result clarifies why it is so hard to to incorporate exact gap counts into many progressive approximate multiple sequence alignment methods We then propose another heuristic algorithm for this problem.

2   Definitions and Problem Description

In this section we give basic definitions and the problem description.

A multiple sequence alignment for sequences $s_1, s_2, \ldots s_K$ is $s'_1, s'_2, \ldots s'_K$ where each s'_i is s_i with null characters inserted and of the same length N. Since in a multiple sequence alignment, every sequence has the same length, we can represent a multiple sequence alignment as an array A with K rows and N columns representing K sequences of length N. For a multiple alignment A, we use A_k to represent the kth sequence, $A[i]$ to represent the ith column, $A_k[i]$ to present the ith element of the kth sequence, and $A[i \cdots j]$ to present column i to column j inclusive.

The goal of multiple sequence alignment is to find the optimal multiple sequence alignment with respect to some objective function measure. The sum-of-pairs (SP) score is the commonly used measure based on the summation of the induced pairwise sequence alignment measures, namely

$$\sum_{1 \leq i \leq K} \sum_{1 \leq j \leq K, j \neq i} D(A_i, A_j),$$

where $D(A_i, A_j)$ is a measure base on edit operations for a pairwise alignment.

A gap in a pairwise alignment is a maximal run of insertions or deletions. We say an alignment measure function uses affine gap costs when a gap of length $x > 0$ costs $a + bx$ for constants a and b where a can be viewed as the gap opening cost and b is the indel cost. The problem of finding optimal multiple alignment is NP-hard whether a is zero or not [11,2,6].

Given two alignments A with K sequences and length N and B with L sequences and length M, an alignment between alignments A and B is A' and B' with same length where A' is A with null columns (of size K) inserted and B' is B with null columns (of size L) inserted.

The sum-of-pair score for an alignment between A and B, (A', B'), is the summation of the induced pairwise sequence alignment measures, namely

$$\sum_{1 \leq i \leq K} \sum_{1 \leq j \leq L} D(A'_i, B'_j),$$

where $D(A'_i, A'_j)$ is a measure base on edit operations for a pairwise alignment.

We note that the alignment (A'_i, B'_j) of two sequences A_i and B_j, induced from the alignment (A', B'), may contain null columns like the following,

$$
\begin{array}{ll}
A'_i: & \texttt{a-----aaa} \\
B'_j: & \texttt{aa-a-aaaa}
\end{array}
$$

The two null columns corresponding to the two null characters in B'_j should be removed before the computation of $D(A'_i, B'_j)$. Therefore, there is actually only one gap in this alignment.

When the gap opening cost $a = 0$, then the optimal alignment between alignments can be determined by the well-known dynamic programming algorithm [9] for the optimal alignment between two sequences by viewing the alignments as sequences of columns. However, when $a > 0$, the method does not work. Actually, we show in this paper that this problem is NP-hard.

3 Hardness Results

In this section, we prove that the problem of aligning two alignments with SP-score and affine gap penalty is NP-Complete.

Given an undirected graph $G = (V, E)$, a partition (V_1, V_2) of vertex set V satisfies $V_1 \cap V_2 = \emptyset$ and $V_1 \cup V_2 = V$. The cut of G associated with (V_1, V_2) is the set of edges each of which has one end point in V_1 and the other end point in V_2.

Max-Cut-3: Given a cubic graph $G = (V, E)$, i.e. every vertex has degree 3, find a partition of V with the maximum associated cut.

It is well known that Max-Cut-3 is MAX SNP-hard [10,12]. We will reduce Max-Cut-3 to our problem. In our reduction, we will construct two multiple alignments from the given graph. Each vertex of the graph would be represented by some columns in each of the two alignments and each edge of the graph will be represented by two sequences in each of the two alignments. The construction will make sure that columns representing the same vertex in the two alignments will be aligned to each other and the way they are aligned corresponds to a partition of the graph. The optimal alignment between these two alignments corresponds to the partition with maximum cut of the graph.

Given a cubic graph $G = (V, E)$, let $n = |V|$ and the vertices of G be $v_1, v_2, \ldots v_n$, then $|E| = 3/2n$. We construct two multiple alignments A and B. The general structure of A and B are $S_0 v'_1 S_1 v'_2 S_2 \ldots S_{n-1} v'_n S_n$ and $S_0 v''_1 S_1 v''_2 S_2 \ldots S_{n-1} v''_n S_n$ where S_i is a "separator" of $s(n)$ columns and v'_i and v''_i are segments corresponding to vertex v_i in G. $s(n)$ should be large enough so that in the optimal alignment, v_i' in A has to be aligned with v_i'' in B for all i. Each v_i' in A has 13 columns and each v_i'' in B has 12 columns. Each edge of G is encoded by 2 sequences in A and 2 sequences in B.

```
----  ...  ----           --           --           --           --
----  ...  ----           --           --           --           --
 ...       ...          ...        ...             ...        ...
----  ...  ----           --           --           --           --
----  ...  ----           --           --           --           --
cccc  ...  cccc     aaaaaaaaaaaaa          aaaaaaaaaaaa
cccc  ...  cccc     -aaaaaaaaaaa-          -aaaaaaaaaa-
 ...       ...          ...        ...             ...        ...
cccc  ...  cccc     aaaaaaaaaaaaa          aaaaaaaaaaaa
cccc  ...  cccc     -aaaaaaaaaaa-          -aaaaaaaaaa-
```

Fig. 1. "Separator" S_i, segment v_i' in A and segment v_i'' in B .

In both alignment A and B there are $3n + 2k(n)$ sequences where $k(n)$ should be sufficiently large so that, together with $s(n)$, in the optimal alignment separators have to be aligned with separators and therefore v_i' in A has to be aligned with v_i'' in B for all i.

For the separator S, the top $3n$ sequences are all spaces and the bottom $2k(n)$ sequences are all c's, see Figure 1. For each v_i' with 13 columns in A, the bottom $2k(n)$ sequences are the interleaving of $aaaaaaaaaaaaa$ and $-aaaaaaaaaaa-$. For the top $3n$ sequences of v_i', columns 3 to 5, columns 6 to 8, and columns 9 to 11 are used to encode three edges incident to v_i. All the other columns are spaces. Similarly for each v_i'' with 12 columns in B, the bottom $2k(n)$ sequences are the interleaving of $aaaaaaaaaaaa$ and $-aaaaaaaaaa-$, For the top $3n$ sequences of v_i'', columns 3 to 4, columns 6 to 7, and columns 9 to 10 are used to encode three edges incident to v_i. All the other columns are spaces. Figure 1 shows v_i' and v_i'' in A and B.

To represent an edge $\{v_i, v_j\}$, where $i < j$, we use two sequences in A and two sequences in B, see Figure 2. For the two sequences in A, we choose three columns from the nine edge encoding columns of v_i' and use aaa and $-aa$ for these three columns, and we also choose three columns from the nine edge encoding columns of v_j' and use $aa-$ and aaa for these three columns. For the two sequences in B, we choose the corresponding two columns from the six edge encoding columns of v_i'' and use $-a$ and $-a$ for these two columns, and we also choose the corresponding two columns from the six edge encoding columns of v_j'' and use $a-$ and $a-$ for these two columns.

In the proofs of this section, we assume that a mismatch has a cost of 2 and an indel or a gap opening has a unit cost.

Lemma 1. *Let $A = S_0 v_1' S_1 \cdots v_n' S_n$ and $B = S_0 v_1'' S_1 \cdots v_n'' S_n$ where $s(n) = O(n^{1.5})$ and $k(n) = O(n^{2.5})$, then in the optimal alignment between A and B, S_i in A has to be aligned with S_i in B and v_i' in A has to be aligned with v_i'' in B for every i.*

```
-----aaa-----------      . . .      . . .      --------aa---------
------aa-----------      . . .      . . .      --------aaa-------

cccaaaaaaaaaaaaaaccc      . . .      . . .      cccaaaaaaaaaaaaaaccc
ccc-aaaaaaaaaaa-ccc       . . .      . . .      ccc-aaaaaaaaaaa-ccc
        . . .
cccaaaaaaaaaaaaaaccc      . . .      . . .      cccaaaaaaaaaaaaaaccc
ccc-aaaaaaaaaaa-ccc       . . .      . . .      ccc-aaaaaaaaaaa-ccc



------a-----------       . . .      . . .      --------a---------
------a-----------       . . .      . . .      --------a---------

cccaaaaaaaaaaaaaccc       . . .      . . .      cccaaaaaaaaaaaaaccc
ccc-aaaaaaaaaa-ccc        . . .      . . .      ccc-aaaaaaaaaa-ccc
        . . .                                          . . .
cccaaaaaaaaaaaaaccc       . . .      . . .      cccaaaaaaaaaaaaaccc
ccc-aaaaaaaaaa-ccc        . . .      . . .      ccc-aaaaaaaaaa-ccc
```

Fig. 2. An edge $\{v_i, v_j\}$ is represented by two sequences in A (top) and two sequences in B (bottom).

Proof. (sketch) For an alignment where S_i is aligned with S_i and v_i' is aligned with v_i'', the cost is bounded by $O(n^6)$ since the cost of aligning S_i to S_i is bounded by $O(n * k(n) * s(n)) = O(n^5)$ and the cost of aligning v_i' to v_i'' is bounded by $O(k(n) * k(n) + n * k(n)) = O(n^5)$. Now if a column of v_i' is aligned with a column of v_j'' or S_j for some $j > i$, then to the left of this alignment point, the number of columns in B that have to be aligned with space columns is $O(s(n))$ and the cost is at least $O((n+k(n)) * k(n) * s(n)) = O(n^{6.5})$. Similarly, v_i' cannot be aligned with v_j'' or S_{j-1} for any $j < i$. Therefore v_i' can only be aligned with v_i'', S_{i-1}, or S_i.

```
. . . aaccc ... ccaaaa ...          . . . aa--cc ... ccaaaa ...
. . . a-cccc ... cc-aaa ...         . . . a---cc ... cc-aaa ...

. . . aaaacc ... ccccaa ...         . . . aaaacc ... cc--aa ...
. . . aaa-cc ... cccc-a ...         . . . aaa-cc ... cc---a ...

           (a)                                 (b)
```

Fig. 3. Separators have to be aligned to separators

We now show that v_i cannot be aligned with S_{i-1} or S_i. If v_i' is aligned with, say, S_{i-1}, then we can transform it into another alignment with better cost where S_{i-1} is aligned completely with S_{i-1}. Consider two bottom sequences from A and two bottom sequences from B in such an alignment, Figure 3 (a) shows a typical situation. When we transform it into Figure 3 (b) the cost decrease is at least 1. Therefore the cost decrease of the whole bottom sequences of A and B is at least $k(n) * k(n) = n^5$ which is large enough to compensate any cost increases, bounded by $O(n * k(n) = n^{3.5})$, from the top sequences of A and B. Therefore in an optimal alignment S_i in A has to be aligned with S_i in B and v_i' in A has to be aligned with v_i'' in B for every i.   □

Theorem 1. *The problem of computing the optimal alignment of two multiple alignments with SP-score and affine gap penalty is NP-complete.*

Proof. (sketch) The problem is clearly in NP, so we just need to prove that the problem is NP-hard. Given a cubic graph $G = (V, E)$, we construct two multiple alignments of $A = Sv_1'Sv_2'S\ldots Sv_n'S$ and $B = Sv_1''Sv_2''S\ldots Sv_n''S$ where S, v_i' and v_i'' are as discussed above.

We now prove that an optimal alignment between A and B corresponds a maximum cut of the G and vice versa.

From Lemma 1, we know that in an optimal alignment S_{i-1} and S_i in A have to be aligned with S_{i-1} and S_i in B, therefore we now consider the optimal alignment between v_i' and v_i''. If we only consider the bottom $2k(n)$ sequences of v_i' and v_i'' with 13 and 12 columns respectively, then it is clear that in an optimal alignment there is no insertion of space columns in v_i' and the insertion of one space column in v_i'' should be either to the left of v_i'' or to the right of v_i''. The reason is that insertion to the left or right has a cost of $11k^2(n)$ whereas the insertion inside v_i'' will have a cost of $16k^2(n)$. Now we consider the top $3n$ sequences in v_i' and v_i''. Except six sequences in v_i' and six sequences in v_i'' encoding edges, all the other sequences are composed of spaces. We do not have to consider these sequences of spaces since there will be no cost change no matter where we insert the space column in v_i''. Since the insertion of a space column inside v_i'' will not decrease the alignment cost between these six sequences and the bottom $2k(n)$ sequences, the only case which may decrease the cost is the pairwise alignments between the six sequences in v_i' and the six sequences in v_i'' encoding edges incident to v_i. However the decrease would be a constant and $5k^2(n) = 16k^2(n) - 11k^2(n)$ is more than enough to compensate that when $k(n)$ is large.

Therefore we know that in the optimal alignment between A and B, for $0 \le i \le n$, S_i has to be aligned with S_i and v_i' has to be aligned with v_i'' with no space column inserted in v_i' and with one space column inserted either to the left or the right of v_i''.

We can now consider the effect of inserting one space column to the left or the right of v_i''. Consider the cost between the two sequences in A and the two sequences in B corresponding to an edge $\{v_i, v_j\}$, there are four cases, see Figure 4.

(a): one space column is inserted to the right of v_i'' and one space column is inserted to the right of v_j''. In this case, the cost is 22 since the alignment cost between the first sequence and the third or fourth sequence is 6 with 3 deletions and 3 gaps and the alignment cost between the second sequence and the third or fourth sequence is 5 with 3 deletions and 2 gaps.

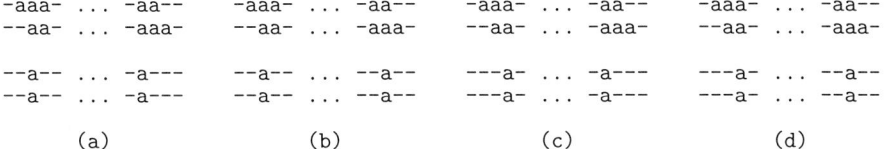

```
-aaa-  ... -aa--     -aaa-  ... -aa--     -aaa-  ... -aa--     -aaa-  ... -aa--
--aa-  ... -aaa-     --aa-  ... -aaa-     --aa-  ... -aaa-     --aa-  ... -aaa-

--a--  ... -a---     --a--  ... --a--     ---a-  ... -a---     ---a-  ... --a--
--a--  ... -a---     --a--  ... --a--     ---a-  ... -a---     ---a-  ... --a--

      (a)                 (b)                 (c)                 (d)
```

Fig. 4. The four cases involving the alignments between the two sequences in A and two sequences in B corresponding to an edge $\{v_i, v_j\}$.

(b): one space column is inserted to the right of v_i'' and one space column is inserted to the left of v_j''. In this case, the cost is 22 since the alignment cost between the first sequence and the third or fourth sequence is 5 with 3 deletions and 2 gaps and the alignment cost between the second sequence and the third or fourth sequence is 5 with 3 deletions and 2 gaps.

(c): one space column is inserted to the left of v_i'' and one space column is inserted to the right of v_j''. In this case, the cost is 22 since the alignment cost between the first sequence and the third or fourth sequence is 5 with 3 deletions and 2 gaps and the alignment cost between the second sequence and the third or fourth sequence is 5 with 3 deletions and 2 gaps.

(d): one space column is inserted to the left of v_i'' and one space column is inserted to the left of v_j''. In this case, the cost is 22 since the alignment cost between the first sequence and the third or fourth sequence is 5 with 3 deletions and 2 gaps and the alignment cost between the second sequence and the third or fourth sequence is 6 with 3 deletions and 3 gaps.

Therefore if these two space columns are inserted at the same side of v_i'' and v_j'' the alignment cost is 22 whereas if they are inserted at different sides the cost alignment is 20. In addition, for any other pair of sequences, one from A and one from B, the alignment costs are the same no matter which side the space columns are inserted. This means that in an optimal alignment between A and B, space columns should be inserted in such a way that the number of sequences in A and B encoding edges $\{v_i, v_j\}$ where space columns are inserted at different sides of v_i'' and v_j'' is maximized.

Now it is clear that given an optimal alignment between A and B, we can construct a partition of V with maximum cut by putting any vertex v_i in V_1 if the space column is inserted to the left of v_i'' and putting any vertex v_i in V_2 if the space column is inserted to the right of v_i''. Conversely given a maximum cut, we can construct an optimal alignment between A and B by aligning S's to S's and v_i' to v_i'' where for any v_i in V_1 the space column is inserted to the left of v_i'' and for any v_i in V_2 the space column is inserted to the right of v_i''.

Therefore the problem of computing the optimal alignment of two multiple alignments with SP-score and affine gap penalty is NP-complete. $\square$

4   A Heuristic Algorithm to Align Two Alignments

When aligning two sequences optimally, using the method proposed by Gotoh [3], one can determine a gap opening by checking the preceding alignment pair. Unfortunately for the alignment of two multiple alignments this method is no longer applicable. In fact, keeping track of any fixed number of preceding columns of the alignment is not enough to determine gap openings for the optimal solution since the problem is NP-hard. Kececioglu and Zhang [7] suggested heuristic suboptimal solutions based on the preceding column alone. Their *optimistic* gap counts count one gap opening in the situation $\frac{-\,-}{-\,X}$ and $\frac{-\,X}{-\,-}$, while their *pessimistic* gap counts do not. While this will not give exact gap counts, the hope is that this will underestimate or overestimate true count rarely.

One observation for this approach is that when counting gap openings one can consider not only the preceding columns but also the preceding suboptimal alignments ending at these columns. The information we need for these alignments is just the number of consecutive null characters at the end of each sequence in these alignments. In this situation, one can count the gap openings for these suboptimal alignments precisely when the necessary information are properly stored. In this section we will present an alternative heuristic solution base on this idea. One can generalize this to consider any fixed number of preceding columns resulting better solutions. This idea is similar to that in [4] with different algorithmic details.

4.1   The Algorithm

Given two alignments A, with K sequences and length N, and B, with L sequences and length M, respectively, following [7] we use $D[i][j]$ to present the value of a suboptimal alignment between $A[1 \ldots i]$ and $B[1 \ldots j]$ which ends at deleting $A[i]$, $I[i][j]$ to represent the value of a suboptimal alignment between $A[1 \ldots i]$ and $B[1 \ldots j]$ which ends at inserting $B[j]$, and $M[i][j]$ to represent the value of a suboptimal alignment between $A[1 \ldots i]$ and $B[1 \ldots j]$ which ends at matching $A[i]$ and $B[j]$.

For a suboptimal alignment ending with a X, $X \in \{D, I, M\}$, operation, we use $glen_A^X(i,j)[s]$ and $glen_B^X(i,j)[t]$ to store the number of consecutive null characters from the end of this alignment for sequence s in $A[1 \cdots i]$ and sequence t in $B[1 \cdots j]$ respectively. With this information, we can count gap openings in these suboptimal alignments precisely. In the following gap is used to represent gap opening cost.

Lemma 2. *For $i \geq 1$, $j \geq 1$, and $X \in \{D, I, M\}$, let $gap_M^X(i,j) = \sum_{s=1}^{K} \sum_{t=1}^{L}$ $(((A_s[i] \neq null) \ \&\& \ (B_t[j] = null) \ \&\& \ (glen_A^X(i-1, j-1)[s] \geq glen_B^X(i-1, j-1)[t])) \ || \ ((A_s[i] = null) \ \&\& \ (B_t[i] \neq null) \ \&\& \ (glen_A^X(i-1, j-1)[s] \leq glen_B^X(i-1, j-1)[t]))),$ then*

$$M[i][j] = \delta(i,j) + \min \begin{cases} D[i-1][j-1] + gap_M^D(i,j) * gap \\ I[i-1][j-1] + gap_M^I(i,j) * gap \\ M[i-1][j-1] + gap_M^M(i,j) * gap \end{cases}$$

If $M[i][j] = \delta(i,j) + X[i-1][j-1] + gap_M^X(i,j) * gap$, where $X \in \{D, I, M\}$, then for $1 \leq s \leq K$ and $1 \leq t \leq L$,

$$glen_A^M(i,j)[s] = \begin{cases} 0 & \text{if } A_s[i] \text{ is a non-null character} \\ glen_A^X(i-1,j-1)[s]+1 & \text{otherwise} \end{cases}$$

$$glen_B^M(i,j)[t] = \begin{cases} 0 & \text{if } B_t[j] \text{ is a non-null character} \\ glen_B^X(i-1,j-1)[t]+1 & \text{otherwise} \end{cases}$$

Proof. (sketch) Consider the suboptimal alignment between $A[1\ldots i]$ and $B[1\ldots j]$ ending with matching $A[i]$ to $B[j]$, the best will come from $D[i-1][j-1]$, $I[i-1][j-1]$, or $M[i-1][j-1]$ with the gap openings created by aligning $A[i]$ to $B[j]$ incorporated.

```
x-   ... -x          x-  ...  --  ... -x        ...... x-  ... -x

y-   ... --           ...... y- ... --         y-  ... --  ... --
     (1)                     (2)                       (3)
```

Fig. 5. Gap openings.

Consider $A_s[i]$ and $B_t[j]$, if both are null characters or both are non-null characters, then there is no gap opening. However if one is a null character and the other is a non-null character, then there may be new gap opening. Assuming that $A_s[i]$ is a non-null character and $B_t[j]$ is a null character and the suboptimal alignment for $A[1\ldots i-1]$ and $B[1\ldots j-1]$ ends at deleting $A[i-1]$, if $glen_A^D(i-1,j-1)[s] \geq glen_B^D(i-1,j-1)[t]$ then there is a new gap opening, see cases (1) and (2) in Figure 5, otherwise there is no new gap opening, see case (3) in Figure 5. Therefore $gap_M^X(i,j)$ is correct.

Since $gap_M^X(i,j)$ is correct, it is easy to see that $M[i][j]$ is correct. Once $M[i][j]$ is calculated, we can compute $glen_A^M(i,j)[s]$ and $glen_B^M(i,j)[t]$ accordingly. If $A_s[i]$ is a non-null character, then $glen_A^M(i,j)[s] = 0$ since the suboptimal alignment ends at $A_s[i]$, otherwise $glen_A^M(i,j)[s] = glen_A^X(i-1,j-1)[s]+1$ since the suboptimal alignment has one more null character than the suboptimal alignment for $A[1\ldots i-1]$ and $B[1\ldots j-1]$. Therefore $glen_A^M(i,j)[s]$ and $glen_B^M(i,j)[t]$ are correct. □

Lemma 3. *For $i \geq 1$, $j \geq 1$, let $gap_D^X(i,j) = \sum_{s=1}^K \sum_{t=1}^L ((A_s[i] \neq null)\&\& (glen_A^X(i-1,j)[s] \geq glen_B^X(i-1,j)[t]))$, then*

$$D[i][j] = \delta(i,0) + \min \begin{cases} D[i-1][j] + gap_D^D(i,j) * gap \\ I[i-1][j] + gap_D^I(i,j) * gap \\ M[i-1][j] + gap_D^M(i,j) * gap \end{cases}$$

If $D[i][j] = \delta(i,0) + X[i-1][j] + gap_D^X(i,j) * gap$, where $X \in \{D, I, M\}$, then for $1 \leq s \leq K$ and $1 \leq t \leq L$,

$$glen_A^D(i,j)[s] = \begin{cases} 0 & \text{if } A_s[i] \text{ is a non-null character} \\ glen_A^X(i-1,j)[s] + 1 & \text{otherwise} \end{cases}$$
$$glen_B^D(i,j)[t] = glen_B^X(i-1,j)[t] + 1$$

Proof. Similar to Lemma 2.

Lemma 4. *For $i \geq 1$, $j \geq 1$, let $gap_I^X(i,j) = \sum_{s=1}^{K} \sum_{t=1}^{L} ((B_t[i] \neq null)\&\& (glen_A^X(i,j-1)[s] \leq glen_B^X(i,j-1)[t]))$, then*

$$I[i][j] = \delta(0,j) + \min \begin{cases} D[i][j-1] + gap_I^D(i,j) * gap \\ I[i][j-1] + gap_I^I(i,j) * gap \\ M[i][j-1] + gap_I^M(i,j) * gap \end{cases}$$

If $I[i][j] = \delta(0,j) + X[i][j-1] + gap_I^X(i,j) * gap$, where $X \in \{D, I, M\}$, then for $1 \leq s \leq K$ and $1 \leq t \leq L$,

$$glen_A^I(i,j)[s] = glen_A^X(i,j-1)[s] + 1$$
$$glen_B^I(i,j)[t] = \begin{cases} 0 & \text{if } B_t[j] \text{ is a non-null character} \\ glen_B^X(i,j-1)[t] + 1 & \text{otherwise} \end{cases}$$

Proof. Similar to Lemma 2.   □

The following lemmas show the boundary conditions. These are derived by considering that at position 0 of each alignment there is a column in which each character is a non-null character.

Lemma 5. *For $i \geq 1$, $j = 0$, let $gap_D^D(i,0) = \sum_{s=1}^{K} \sum_{t=1}^{L} ((A_s[i] \neq null)\&\& (glen_A^D(i-1,0)[s] \geq glen_B^D(i-1,0)[t]))$, then*

$$D[i][0] = \delta(i,0) + D(i-1)[0] + gap_D^D(i,0) * gap$$
$$I[i][0] = M[i][0] = D[i][0];$$
$$glen_A^D(i,0)[s] = \begin{cases} 0 & \text{if } A_s[i] \text{ is a non-null character} \\ glen_A^D(i-1,0)[s] + 1 & \text{otherwise} \end{cases}$$
$$glen_B^D(i,0)[t] = glen_B^D(i-1,0)[t] + 1$$
$$glen_A^I(i,0)[s] = glen_A^M(i,0)[s] = glen_A^D(i,0)[s]$$
$$glen_B^I(i,0)[t] = glen_B^M(i,0)[t] = glen_B^D(i,0)[t]$$

Lemma 6. *For $i = 0$, $j \geq 1$, let $gap_I^I(0,j) = \sum_{s=1}^{K} \sum_{t=1}^{L} ((B_t[i] \neq null)\&\& (glen_A^I(0,j-1)[s] \leq glen_B^I(0,j-1)[t]))$, then*

$$I[0][j] = \delta(0,j) + I[0][j-1] + gap_I^I(0,j) * gap$$
$$D[0][j] = M[0][j] = I[0][j];$$
$$glen_A^I(0,j)[s] = glen_A^I(0,j-1)[s] + 1$$
$$glen_B^I(0,j)[t] = \begin{cases} 0 & \text{if } B_t[j] \text{ is a non-null character} \\ glen_B^I(0,j-1)[t] + 1 & \text{otherwise} \end{cases}$$
$$glen_A^D(0,j)[s] = glen_A^M(0,j)[s] = glen_A^I(0,j)[s]$$
$$glen_B^D(0,j)[t] = glen_B^M(0,j)[t] = glen_B^I(0,j)[t]$$

Lemma 7.

$$D[0][0] = I[0][0] = M[0][0]$$

$$glen_A^I(0,0)[s] = glen_A^D(0,0)[s] = glen_A^M(0,0)[s] = 0;$$

$$glen_B^I(0,0)[t] = glen_B^D(0,0)[t] = glen_B^M(0,0)[t] = 0;$$

4.2   Complexities

The time complexity of the algorithm depends on how to calculate $\delta(A[i], B[j])$, $glen_A^X(i,j)[s]$ and $glen_B^X(i,j)[t]$, and $gap_X^Y(i,j)$, where $X, Y \in \{D, I, M\}$.

For the computation of $\delta(A[i], B[j])$, we can transform the columns involved into "profile" [4] and use constant time assuming that the alphabet size of the input sequences is a constant. The calculation of $glen_A^X(i,j)[s]$ and $glen_B^X(i,j)[t]$ takes $O(K + L)$ time. A direct calculation of $gap_X^Y(i,j)$ needs $O(KL)$ time. However this can be reduced to $O(K + L)$. We now consider how to compute $gap_D^X(i,j)$ and the computation of $gap_I^X(i,j)$ and $gap_M^X(i,j)$ are similar. Recall that

$$gap_D^X(i,j) = \sum_{s=1}^{K}\sum_{t=1}^{L}((A_s[i] \neq null) \text{ \&\& } (glen_A^X(i-1,j)[s] \geq glen_B^X(i-1,j)[t])).$$

Therefore for each non-null character $A_s[i]$ we only need to know the number of t such that $glen_A^X(i-1,j)[s] \geq glen_B^X(i-1,j)[t])$. If we sort $glen_B^X(i-1,j)[t])$ in increasing order, then it is easy to calculate this number. In fact we can also sort $glen_A^X(i-1,j)[s]$ in increasing order for those s such that $A_s[i]$ is a non-null character. Once we have these two sorted lists, a procedure similar to merging two sorted sequences can be applied to determine $gap_D^X(i,j)$. The time complexity of this process is $O(K \log(K) + L \log(L))$. In fact, it is not hard to further improve this to $O(K+L)$. Therefore the time complexity of the algorithm is $O(NM(K + L))$ which is better than that in [4].

The extra space needed to store $glen_A^X(i,j)[s]$ and $glen_B^X(i,j)[t]$ is $O(\min\{N, M\}(K + L))$ since we only need to keep two rows or two columns of these values. Therefore the space complexity is $O(NK + ML + NM)$.

5   Conclusions

We have proved that the problem of aligning two alignments under sum-of-pairs objective function with affine gap penalty is NP-complete. We present an efficient heuristic algorithm for this problem. Intuitively this algorithm should be better than that of Kececioglu and Zhang [7] with optimistic or pessimistic gap counts since it uses all the information available based on the last two columns at each location. A computational study comparing this algorithm with optimistic and pessimistic gap counts is under way.

References

1. E.L. Anson and G. Myers. Realigner: a program for refining dna sequence multi-alignments. In *Proceedings of the First ACM conference on Computational Molecular Biology*, pages 9–13, 1997.
2. P. Bonizzoni and G. Della Vedova. The complexity of multiple sequence alignment with sp-score that is a metric. *Theorectical Computer Science*, 259(1):63–79, 2001.
3. O. Gotoh. An improved algorithm for matching biological sequences. *Journal of Molecular Biology*, 162:705–708, 1982.
4. O. Gotoh. Optimal alignment between groups of sequences and its application to multiple sequence alignment. *Computer Application in the Biosciences*, 9(3):361–370, 1993.
5. D. Gusfield. Efficient methods for multiple sequence alignment with guaranteed error bounds. *Bulletin of Mathematical Biology*, 55:141–154, 1993.
6. W. Just. Computational complexity of multiple sequence alignment with sp-score. *Journal of computational biology*, 8(6):615–623, 2001.
7. J. D. Kececioglu and W. Zhang. Aligning alignments. In *Proceedings of the Ninth Symposium on Combinatorial Pattern Matching*, Lecture Notes in Computer Science 1448, pages 189–208. Springer-Verlag, 1998.
8. G. Myers, S. Selznick, Z. Zhang, and W. Miller. Progressive multiple alignment with constraints. In *Proceedings of the First ACM conference on Computational Molecular Biology*, pages 220–225, 1997.
9. S.B. Needleman and C.D. Wunsch. A general method applicable to the search for similarities in the amino acid sequence of two proteins. *Journal of Molecular Biology*, 48:443–453, 1970.
10. C.H. Papadimitriou and M. Yannakakis. Optimization, approximation, and complexity classes. *J. Comput. System Sciences*, 43:425–440, 1991.
11. L. Wang and T. Jiang. On the complexity of multiple sequence alignment. *Journal of computational biology*, 1(4):337–448, 1994.
12. Z. Wang and K. Zhang. Alignment between rna structures. In *Proceedings of the 26th International Symposium on Mathematical Foundations of Computer Science*, Lecture Notes in Computer Science 2136, pages 690–702. Springer-Verlag, 2001.

An Effective Algorithm for the Peptide *De Novo* Sequencing from MS/MS Spectrum

Bin Ma[1], Kaizhong Zhang[1], and Chengzhi Liang[2]

[1] Department of Computer Science, University of Western Ontario
London, ON, Canada N6A 5B7. {bma,kzhang}@csd.uwo.ca
[2] Bioinformatics Solutions Inc., 145 Columbia St. West, Suite 2B
Waterloo, ON, Canada N2L 3L2. cliang@BioinformaticsSolutions.com

Abstract. The determination of the amino acid sequence of a peptide from its MS/MS spectrum is an important task in proteomics. The determination without the help of a protein database is called the *de novo* sequencing, which is especially useful in the identification of unknown proteins. Many studies on the *de novo* sequencing problem have been done but none proves to be practical. In this paper, we define a new model for this problem, and provide a sophisticated dynamic programming algorithm to solve it. Experiments on real MS/MS data demonstrated that the algorithm works very well on QTof MS/MS data.

1   Introduction

Tandem mass spectrometry (MS/MS) now plays a very important role in protein identification due to its fastness and high sensitivity. In an MS/MS experiment, proteins are digested into peptides. Those peptides are charged and selected by a mass analyzer according to their mass to charge ratio (also called m/z value). Usually, different peptides of the proteins have different m/z values. Therefore, we can assume that the mass analyzer selects many copies of the same peptide. The selected peptides are then fragmented and the resulting fragment ions are measured again by a second mass analyzer to generate an MS/MS spectrum.

Figure 1 illustrates how a peptide with four amino acids fragments into different fragment ions. In an MS/MS experiment, the peptide in Figure 1 usually can fragment into two pieces at any of the several labeled locations, and generate six types of fragment ions: a_i, b_i, c_i, x_i, y_i, z_i ($i = 1, 2, 3$).

The MS/MS spectrum of a peptide consists of many peaks (see Figure 2), each of which is presumably generated by many copies of one fragment ion. The position of the peak indicates the mass to charge ratio of the corresponding fragment ion, and the height of the peak indicates the relative intensity of the fragment ion. In general, there are 20 different types of amino acids, of which most have distinct masses to each other. Consequently, different peptides usually produce different MS/MS spectra. It is thus possible, and now a common practice, to use the spectrum of a peptide to determine its sequence. This step of sequence determination is an indispensable extension of MS/MS lab work for peptide sequencing.

R. Baeza-Yates et al. (Eds.): CPM 2003, LNCS 2676, pp. 266–277, 2003.

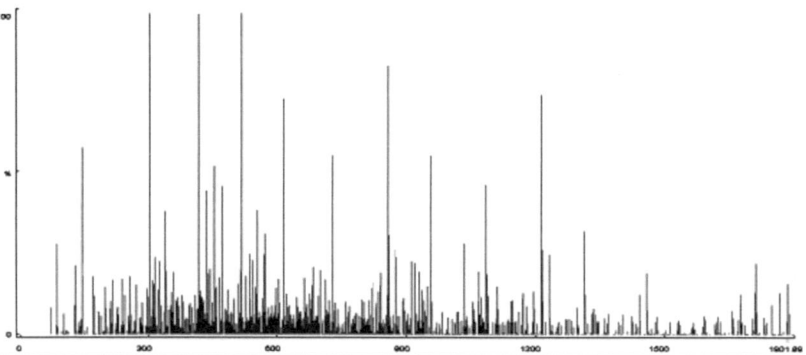

Fig. 1. Four amino acids are linked by the C-N bonds to form a peptide. The fragmentation can occur at any of the sites labeled by the dashed vertical lines. The resulting fragment ions with the N-terminus are the a-, b-, and c-ions, and the ions with the C-terminus are the x-, y-, and z-ions.

Fig. 2. An MS/MS spectrum.

One method to determine the peptide sequence from the MS/MS spectrum involves searching protein databases for a peptide whose hypothetical spectrum matches the observed spectrum [4,12,13,14,20]. However, the database search method is not always successful because the target peptide sequences may be absent in the databases. Therefore, the *de novo* sequencing, which determines the peptide sequence from the MS/MS spectrum without the help of a protein database, is very important to the study of novel proteins.

1.1   Related Work

There have been attempts to solve the *de novo* sequencing problem. Sakurai *et al.* [16] used a method which generates all possible peptide sequences and compares each of them with the spectrum. The method is only feasible for very short peptides. Much research has been done on *prefix pruning* to speed up the search [6,8,9,17,21,22]. But the prefix pruning cannot find the correct sequence if its prefixes are poorly represented in the spectrum [3].

Another approach to *de novo* sequencing generates a *spectrum graph* for an MS/MS spectrum [1,2,3,5,7,19], and then conduct computation on the graph.

Each peak in the spectrum generates a few of nodes in the spectrum graph, corresponding to the different types of ions that may produce the peak. Each edge in the graph indicates that the mass difference of the two adjacent nodes is approximately the mass of an amino acid, and the edge is labeled with the amino acid. The *de novo* sequencing problem is therefore reduced to the finding of the longest path in the graph.

Spectrum graph approach has two major difficulties. First, when all the ion types, $a_i, b_i, c_i, x_i, y_i, z_i$, are missing for one or more consecutive i, then there is no path in the graph corresponding to the correct sequence.

Secondly, a peak may be interpreted by two different ions. In spectrum graph approach the score of a path is the sum of the scores of the nodes in the path. Consequently, if a peak has very high intensity, then there is a tendency that the longest path will include the two nodes corresponding to the two different interpretations of the same peak. This usually (but not always) indicates a wrong sequence. To avoid this problem, one way is to forbidden the simultaneous occurrences of pairs of nodes corresponding to the same peak [2,3]. However, in practice, two different ions are possible to generate the same peak. The forbidden pairs approach will exclude the correct solutions for those cases. Consequently, We feel that a better way would be changing the score definition. That is, if both of the two nodes corresponding to the same peak occur in the path, then only the score of one node will be added into the score of the path. This eliminates the tendency. The method we use in this paper is analogous to the latter way.

In this paper, we present a new model for *de novo* sequencing problem which overcomes both difficulties of the spectrum graph approach.

1.2   Our Contributions

Our contributions with this paper are the following:

1. We introduce a new model for the *de novo* sequencing problem. The new model fits into the paradigm of the *Peptide Sequencing Problem* defined in [14]. However, we introduce a new scheme to define the score function. Our model accounts for most of the ion types that have been observed in practice. Also, our model deals with the mass overlap of different ions in a more reasonable manner.
2. We provide an efficient algorithm to find the optimal solution of the *de novo* sequencing problem with our model. The algorithm tolerants the mass errors and the missing of ions in the spectra. Because the dealing of the mass overlaps of different ions, the proof of the correctness of the algorithm is fairly complicated.

We implemented our algorithm into a Java program and did experiments on real data, which demonstrated that our algorithm performed very well for QTof MS/MS data. However, it is not the focus of the paper to discuss the software implementation. Instead, we only present the basic model and the algorithm.

Also, for the simplicity of presentation, we use a very simple score function conforming our scheme of score functions. But more complex score functions can be used without major changes to our algorithm. In fact, the software program we have implemented used a more sophisticated score function.

The rest of the paper is organized as follows: Section 2 gives our model to the *de novo* sequencing problem; Section 3 gives the algorithm; and Section 4 discusses some experimental results.

2   Modeling the *De novo* Sequencing Problem

Let $\mathcal{M}$ be a spectrum. For any peptide sequence P, because we know the precise mass of each atom, we are able to compute the masses of all possible fragment ions of P. Consequently, we can find all the peaks in $\mathcal{M}$ whose positions match these fragment ion masses. Intuitively, the more and higher peaks are matched, the more likely P is the correct sequence for the peptide that generates $\mathcal{M}$. In this section we will formulate this intuition and give a new model to the *de novo* sequencing problem.

2.1   The Ion Masses of a Peptide

In the MS/MS experiment, an ion can be charged with different charges, and therefore generates a few different peaks. Fortunately, there are standard methods to preprocess an MS/MS spectrum to convert all the peaks for multiply charged ions to their singly charged equivalents [18]. Therefore, in this paper we assume all the ions have charge one. Consequently, the mass to charge ratio of an ion is equal to the mass of the ion. In this section we will examine how to compute all the fragment ion masses of a given peptide.

We denote the alphabet with the 20 amino acids by Σ. For an amino acid $a \in \Sigma$, we use $\|a\|$ to denote the mass of its residue (i.e., the amino acid losing a water). The masses of all the amino acid residues can be found in [18]. Here, we only note that $\max_{a \in \Sigma} \|a\| = 186.08$ Dalton and

$$\min_{a \in \Sigma} \|a\| = 57.02 \text{ Dalton}. \tag{1}$$

It is also noteworthy that the masses of the amino acid I and L are identical. Therefore, they are not distinguishable using usual MS/MS spectrometry. Following the common practice, we ignore this problem and simply treat I and L as the same amino acid.

For $P = a_1 a_2 \ldots a_k$ being a string of amino acids, define $\|P\| = \sum_{1 \le j \le k} \|a_j\|$. The actual mass of the peptide with sequence P is then $18 + \|P\|$, because of an extra H_2O in the peptide.

From Figure 1, the mass of the b-ion of P with i amino acids, denoted by b_i, can be computed with $b_i = 1 + \sum_{1 \le j \le i} \|a_j\|$. Similarly, the mass of the y-ion of P with i amino acids, denoted by y_i, can be computed with $y_i = 19 + \sum_{n-i+1 \le j \le n} \|a_j\|$.[1] Let $A = a_1 a_2 \ldots a_n$ be a string of amino acids. We

---

[1] Because the ionization process adds protons to the ions, the actual compositions of the ions are slightly different from those shown in Figure 1.

introduce two notations $\|A\|_b = 1 + \sum_{1 \leq j \leq n} \|a_i\|$, and $\|A\|_y = 19 + \sum_{1 \leq j \leq n} \|a_i\|$. Therefore, $b_i = \|a_1 a_2 \ldots a_i\|_b$ and $y_i = \|a_{k-i+1} \ldots a_{k-1} a_k\|_y$, $(1 \leq i < k)$. Clearly,

$$y_{k-i} + b_i = 20 + \|P\|, \quad 1 \leq i < k. \tag{2}$$

This fact will be used in the design of our algorithm in Section 3.

Let x be the mass of a b-ion. The masses of the a-ions and c-ions with the same number of amino acids are $x - 28$ and $x + 17$, respectively. The b-ion may lose a water or an ammonia to form ions with masses $x - 18$ and $x - 17$, respectively. So, we use a notation $B(x)$, defined as follows,

$$B(x) = \{x - 28, x - 18, x - 17, x, x + 17\}, \tag{3}$$

to denote the set of masses of the ions related to the b-ion with mass x. Similarly, for each y-ion with mass x, the following set,

$$Y(x) = \{x - 18, x - 17, x, x + 26\}, \tag{4}$$

is the set of masses of the ions related to the y-ion with mass x. The reason that $Y(x)$ has one fewer element than $B(x)$ is that both the mass of the y-ion losing an ammonia and the mass of the corresponding z-ion are $x - 17$.

Therefore, hypothetically, the spectrum of the peptide P have a peak at each mass value in

$$S(P) = \bigcup_{i=1}^{n-1} (B(b_i) \cup Y(y_i)). \tag{5}$$

2.2 *De novo* Sequencing Problem

Most tandem spectrometers can output an MS/MS spectrum $\mathcal{M} = \{(x_i, h_i) \mid 1 \leq i \leq n\}$, where (x_i, h_i) denotes the i-th peak at position x_i and with height h_i. Together with $\mathcal{M}$, the mass of the peptide P can also be measured. Let $M = \|P\| + 20$. We want to derive the sequence P from $\mathcal{M}$ and M.

The masses given by the spectrometer are not accurate. Depending on different spectrometers, the maximum error varies from ± 0.01 Dalton to ± 0.5 Dalton. Let $\delta > 0$ be the maximum error of the mass spectrometer that we use. We assume that $\delta < 0.5$ throughout this paper. Let S be a set of masses. A peak $(x, h) \in \mathcal{M}$ is *explained* by S, if there is $y \in S$, such that $|x - y| \leq \delta$. The subset of peaks in $\mathcal{M}$, explained by S, is denoted by $\overline{S}$. That is,

$$\overline{S} = \{(x_i, h_i) \in \mathcal{M} \mid \text{there is } y \in S \text{ s.t. } |y - x_i| \leq \delta\}.^2$$

Let $S(P)$ be the set of the mass values of all possible ions of P. $S(P)$ can be computed by (5). Then $\overline{S(P)}$ consists of all the peaks in $\mathcal{M}$ that can be explained

---

[2] $\overline{S}$ depends not only on S, but also on $\mathcal{M}$ and δ. However, because $\mathcal{M}$ and δ are given and fixed for an instance of our problem, we use the notation $\overline{S}$ for similicity.

by the ions of P. Intuitively, the more and higher peaks $\overline{\mathcal{S}(P)}$ contains, the more likely $\mathcal{M}$ is the mass spectrum of P. For any peak list $\mathcal{M}$, we define

$$h(\mathcal{M}) = \sum_{(x,h)\in\mathcal{M}} h.$$

Then the *de novo* sequencing problem is defined as follows

De novo sequencing   Given a mass spectrum $\mathcal{M}$, a positive number M, and an error bound $\delta > 0$, to construct a peptide P, so that $|\|P\| + 20 - M| \leq \delta$ and $h(\overline{\mathcal{S}(P)})$ is maximized.

Without further discussion, we note that without any major modifications to our algorithm, the score function $h(\overline{\mathcal{S}(P)})$ can be broaden. One possible scheme of score functions is $\sum_{q \in \overline{\mathcal{S}(P)}} w(q)$, where $w(q)$ is the "weight" of the peak q. In a real implementation of our algorithm, $w(q)$ can involve many factors such as the height of q, the type of ion that explains q, and the mass error between the ion and q. However, for the simplicity of presentation, we simply set $w(q)$ to be the height of q in this paper.

3   An Algorithm for *De novo* Sequencing

There are two major difficulties of the *de novo* sequencing problem. First, each fragmentation may produce a pair of ions. When the fragmentation happens closely to the N- or C-terminus, one of the two ions has relatively large mass and the other is small. Therefore, both ends of the spectrum must be considered at the same time, in order to evaluate the score contribution of the pair of ions caused by the same fragmentation. Secondly, the types of the peaks are unknown, and a peak might be matched by zero, one, or two different types of ions. In the case a peak is matched by two ions, the height of the peak can only be counted once. (An algorithm that counts the height twice has the tendency to match the highest peaks more than once, instead of matching more peaks. We have observed this fact in many MS/MS data sets.) Therefore, the algorithm should know whether a peak has already been matched, before it can evaluate the match between the peak and a new ion.

Because of the difficulties, a straightforward dynamic programming approach, which attempts to construct the optimal peptide from one terminus to the other, does not work. In this section, we give a much more sophisticated dynamic programming algorithm for the *de novo* sequencing problem. Our algorithm gradually construct optimal pairs of prefixes and suffixes in a carefully designated pathway, until the prefix and the suffix become sufficient long to form the optimal solution. Before we can present the algorithm, we need some definitions.

Let $\mathcal{M} = \{(x_i, h_i) | 1 \leq i \leq n\}$ be the given spectrum and $M = \|P\| + 20$. Let $A = a_1 a_2 \ldots a_k$ be a string of amino acids. If A is a prefix (at the N-terminus) of $P = a_1 a_2 \ldots a_n$, then the mass of the b-ion produced by the fragmentation between a_i and a_{i+1} is $\|a_1 a_2 \ldots a_i\|_b$. From (2), the mass of the y-ion caused by the same fragmentation is $M - \|a_1 a_2 \ldots a_i\|_b$. If use $\mathcal{S}_N(A)$ to denote the set of masses of all the ions caused by the fragmentations between a_i and a_{i+1} $(1 \le i \le k)$, then from (3) and (4),

$$S_N(A) = \bigcup_{i=1}^{k} [B(\|a_1 a_2 \dots a_i\|_b) \cup Y(M - \|a_1 a_2 \dots a_i\|_b)].$$

Similarly, suppose that $A' = a'_k \dots a'_2 a'_1$ is a suffix (at the C-terminus) of a peptide $P' = a'_n \dots a'_2 a'_1$ and $M = \|P'\| + 20$. We use $S_C(A)$ to denote the set of masses of all the ions caused by the fragmentation between a'_{i+1} and a'_i $(1 \le i \le k)$. It is easy to verify that

$$S_C(A') = \bigcup_{i=1}^{k} [Y(\|a'_i \dots a'_2 a'_1\|_y) \cup B(M - \|a'_i \dots a'_2 a'_1\|_y)].$$

Let peptide P be the optimal solution. For any string A, A' and an amino acid a, such that $P = AaA'$, the following fact is obvious.

$$S(P) = S_N(A) \cup S_C(A'). \tag{6}$$

This suggests us to reduce the *de novo* sequencing problem to the problem of finding the appropriate prefix A and suffix A'. There are $l(P)$ different ways to divide P into the form AaA', where $l(P)$ denotes the length of P. In the rest of this section, we will compute a specific division so that (A, A') is a *chummy pair*, defined as follows.

Definition 1. *For any string of amino acids* $s = s_1 s_2 \dots s_n$, $s_{\gg}$ *denotes the length* $n-1$ *string* $s_1 s_2 \dots s_{n-1}$ *and* $s_{\ll}$ *denotes the length* $n-1$ *string* $s_2 s_3 \dots s_n$. *A string pair* (A, A') *is called a chummy pair, if*

$$\|A\|_b + \|A'\|_y < M$$

and either of the following two inequalities holds:

$$\|A_{\gg}\|_b < \|A'\|_y \le \|A\|_b \tag{7}$$
$$\|A'_{\ll}\|_y \le \|A\|_b < \|A'\|_y \tag{8}$$

Figure 3 illustrates two chummy pairs (A, A') and (Aa, A'). Both (A, A') and (Aa, A') are such that (8) holds. Let Z be any prefix of A such that $Z \ne A$. Let Z' be any suffix of A' such that $Z' \ne A'$. From the figure we can see that $B(\|Aa\|_b)$ is apart from both $Y(\|Z'\|_y)$ and $B(\|Z\|_b)$. In fact, from the definition of $B(\cdot)$ and (1), we know that the distance between $B(\|Z\|_b)$ and $B(\|Aa\|_b)$ is at least $2 \times \min_{a \in \Sigma} \|a\| - 45 > 69$. Because $\delta < 0.5$, there is no peak in $\mathcal{M}$ that can be explained by both $B(\|A_{\gg}\|_b)$ and $B(\|Aa\|_b)$. In other words,

$$\overline{B(\|Aa\|_b)} \cap \overline{B(\|Z\|_b)} = \emptyset. \tag{9}$$

Because (8) holds, $\|Z'\|_y \le \|A'_{\ll}\|_y \le \|A\|_b$. From Figure 3, we can similarly prove that

$$\overline{B(\|Aa\|_b)} \cap \overline{Y(\|Z'\|_y)} = \emptyset. \tag{10}$$

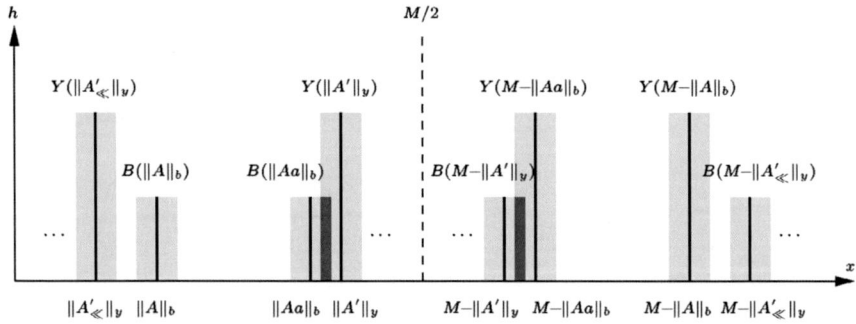

Fig. 3. An illustration to the relationship of chummy pairs. Each solid vertical line indicates a hypothetical peak. We draw the hypothetical y-ion peaks higher to reflect that the y-ions are the most often observed ions in the spectrum. The shadowed areas $Y(x)$ and $B(x)$ indicate that there are some other hypothetical peaks nearby x (See (4) and (3)). As the figure shows, both (A, A') and (Aa, A'), but not $(Aa, A'_{\ll})$, are chummy pairs.

Also, because (Aa, A') is a chummy pair, $\|Aa\| + \|A'\| < M$. Therefore, it is easy to prove that both $B(\|Aa\|_b) \cap B(M - \|Z'\|_y)$ and $B(\|Aa\|_b) \cap Y(M - \|Z\|_b)$ are empty sets. Combined with Formula (9) and (10), we conclude that

$$\overline{B(\|Aa\|_b) \cap \overline{S_N(A_{\gg}) \cup S_C(A'_{\ll})}} = \emptyset. \tag{11}$$

The same arguments can be applied to $Y(M - \|Aa\|_b)$ and get the conclusion that

$$\overline{Y(M - \|Aa\|_b) \cap \overline{S_N(A_{\gg}) \cup S_C(A'_{\ll})}} = \emptyset. \tag{12}$$

The following lemma reveals the reason we investigate chummy pairs. That is, we can compute $h(\overline{S_N(Aa) \cup S_C(A')})$ from $h(\overline{S_N(A) \cup S_C(A')})$.

Lemma 1. *Let (A, A') be a chummy pair and a be an amino acid and*

$$f(u, v, w) = h\left(\overline{B(u) \cup Y(M - u)} \setminus \overline{B(v) \cup Y(M - v) \cup Y(w) \cup B(M - w)}\right).$$

Then (i) if (Aa, A') is a chummy pair, then

$$h\left(\overline{S_N(Aa) \cup S_C(A')}\right) = h\left(\overline{S_N(A) \cup S_C(A')}\right) + f(\|Aa\|_b, \|A\|_b, \|A'\|_y).$$

(ii) if (A, aA') is a chummy pair, then

$$h\left(\overline{S_N(A) \cup S_C(aA')}\right) = h\left(\overline{S_N(A) \cup S_C(A')}\right) + f(M - \|aA'\|_y, M - \|A'\|_y, M - \|A\|_b).$$

Proof. We only prove (i). The proof of (ii) is similar therefore omitted.

Let $u = \|Aa\|_b$, $v = \|A\|_b$, and $w = \|A'\|_y$. It suffices to prove the following two equalities:

$$\overline{S_N(Aa) \cup S_C(A')} = \overline{S_N(A) \cup S_C(A') \cup B(u) \cup Y(M - u)}, \tag{13}$$

and

$$\emptyset = \overline{\mathcal{S}_N(A) \cup \mathcal{S}_C(A')} \cap$$
$$\left(\overline{B(u) \cup Y(M-u)} \setminus \overline{B(v) \cup Y(M-v) \cup Y(w) \cup B(M-w)} \right) \quad (14)$$

(13) is true because of the definition of $\mathcal{S}_N(\cdot)$ and $\mathcal{S}_C(\cdot)$. (14) is true because of (11) and (12).   □

Lemma 2-4 give more properties of chummy pairs, which are also useful in our algorithm.

Lemma 2. *Let (A, A') be a chummy pair. Let a be any letter such that $\|A\|_b + \|A'\|_y + \|a\| < M$. Then*
   (i) If $\|A\|_b < \|A'\|_y$, then (Aa, A') is a chummy pair and (A, aA') is not;
   (ii) If $\|A'\|_y \leq \|A\|_b$, then (A, aA') is a chummy pair and (Aa, A') is not;

Proof. (i) Because (A, A') is a chummy pair, $\|A\|_b < \|A'\|_y$ implies that $\|A'_{\ll}\|_y \leq \|A\|_b < \|A'\|_y$. If $\|Aa\|_b < \|A'\|_y$, then (Aa, A') is a chummy pair because $\|A'_{\ll}\|_y \leq \|A\|_b < \|Aa\|_b < \|A'\|_y$. If $\|A'\|_y \leq \|Aa\|_b$, still (Aa, A') is a chummy pair because $\|A\|_b < \|A'\|_y \leq \|Aa\|_b$. (A, aA') is not a chummy pair because $\|A\|_b < \|A'\|_y \leq \|aA'\|_y$.
   The proof of (ii) is very similar to (i). We leave it to the readers.   □

Lemma 3. *Let (A, A') be a chummy pair. Then either $(A_{\gg}, A')$ or $(A, A'_{\ll})$, but not both, are chummy pairs.*

Proof. If $\|A_{\gg}\|_b < \|A'_{\ll}\|_y$, it is easy to check that no matter which of (7) and (8) is true, we always have $\|A_{\gg}\|_b < \|A'_{\ll}\|_y \leq \|A\|_b$. Therefore, $(A, A'_{\ll})$ is a chummy pair. $(A_{\gg}, A')$ is not a chummy pair because $\|A_{\gg}\|_b < \|A'_{\ll}\|_y < \|A'\|_y$, which conflicts both (7) and (8).
   For the same reason, if $\|A'_{\ll}\|_y \leq \|A_{\gg}\|_b$, then $(A_{\gg}, A')$ is a chummy pair but $(A, A'_{\ll})$ is not. We leave the proof to the readers.   □

Lemma 4. *Let P be the optimal solution. Then there is a chummy pair (A, A') and a letter a, so that $P = AaA'$.*

Proof. Suppose $P = a_1 a_2 \dots a_m$. Let $l(s)$ denote the length of a string s. From Lemma 2, the desired A and A' can be found by the following procedure:
   1. Let A and A' be empty strings.
   2. for i from 1 to $m-1$ do
      if $\|A\|_b < \|A'\|_y$ then let $A = Aa_{l(A)+1}$.
      else let $A' = a_{m-l(A')}A'$.
     □

From Lemma 4 and (6), in order to find the optimal solution, it suffices to find a chummy pair (A, A'), so that

1. there is a letter a, such that $|\|A\|_b + \|A'\|_y + \|a\| - M| \leq \delta$.
2. $h(\overline{\mathcal{S}_N(A) \cup \mathcal{S}_C(A')})$ is maximized.

Algorithm Sandwich

Input   A peak list $\mathcal{M}$, a mass value M, an error bound δ, and a calibration Δ.

Output   A peptide P such that $h(\overline{S(P)})$ is maximized and $|\,\|P\| + 20 - M| \leq \delta$.

1. Initialize all $DP[i, j] = -\infty$; Let $DP[1, 19] = 0$

2. for x from 1 to $M/2 + \max_{a \in \Sigma} \|a\|$ step Δ do

3.   for y from $x - \max_{a \in \Sigma} \|a\|$ to $\min(x + \max_{a \in \Sigma} \|a\|, M - x)$ step Δ do

4.   for $a \in \Sigma$ such that $x + y + \|a\| < M$

5.   if $x < y$ then

6.   let $DP[x + \|a\|, y] = \max \begin{cases} DP[x + \|a\|, y], \\ DP[x, y] + f(x + \|a\|, x, y). \end{cases}$

7.   else

8.   let $DP[x, y + \|a\|] = \max \begin{cases} DP[x, y + \|a\|], \\ DP[x, y] + f(M - y - \|a\|, M - y, M - x). \end{cases}$

9. Compute the best $DP[x, y]$ for all x, y and a satisfying $|x + y + \|a\| - M| \leq \delta$.

10. Compute the best A, A', a using backtracking, and output AaA'.

Fig. 4. Our algorithm for *de novo* sequencing, where $f(u, v, w)$ is defined in Lemma 1.

For any two positive numbers x and y such that $x + y \leq M$, let $DP(x, y)$ be the maximum value of $h(\overline{S_1(A) \cup S_2(A')})$ for all chummy pairs (A, A') such that $\|A\|_b = x$ and $\|A'\|_y = y$. Let $\Delta > 0$ be the finest calibration of the spectrometer, e.g. $\Delta = 0.005$ Dalton. Let $\delta > 0$ be the mass error tolerance, e.g. $\delta = 0.1$. From Lemma 1-4, it is easy to see that the algorithm in Figure 4 computes $DP(x, y)$ gradually and outputs the optimal solution.

Theorem 1. *Algorithm Sandwich computes the optimal solution of the* de novo *sequencing problem in* $O\left(\frac{M}{\Delta} \times \frac{\delta}{\Delta} \times \frac{\max_{a \in \Sigma} \|a\|}{\Delta}\right)$ *time.*

Proof. We first prove that the $DP[x, y]$ computed by the algorithm is equal to the $DP(x, y)$ defined by us. This can be proven by an induction on $x + y$. Clearly the claim is true for $x + y \leq 20$.

For any x, y such that $DP(x, y) \geq 0$, there is a chummy pair (A, A') such that $x = \|A\|_b$, $y = \|A'\|_y$, and $DP(x, y) = h(\overline{S_N(A) \cup S_C(A')})$. From Lemma 3, without loss of generality, we assume that $(A_\gg, A')$ is a chummy pair. Let $x_0 = \|A_\gg\|_b$. Due to Lemma 1 and the optimality of (A, A'), $(A_\gg, A')$ must also maximize $h(\overline{S_N(A_\gg) \cup S_C(A')})$. That is, $DP(x_0, y) = h(\overline{S_N(A_\gg) \cup S_C(A')})$. The induction hypothesis confirms that $DP(x_0, y) = DP[x_0, y]$. Therefore, we can use the method provided in Lemma 1 to compute $DP(x, y)$ from $DP[x_0, y]$. Clearly the same computation has been done in the algorithm to compute $DP[x, y]$. Therefore, $DP(x, y) = DP[x, y]$. From Lemma 4 and (6), we can conclude that the sequence AaA' output at line 10 is the optimal peptide sequence.

Because there are at most $O\left(\frac{\delta}{\Delta}\right)$ peaks in $\mathcal{M}$ explained by a single mass, $f(\cdot, \cdot, \cdot)$ in lines 6 and 8 can be computed in time $O\left(\frac{\delta}{\Delta}\right)$. The time complexity of the algorithm can then be proved straightforwardly. $\qquad\square$

4   Implementation and Experiments

We have implemented the algorithm into a Java program, and tested it with thirteen MS/MS spectra that were obtained from a biochemistry research lab in the University of Western Ontario. A technician with no prior knowledge to the software picked up the thirteen spectra, of which the sequences were determined previously using manual or semi-automated methods and considered reliable. However, the quality of the thirteen spectra are not great. Commercial *de novo* sequencing software failed to determine the sequences of most of the thirteen spectra. Our program, however, successfully determined eleven of the thirteen sequences. For the other two spectra, our program produced slightly different sequences. More specifically, it output `ADVNDNEEGFFSAR` instead of `EGVNDNEEGFFSAR` and `VNMEVEQVQLVVDGVK` instead of `LGSSEVEQVQLVVDGVK`. We note that the mistakes at the N-terminuses in both cases are due to the bad quality of the spectra at the high-mass ends.

Besides of the Sandwich algorithm, the program also involves more techniques in handling the mass spectrometer errors and optimizing the scoring method. More information about the program can be found in [11]. The program has been improved and integrated into a commercial software package, *PEAKS*. The free on-line version of *PEAKS* can be found at
http://www.BioinformaticsSolutions.com/.

Acknowledgement. BM was supported partially by NSERC research grant RGP0238748. KZ was supported partially by NSERC research grant OGP0046373 and a SharcNet research fellowship. CL was supported by Bioinformatics Solutions Inc. We thank Dr. Ming Li for suggesting the research problem. We thank Drs. Gilles Lajoie and Amanda Doherty-Kirby for providing the test data sets.

References

1. Bartels, C. 1990. Fast algorithm for peptide sequencing by mass spectroscopy. *Biomed. Environ. Mass Spectrom* 19, 363–368.
2. Chen, T., Kao, M-Y., Tepel, M., Rush J., and Church, G. 2001. A Dynamic Programming Approach to *de novo* Peptide Sequencing via Tandem Mass Spectrometry. *J. Comp. Biology* 8(3), 325-337.
3. Dančík, V., Addona, T., Clauser, K., Vath, J., and Pevzner, P. 1999. De novo protein sequencing via tandem mass-spectrometry. *J. Comp. Biology* 6, 327–341.
4. Eng, J.K., McCormack, A.L., and Yates, J.R. 1994. An approach to correlate tandem mass spectral data pf peptides with amino acid sequences in a protein database. *J. Am. Soc. Mass Spectrom* 5, 976–989.
5. Fernández de Cossío, J., Gonzales, J., and Besada, V. 1995. A computer program to aid the sequencing of peptides in collision-activated decomposition experiments. *CABIOS* 11(4), 427–434.
6. Hamm, C.W., Wilson, W.E., and Harvan, D.J. 1986. Peptide sequencing program. *CABIOS* 2, 365.

7. Hines, W.M., Falick, A.M., Burlingame, A.L., and Gibson, B.W. 1992. Pattern-based algorithm for peptide sequencing from tandem high energy collision-induced dissociation mass spectra. *J. Am. Sco. Mass. Spectrom.* 3, 326–336.

8. Ishikawa K., and Niva, Y. 1986. Computer-aided peptide sequencing by fast atom bombardment mass spectrometry. *Biomed. Environ. Mass Spectrom.* 13, 373–380.

9. Johnson, R.J., and Biemann, K. 1989. Computer program (seqpep) to aid the interpretation of high-energy collision tandem mass spectra of peptides. *Biomed. Environ. Mass. Spectrom.* 18, 945–957.

10. Johnson, R.S., Martin, S.A., Biemann, K., Stults, J.T., and Watson, J.T. 1987. Novel fragmentation process of peptides by collision-induced decomposition in a tandem mass spectrometer: differentiation of leucine and isoleucine. *Anal. Chem.* 59(21), 2621–5.

11. Ma, B., Zhang, K., Lajoie, G., Doherty-Kirby, A., Liang, C., and Li, M. 2002. A powerful software tool for the *de novo* sequencing of peptides from MS/MS data. *50th ASMS Conference on Mass Spectrometry and Allied Topics.*

12. Mann, M., and Wilm, M. 1994. Error-tolerant identification of peptides in sequence databases by peptide sequence tags. *Anal. Chem.* 66, 4390–4399.

13. Perkins, D.N., Pappin, D.J.C., Creasy, D.M., and Cottrell, J.S. 1999. Probability-based protein identification by searching sequence database using mass spectrometry data. *Electrophoresis* 20, 3551–3567.

14. Pevzner, P.A., Dančĭk, V., and Tang, C. 2000. Mutation Tolerant Protein Identification by Mass Spectrometry. *Journal of Computational Biology* 6, 777–787.

15. Roepstorff, P., and Fohlman J. 1984. Proposal for a common nomenclature for sequence ions in mass spectra of peptides. *Biomed Mass Spectrom* 11(11), 601.

16. Sakurai, T., Matsuo, T., Matsuda, H., and Katakuse, I. 1984. Paas3: A computer program to determine probable sequence of peptides from mass spectrometric data. *Biomed. Mass spectrum* 11(8), 396–399.

17. Siegel, M.M., and Bauman, N. 1988. An efficient algorithm for sequencing peptides using fast atom bombardment mass spectral data. *Biomed. Environ. Mass Spectrom* 15, 333–343.

18. Snyder, A.P. 2000. Interpreting Protein Mass Spectra: A Comprehensive Resource. *Oxford University Press.*

19. Taylor, J.A., and Johnson, R.S. 1997. Sequence Database Searches via *de novo* peptide sequencing by tandem mass spectrometry. *Rapid Communications in Mass Spectrometry* 11, 1067–1075.

20. Yates, J.R.I., Eng, J.K., McCormack, A.L., and Schieltz, D. 1995. Method to correlate tandem mass spectra of modified peptides to amino acid sequences in the protein database. *Analytical Chemistry* 67, 1426–36.

21. Yates, J.R., Griffin, P.R., Hood, L.E., and Zhou, J.X. 1991. Computer aided interpretation of low energy MS/MS mass spectra of peptides, 477–485. in J.J. Villafranca ed., *Techniques in Protein Chmistry II*, Academic Press, San Diego.

22. Zidarov, D., Thibault, P., Evans, M.J., and Bertrand, M.J. 1990. Determination of the primary structure of peptidesusing fast atom bombardment mass spectrometry. *Biomed. Environ. Mass Spectrom* 19, 13–16.

Pattern Discovery in RNA Secondary Structure Using Affix Trees

Giancarlo Mauri and Giulio Pavesi

Dept. of Computer Science, Systems and Communication
University of Milan–Bicocca
Via Bicocca degli Arcimboldi 8
20126 Milan, Italy
Phone: +39 02 6448 7864, Fax: +39 02 6448 7839
{mauri,pavesi}@disco.unimib.it

Abstract. We present an algorithm for finding common secondary structure motifs in a set of unaligned RNA sequences. The basic version of the algorithm takes as input a set of strings representing the secondary structure of the sequences, enumerates a set of candidate secondary structure patterns, and finally reports all those patterns that appear, possibly with variations, in all or most of the sequences of the set. By considering structural information only, the algorithm can be applied to cases where the input sequences do not present any significant similarity. However, sequence information can be added to the algorithm at different levels. Patterns describing RNA secondary structure elements present a peculiar symmetric layout that makes affix trees a suitable indexing structure that significantly accelerates the searching process, by permitting bidirectional search from the middle to the outside of patterns. In case the secondary structure of the input sequences is not available, we show how the algorithm can deal with the uncertainty deriving from prediction methods, or can predict the structure by itself on the fly while searching for patterns, again taking advantage of the information contained in the affix tree built for the sequences. Finally, we present some case studies where the algorithm was able to detect experimentally known RNA stem–loop motifs, either by using predicted structures, or by folding the sequences by itself.

1 Introduction

In molecular biology, sequence similarity often implies structural, and consequently functional similarity. In some cases, however, molecules of related function do not present any significant similarity at sequence level, but anyway form similar structures. This is the case, for example, of RNA. While little or no similarity can be detected among different RNA sequences, these sequences can anyway fold into structures that present similar elements, that in turn play the same biological function. Thus, in order to obtain biologically meaningful results, it is essential to compare and to find motifs also in the structures formed

R. Baeza-Yates et al. (Eds.): CPM 2003, LNCS 2676, pp. 278–294, 2003.

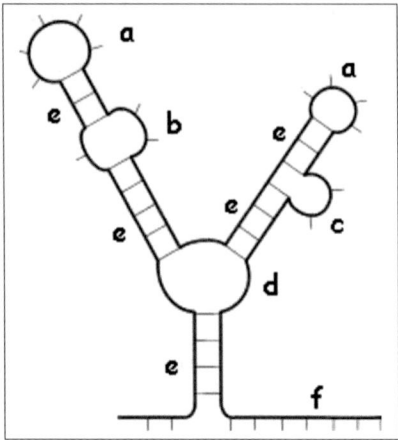

Fig. 1. Decomposition of RNA secondary structure in hairpin loops (a), internal loops (b), bulges (c), multi–loops (d), stacks (e), external elements (f).

by the sequences, and the first step is to take into account the *secondary structure* associated with them.

RNA (ribonucleic acid) molecules are composed of four different nucleotides (bases), guanine, adenine, cytosine and uracil (that replaces thymine, found in DNA), and are usually denoted by strings over the alphabet $\Sigma_{RNA} = \{$A,C,G,U$\}$. In living organisms, RNA is synthesized using one of the two DNA strands as a template. Once a RNA molecule has been copied from DNA, it starts to fold on itself forming a three–dimensional structure. Nucleotides composing the sequence bind to each other in different ways. The complementary nucleotides C-G and A-U form stable base pairs with each other through the creation of hydrogen bonds between donor and acceptor sites on the bases. These are called *Watson–Crick* base pairs, since they are the same bonds that hold the two strands of a DNA molecule together. In addition, some other base pairings are found in RNA structures, like the weaker G-U pair (*wobble pair*) where bases bond in a skewed fashion. All of these are called *canonical base pairs*. Other non–canonical pairs can occur, some of which are stable (like the G-A pair).

RNA secondary structure is usually represented by the list of the bonds taking place between the bases forming the sequence, with the additional constraints that no nucleotide takes part in more than one bond (base pair), and base pairs never cross: if nucleotide in position i of the sequence is paired with nucleotide $j > i$, and nucleotide $k > i$ is paired with $l > k$, then either $i < j < k < l$ or $i < k < l < j$, but never $i < k < j < l$. In this way, RNA secondary structure can be decomposed in blocks belonging to three different classes:

1. A *stack* consists of nested, consecutive base pairs, $(i,j),(i+1,j-1),\ldots,(i+k,j-k)$. The length of the stack is $k+1$. Pair (i,j) is the *terminal* base pair of the stack, while pair $(i+k,j-k)$ is the *closing* base pair.

2. A *loop* is defined as all unpaired nucleotides immediately interior to a base pair.
3. An *external nucleotide* is any unpaired nucleotide not contained in a loop. Consecutive external nucleotides are called *external elements*.

Any secondary structure can be uniquely decomposed into external elements, loops and stacks. Loops are also characterized in different ways. The *degree* of a loop is defined as $k + 1$, where k is the number of terminal base pairs that are internal to the closing base pair.

1. Loops of degree 1 are called *hairpin loops*. In biological circumstances, the minimum size of a hairpin loop is three nucleotides.
2. Loops of degree 2 are divided into two groups. Consider an external base pair (i, j) and an internal base pair $(i + p, j - q)$, where p and q are non–negative integers and nucleotides $i+1,\ldots, i+p-1$ and $j-1,\ldots, j-q+1$ are unpaired:
   a) A *bulge* is defined as the situation where p or q is equal to zero.
   b) An *internal loop* occurs when both p and q are greater than zero.
3. Loops of degree 3 or more are known as multi-loops.

An example is shown in Fig. 1. Since base pairs cannot cross, the secondary structure of a RNA sequence can be expressed with a string over the bracket alphabet $\Sigma_B = \{(, .,)\}$. A base pair between nucleotides i and j is denoted by an open bracket in position i and a close bracket in position j, and it is always possible to determine for each open bracket which is the corresponding close one. Unpaired nucleotides are indicated by a dot. In the bracket alphabet, the structure shown in Fig. 1 is denoted as:

$$..((((..(((((.(((.....)))..)))))..(((((((....)))..))))..))))......$$

Once believed to be a mere mediator of the genetic code, RNA is nowadays regarded as a key player in many biological processes, like in post–translational gene regulation [1,2]. That is, once a messenger RNA strand has been synthesized by using one of the two DNA strand as a template, its actual translation into a protein is further regulated by the formation of particular structural elements in the so–called *untranslated regions* (UTR), that are located immediately before (5' UTR) and after (3' UTR) the coding part of the messenger sequence that is actually translated into a protein. Moreover, many relevant functions of this kind have been associated with small but well conserved structural motifs, found in different sequences of related function, that can be described by considering only the secondary structure formed by the sequence. Thus, given a set of RNA sequences of related function (as for example messenger RNA for homologous proteins), finding similar elements in their secondary structure could provide significant insights on which parts of the molecules are more likely to be responsible for their function.

So far, the problem of finding common motifs in RNA secondary structure has been coupled with the prediction of the structure itself. Usually, the input sequences are aligned beforehand, and a common secondary structure is predicted for the alignment [3,4,5]. Differences in the structures associated with the sequences correspond to gaps and insertions in the alignment. This approach functions less well when the structural similarity is local and limited to small functional elements, or sequence similarity is not sufficient to produce a reliable alignment. To our knowledge, the only method available that finds similar elements in RNA secondary structure, either local or global, in the absence of significant (or detectable) sequence similarity (that is, without pre–aligning the sequences) or any prior knowledge about the structure of the motif is the FOLDALIGN algorithm [6,7]. The algorithm takes a set of RNA sequences, and tries to align them and at the same time to predict shared secondary structure elements by a double dynamic programming approach. The main drawback of FOLDALIGN lies in its time complexity, exponential in the number of input sequences. Even if reduced, with the introduction of some heuristics, to about $O(n^4 k^4)$ for k sequences of length n, it limits the applicability of the method to small sets or short sequences (no longer than 200–300 nucleotides). The algorithm we present, instead, works directly on the structure associated with the sequences, either known experimentally or predicted. In the latter case, we show how to deal with the uncertainty deriving from predictors. Moreover, we also introduce an approach similar to FOLDALIGN, where the secondary structure of common motifs is predicted on the fly during the search by looking for parts of the input sequences that can fold into similar structures.

2 Comparing Structural Elements

Let us assume for the moment that the secondary structure of the input sequences is already somehow available. Then, given the bracket alphabet, it is straightforward to see how the problem of finding common motifs in a set of RNA structures can be recast as finding motifs in a set of strings. However, a given structural element can appear in different forms, while keeping its biological function intact. Thus, we first of all have to introduce some distance measures suitable for the comparison of different structural parts.

The basic (or, at least, the most biologically interesting) feature of RNA secondary structure is the *stem–loop* motif (also called *hairpin*). It is made of a stack of paired nucleotides, possibly interrupted by bulges or internal loops, surmounted by a hairpin loop. If we compare two stem–loop structures we see that they can differ in various ways:

1. One may present one or more internal loops or bulges along the stem that the second one does not have, or the two stems may have internal loops or bulges in different positions; we will define *bulge distance*, denoted by d_b the total number of nucleotides, in either structure, forming bulges or internal loops that are not matched in the other one;
2. The two hairpin loops can be of different size; we will call *hairpin distance*, denoted by d_h, the difference in size of the two loops;
3. The two stems can present internal loops or bulges at the same position, but of different size; we will call *internal loop distance*, denoted by d_i, the

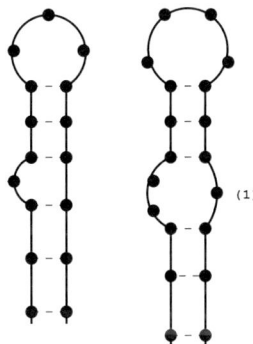

Fig. 2. Two hairpins having $d_b = 1$ (bulge formed by nucleotide (1) on the right one does not appear on the left one), $d_i = 1$, and $d_h = 1$. Dashes indicate bonds between nucleotides.

sum of the size difference of bulges and/or internal loops located at the same positions along the stems.

An example of the various distances is shown in Fig. 2. Now, to decide whether two stem–loop structures can be considered similar, we can set suitable threshold values τ_b, τ_i, and τ_h, respectively, for the distance measures just defined. Moreover, given a string S on Σ_B, we say that a pattern P representing a stem–loop structure occurs in S if there exists a substring of S whose distances from P do not exceed the corresponding thresholds.

However, to accelerate the operations needed for the comparisons, we can resort to a representation of the structure even more space efficient than the bracket alphabet. The idea is to replace any stretch of unpaired nucleotides (dots) with a unique symbol denoting its type (hairpin, internal loop or bulge, multi–loop, external element) and its size. For sake of clarity suppose that:

1. Hairpin loops are denoted with an uppercase letter. Letter A will denote a hairpin of size one, B of size two, and so on.
2. Internal loops, bulges, and multi–loops will be denoted by a lowercase letter, a for an element of size one, b for size two, and so on.
3. External elements are denoted by an integer representing their size.

Notice that two non–bracket characters cannot be adjacent in the compressed string. Secondary structure strings are thus defined over the alphabet $\Sigma_C = \{(,), u_1, \ldots, u_{m_1}, l_1, \ldots, l_{m_2}, i_1, \ldots, i_{m_3}\}$, where m_1, m_2, and m_3 are the maximum loop sizes (for each type) found in the original bracket string. Starting from a secondary structure in bracket notation, the compressed expression can be always determined in a unique way. With the compressed notation we are able to compare loops of different size by just comparing two characters, and moreover we can determine immediately if we are comparing two elements of the same type (in our definition of error, we always compare elements of the same type, treating bulges, internal loops and multi–loops as elements of the same type). For example, the structure of Fig. 1 thus becomes:

$$[2]((((b(((((a(((E)))b)))))b((((((D)))b))))b))))[2]$$

and the two structures of Fig 2 are denoted on Σ_C as $(((a((C))))))$ and $(((b((D))a)))$, respectively. From now on, we will denote with $size(e)$ the number of unpaired nucleotides forming an element denoted by symbol e, and with $D(e_1, e_2)$ the size difference of two elements e_1 and e_2 of the same type.

3   The Algorithm

Given a set of secondary structure strings S in the compressed bracket alphabet Σ_C, the basic idea of our algorithm is the following. First of all, we define suitable distance threshold values. These values might also depend on the size of the stem (for example, we might allow one unmatched bulge in a stem of four base pairs, two in a stem of eight base pairs, and so on). Then, we search exhaustively for all patterns on Σ_C representing stem–loop structures, containing also bulges and/or internal loops. For each pattern, we check whether it appears (according to distance thresholds) in at least q strings of S, and report all patterns that occur in at least q strings. A priori, the set of candidate patterns to be searched is infinite, if we do not require a pattern to appear exactly at least once in the structures. However, the exhaustive search for a large set of patterns can be made more efficient if the strings are first organized in a suitable text–indexing structure. Moreover, text indexing structures permit to implement the search *recursively*, pruning in a dramatic way the search space.

In this case, also, patterns present a peculiar symmetric layout. That is, they present a single uppercase letter around the middle, with two stretches of open and close brackets to the left and to the right, respectively (possibly interspersed by lowercase letters). As we will show in the next section, this feature makes *affix trees* particularly well suited for the problem.

3.1   Affix Trees

Let Σ be an alphabet, and Σ^* the set of strings over Σ. Given a string $S = s_1 \ldots s_n \in \Sigma^*$, we denote with $|S|$ its length, and with S^{-1} the *reverse* of S, that is, $S^{-1} = s_n \ldots s_1$. If $S = \alpha\beta\gamma$, with $\alpha, \beta, \gamma \in \Sigma^*$, then α is a prefix of S, and γ is a suffix of S.

Definition 1. *An* affix tree *[8] $\mathcal{T}$ for a string $S = s_1 \ldots s_n$ is a rooted directed tree $\mathcal{A}(S) = (V, E)$ such that:*

1. *V is the set of nodes, E is the set of edges;*
2. *$E = E_s \cup E_p$, and $E_s \cap E_p = \emptyset$. That is, the edges of $\mathcal{A}(S)$ are divided in two disjoint subsets, that we will call* suffix *and* prefix *edges;*
3. *Suffix edges are labeled with non–empty substrings of S;*

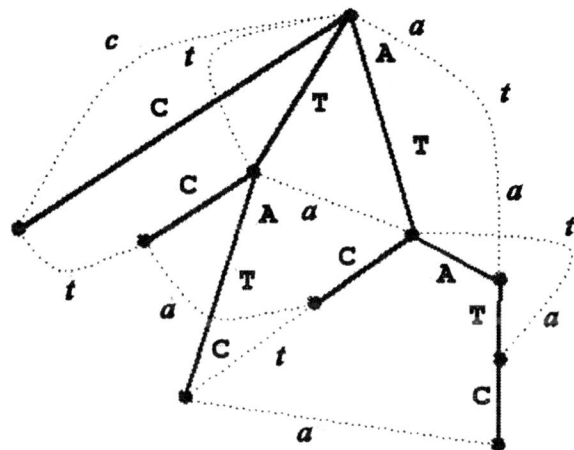

Fig. 3. Affix tree for string ATATC. Suffix edges are solid, prefix edges are dotted. Labels of prefix edges are in italic.

4. *Prefix edges are labeled with non–empty substrings of S^{-1};*
5. *Two edges of the same type leaving the same node cannot have labels starting with the same character;*
6. *On each path starting from the root of the tree and following suffix edges only, the concatenation of the edge labels spells out a substring of S;*
7. *On each path starting from the root of the tree and following prefix edges only, the concatenation of the edge labels spells out a substring of S^{-1};*

In $\mathcal{A}(S)$, *every* substring of S is spelled out by a unique path starting from the root and following suffix edges only, while every substring of S^{-1} is spelled by a unique path following prefix edges only. An example is shown in Fig. 3.

Now, let p be a node of the tree, and α be the concatenation of the labels of suffix edges from the root to p. The path from the root to p following prefix edges will spell α^{-1}. Let $p \to q$ be a prefix edge leaving p and entering node q, and $\gamma = \gamma_1...\gamma_l$ be its label. Then, the suffix path from the root to q (if it exists) spells out substring $\gamma^{-1}\alpha$ of S. An example can be seen in Fig. 3. If we consider the node whose label is AT, and follow the dotted prefix edge labeled ta, we reach a node whose label on suffix edges is ATAT. This property is essential for our method, since once a substring of S has been located in $\mathcal{A}(S)$, we can expand it both by adding characters to its right end (by following suffix edges) and to the left end, by following prefix edges. If $|\gamma| > 1$, then substring $\gamma_1\alpha$ occurs in the string *always* preceded by substring $\gamma_l...\gamma_2$. If no prefix edge leaves node p, then substring α starts only at position 1 (is a prefix) of S. Conversely, if no suffix edge leaves node p, then α occurs only as a suffix of S. The same argument holds for the substrings of S^{-1}.

Perhaps, the most amazing feature of affix trees is that they can be built in linear time, and take linear space [8]. An analogous structure can be built for a set of k strings, following for example the construction method of [9] for suffix trees. Moreover, each node of the tree can be annotated with a k–bit string, where the i–th bit is set iff the corresponding substring (spelled by suffix edges) occurs in the i–th string of the set.

Now, let $\mathcal{S} = \{S_1, \ldots, S_k\}$ be a set of secondary structure strings over Σ_C, and $\mathcal{A}(\mathcal{S})$ the corresponding affix tree. In order to find the approximate occurrences of a stem–loop pattern $P = p_1 \ldots p_m$ in $\mathcal{S}$, having set τ_h, τ_i, and τ_b, we proceed by first matching in the tree the central uppercase character p_i on each suffix edge leaving the root, then we proceed with the other characters, matching characters $p_{i+1}...p_m$ on suffix edges, and characters $p_{i-1}...p_1$ on prefix edges. On each path, we keep count of the errors associated with the substring spelled by the path itself. Whenever one of the errors exceeds the corresponding threshold, we discard the path. Once all the characters of P have been matched, the surviving paths (if any) spell out the approximate occurrences of P in $\mathcal{S}$. In other words:

1. Match the central uppercase character u_i with the first character u_j on the labels of each of the suffix edges leaving the root. Save those edges such that $D(u_i, u_j) \leq \tau_h$.
2. (Forward match). Match the close bracket following u_i with the next character on each of the edges saved at the previous step. Keep the surviving edges (no mismatches allowed).
3. (Backward Match). On each of the surviving paths, match the open bracket preceding u_i in the pattern:
   a) Go to the node entered by the edge the path ended on. Let c_s be the number of characters of the edge label we skip.
   b) Traverse the prefix edge whose label starts with character ((if it exists). Move backward toward the root (on suffix edges) by c_s characters. Let (γ be the label of the prefix edge. We have two cases:
      i. $|\gamma| = 0$. The point reached corresponds to the endpoint of the path corresponding to an approximate occurrence of (u_i). The logical OR of the bitstrings of the nodes entered by the prefix edges tells us which strings pattern (u_i) appears in.
      ii. $|\gamma| > 0$. The endpoint we reached actually corresponds to substring $\gamma^{-1}(u_i)$. Since (u_i) and $\gamma^{-1}(u_i)$ occur exactly at the same positions in the strings, the bitstring resulting from the OR will be the same. And, also, all paths starting from the endpoints of (u_i) and $\gamma^{-1}(u_i)$ will have the same suffix edge labels. The only difference is that in the next backward matches on this path we will use the same prefix link with label (γ, without traversing it but only matching characters, keeping the endpoints reached at the previous forward match, until all the characters of γ have been used. Then, we proceed as usual. In general, we will follow these steps for any prefix edge with label $a\gamma$, with $a \in \Sigma_C$, and $|\gamma| > 0$.

    c) If at least q bits are set in the resulting bitstring, we match the next forward character from the endpoints we have reached, and so on.

We proceed in this way for all the characters of P, alternating a forward match with a backward match. Each time a path reaches a node, it forks on each of the edges of the same type (suffix or prefix, according to the type of match) leaving it. In case we meet a lowercase character in the pattern, we match the character, and immediately after it the bracket following (or preceding) it. Along the stem, we can encounter different errors:

1. We are matching an open (close) bracket in P, and we encounter a lowercase letter l_i. We have found an internal loop (bulge) not present in the pattern. We set $d_b = d_b + size(l_i)$. If $d_b \leq \tau_b$, and if at least one path continues with an open (close) bracket, we keep this path, considering the open (close) bracket of the pattern to be matched by the one immediately following the lowercase letter in the path. Otherwise, we discard the path.
2. We are matching a lowercase letter l_i, and we encounter an open (close) bracket. If $d_b = d_b + size(l_i) \leq \tau_b$ we will proceed with the next character of P exactly from the same point we are on now, setting $d_b = d_b + size(l_i)$.
3. We are matching a lowercase letter l_i, and we encounter a different lowercase letter l_j. We set $d_l = d_l + D(l_i, l_j)$. If $d_l \leq \tau_l$ we continue, otherwise we discard the path.
4. Whenever we try to match any other pair of characters, we discard the path.

    As we mentioned, each time new characters are been added to a pattern P, we can determine whether P appears in at least q input sequences by employing the bit strings annotating the nodes of the affix tree. We perform the logical OR of the bit strings associated with the nodes entered by the edges where each path ends. If at least q bits are set to one in the resulting bit string, then P appears in at least q input strings and we continue expanding it. Otherwise we can stop, since no other pattern having P as a core (and thus deriving from every possible expansion of P) will appear in q sequences.

    As already demonstrated with suffix trees in case of motifs in nucleotide and protein sequences [10,11,12], affix trees permit to implement *recursively* the exhaustive search for patterns. The pseudo–code of the algorithm is shown in Fig. 4. We start with a pattern made of a single uppercase character (the hairpin loop), and try to expand it in both directions, by adding pairs of brackets and lowercase letters. Whenever a pattern P does not occur in at least q sequences, we stop expanding it. Then, we continue with the pattern we had used as a starting point in the expansion that had generated P, expanding it in the remaining possible ways. The distance thresholds can be also defined dynamically, for example according to the number of base pairs in the pattern. In this way, we also sort of spread unmatched bulges and loops along the stem. As another side effect, we do not need to know the length of the motif we are looking for beforehand, nor the size of any of its components, but only that it has a stem–loop overall structure. The expansions will stop at the longest motif that satisfies the input constraints.

```
 1.  Procedure Expand(Paths P, char f, char b, pattern S)
 2.  P' = P;
 3.  If(f ≠ null)
 4.      P' = forwardmatch(P', f);
 5.  end If
 6.  If(b ≠ null) and (Checkbits(P' ≥ q))
 7.      P' = backwardmatch(P', b);
 8.  end If
 9.  If (Checkbits(P' ≥ q))
10.      S = bSf;
11.      report(S);
12.      If f ∉ Lowercasechars
13.        For all a' ∈ Lowercasechars
14.          Expand(P', a', null, S);
15.          If b ∉ Lowercasechars
16.            For all b' ∈ Lowercasechars
17.              Expand(P', a', b', S);
18.            end For
19.          end If
20.        end For
21.      end If
22.      If b ∉ Lowercasechars
23.        For all a' ∈ Lowercasechars
24.          Expand(P', null, a', S);
25.        end For
26.      end If
27.      Expand(P', ), (, S);
28.  end If
29.  return
```

Fig. 4. The pseudo–code of the recursive procedure expand. *Checkbits(P)* returns the number of bits set in the string resulting from the logical OR of the bitstrings associated with the paths in P. Forwardmatch(P, f) returns the pointers to the endpoints of valid paths obtained by matching f on the next symbol on suffix edges from the endpoints of paths P, associating with each one the corresponding distances. Backwardmatch does the same, using prefix edges as explained in Section 3. The algorithm starts by calling **Expand**(root, u_i, null, ϵ) for each uppercase character u_i.

3.2   Complexity

Given a set of k strings $\mathcal{S}$ of overall size N, building the annotated affix tree takes $O(kN)$ time and $O(|\Sigma_C|kN)$ space if the edges are implemented with a transition matrix, otherwise it takes $O(kN \log |\Sigma_C|)$ time and $O(kN)$ space if edges are implemented with adjacency lists. The complexity of the searching phase is, as in [10,11,12], $O(\mathcal{V}(m)kN)$, where m is the length of the longest pattern found, and $\mathcal{V}(m)$ is the maximum number of valid occurrences a pattern of length m can have in the strings. The latter term clearly depends on τ_b, τ_h, and τ_i. A rough upper bound is given by $\mathcal{V}(m) \leq 2 \cdot \tau_h \cdot (2 \cdot \tau_i)^i \cdot (m-1)^{\tau_b}$, where i is the number of lowercase characters in the pattern. Thus, the algorithm is linear in the input size, and exponential only in τ_b. The k factor derives from the time needed to perform the OR of the bit strings.

4   Introducing Sequence Similarity

So far, we have been looking only for structural similarity among various sequences. However, as we will see in the following sections, different RNA molecules of analogous function often present both structural and sequence similarity in the conserved parts. The latter is usually found in unpaired nucleotides (either in the hairpin loop or in internal loops), that are thus free to bind to other parts of the sequence, forming more complex tertiary structures, or to other molecules like proteins. Even if the basic idea of our method considers structures only, it can be integrated with sequence information in different ways. One possible approach could be to represent explicitly the unpaired nucleotides, instead of using dots or letters, and consider also matches and mismatches between nucleotides while adding unpaired elements to a structure. One drawback of this idea is that in this way the search would be significantly slower. But the main problem is that often it is far from easy to determine a priori what kind of sequence similarity one might expect in the conserved structures, and define, for example, how two hairpin loops can differ in their nucleotide composition to be considered similar.

Therefore, we introduced sequence similarity as an additional post–processing step, starting from the motifs reported by the algorithm according to the input constraints. The motivation is that if more than a single occurrence per sequence is reported for a motif, then just one of those could be its real occurrence, while the other are spurious and correspond to elements unrelated to it. But, we can also expect real functional motifs to share some sequence elements.

We introduced a greedy approach similar to the pattern–discovery method Consensus [13,14]. If all motif occurrences share exactly the same structure, we first align all the unpaired nucleotides (in internal or hairpin loops) of each instance in the first sequence, with those of each of the instances appearing in the second one. For each pair, we build a profile, as in [13,14], associating with it a score based on the information content. We keep the h highest scoring profiles. Then, we extend the profiles with the instances of the third sequence, trying every possible combination, and scoring each one. Once again, we keep only the h highest scoring profiles. We proceed in this way for each of the k sequences. At the end, we obtain h profiles, built taking one motif occurrence for each sequence, in which the most similar ones, considering sequence information only, have been grouped together. Finally, we report as real occurrences of a motif those that were used to build the highest scoring profile.

In case also the structural similarity required to the algorithm is not very stringent, that is, the motifs present structural differences in their occurrences, we can compare different candidate motif occurrences by aligning nucleotides and the structure associated with the motif occurrences simultaneously, employing a scoring function like the one used in FOLDALIGN. The difference is that, instead of aligning and comparing whole sequences, in this case the regions to compare, as well as the structure associated with them, have already been selected, with a few occurrences in each input sequence.

In this way, it is also possible to divide into separate groups occurrences that on a structural basis were assigned to the same motif. For example, suppose that a set of sequences contains two distinct motifs, corresponding to two different alignments, both having a four nucleotide loop, with different nucleotide composition, thus performing a different biological function. By considering only structural similarity, the algorithm will report their occurrences as belonging to the same motif. Instead, if sequence similarity in the loop can be used to discriminate one motif from the other, then the algorithm at the end of the post–processing step will be able to report two motifs instead of one, since two profiles will be built with two disjoint sets of occurrences. The same argument holds also for structural similarity. Moreover, in case more than a single motif is reported, the score of the alignment (of both structure and sequence) of its occurrences can be used as a measure of significance to rank the motifs output according to their conservation.

5   Dealing with Predictors

When the secondary structure of the input sequences is not available, the natural choice is to use a tool for the prediction of RNA secondary structure. Today's state of the art software for the prediction of RNA secondary structure, like mFold [15] and RNAfold [16], finds the structure that minimizes a free energy function based on the base pairings occurring in the structure as well as some additional parameters. The problem is that sometimes the structure of minimal energy does not correspond to the biologically correct one, that instead might be located on a local minimum of the free energy function. However, as an additional feature, the two methods just mentioned can output also a number of sub–optimal structures, whose associated energy is close to the minimal one. In particular, RNAsubopt, part of the Vienna RNA package based on the RNAfold algorithm, outputs *all* the structures within a given energy range ΔE from the optimal one [17]. The idea is that a motif could occur, if not in the structure of minimal energy, at least in some or most of the sub–optimal ones.

Thus, instead of working on just one structure per sequence, we input the A structures of minimal energy that have been predicted for each one. Then, we run the algorithm as usual, and we check whether patterns appear in at least one alternative structure of at least q sequences. Clearly, the more alternative structures contain the motif, the more reliable can be considered the presence of the motif in a structure (at least, according to predictors). The preliminary construction of the affix tree permits to handle efficiently the hundreds–fold increase in the number of input strings. However, we have to consider an additional problem. As the length of the sequence increases, many sub–optimal structures tend to be very similar to the optimal one, with only a few base pairings of difference. Thus, if we simply take the best A structures predicted, it might happen that most of them are just close relatives of the optimal one, differing just in a few base pairings. To overcome this problem, we chose to *cluster* the alternative structures in groups, according to their similarity. This can be done, for example, by using Ward's variance method [18], where at the beginning each structure is a cluster by itself, and two clusters are merged together to minimize the associated increase in variance. In this way, we can consider a much larger set of alternative structures, by clustering the redundant ones together. One representative per cluster (the one of lowest energy) is input to the algorithm.

5.1 Experimental Results

The Iron Responsive Element (IRE) is a particular hairpin structure located in the 5' untranslated region or in the 3' untranslated region of various messenger RNAs coding for proteins involved in cellular iron metabolism, like ferritin and transferrin. Two alternative IRE consensus structures have been found: Some IREs present an unpaired nucleotide (usually cytosine) on the stem, whereas in others the cytosine nucleotide and two additional bases seem to oppose one free 3' nucleotide [19]:

```
(((((.(((((.....)))))))))) 
NNNNNCNNNNNCAGWGHNNNNNNNNNN
(((...(((((......))))).)))
NNNNNCNNNNNCAGWGHNNNNNNNNNN
```

where W = A,U, and H = not G. The lower stem can be of variable length. As we can see, there is no conservation in the nucleotides forming the stem, but only in those forming the loop.

We took a sample of 20 5'UTR ferritin sequences (ranging in length from 100 to 700 nucleotides) known to contain a known functional IRE. We folded them with RNAsubopt with $\Delta E = 4.5$. Instead of just taking all the suboptimal structures, we formed 200 candidate clusters for each sequence. We also set $\tau_b = 1$, $\tau_h = 0$, and $\tau_i = 2$, in order to be able to detect the two alternative forms of the motif. Pattern $(((.(((((. . . . .))))))))$ (in this form, or with three unpaired nucleotides opposing a bulge) was found to appear in 19 sequences out of 20 and the motif with a single bulge in 16. In the missing sequence the motif simply had not been predicted in any alternative structure. The same test set (limited to the sequences shorter than 300 nucleotides) had been used in [7], where FOLDALIGN was reported to take 10–12 hours to detect the same motif. In our tests, our method (folding included) always took less than 15 minutes on a Pentium III-class PC with 128 Mb of RAM, with the actual pattern discovery stage taking a few seconds.

6 Avoiding Predictors

In the test on the Iron Responsive Element, we could not find the motif in a sequence simply because the predictor failed to generate a structure that contained it. But, experimental evidence proves the fact that the motif actually appears in the sequence. Thus, if we do not trust prediction algorithms, another idea could be to work directly on *unfolded* sequences, and try to predict stem–loop structures *on the fly* while searching for patterns by looking for complementary base pairs.

For the moment, suppose we consider only stem–loop elements without bulges or internal loops. We build the affix tree, but this time for the sequences. We find similar sequence patterns for candidate hairpin loops (or consider a valid hairpin candidate every substring of a given length that cannot contain base pairs) starting from the suffix edges leaving the root of the affix tree; then we proceed with forward/backward expansions as usual, but now, instead of brackets, we find in the affix tree letters representing nucleotides. Thus, now we have to check whether the forward–backward pairs of nucleotides we encounter on the paths at each expansion can form a valid base pair, for example a canonical base pair. If they do, we keep the path: otherwise, we discard the path. In practice this corresponds to looking for conserved motifs in the sequences, but instead of considering directly sequence similarity we take into account which patterns of complementary base pairs are conserved in the sequences, potentially corresponding to the same secondary structure element. And, to discover these patterns, we take advantage of the information contained in the affix tree, that allow us to expand a given pattern bidirectionally looking for complementary base pairings.

The first example we used for our tests is Histone 3'–UTR mRNA. Metazoan histone 3'–UTR mRNAs contain a highly conserved stem–loop structure with a six base stem and a four base loop. This stem–loop structure plays a different role in the nucleus and the cytoplasm [20], and is usually represented as:

$$((((((\ldots.))))))$$
GGYYYUHURHARRRCC

where H = not G, Y = C,T, and R = A,G. As in the previous example, we took a sample of 20 sequences of 3'–UTR mRNA experimentally known to contain this motif, and built the affix tree, this time for the unfolded sequences. Then we considered as valid hairpin loops all the substrings of length four appearing in the tree, and from the corresponding positions we proceeded with forward and backward expansions considering valid base pairs only Watson–Crick pairs (A–U and C–G). In this way, we found the motif in 19 sequences; by allowing also at most one wobble pair (G–U, normally found in RNA but less stable) we discovered the motif also in the missing sequence. Remarkably enough, without false positives in any sequence.

We can add unpaired nucleotides to the stem of the structure by modifying the *forwardmatch* and *backwardmatch* procedures shown in Fig. 4 accordingly. That is, to include a bulge or an internal loop on the left side of the structure, we just have to follow prefix edges for a number of characters matching the size of the corresponding elements, regardless of which nucleotides we find. Analogously, we follow suffix edges only for adding unpaired nucleotides on the right side of the stem. The two procedures can also be extended in order to consider differences in the size and position of internal loops and bulges as in the original algorithm.

However, when bulges and/or internal loops are included in the prediction, a given substring can fold in different ways, that is, different pairings can be associated with it. Thus, we have also to choose "on the fly" whether or not to accept a potential stem–loop prediction, in other words, if the base pairings

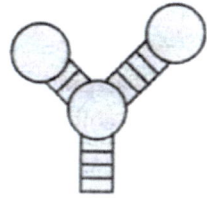

Fig. 5. A schematic view of the Y–shaped IRES secondary structure appearing in viral RNA and cellular mRNA [23].

found on the paths of the affix tree can be considered a reliable prediction for the secondary structure formed by the associated substring. The criterion we adopted to validate the predictions is very simple: if a given stem–loop structure is associated with a given substring, it also has to be the structure of *minimal* energy that the corresponding region can form, according to thermodynamical rules usually employed by secondary structure prediction algorithms [21,22]. Thus, before reporting a motif, we check whether its occurrences are of minimal energy: if the surviving ones satisfy this constraint for at least q sequences we report the motif itself. Notice that we perform this control only before reporting a motif, but anyway expand it regardless of the energy associated with it. In fact, the formation of a stem–loop structure as we build it goes through some intermediate states not corresponding to the optimal one. When we re–ran the IRE benchmark, with the same parameters of the previous test, but predicting structures on the fly, the algorithm reported a large set of potential motifs, with different hairpin loop sizes, some of which corresponded to overlapping regions in the sequences. But, when we added the energy constraint the *only* motif reported was the IRE, without false positives, and all its occurrences were reported, including the one that had been missed by the predictors. Moreover, the execution time, including the construction of the affix tree and the energy validation, was much faster, taking just a few seconds, on the same machine of the previous test.

7   Conclusions and Current Work

We have presented a novel pattern discovery algorithm, based on affix trees, that finds common secondary structure elements in a set of RNA sequences. The algorithm takes as input the secondary structure of the sequences, if the latter is known in advance, or a set of alternative predictions for each one. Experimental results have shown that the algorithm is able to detect known structural motifs with a significant time improvement over existing methods. Although dependent on the output of structure prediction algorithms, our method can anyway provide significant hints to researchers on which parts of a set of functionally related sequences are more likely to play an active role in their function, and which structure is associated with them.

Moreover, we have also shown how it is possible to make the algorithm work on the unfolded sequences only, by predicting structural elements while searching for sequence motifs sharing the same patterns of base pairs. Perhaps the more encouraging results came from this idea, that we are now validating with other real biological instances to ensure the reliability of the predictions and the criteria adopted. We are also extending both approaches to the detection of motifs more complex than single stem–loop elements, like Y–shaped elements, as shown in Fig. 5. The idea is to detect single hairpins first, then combine those motifs that appear to be adjacent in the sequences in more complex structures. We will present the results of these works in the near future.

References

1. Gesteland, R., Cech, T., Atkins, J., (eds.): The RNA World. Cold Spring Harbor Laboratory Press, New York (1999)
2. Simons, R., Grumberg-Magnago, M., (eds.): RNA Structure and Function. Cold Spring Harbor Laboratory Press, New York (1998)
3. Fox, G., Woese, C.: 5s rna secondary structure. Nature **256** (1975) 505–507
4. Westhof, E., Auffinger, E., Gaspin, C.: Dna and rna structure prediction. In: DNA – Protein Sequence Analysis, Oxford (1996) 255–278
5. Stephan, W., Parsch, J., Braverman, J.: Comparative sequence analysis and patterns of covariation in rna secondary structures. Genetics **154** (2000) 909–921
6. Gorodkin, J., Heyer, L., Stormo, G.: Finding common sequence and structure motifs in a set of rna sequences. Nucleic Acids Research **25** (1997) 3724–3732
7. Gorodkin, J., Stricklin, S., Stormo, G.: Discovering common stem–loop motifs in unaligned rna sequences. Nucleic Acids Research **29** (2001) 2135–2144
8. Maass, M.: Linear bidirectional on-line construction of affix trees. Proc. of CPM 2000, Lecture Notes in Computer Science **1848** (2000) 320–334
9. Gusfield, D.: Algorithms on Strings, Trees and Sequences: Computer Science and Computational Biology. Cambridge University Press, New York (1997)
10. Marsan, L., Sagot, M.: Algorithms for extracting structured motifs using a suffix tree with application to promoter and regulatory site consensus identification. Journal of Computational Biology **7** (2000) 345–360
11. Sagot, M.: Spelling approximate repeated or common motifs using a suffix tree. Lecture Notes in Computer Science **1380** (1998) 111–127
12. Pavesi, G., Mauri, G., Pesole, G.: An algorithm for finding signals of unknown length in dna sequences. Proc. of ISMB '01, Bioinformatics **17** (2001) S207–S214
13. Hertz, G., Hartzell, G., Stormo, G.: Identification of consensus patterns in unaligned dna sequences known to be functionally related. Comput.Appl.Biosci. **6** (1990) 81–92
14. Hertz, G., Stormo, G.: Identifying dna and protein patterns with statistically significant alignment of multiple sequences. Bioinformatics **15** (1999) 563–577
15. Zucker, M., Matthews, D.H., Turner, D.H.: Algorithms and thermodynamics for rna secondary structure prediction: a practical guide. In: RNA Biochemistry and Biotechnology, NATO ASI Series, Kluwer Academic Publishers (1999) 11–43
16. Hofacker, I., Fontana, W., Stadler, P., Bonhoeffer, S., Tacker, M., Schuster, P.: Fast folding and comparison of rna secondary structures. Monatshefte f Chemie **125** (1994) 167–188

17. Wuchty, S., Fontana, W., Schuster, P.: Complete suboptimal folding of rna and the stability of secondary structures. Biopolymers **49** (1999) 145–165
18. Ward, J.: Hierarchical grouping to optimize an objective function. Journal of the American Statistical Association **58** (1963) 236–244
19. Hentze, M., Kuhn, L.: Molecular control of vertebrate iron metabolism: mrna based regulatory circuits operated by iron, nitric oxide and oxidative stress. Proc. Natl. Acad. Sci. USA **93** (1996) 8175–8182
20. Williams, A., Marzluff, W.: The sequence of the stem and flanking sequences at the 3' end of histone mrna are critical determinants for the binding of the stem–loop binding protein. Nucleic Acids Research **23** (1996) 654–662
21. Walter, A., Turner, D., Kim, J., Lyttle, M., Muller, P., Mathews, D., Zuker, M.: Coaxial stacking of helices enhances binding of oligoribonucleotides. PNAS **91** (1994) 9218–9222
22. Mathews, D., Sabina, J., Zucker, M., Turner, D.: Expanded sequence dependence of thermodynamic parameters provides robust prediction of rna secondary structure. Journal of Molecular Biology **288** (1999) 911–940
23. Pain, V.: Initiation of protein synthesis in eukaryotic cells. Eur. J. Biochem. **236** (1996) 747–771

More Efficient Left-to-Right Pattern Matching in Non-sequential Equational Programs

Nadia Nedjah and Luiza de Macedo Mourelle

Department of Systems Engineering and Computation,
Faculty of Engineering, State University of Rio de Janeiro,
Rua São Francisco Xavier, 524, Maracanã,
Rio de Janeiro, RJ, Brazil
{nadia, ldmm}@eng.uerj.br
http://www.eng.uerj.br/ldmm/index.html

Abstract. Pattern matching is a fundamental feature in many applications such as functional programming, logic programming, theorem proving, term rewriting and rule-based expert systems. Usually, patterns are pre-processed into a deterministic finite automaton. Using such an automaton allows one to determine the matched pattern(s) by a single scan of the input term. The matching automaton is typically based on left-to-right traversal of patterns. In this paper, we propose a method to build such an automaton. Then, we propose an incremental method to build a deterministic concise automaton for non-necessarily sequential rewriting systems. With ambiguous patterns a subject term may be an instance of more than one pattern. To select the pattern to use, a priority rule is usually engaged. The pre-processing of the patterns adds new patterns, which are instances of the original ones. When the original patterns are ambiguous, some of the instances supplied may be irrelevant for the matching process. They may cause an unnecessary increase in the space requirements of the automaton and may also reduce the time efficiency of the matching process. Here, we devise a new pre-processing operation that recognises and avoids such irrelevant instances. Hence improves space and time requirements for the matching automaton.

1 Introduction

Pattern matching is a corner-stone operation in several applications such as functional, equational and logic programming [5], [16], theorem proving [4] and rule-based expert systems [3]. With ambiguous patterns, an input term may be an instance of more than one pattern. Usually, patterns are partially ordered using priorities. Notice that pattern matching techniques usually fall into two categories:

- Root matching techniques determine whether a given subject term is an instance of a pattern in a given set of patterns,
- Complete matching techniques determine whether the subject term contains a subterm (including the term itself) that is an instance of a pattern in the pattern set.

R. Baeza-Yates et al. (Eds.): CPM 2003, LNCS 2676, pp. 295–314, 2003.

Thus, complete matching subsumes root matching and root matching may be used to implement complete matching (by a recursive descent into the subject term). Throughout this paper, we only deal with root matching. Pattern matching automata have been studied for over a decade. It can be achieved as in lexical analysis by using a finite automaton [2], [7], [8], [11-15], [17]. Gräf [7] and Christian [2] construct deterministic matching automata for unambiguous patterns based on the left-to-right traversal order. In functional programming, Augustsson [1] and Wadler [19] describe matching techniques that are also based on left-to-right traversal of terms but allow prioritised overlapping patterns. Although these methods are economical in terms of space usage, they may re-examine symbols in the input term. In the worst case, they can degenerate to the naive method of checking the subject term against each pattern individually. In contrast, Christian's [2] and Gräf's [7] methods avoid symbol re-examination at the cost of increased space requirements. In order to avoid backtracking over symbols already examined, like Gräf's our method introduces new patterns. These correspond to overlaps in the scanned prefixes of original patterns. When patterns overlap, some of the added patterns may be irrelevant to the matching process. The method proposed here improves Gräf's in the sense that it introduces only a subset of the patterns that his method adds. This improves both space and time requirements as we will show later. Sekar [17] uses the notion of irrelevant patterns to compute traversal orders of pattern matching. His algorithm eliminates a pattern π whenever a match for π implies a match for a pattern of higher priority than π. In contrast with Sekar's method, we do not introduce irrelevant patterns at once.

In this paper, we focus on avoiding the introduction of irrelevant patterns while constructing matching automata. This results in a more efficient pattern-matcher. First, we introduce a method for generating a deterministic tree matching automaton for a given pattern set. Then, we show how these automata can be constructed incrementally, which is suitable for some application such as Knuth-and-Bendix problem. Subsequently, We prove that the incremental construction preserves the determinism of the constructed automata. Although the generated automaton is efficient since it avoids symbol re-examination, it may contain unnecessary branches. As we shall see, the main reason for this is the presence of ambiguous patterns with more general patterns having higher priority. Here, we modify that method so that only relevant patterns are added. A smaller and more efficient automaton is thereby obtained.

2 Preliminaries

In this section, we recall the notation and concepts that will be used in the rest of the paper. Symbols in a term are either function or variable symbols. The non-empty set of function symbols F = a, b, f, g, ... is ranked i.e., every function symbol f in F has an arity which is the number of its arguments and is denoted #f. A term is either a constant, a variable or has the form $f t_1 t_2 \ldots t_{\#f}$ where each t_i, $1 \leq i \leq \#f$, is itself a term. We represent terms using their corresponding abstract tree. We abbreviate terms by removing the usual parentheses and commas. This is unambiguous in our examples since the function arities will be kept unchanged throughout, namely $\#f = 3$, $\#g = 1$, $\#a = \#b = 0$. Variable occurrences are replaced by ω, a meta-symbol which is used since the actual symbols are irrelevant here. A term containing no variables is said to be a *ground term*. We generally assume that patterns are *linear* terms, i.e. each variable symbol can occur at most once in them. Pattern sets will be denoted by L and patterns by π_1, π_2, ..., or simply by π. A term t is said to be an *instance* of a (linear) pattern π if t can be obtained from π by replacing the variables of π by corresponding sub-terms of t. If term t is an instance of pattern π then we denote this by $t \triangleright \pi$.

Definition 1. A matching item is a triple $r{:}\alpha\bullet\beta$ where $\alpha\beta$ is a term and r is a rule label. The label identifies the origin of the term $\alpha\beta$ and hence, in a term rewriting system, the rewrite rule which has to be applied when $\alpha\beta$ is matched. The label is not written explicitly below except where necessary. The meta-symbol $\bullet$ is called the *matching dot*, α and β are called the *prefix* and *suffix* respectively. A final matching item is one of the form $\alpha\bullet$.

Throughout this paper left-to-right traversal order is used. So the matching item $\bullet\beta$ represents the initial state prior to matching the pattern β. In general, the matching item $\alpha\bullet\beta$ denotes that the symbols in α have been matched and those in β have not yet been recognised. Finally, the matching item $\alpha\bullet$ is reached on successfully matching the whole pattern α.

Definition 2. A set of matching items in which all the items have the same prefix is called a *matching set*. A matching set in which all the items have an empty prefix is called an *initial* matching set whereas a matching set in which all the items have an empty suffix is called a *final* matching set.

Definition 3. For a set L of pattern suffixes and any symbol s, let L/s denote the set of pattern suffixes obtained by removing the initial symbol s from those members of L which commence with s and excluding the other members of L. Then, for $f \in F$ define L_ω and L_f by:

$$L_\omega = L\backslash\omega$$

$$L_f = \begin{cases} L\backslash f \cup \omega^{\#f} L\backslash\omega & if\, L = \emptyset \\ \\ L = \emptyset & otherwise \end{cases} \tag{1}$$

where $\omega \# f$ denotes a string of $\#f$ symbols ω and $\emptyset$ is the empty set. The closure $\overline{L}$ of a pattern set L is then defined recursively by Gräf [7] as follows:

$$\overline{L} = \begin{cases} L & if\, L = \{\epsilon\}\, or\, L = \emptyset \\ L = \bigcup_{s \in F \cup \{\omega\}} s\overline{L_s} & otherwise \end{cases} \qquad (2)$$

Roughly speaking, with two item suffixes of the form $f\alpha$ and $\omega\beta$ we always add the suffix $f\omega^{\#f}\beta$ in order to postpone by one more symbol the decision between these two patterns. Otherwise backtracking might be required to match $\omega\beta$ if input f leads to failure to match $f\alpha$.

3   Deterministic Matching Automata

In this section, we briefly describe a practical method to construct a deterministic tree matching automaton for a prioritised ambiguous pattern set [11], [12]. The pattern set L is extended to its closure $\overline{L}$ while generating the matching automaton.

The automaton is represented by the 4-tuple $\langle S_0, S, Q, \delta \rangle$ where S is the state set, $S_0 \in S$ is the initial state, $Q \subseteq S$ is the final state set and δ is the state transition function. The states are labelled by matching sets, which consist of original patterns whose prefixes match the current input prefix, together with extra instances of the patterns, which are added to avoid backtracking in reading the input. In particular, the matching set for S_0 contains the initial matching items formed from the original patterns and labelled by the rules associated with them. Transitions are considered according to the symbol at the matching position, i.e. that immediately after the matching dot. For each symbol $s \in F \cup \{\omega\}$ and state with matching set M, a new state with matching set $\delta(M, s)$ is derived using the composition of the functions *accept* and *close* defined in the equations below:

$$accept(M, s) = \{r : \alpha \bullet \beta \mid r : \alpha \bullet \beta \in M\} \qquad (3)$$

$$close(M) = M \cup \{r : \alpha \bullet f\omega^{\#f}\mu \mid r : \alpha \bullet \omega\mu \in M\ \&\ \exists q : \alpha \bullet f\lambda \in M,\ f \in F\} \qquad (4)$$

$$\delta(M, s) = close\,(accept(M, s)) \qquad (5)$$

The items obtained by recognising the symbols in those patterns of M where s is the next symbol form the set $accept(M, s)$, which is called the *kernel* of $\delta(M, s)$. However, the set $\delta(M, s)$ may contain more items. The presence of two items $\alpha \bullet \omega\mu$ and $\alpha \bullet f\lambda$ in M creates a non-deterministic situation since the variable ω could be matched by a term having f as head symbol. The item $\alpha \bullet f\omega^{\#f}\mu$ is added to remove this non-determinism and avoid backtracking. The transition function thus implements simply the main step in the closure operation described by Gräf [6] and set out in the previous section. Hence the pattern set resulting from the automaton construction using the transition function of equation (5) coincides with the closure operation of Definition 3. The item labels simply keep account of the originating pattern for when a successful match is achieved. As we deal here with root matching, every failure transition ends up in a single global failure state. Non-determinism is worst where the input can end up matching the whole of two different patterns. Then we need a priority rule to determine which pattern to select.

Definition 4. A pattern set L is said *ambiguous* if there is a ground term that is an instance of at least two distinct patterns in L. Otherwise, L is *non-ambiguous*.

Definition 5. A *priority rule* is a partial ordering on patterns such that if π_1 and π_2 are distinct ambiguous patterns then either π_1 has higher priority than π_2 or π_2 has higher priority than π_1. In the latter case, we write $\pi_1 \prec \pi_2$.

When a final state is reached, if several patterns have been successfully matched, then the priority rule is engaged to select the one of highest priority. An example is the textual priority rule which is used in the majority of functional languages [1], [9], [10], [18]. Among the matched patterns, the rule chooses the pattern that appears first in the text. Whatever rule is used, we will apply the word *match* only to the pattern of highest priority, which is matched.

Definition 6. For a prioritised pattern set L and pattern $\pi \in L$, the term t is said to *match* $\pi \in L$ if, and only if, t is an *instance* of π but not instance of any other pattern in L of higher priority than π.

4   Building Deterministic Matching Automata

A pattern set is compiled into a deterministic finite matching automaton using the algorithm $BuildAutomaton(S, S_0, Q, \Delta)$ below. This automaton has S as the set of states that includes the initial state S_0. The set of final states is represented by Q while Δ represents the set of valid state transitions of the automaton.

Procedure BuildAutomaton(S, S$_0$, Q, Δ);
Begin
    S := Δ := Q := $\emptyset$; S$_0$:= CloseSet($\bullet\pi \, |\pi \in \Pi$);
    For each state s $\in$ S Do
        For each symbol $\sigma \in$ F Do
            If AcceptSymbol(s, σ) $\neq \emptyset$ Then
                s' := CloseSet(AcceptSymbol(s, σ));
                S := S $\cup$ { s'};
                Δ := $\Delta\cup$ (s, σ) $\rightarrow$ s';
        For each state s $\in$ S Do
            If $\pi\bullet \in$ s Then Q := Q $\cup$ s;
End.

The state transitions of the matching automaton are obtained by composing the two functions $AcceptSymbol(s, \sigma)$ of Equation (3) and $CloseSet(s)$ of Equation (4). The former function simply accepts the symbol σ that comes immediately after the matching dot in the matching items of the state s while the latter yields the necessary matching items to guarantee that the matching process deterministic.

Procedure AcceptSymbol(s, σ);
Begin
    $s' := \emptyset$;
    For each item $\alpha \bullet \beta \in s$ Do
        $s' := s' \cup \{\alpha\sigma \bullet \beta\}$;
    Return s';
End.

Notice that the presence of the items $\alpha \bullet \omega\beta$ together with the items $\alpha \bullet f\mu$ in the same matching state creates a non-deterministic situation for a pattern-matcher since ω can be substituted with a term having f as head symbol. The items $\alpha \bullet f\omega^{\#f}\beta$ are added to remove such non-determinism and avoid back-tracking. For instance, let $\Pi = \{f\omega wa, fcwc\}$ and let s be the matching state obtained after accepting the root symbol f so $s = \{f \bullet \omega wa, f \bullet cwc\} \cup \{f \bullet cwa\}$. The item $f \bullet cwa$ is added because a target term with the prefix fc could match the pattern $f\omega wa$ too if the last argument of f were a rather than c. So supplying the instance $fcwa$ would allow the pattern-matcher to decide deterministically which option to take. Without this new item, the pattern-matcher would need to backtrack to the first argument of f if the option offered by $f\omega wa$ were taken and a symbol c encountered as the last argument of f in the target term. (For more details and formal proofs see [9].) In the following, we will call $Kernel$ of a state s the set of matching items q such that $s = CloseSet(q)$.

Procedure CloseSet(s);
Begin
    For each pair of items $\alpha \bullet \omega\beta \in s$, $\alpha\bullet f\mu \in s \mid f \in F$ Do
        $s := s \cup\{\alpha\bullet f\omega^{\#f}\beta\}$;
    Return s;
End.

Example 1. Let $L = \{1 : faww, 2 : fwaa, 3 : fwba, 4 : fg\omega g\omega b\}$ be the pattern set where $\#f - 3$, $\#g = 1$ and $\#a = \#b = 0$, as throughout this paper. Assuming a textual priority rule, the matching automaton for L is given in Fig. 1. Transitions corresponding to failures are omitted. Each state is labelled with its matching set. In the construction process, each new item is associated with the rule from which it is directly derived and whose pattern it is known to match. So, an added item $a \bullet f\omega^{\#f}b$ is associated with the same rule as is its parent $a\bullet\omega b$. At the final nodes, whatever item is matched, the matching rule of highest priority is chosen. This rule may be different from the one inherited by the item at that node. When this happens, it indicates what we call irrelevancy in the next section. During pattern matching, a ω-transition is only taken when there is no other available transition, which accepts the current symbol. The automaton can be used to drive pattern matching with any chosen term rewriting strategy.

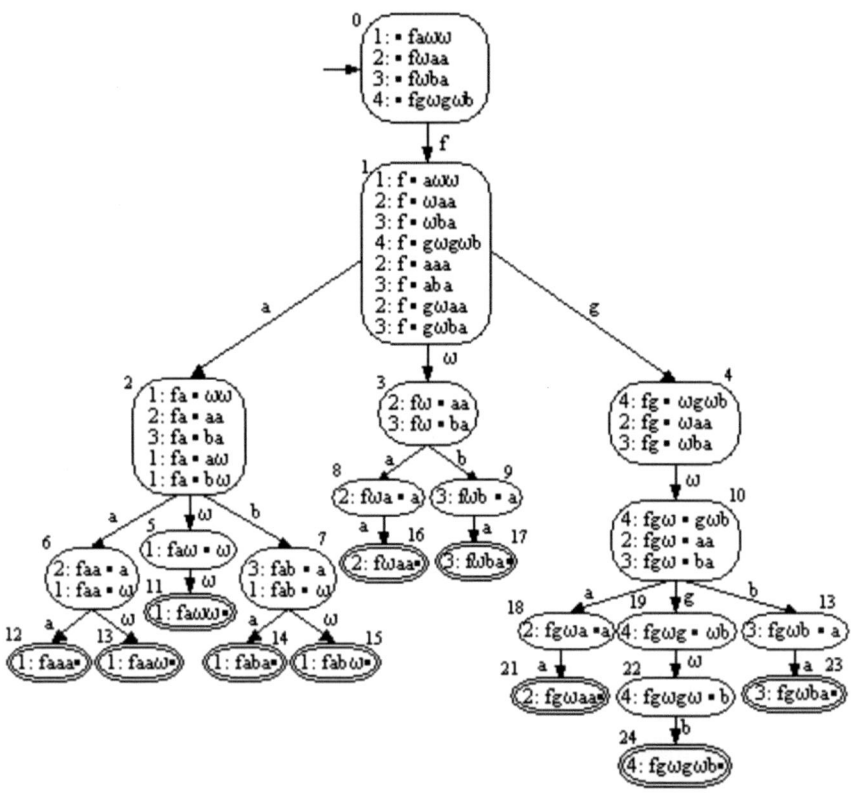

Fig. 1. Matching automaton for $\{1 : f a \omega \omega, \ 2 : f \omega a a, \ 3 : f \omega b a, \ 4 : f g \omega g \omega b\}$

5   Incremental Construction

It is possible to construct the matching automaton returned by the procedure *BuildAutomaton* of Section 4 incrementally, i.e. by progressively adding and/or deleting patterns. Incremental construction of term matching automata is important in practical applications in which pattern sets change dynamically. Examples of such applications include term rewrite systems interpreters and the

Knuth-Bendix completion algorithm. In the latter application rewrite rules are added and removed at each step of the completion process which attempts to normalise overlaps between rules, i.e. critical pairs. Incremental techniques have only been considered in [16] for a very special case of patterns.

The general idea for the incremental construction here is to update the matching item sets with each change in the pattern set accordingly. When items are added to or removed from the kernel of a state, starting from the initial state of the automaton, the set of matching items of the state is recomputed and the successor states are updated recursively. Notice that in the process, parts of the matching automaton that are not concerned with the change are kept unchanged. Also, notice that it is possible that new states are added when a new pattern is added or old states can be removed when removing an existing pattern.

Let $A = (S, s_0, Q, \Delta)$ be a deterministic matching automaton for a pattern set Π and an alphabet F. Observe that the pattern set can be an empty set which will yield the simplest matching automaton $A_0 = (\{s_0\}, s_0, \{s_0\}, \emptyset)$. This particular matching automaton accepts the language $L(A_0) = \emptyset$. So, when adding a pattern π starting off from the initial state we must add the item $\bullet\pi$ to the set $Kernel(s_0)$ and then re-compute the closure of the set $Kernel(s_0) \cup \{\bullet\pi\}$. This may yield new kernel items in the successor states of s_0 and so, we must recursively introduce new items in the successors states. It is possible that the new items induce new transitions and corresponding new states. This recursive process can be implemented using three new procedures:

- $AddPattern(\pi)$ which simply instigates the addition of the item $\bullet\pi$ to the initial state of the automaton.
- $AddItems(s, I)$ which adds a set of items to the kernel of the state s and updates the successor states of s recursively.
- $NewState(I)$ which constructs a new state with the set I as its kernel and recursively all possible new successor states, employing the method used earlier in Section 3 to construct deterministic matching automata.

Procedure AddPattern(π);
Begin
    s_0 := AddItems(s_0, $\bullet\pi$);
End.

The functions are designed such that they return the modified or newly constructed states. Note that the procedures are applied in the environment of the given matching automaton A.

When the item set provided to $AddItems$ is empty, the state s is kept unchanged. Otherwise, the kernel of s together with the items in I are closed and the items of s replaced by those freshly obtained from the closure operation. For each of the matching symbols (i.e., symbols appearing immediately after the matching dot) of the matching item set obtained, we deduce the set I' of items that should be added to each of the existing successors of state s. If the transition for the matching symbol considered exists already in the automaton, I' is simply added (using the procedure *AddItems*) to the state successor. Otherwise, a new state is generated, using the procedure *NewState*, and the new state transition is added to the set of transitions δ.

Procedure AddItems(s, I);
Begin
    If $I = \emptyset$ Then Return s;
    $s_1 :=$ CloseSet(Kernel(s) $\cup$ I); S $:=$ S $\setminus$ {s} $\cup$ {s_1};
    If s $\in$ Q Then Q $:=$ Q $\setminus$ {s} $\cup$ {s_1};
        For each symbol $\sigma \in$ F such that AcceptSymbol(s_1, σ) $\neq \emptyset$ Do
            $I' :=$ AcceptSymbol(s_1, σ)$\setminus$ AcceptSymbol(s, σ);
            If (s, σ) $\rightarrow$ s' $\in \Delta$ Then
                $s'_1 :=$ AddItems(s', I'); $\Delta := \Delta\setminus\{(s,\sigma) \rightarrow s'\}\cup \{(s_1,\sigma) \rightarrow s'_1\}$;
            Else $s'_1 :=$ NewState(I');
                $\Delta := \Delta\cup \{(s_1,\sigma) \rightarrow s'_1\}$;
    Return s_1;
End.

Procedure NewState(I);
Begin
    s $:=$ CloseSet(I);
    S $:=$ S $\cup$ {s};
    If s $\in$ Q Then Q $:=$ Q $\cup$ {s_1};
    For each symbol $\sigma \in$ F such that AcceptSymbol(s, σ) $\neq \emptyset$ Do
        $s'_1 :=$ NewState(I');
        $\Delta := \Delta\cup \{(s, \sigma) \rightarrow s'\}$;
    Return s;
End.

Deletion of patterns is similar to their inclusion in a matching automaton. As before, let $A = (S, s_0, Q, \Delta)$ be a deterministic matching automaton for a pattern set Π and an alphabet F. Updating the automaton A to accept the language $L(\Pi)\setminus\{\pi\}$ only with $\pi \in \Pi$ starts off by removing the item $\bullet\pi$ from the kernel of s_0 and carries on updating all the successor states accordingly. As for the addition of patterns, this recursive process can be implemented using three new procedures:

- *DeletePattern*(π) which simply instigates the deletion of the item $\bullet\pi$ from the initial state of the automaton.
- *DeleteItems*(s, I) which removes a set of items from the kernel of the state s and updates the successor states of s recursively.
- *DeleteState*(I) which discards an existing state s and all its possible successor states recursively.

As before, the functions are designed so that they return the modified states. Note that the procedures are applied to a given matching automaton A.

Procedure DeletePattern(p);
Begin
    s_0 := DeleteItems(s_0, $\{\bullet\pi\}$);
End.

Procedure DeleteItems(s, I);
Begin
    If I = $\emptyset$ Then Return s;
    s_1 := CloseSet(Kernel(s) $\cup$ I);
    S := S $\setminus$ $\{s\}$ $\cup$ $\{s_1\}$;
    If $s_1 \in$ Q Then Q := Q $\setminus$ $\{s\}$ $\cup$ $\{s_1\}$;
        For each transition (s, σ) $\rightarrow$ s' $\in \delta$ Do
            I' := AcceptSymbol(s, σ)$\setminus$ AcceptSymbol(s_1, σ);
            If AcceptSymbol(s_1, σ) $\neq \emptyset$ Then
                s_1' := DeleteItems(s', I');
                Δ := $\Delta\setminus\{\Delta(s, \sigma) \rightarrow s'\}\cup\{(s_1, \sigma) \rightarrow s_1'\}$;
            Else
                DeleteState(s');
                Δ := $\Delta\setminus$ $\{(s, \sigma) \rightarrow s'\}$;
        Return s_1;
End.

When the item set provided to *DeleteItems* is empty, the state s is kept unchanged. Otherwise, the kernel of s together with the items of I are closed and the items of s replaced by those freshly obtained from the closure operation. For each state transition of the automaton, we deduce the set I' of items that should be discarded from each of the existing successors of state s. If the transition for the matching symbol considered is still needed in the automaton, I' is simply removed (using the procedure *DeleteItems*) to the state successor. Otherwise, the state successor is deleted, using the procedure *DeleteState*, and the state transition is eliminated from the set of transitions Δ. Note that if the original pattern set Π is the singleton $\{\pi\}$ (i.e., $\Pi\setminus\{\pi\} = \emptyset$), then *DeletePattern*($\pi$) applied to the automaton A resumes into the automaton $A_0 = (\{s_0\}, s_0, \{s_0\}, \emptyset)$.

Procedure DeleteState(s);
Begin
    S := S $\setminus$ $\{s\}$; Q := Q $\setminus$ $\{s\}$ $\cup$ $\{s_1\}$;
        For each transition (s, σ) $\rightarrow$ s' $\in \Delta$ Do
            Δ := $\Delta\setminus$ $\{(s, \sigma) \rightarrow s'\}$;
            DeleteState(s');
End.

Some further optimisations are possible. In the functions *AddItems* respectively *DeleteItems*, we do not have to recompute the changed states $s_1 = CloseSet(Kernel(s) \cup I)$ respectively $s_1 = CloseSet(Kernel(s) \backslash I)$ each time from nothing. Since s is already closed, to determine $CloseSet(Kernel(s) \cup I)$ it suffices to add to s the items of the set $CloseSet(I)$ and those items $\alpha \bullet f\omega^{\#f}\mu$ induced by pairs $(\alpha \bullet \omega\mu, \alpha \bullet f\mu') \in (Kernel(s) \cup I) \times (Kernel(s) \cup I)$. Likewise, $CloseSet(Kernel(s) \backslash I)$ can be determined by removing from s all items of I and those closure items $\alpha \bullet f\omega^{\#f}\mu$ which are not induced by a pair $(\alpha \bullet \omega\mu, \alpha \bullet f\mu') \in (Kernel(s) \backslash I) \times (Kernel(s) \backslash I)$.

For simplicity, we have considered adding and removing single patterns. However, the extension to add and remove a set of patterns is straightforward. Instead of adding respectively deleting the singleton $\{\bullet\pi\}$ to respectively from the automaton A, we can also use arbitrary sets $\{\bullet\pi | \pi \in Q\}$. So, the procedure that adds and that which removes a set of patterns P from the automaton A are given below.

Procedure AddPatterns(P);
Begin
    For each pattern $\pi \in$ P Do
        $s_0 :=$ AddItems($s_0, \{\bullet\pi\}$);
End.

Procedure DeletePatterns(P);
Begin
    For each pattern $\pi \in$ P Do
        $s_0 :=$ DeleteItems($s_0, \{\bullet\pi\}$);
End.

Lemma 1. Let $A = (S, s_0, Q, \Delta)$ be the deterministic matching automaton for a pattern set Π with the alphabet F and let P be another set of patterns. $AddPatterns(P)$ applied to A yields the deterministic matching automaton $A' = (S', s_0', Q', \Delta')$ for $\Pi \cup P$.

Proof. Let $A_1 = (S_1, s_{01}, Q_1, \Delta_1)$ be the automaton obtained by applying the procedure $AddPatterns(P)$ to A. Then we want to show that $A_1 = A'$. It is easy to verify that $AddPatterns$ produces neither garbage states (i.e., unreachable states from s_{01}) nor garbage transitions (i.e., transitions $(s, \sigma) \rightarrow s' \in \Delta$ such that either $s \ni Sors' \ni S$). Furthermore, we can easily check that Q_1 consists exactly of those states of S_1 that contains at least one final item. So, we need to show that: *(i)* s_{01} and s_0' coincide; *(ii)* each successor state s_1' of any matching state $s_1 \in S_1$ coincides with $closeSet(AcceptSymbol(s_1, \sigma))$ for some $\sigma \in F$; *(iii)* for each state $s_1 \in S_1$ and symbol $\sigma \in F$ such that $AcceptSymbol(s_1, \sigma) \neq \emptyset, s_1' = closeSet(AcceptSymbol(s_1, \sigma)) \in S_1$ and $(s_1, \sigma) \rightarrow s_1' \in \Delta_1$. If A_1 meets these three conditions, then clearly A_1 must be the deterministic matching automaton A' for the pattern set $\Pi \cup P$.

– If $P = \emptyset$ then $s_{01} = AddItems(s_0, \emptyset) = s_0 = s_0'$. Otherwise, $s_{01} = AddItems($ $s_0, \{\bullet\pi | \pi \in P\})$ which is nothing but $CloseSet(Kernel(s_0) \cup \{\bullet\pi | \pi \in P\})$. So, $s_{01} = CloseSet(\{\bullet\pi | \pi \in \Pi \cup P\})$ which is s_0'. Before tackling the other two conditions, let us make the following useful observation: for each state $s_1 \in S_1, s_1$ is obtained by either $AddItems(s, I)$ or $NewState(I)$ for some state $s \in S$ and some non-empty set of items I or s_1 is already in S and therefore all successors of s_1 in A are also in A_1 together with the corresponding transitions. (This is the case in which $s_1 = s \in S$ is left unchanged by the $AddItems$ procedure which needs no further proving.)

– Let us first suppose that *(a)* $s_1 = AddItems(s, I)$. Let s_1' be a successor state of s_1 with $(s_1, s) \rightarrow s_1' \in \Delta_1$, and let $I' = AcceptSymbol(s_1, s) \backslash$ $AcceptSymbol(s, \sigma)$. Then depending on whether the transition for the symbol σ already exists in the original automaton A, we have the following couple of sub-cases:

- First, suppose that $(s, \sigma) \rightarrow s' \in \Delta$ $(\Rightarrow s' = CloseSet(AcceptSymbol(s,$ $\sigma)))$ and $s_1' = AddItems(s', I')$. If $I' = \emptyset$, then $AcceptSymbol(s_1, \sigma)$ $= Accept\text{-}Symbol(s, \sigma)$ and $s_1' = AddItems(s', \emptyset) = s'$. As s' equals $CloseSet(AcceptSymbol(s, \sigma))$ and $AcceptSymbol(s_1, \sigma) = AcceptSymbol$ (s, σ), then s_1' is nothing but $CloseSet(AcceptSymbol(s_1, \sigma))$. Otherwise, $s_1' = AddItems(s', I') = CloseSet(Kernel(s') \cup I')$. So, substituting $Kernel(s')$ and I' by $CloseSet(AcceptSymbol(s, \sigma)$ and $AcceptSymbol$ $(s_1, \sigma) \backslash AcceptSymbol(s, \sigma))$ respectively yields $s_1' = CloseSet(Accept\text{-}Symbol(s_1, \sigma))$.

- Now, suppose that $\delta(s, \sigma) \rightarrow s' \in \Delta$ i.e., $AcceptSymbol(s_1, \sigma) = \emptyset$, then $I' = AcceptSymbol(s_1, \sigma)$ and $s_1' = NewState(I') = NewState$ $(AcceptSymbol(s_1, \sigma)) = CloseSet(AcceptSymbol(s_1, \sigma))$.

Now, let us consider the case *(b)* in which $s_1 = NewState(I)$. Then each successor state s_1' of s has the form $s_1' = NewState(AcceptSymbol(s_1, \sigma))$ which is $CloseSet(AcceptSymbol(s_1, \sigma))$ for some $\sigma \in F$ with $AcceptSymbol(s, \sigma)$ $\neq \emptyset$.

– Let $s_1 \in S_1$ and symbol $\sigma \in F$ such that $AcceptSymbol(s_1, \sigma) \neq \emptyset$. Then depending on whether $s_1 = AddItems(s, I)$ or $s_1 = NewState(I)$, we have the following couple of sub-cases:

- Suppose that $s_1 = AddItems(s, I)$ with $I \neq \emptyset$. Since $AcceptSymbol(s_1, \sigma) \neq \emptyset$, then either $s_1' = CloseSet(AcceptSymbol(s_1, \sigma)) = AddItems(s', I') \in S_1$ or $s_1' = CloseSet(AcceptSymbol(s, \sigma)) = NewState(I') \in S_1$ where $I' = AcceptSymbol(s_1, \sigma) \backslash AcceptSymbol(s, \sigma)$. In either cases we have that $(s_1, \sigma) \rightarrow s_1' \in \Delta_1$.

- Now, suppose that $s_1 = NewState(I)$ with $I \neq \emptyset$. Since $AcceptSymbol$ $(s_1, \sigma) \neq \emptyset$, then $s'_1 = CloseSet(AcceptSymbol(s, \sigma)) = NewState(I') \in S_1$ where $I' = AcceptSymbol(s_1, \sigma)$ and so we have that $(s_1, \sigma) \to s'_1 \in \Delta_1$. $\diamond$

Lemma 2. Let $A = (S, s_0, Q, \Delta)$ be the deterministic matching automaton for a pattern set Π with the alphabet F and let P be another set of patterns. $DeletePatterns(Q)$ applied to A yields the deterministic matching automaton $A' = (S', s'_0, Q', \Delta')$ for $\Pi \backslash Q$.

Proof. Let $A_1 = (S_1, s_{01}, Q_1, \Delta_1)$ be the automaton obtained by applying the procedure $DeletePatterns(P)$ to A. As in the proof of Lemma 1, it is easy to check that A_1 contains no garbage states and transitions and that Q_1 is the proper set of final states for A_1. As in the previous proof we verify the two conditions *(i)* and *(ii)* knowing that here for each state $s_1 \in S_1$, either $s_1 = DeleteItems(s, I)$ for some $s \in S$ and a nonempty item set I or $s_1 \in S$ and then all successors of s_1 in A are in A_1 together with all the corresponding transitions. Clearly, the latter case does not need any further proving.

- If $P = \emptyset$ then $s_{01} = DeleteItems(s_0, \emptyset) = s_0 = s'_0$. Otherwise, $s_{01} = DeleteItems(s_0, \{\bullet\pi | \pi \in P\})$ which is nothing but $CloseSet(Kernel(s_0) \backslash \{\bullet\pi | \pi \in P\})$. So, $s_{01} = CloseSet(\{\bullet\pi | \pi \in \Pi \backslash P\})$ which is s'_0.

- Let $s_1 = DeleteItems(s, I)$ with $I \neq \emptyset$. Let s'_1 be a successor state of s_1 with $(s_1, \sigma) \to s'_1 \in \Delta_1$, and let $I' = AcceptSymbol(s, \sigma) \backslash AcceptSymbol(s_1, \sigma)$. If $I' = \emptyset$, then $AcceptSymbol(s, \sigma) = AcceptSymbol(s_1, \sigma)$ and $s'_1 = Delete- Items(s', \emptyset) = s' = CloseSet(AcceptSymbol(s, \sigma)) = CloseSet(AcceptSym- bol(s_1, \sigma))$. Otherwise, $s'_1 = DeleteItems(s', I') = CloseSet(Kernel(s') \backslash I')$. By substituting I' by its value we have $s'_1 = CloseSet(AcceptSymbol(s, \sigma) \backslash (CloseSet(AcceptSymbol(s_1, \sigma)))) = CloseSet(AcceptSymbol(s_1, \sigma)))$.

- Let $s_1 = DeleteItems(s, I)$ with $I \neq \emptyset$ and $AcceptSymbol(s_1, \sigma) \neq \emptyset$, then $(s, \sigma) \to s' \in \Delta_1$ for some $s' \in S$ and $s'_1 = CloseSet(AcceptSymbol(s_1, \sigma)) = DeleteItems(s', I') \in S_1$ where $I' = AcceptSymbol(s, \sigma) \backslash AcceptSymbol(s_1, \sigma)$. $\diamond$

6 Reduced Matching Automata for Overlapping Patterns

Now, we observe that the close function of Equation (4) may add more items than it needs when the patterns are ambiguous. For example, in Fig. 1 consider the items $2 : f \bullet aaa$ and $3 : f \bullet aba$ which function *close* adds to state 1. Every term matching these items will instead be eventually associated with derivatives of the higher priority item $1 : f \bullet a\omega\omega$ also of state 1 and so the two items could have safely been omitted. We return to this example again after introducing some necessary definitions.

A *position* in a term is a path specification, which identifies a node in the abstract tree of the term. Positions are specified here using a list of positive integers. The empty list Λ denotes the position of the root of the abstract tree and the position $p.k$ $(k \leq 1)$ denotes the root of the kth argument of the function symbol at position p. For a term t and a position p in this term, we denote by $t[p]$ the symbol at position p in t.

Definition 7. A term t is said to be *more general* than term t' at position p if, and only if, the symbol $t[p]$ is ω, $t'[p]$ is a function symbol and the prefixes of t and t' ending immediately before p are the same. Without too much confusion, we hope, we will also say t is initially more general than t', if t is more general than t' at the first position for which the symbols of t and t' differ.

The function *close* adds new patterns to which the priority rule of the original rule set must be extended. The following definition enables us to associate a unique item amongst all those which pattern-match a given term: if there is a unique pattern-matched item whose rule has highest priority then that is chosen; otherwise, when there are several pattern-matched items associated with rules of maximal priority, those items must all derive from the same original pattern and we can select the initially most general. (Any other uniquely defined choice would also be acceptable, but this is the most convenient in what follows.)

Definition 8. Item $r : \alpha \bullet \beta$ has higher priority than item $r' : \alpha' \bullet \beta'$ if the original pattern of r has higher priority than that of r' or $r = r'$ and $\alpha\beta$ is initially more general than $\alpha'\beta'$. For a matching set M and item $r : \alpha \bullet \beta \in M$, a term t is said to match $r : \alpha \bullet \beta \in M$ if, and only if, t is an instance of $\alpha\beta$ but not an instance of any other item in M of higher priority.

Although the pattern $\alpha\beta$ of the item $r : \alpha \bullet \beta$ will always match the pattern of rule r, it may match a pattern of higher priority. This could have been used in defining a priority rule on all terms and hence on items, but this is computationally more expensive, and unnecessary here. It is now possible to determine which patterns are useful for *close* to include. Indeed, we can start by considering the usefulness of each pattern in the initial pattern set.

Definition 9. Suppose $L \cup \{\pi\}$ is a prioritised pattern set. Then π is said to be *relevant* for L if there is a term that matches π in $L \cup \{\pi\}$ in the sense of Definition 6. Otherwise, π is *irrelevant* for L. Similarly, an item π is *relevant* for (the matching set) M if there is a term that deterministically matches π in $L \cup \{\pi\}$ in the sense of Definition 8.

Clearly, any term that matches an element of a pattern set, respectively item of a matching set, will still have that property even when an irrelevant pattern, respectively item, is removed. We can therefore immediately prune irrelevant patterns one by one from the initial pattern set until every remaining pattern is relevant to the remaining pattern set, and do the same for each matching set generated by *close*.

The function *close* of Equation (4) may certainly supply items that are irrelevant for subsequent matching. This may happen when the original pattern set contains ambiguous patterns with more general ones having lower priorities. For instance, in Example 1, the original patterns $faww$ and $fwaa$ are ambiguous as $fwaa$ is more general than $faww$ at position 1 yet $faww$ has higher priority. The close function supplies the items $2 : f \bullet aaa$, $3 : f \bullet aba$ and two others to state 1. Then accepting symbol a would yield a superset of $\{1 : fa \bullet ww, 2 : fa \bullet aa, 3 : fa \bullet ba\}$. At this stage, based only on the item $fa \bullet ww$ a match for $faww$ can be announced and hence $fa \bullet aa$ and $fa \bullet ba$ are redundant, and indeed irrelevant under the definition above. Note also that the items $1 : fa \bullet aw$ and $1 : fa \bullet bw$ are similarly irrelevant for the matching set of state 2 in this case, because the item $1 : fa \bullet ww$ has higher priority due to its initially greater generality (at position 2).

Since the relevance of items may depend on the order in which items are added to a matching set M to form $close(M)$, we need to be careful about re-defining *close* to exclude irrelevant items; the result may depend on this order. The new, improved function is $close'$ defined (non-uniquely) from the initial *close* by Equation (6). It seems best to consider items for inclusion using an ordering which preserves decreasing priorities. So, highest priority items will be added first. This ensures that items already added to the partially created $close'(M)$ never subsequently become irrelevant.

$$close'(M) = \text{any maximal subset } S \text{ of } close(M)|$$
$$\text{if } \pi \in close(M) \backslash S \text{ then p is irrelevant for } S \qquad (6)$$

Finally, we consider a special case where the revised specification for close can be computed more easily. For this new definition, we assume that there is at least one function symbol in F that does not occur in any original pattern. This definition is given in Equation (6) where the first line duplicates the conditions of the initial *close* and the subsequent lines add an extra condition to exclude some irrelevant items. Roughly speaking, this condition says that any potentially added item $\alpha \bullet f\omega^{\#f}\beta$ must contribute new terms, which are not already covered by patterns in M with higher priority. However, in the general case when *close* is computed iteratively an added item may actually be covered by a previously added item, superseded by a subsequent item of higher priority or even covered by a number of more specific items.

$$close''(M) = M \cup \{r : \alpha \bullet f\omega^{\#f}\beta | r : \alpha \bullet \omega\beta \in M, \exists r' : \alpha \bullet f\beta' \in M,$$
$$\forall r'' : \alpha \bullet f\beta'' \in M, (\text{if } r : \alpha \bullet \omega \prec r'' : \alpha \bullet \beta''$$
$$\text{then } \neg(\alpha f\omega^{\#f}\beta \triangleright \alpha f\beta'') \text{ else } \beta = \omega \cdots \omega)\} \qquad (7)$$

Theorem 1. Assuming that there is at least one function symbol which does not occur in any pattern, then all items supplied by the function $close''$ are relevant.

Proof. We proceed by induction on matching sets. Assuming that all original patterns are relevant, then all items in the initial matching set are relevant. For the general case which unfolds in two sub-cases, let $M = \delta(N, \sigma), \alpha \bullet \beta \in M$ and $s \in F \cup \{\omega\}$:

- When $\alpha \bullet \beta \in kernel(M)$, i.e. the item set $accept(N, s)$, then it is clear that $\alpha \bullet \beta$ is relevant for M as by induction hypothesis, the item $\alpha \bullet s\beta$ is relevant for the matching set N.
- Suppose that $\alpha \bullet \beta = \alpha \bullet f\omega^{\#f}\beta \ni kernel(M)$. Then there are some items $\alpha \bullet \omega\beta'$ and $\alpha \bullet f\beta''$ in $kernel(M)$. Depending on the priorities of these patterns, there are two cases to consider: first, assume that $\alpha \bullet f\beta'' \prec \alpha \bullet \omega\beta'$. Then when $\beta' \neq \omega \ldots \omega$, the item $\alpha \bullet f\omega^{\#f}\beta'$ is always relevant for M because it is needed to deterministically match either $\alpha \bullet f\beta'$ or $\alpha \bullet f\omega^{\#f}\beta'$. If β' is a sequence of ωs, the item $\alpha \bullet \beta = \alpha \bullet f\omega \ldots \omega$ is not relevant for M since a match for the item $\alpha \bullet \omega\beta'$ is already determined (at this matching set M);

Now, suppose that $\alpha \bullet f\beta_1'', \ldots, \alpha \bullet f\beta_n''$ are all the items in $kernel(M)$ such that $\alpha\omega\beta' \prec \alpha \bullet f\beta_i'', 1 \leq i \leq n$ and $\alpha \bullet \omega\beta_1', \ldots \alpha \bullet \omega\beta_m'$ are all the items in $kernel(M)$ such that $\alpha\omega\beta' \prec \alpha \bullet \omega\beta_k', 1 \leq k \leq m$. The $close''$ function would add the item $j = \alpha \bullet f\omega^{\#f}\beta'$ if the term $t_j = \alpha f\omega^{\#f}\beta'$ were not an instance of any term $\alpha f\beta_i''$. We need to show that if t_j is not an instance of any $\alpha f\beta_i''$ then there is a term t that matches t_j i.e., t is an instance of t_j, t is not an instance of any of the terms $\alpha f\beta_i''$ and t is not an instance of any of the so far added items (by $close''$). Here, we need only to consider added items having an f at the matching position of M and higher priority than $\alpha \bullet \beta$, i.e., any item $\alpha \bullet f\omega^{\#f}\beta_k'$ already in M. Using the induction hypothesis, $\alpha\omega\beta'$ is not an instance of any of the terms $\alpha\omega\beta_k'$. So, $\alpha f\omega^{\#f}\beta'$ is not an instance of any of the terms $\alpha f\omega^{\#f}\beta_k'$.

Now we construct a term t that matches $\alpha \bullet f\omega^{\#f}\beta'$. For this purpose, let g be a function symbol that does not occur in any original pattern and t the term obtained from t_j by substituting all the ωs by the term $g\omega^{\#g}$. The fact that t_j is not an instance of any $\alpha f\beta_i''$ and $\alpha f\omega^{\#f}\beta_k'$ means that for each of these terms there is a position p_i, respectively p_k, such that $\alpha f\beta_i''[p_i]$, respectively $\alpha f\omega^{\#f}\beta_k'[p_k]$ is a function symbol and $\alpha f\beta_i''[p_i] \neq t_j[p_i]$, respectively $\alpha f\omega^{\#f}\beta_k'[p_k] \neq t_j[p_k]$. t_j cannot have a function symbol at position p_i and p_k because $\alpha f\beta_i''$ and $\alpha\omega\beta_k'$ overlap with $\alpha\omega\beta$ and so are $\alpha f\beta_i''$ and $\alpha f\omega^{\#f}\beta_k'$ with t_j. Then $t_j[p_i]$ and $t_j[p_k]$ is an ω symbol and so $t[p_i] = t[p_k] = g$. Hence, t cannot be an instance of any of the terms $\alpha f\beta_i''$ or $\alpha f\omega^{\#f}\beta_k'$ because g does not occur in any original pattern. ◇

Example 2. Using any of the improved closure functions, the automaton corresponding to $L = \{1 : fa\omega\omega, 2 : f\omega aa, 3 : f\omega ba, 4 : fg\omega g\omega b\}$ is given in Fig. 2. As usual, transitions corresponding to failure are omitted. Notice that for the same pattern set, the automaton of Fig. 1 has six more states, namely states 6, 7, 12, 13, 14 and 15. Pattern matching for the terms *faaa* and *faba* using the automaton of Fig. 1 requires four symbol examinations whereas by using the automaton of Fig. 2 only two symbols need to be examined as ωs match any term. Thus, using the new function *close* in this example, not only does the automaton have fewer states but it also allows pattern matching to be performed more quickly.

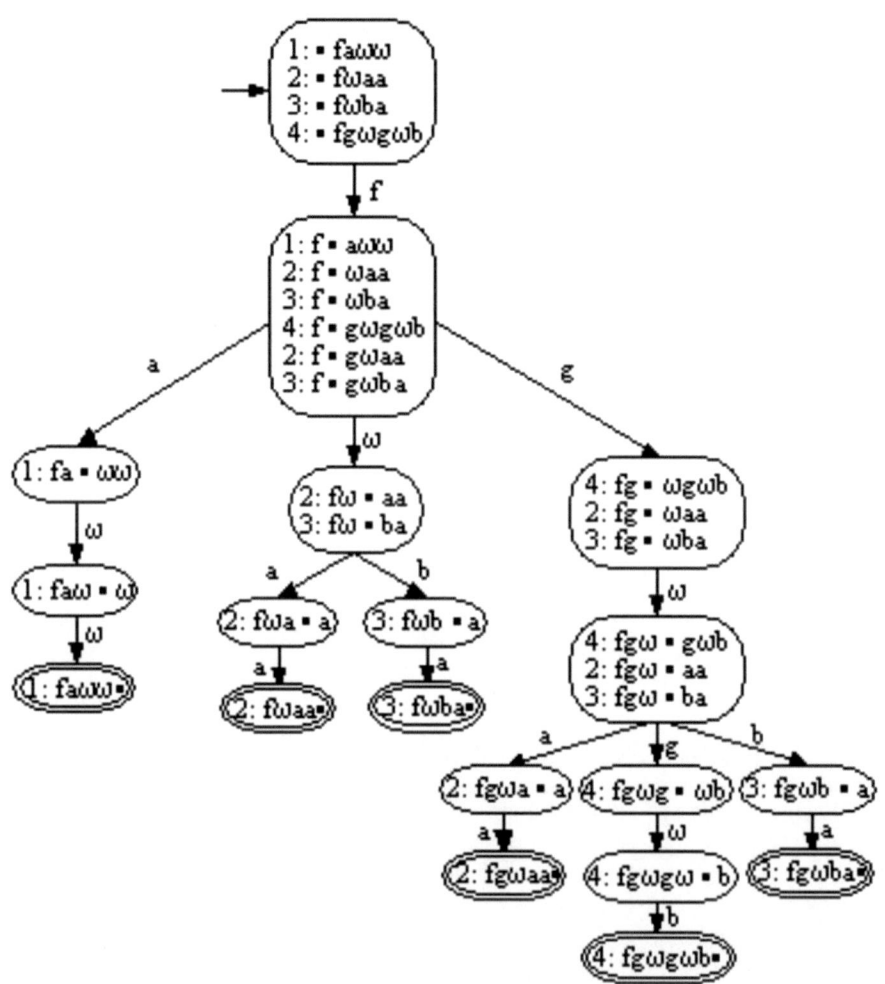

Fig. 2. Matching automaton using a revised function for *close*

Now, we need to develop a procedure that implements the new closure function and so will substitute procedure *ClosetSet*. This is a straightforward implementation of function *close''*, which is given below. It is clear that for the incremental construction, all the previously given procedures will be kept unchanged.

Procedure NewCloseSet(s);
Begin
    For each pair of items $\alpha \bullet \omega \beta \in$ s, $\alpha \bullet f \beta' \in$ s | f $\in$ F Do
        For each item $\alpha \bullet$ f $\beta'' \in$ s $|\beta'' \neq \beta'$ Do
            If (r:$\alpha \bullet \omega \prec$ r'':$\alpha \bullet \beta''$) and ($\alpha f \omega^{\#f} \beta \rhd \alpha f \beta''$ or $\beta = \omega \ldots \omega$) then
                s := s $\cup \{\alpha \bullet$ f$\omega^{\#f} \beta\}$;
    Return s;
End.

7 Evaluation

The pattern matching automata for some known problems were used as benchmarks to evaluate the improvement proposed. These problems were first used by Christian [2] to evaluate his system HIPER. The *Kbl* benchmark is the ordinary three-axiom group completion problem. The *Comm* benchmark is the commutator theorem for groups. The *Ring* problem is to show that if $x^2 = x$ is a ring then the ring is commutative. Finally, the *Groupl* problem is to derive a complete set of reductions for Highman's single-law axiomatisation of groups using division.

For each of the benchmarks, we built the both matching automata, i.e. using *close* and *close''*, and obtained the number of necessary states. This should provide an idea about the size of the automaton. Furthermore, we obtained the evaluation times of a given subject term under our rewriting machine (for details see [11]) using both matching automata as well as the evaluation times of the same subject terms under the system HIPER. Evaluation times under HIPER are for Sun 4 while times for our implementation are for MicroSPARC I. The space and time requirements are given in Table 1.

Table 1. Space and time requirements for miscellaneous benchmarks

Benchmark	Number of states *close*	Number of states *close''*	Time (s) *close*	Time (s) *close''*	Time (s) *HIPER*
Kbl	49	38	0.079	0.052	0.067
Comm	153	101	1.229	0.811	1.023
Ring	256	170	2.060	1.160	2.830
Groupl	487	392	1.880	0.980	2.000

8 Conclusion

In this paper, we started off by giving a description of our technique to compile pattern sets into deterministic finite matching automata (i.e., no backtracking over the symbols of the subject terms is needed to perform pattern matching). The construction procedure of such automata was provided. (Several examples can be found in [11], [13].)

A part from the important characteristic of matching determinism, another major improvement over most known term matching techniques is that here deterministic matching automata can easily be constructed incrementally. As we pointed out, this feature is very desirable for applications in which pattern sets change dynamically. Then, we outlined in details the incremental construction procedure of deterministic automata. We also proved that the automata obtained from such incremental construction procedure coincide with those obtained from the initial construction technique (Lemma1 and Lemma 2).

Despite the fact the automata thus yileded are deterministic and so efficient, we showed that the obtained automaton include unnecessary subautomata when the patterns are ambiguous. In the main body of the paper, we identified those subautomata by considering relevant patterns/items. By avoiding irrelevant patterns/items, we modified the method so that only necessary subautomata are included in the matching overall automaton. An example showed how the matching automaton obtained can improve both the space and time requirements. Finally, in Theorem 1, we proved that assuming there is at least one function symbol which does not occur in any pattern, then all items supplied are relevant, and hence the thus yield matching automaton does not include any unnecessary subautomata.

References

1. Augustsson A., A Compiler for lazy ML, Proc. Conference on Lisp and Functional Programming, ACM, pp. 218–227, 1984
2. Christian J., Flatterms, Discrimination nets and fast term rewriting, Journal of Automated Reasoning, vol. 10, pp. 95–113, 1993
3. Cooper, D. and Wogrin, N. Rule-based programming with OPS5, Morgan Kaufmann, San Francisco, 1988
4. Dershowitz, N., Jouannaud, J.P., Rewrite systems, Handbook of Theoretical Computer Science, vol. 2, chap. 6, Elsevier Science, 1990
5. Field, A.J., Harrison, P.G., Functional programming, International Computer Science Series, 1988
6. Goguen, J.A., Winkler, T., Introducing OBJ3, Technical report SRI-CSL-88-9, Computer Science Laboratory, SRI International, 1998
7. Gräf, A., Left-to-right tree pattern matching, Proc. Conference on Rewriting Techniques and Applications, Lecture Notes in Computer Science, Springer-Verlag, vol. 488, pp. 323–334, 1991
8. Hoffman, C.M., O'Donnell, M.J., Pattern matching in trees, Journal of ACM, vol. 29, no. 1, pp. 68–95, 1982

9. Hudak, P., al., Report on the programming language Haskell: a Non-Strict, Purely Functional Language, Sigplan Notices, Section S, May 1992

10. Laville, A., Comparison of priority rules in pattern matching and term rewriting, Journal of Symbolic Computation, no. 11, pp. 321–347, 1991

11. Nedjah, N., Walter, C.D., Eldridge, S.E., Optimal left-to-right pattern matching automata, Proc. Conference on Algebraic and Logic Programming, Southampton, UK, Lecture Notes in Computer Science, M. Hanus, J. Heering and K. Meinke (Eds.), Springer-Verlag, vol. 1298, pp. 273–285, 1997

12. Nedjah, N., Postponing redex contractions in equational programs, Proc. Symposium on Functional and Logic Programming, Kyoto, Japan, M. Sato and Y. Toyama (Eds.), World Scientific, pp. 40–60, 1998

13. Nedjah, N., Walter, C.D., Eldridge, S.E., Efficient automata-driven pattern matching for equational programs, Software-Practice and Experience, vol. 29, n° 9, pp. 793–813, John Wiley, 1999

14. Nedjah, N., Mourelle, L.M., Dynamic deterministic pattern matching, Proc. Computing: the Australasian Theory Symposium, Canberra, Australia, Elsevier Science, D.A. Wolfram (Ed.), vol. 31, 2000

15. Nedjah, N., Mourelle, L.M., Improving time, space and termination in term rewriting-based programming, Proc. International Conference on Industrial & Engineering Applications of Artificial Intelligence & Expert Systems, Budapest, Hungary, Lecture Notes in Computer Science, L. Monostori, J. Váncsa and A. M. Ali (Eds.), Springer-Verlag, vol. 2070, pp. 880–890, June 2001

16. O'Donnell, M.J, Equational logic as programming language, MIT Press, 1985

17. Sekar, R.C., Ramesh, R. and Ramakrishnan, I.V., Adaptive pattern matching, SIAM Journal, vol. 24, no. 5, pp. 1207–1234, 1995

18. Turner, D.A., Miranda: a Non strict functional language with polymorphic Types, Proc. Conference on Lisp and Functional Languages, ACM, pp. 1–16, 1985

19. Wadler, P., Efficient compilation of pattern matching, In "The Implementation of Functional Programming Languages", In S. L. Peyton-Jones, Prentice-Hall International, pp. 78–103, 1987

Complexities of the Centre and Median String Problems

François Nicolas and Eric Rivals

L.I.R.M.M., CNRS U.M.R. 5506
161 rue Ada, F-34392 Montpellier Cedex 5, France
{nicolas, rivals}@lirmm.fr

Abstract. Given a finite set of strings, the MEDIAN STRING problem consists in finding a string that minimizes the sum of the distances to the strings in the set. Approximations of the median string are used in a very broad range of applications where one needs a representative string that summarizes common information to the strings of the set. It is the case in Classification, in Speech and Pattern Recognition, and in Computational Biology. In the latter, MEDIAN STRING is related to the key problem of Multiple Alignment. In the recent literature, one finds a theorem stating the NP-completeness of the MEDIAN STRING for unbounded alphabets. However, in the above mentioned areas, the alphabet is often finite. Thus, it remains a crucial question whether the MEDIAN STRING problem is NP-complete for finite and even binary alphabets. In this work, we provide an answer to this question and also give the complexity of the related CENTRE STRING problem. Moreover, we study the parametrized complexity of both problems with respect to the number of input strings.

1 Introduction

Given an alphabet Σ, a set W of strings over Σ, and the Levenshtein distance between strings, the problem of finding a string over Σ that minimizes the sum of distances to the strings of W is called the MEDIAN STRING problem. Alternative terminologies include the GENERALIZED MEDIAN STRING problem [1], the STAR ALIGNMENT problem [13] and also the STEINER STRING problem [7].

The MEDIAN STRING problem is of major significance in several areas of research: Pattern Recognition, Speech Recognition and Computational Biology. Its importance is reflected by the wide use of a polynomial time approximation, the *set median* string ([9,18,17,7,6,8]). In this restricted version of the problem, the solution string must be taken in the input set W (it is also termed the "center string" in [7, p. 349]). One class of applications, encountered in all three areas, looks for a string (or a language) that models the input set of strings. In other words, this string summarizes the information shared by strings of W. Depending on the application, it then serves as an index for W (in databases and data mining), as a pattern that is searched for in longer texts (in computational biology [7,19]) or used for classification purposes (in speech recognition [9], classification [7] and computational biology [7,19]).

R. Baeza-Yates et al. (Eds.): CPM 2003, LNCS 2676, pp. 315–327, 2003.

In [1], it is shown that CENTRE STRING and MEDIAN STRING are NP-hard for alphabets of size at least 4 and for unbounded alphabets, respectively. In [21], it is shown that MEDIAN STRING is NP-hard for alphabet of size 7 and a particular weighted edit distance. In many practical situations, the alphabet is of fixed constant size. In computational biology, the DNA and protein alphabets are respectively of size 4 and 20. However, other alphabet sizes are also used. Indeed, for some applications, one needs to encode the DNA or protein sequences on a binary alphabet that expresses only a binary property of the molecule, e.g., hydrophoby. For instance, it is the case in some protocols to identify similar DNA sequences [22]. The important practical question is whether CENTRE STRING and MEDIAN STRING are NP-hard for finite and even binary alphabets. In the above-mentioned article, these questions remain open [1, p. 48]. These conjectures are solved in this paper. Additionnally, an interesting issue concerns the existence of fast exact algorithms when the number of input strings is fixed. We provide an answer to this issue for both CENTRE STRING and MEDIAN STRING.

1.1   Definitions

We denote by $\mathbb{N}$ the set of non negative integers and by $\mathbb{N}^*$ the set of positive integers. For all $n \in \mathbb{N}^*$, we denote by $\mathbb{N}_n^*$ the set $\{1, 2, \dots, n\}$ and for every finite set X we denote by $\#X$ the cardinality of X.

Words. An *alphabet* is a non empty set of *letters*. In the sequel, Σ always denotes an alphabet. A *word* over Σ is a finite sequence of elements of Σ. The set of all words over Σ is denoted by $\Sigma^\star$. A *language* over Σ is any subset of $\Sigma^\star$. The empty sequence, denoted by ε, is called the *empty word*. For a word w, $|w|$ denotes the length of w. For all $a \in \Sigma$, $|w|_a$ denotes the number of occurences of the letter a in w. For all $i \in \mathbb{N}_{|w|}^*$, $w[i]$ denotes the i-th letter of w. Given two words x and y, we denote by xy the *concatenation* of x and y. For all $n \in \mathbb{N}$, we denote by x^n the *n-th power* of x that is, the concatenation of n copies of x (note that $x^0 = \varepsilon$). For all $L \subseteq \Sigma^\star$ and for all $w \in \Sigma^\star$, we denote $Lw := \{xw : x \in L\}$.

Edit Distance. Let x, y be words over Σ. The *edit distance* between x and y is the smallest number of single letter deletions, insertions and substitutions needed to transform x into y. It is also called LEVENSHTEIN *distance* [11] and denoted by $\mathrm{Lev}(x, y)$. For exemple, we have, for all $n \in \mathbb{N}^*$, $\mathrm{Lev}\left((01)^n, (10)^n\right) = 2$. WAGNER & FISHER's algorithm [23] compute the edit distance $\mathrm{Lev}(x, y)$ in polynomial time $\mathrm{O}\left(|x|\,|y|\right)$.

Radius, Centre String and Median String.

Definition 1. *For all languages W over Σ, we denote:*

$$\mathcal{R}(W) := \inf_{\gamma \in \Sigma^\star} \left(\sup_{w \in W} \mathrm{Lev}(\gamma, w) \right)$$

$$S(W) := \inf_{\mu \in \Sigma^\star} \left(\sum_{w \in W} \mathrm{Lev}(\mu, w) \right)$$

and we call $\mathcal{R}(W)$ *the* radius *of* W.

A centre *of* W *is a word* γ *over* Σ *such that* $\sup_{w \in W} \mathrm{Lev}(\gamma, w) = \mathcal{R}(W)$.

A median *of* W *is a word* μ *over* Σ *such that* $\sum_{w \in W} \mathrm{Lev}(\mu, w) = S(W)$.

Our goal is to prove the intractability of the following two problems.

Definition 2. *The* CENTRE STRING *(resp.* MEDIAN STRING*) problem is the decision problem: given a non empty finite language* W *over* Σ *and* $K \in \mathbb{N}$, *is* $\mathcal{R}(W) \leq K$ *(resp.* $S(W) \leq K$*)?*

1.2   Related Works

Related problems. Computational biology exhibits numerous problems related to MEDIAN STRING. In the more studied ones, the computationally less demanding HAMMING distance replaces the edit distance. One often uses a closest representative of a set of constant length strings that share a biological function. Its computation is known to be NP-hard and is called the CONSENSUS STRING problem (see [19], [13] and references therein). The CONSENSUS PATTERNS and its variants, like the CLOSEST SUBSTRING problem, aim at finding common substrings of a given length in a set of strings, and a model for them. Li et al. [13,14,12] exhibit PTAS for all of these, while [5,4] give exact polynomial time algorithms for some special cases and studied their parameterized complexities. When the Levenshtein distance is used, finding common substrings is termed pattern discovery or motif extraction (see [16,19]).

Another interesting problem is the DISTINGUISHING STRING SELECTION problem. Given two sets, one of "positive" and the other of "negative" example strings, one has to find a string that is close to the positive, and far from the negative strings (see [10,5,2]).

MEDIAN STRING is also important because of its relation with the *Multiple Alignment* problems. Indeed, once given a median string, one can compute an approximate multiple alignment from the pairwise alignments between the median and any string in the input set [7]. Thus, an algorithm for the set median string is used by several authors as an approximation of the MEDIAN STRING. First, Gusfield [6] provides an approximation algorithm for the SUM-OF-PAIRS MULTIPLE ALIGNMENT problem. In this problem, one wishes to minimise the sum of all pairwise alignment costs, hence the name Sum-of-Pairs. Second, Jiang et al. [8] also give an approximation for the TREE MULTIPLE ALIGNMENT problem. Given a tree and sequences associated with its leaves, one has to find sequences for the internal nodes, such that the sum of distances between adjacent strings/nodes over all edges is minimal. They show that associating the set median string to each internal node provides a good approximation scheme. This result was further improved in [24].

Known results. CENTRE STRING and MEDIAN STRING are polynomial in some trivial setup, e.g., with two sequences. Hence, we can deduce from WAGNER & FISCHER's algorithm a dynamic programming algorithm that computes for every non empty, finite language W over Σ a median and a centre of W in $O\left(\prod_{w \in W} |w|\right)$ time. Thus, for all fixed $n \in \mathbb{N}^*$, the restrictions of CENTRE STRING and MEDIAN STRING to the instances such that $\#W = n$ are polynomial.

In [13], a theorem states that the STAR c-ALIGNMENT problem is NP-hard but no proof is given. The STAR c-ALIGNMENT problem consists in the MEDIAN STRING problem where the number of gaps between any two sequences is constrained to be at most c. Nevertheless, it would not imply that MEDIAN STRING is NP-hard in its general setup. [1]

In [1], it is shown that if Σ is unbounded (resp. $\#\Sigma$ is at least 4) then MEDIAN STRING (resp. CENTRE STRING) is NP-complete. Above, we argue already that especially the NP-completeness of MEDIAN STRING for finite alphabet is an important conjecture. In this work, we demonstrate that both problems are NP-complete even if Σ is binary. Both proofs consist in reducing a well known NP-complete problem, LONGEST COMMON SUBSEQUENCE (LCS), to CENTRE STRING and MEDIAN STRING.

Note that the main difficulty of the NP-completess proof of MEDIAN STRING is that in the instances (W, K), W is a *set* and not a *family*. Hence we do not allow repetitions of words in W, e.g., $\mathcal{S}(\{\varepsilon, \varepsilon, 01011000, 10, 10\}) = \mathcal{S}(\{\varepsilon, 01011000, 10\})$. Otherwise, a little modification of the proof in [1] shows the NP-completeness of MEDIAN STRING problem for families of words.

We also demonstrate that both CENTRE STRING and MEDIAN STRING are hard in the sense of parametrized complexity with respect to the number of strings. These are important results from a practical point of view since they imply that the existence of an exact polynomial time algorithm is unprobable even if the number of strings is fixed.

Organisation of the paper. We conclude this section with some definitions about parameterized complexity and some known results about the LCS problem. In Section 2, we prove that CENTRE STRING is NP-complete and $W[1]$-hard. In Section 3 we prove that MEDIAN STRING is NP-complete and $W[1]$-hard. We conlude the paper in Section 4 with some open problems.

1.3   Parameterized Complexity

We give a short introduction to parameterized complexity and the $W[1]$-class (see [3] for a definition of the whole W-hierarchy).

Let $L, L' \subseteq \{0, 1\}^* \times \mathbb{N}$ be two parameterized binary languages.

We say that L is *fixed parameter tractable* if there exists an algorithm that decides for all $(x, k) \in \{0, 1\}^* \times \mathbb{N}$ wether $(x, k) \in L$ in time $f(k) |x|^c$ where

---

[1] The authors mention on p. 172 that MEDIAN STRING is NP-hard but without any reference nor proof.

$f : \mathbb{N} \to \mathbb{N}$ is an arbitrary fonction and c an integer constant. We denote FPT the set of all fixed parameter tractable parameterized languages.

We says that L *reduces to L' by a standard parameterized (many to one) reduction* if there are functions $f, m : \mathbb{N} \to \mathbb{N}$, $M : \{0,1\}^{\star} \times \mathbb{N} \to \{0,1\}^{\star}$ and a constant $c \in \mathbb{N}$ such that for all $(x, k) \in \{0,1\}^{\star} \times \mathbb{N} : M(x, k)$ is computable in time $f(k)\,|x|^{c}$ and $(M(x, k), m(k)) \in L'$ iff $(x, k) \in L$.

We say that a parameterized language L belongs to $W[1]$ if there exists a standard parameterized reduction from the k-STEP HALTING problem[2] to L. A language L is $W[1]$-*hard* if there exists a standard parameterized reduction from L to the k-STEP HALTING problem.

The k-STEP HALTING problem is the parameterized analog of the TURING MACHINE ACCEPTANCE problem, which is the basic generic NP-complete problem. The conjecture FPT $\neq W[1]$ is to parameterized complexity what P $\neq$ NP is to classical computational complexity. Hence, from a practival point of view, $W[1]$-hardness gives a concrete indication that a paramerized problem is fixed parameter untracktable.

1.4   The Longest Common Subsequence Problem

Let w be a word. A *subword* of w is any word obtained from w by deleting between 0 and $|w|$ letters. We denote by $\mathrm{Sub}(w)$ the set of all subwords of w. For every language L, we denote by $\mathrm{CSub}(L)$ the set of all the words which are common subwords of all the words in L and by $\mathrm{lcs}(L)$ the length of the longest words in $\mathrm{CSub}(L)$. Formally, we have:

$$\mathrm{CSub}(L) = \bigcap_{x \in L} \mathrm{Sub}(x) \quad \text{and} \quad \mathrm{lcs}(L) = \max_{s \in \mathrm{CSub}(L)} |s| \ .$$

For example, for all $n \in \mathbb{N}$, we have, $\mathrm{CSub}\left(\{0^{n}1^{n}, 1^{n}0^{n}\}\right) = \bigcup_{i=0}^{n} \left\{0^{i}, 1^{i}\right\}$ and therefore $\mathrm{lcs}\left(\{0^{n}1^{n}, 1^{n}0^{n}\}\right) = n$.

Definition 3 (Longest Common Subsequence problem (LCS)).
Given a non empty finite language L over Σ and $k \in \mathbb{N}$, is $\mathrm{lcs}(L) \geq k$?

The intractability of LCS was studied firstly by MAIER [15], and later by PIETRZAK [20] who slightly improved MAIER's results in terms of parameterized complexity :

Theorem 1 (MAIER). *If $\#\Sigma$ is at least 2, then LCS is NP-complete.*

Theorem 2 (PIETRZAK). *If $\#\Sigma$ is at least 2, then LCS problem parameterized in $\#L$ is $W[1]$-hard.*

---

[2] also known as SHORT TURING MACHINE ACCEPTANCE problem

2   NP-**Completeness of** CENTRE STRING

In order to reduce LCS to CENTRE STRING we introduce, like in [1], the following intermediate problem, LCS0, which consists in the restriction of LCS to the
instances in which strings have length $2k$, i.e., such that $L \subseteq \Sigma^{2k}$.

Before stating our theorems, we need the following lemma. In substance, it
says that if one concatenates a letter a to all words in a language L then the lcs
of L increases by one. Indeed, by doing this, one "adds" an a to any maximal
common subword of L (one changes $\mathrm{CSub}(L)$ into $\mathrm{CSub}(L) \cup \mathrm{CSub}(L)a$). Thus,
the lcs increases by one. The formal proof is left to the reader.

Lemma 1. *For every language L and for every letter a, we have* $\mathrm{lcs}(La) = \mathrm{lcs}(L) + 1$.

It is shown in [1] that if $\#\Sigma$ is at least 4 then LCS0 is NP-complete (note
that the proved result is stronger than the one stated in their proposition). We
improve this result.

Theorem 3. *The* LCS0 *problem is* NP-*hard even if* Σ *is binary. Moreover, the*
LCS0 *problem parameterized in* $\#L$ *is* $W[1]$-*hard.*

Proof. Suppose that Σ is the binary alphabet $\{0, 1\}$. By Theorem 2, it is sufficient to reduce LCS (parameterized in $\#L$) to LCS0 (parameterized in $\#L$).
Let (L, k) be an instance of LCS, L being a non empty finite language and k a
positive integer. We construct $\left(\tilde{L}, \tilde{k} \right)$ such that it is an instance of LCS0. Let

$$n := \max_{x \in L} |x|, \quad N := 2k + n, \quad \tilde{k} := k + n,$$
$$L' := \bigcup_{x \in L} \left\{ x0^{N-|x|}, x1^{N-|x|} \right\} \quad \text{and} \quad \tilde{L} := L'0^n.$$

We have $L' \subseteq \{0, 1\}^N$. Therefore, $\tilde{L}$ is a subset of $\{0, 1\}^{2\tilde{k}}$ and $\left(\tilde{L}, \tilde{k} \right)$ is an
instance of LCS0. The transformation of an instance (L, k) of LCS into the instance
$\left(\tilde{L}, \tilde{k} \right)$ of LCS0 is polynomial and parameter preserving ($\#\tilde{L} = \#L' = 2\#L$). It
remains to prove that

$$\mathrm{lcs}(L) \geq k \iff \mathrm{lcs}\left(\tilde{L} \right) \geq \tilde{k}. \tag{1}$$

Note that for all words u, v, w, $\mathrm{Sub}(wu) \cap \mathrm{Sub}(wv) = \mathrm{Sub}(w)$ if and only if
u and v do not share any letter. We have

$$\mathrm{CSub}(L') = \bigcap_{x \in L} \underbrace{\mathrm{Sub}\left(x0^{N-|x|} \right) \cap \mathrm{Sub}\left(x1^{N-|x|} \right)}_{\mathrm{Sub}(x)} = \mathrm{CSub}(L)$$

and therefore $\mathrm{lcs}(L) = \mathrm{lcs}(L')$ (the polynomial transformation of (L, k) into
(L', k) shows that the restriction of LCS to the instances such that all words in
L share the same length is NP-complete).

On the other end, Lemma 1 assures that $\text{lcs}\left(\tilde{L}\right) = \text{lcs}(L') + n$ and therefore:

$$\text{lcs}\left(\tilde{L}\right) = \text{lcs}(L) + n$$

which implies (1). Moreover, as our reduction preserves the parameter $\#L$, the $W[1]$-hardness of LCS0 follows from the $W[1]$-hardness of LCS.   □

Now, we have to link the edit distance and the notion of subword to complete the reduction of LCS0 to CENTRE STRING.

Lemma 2. *For all x, $y \in \Sigma^\star$ we have:*

1. $\text{Lev}(x, y) \geq |x| - |y|$,
2. $\text{Lev}(x, y) = |x| - |y|$ *if and only if y is a subword of x.*

Proof. Let $x, y \in \Sigma^\star$ and w.l.o.g. assume x is longer than y. The first statement says that the edit distance is larger than or equal to the length difference of x and y. Clearly, any transformation of x into y has to delete $|x| - |y|$ supernumerary symbols. The second statement says that the equality holds iff y is a subword of x. Again, once the transformation has deleted the $|x| - |y|$ supernumerary symbols, if the resulting subword is y, it means that y is a subword of x, and conversely.   □

Theorem 4. *The CENTRE STRING problem is NP-complete even if Σ is binary. Moreover, CENTRE STRING problem parameterized in $\#W$ is $W[1]$-hard.*

Proof. The proof is the same as in [1]. It consists in reducing LCS0 to CENTRE STRING: we transform an instance (L, k) of LCS0 in the instance $(W, K) := (L \cup \{\varepsilon\}, k)$ of CENTRE STRING. The transformation is clearly polynomial and parameter preserving ($\#W \in \{\#L, \#L + 1\}$) and to check the equivalence $\text{lcs}(L) \geq k \iff \mathcal{R}(W) \leq K$, we only need the properties of Lev stated in Lemma 2.

Suppose that Σ is binary. Since in this case LCS0 is NP-complete according to Theorem 3, our reduction shows that CENTRE STRING (parameterized in $\#W$) is $W[1]$-hard too.   □

3   NP-Completeness of MEDIAN STRING

In order to reduce LCS to MEDIAN STRING, we need to link edit distance and subwords by a tighter inequality than the one provided by Lemma 2. Let x, $y \in \Sigma^\star$ and w.l.o.g. assume $|x| \geq |y|$. The lemma shows that any transformation of x into y contains at least as much operations as the difference between the lengths of x and of its longest common subwords with y. An explanation is as follows. Consider the positions of x that do not belong to a fixed maximal common subword of x and y. All these are either supernumerary and have to be deleted, or differ from the corresponding position in y and need to be substituted.

Lemma 3. *For all x, $y \in \Sigma^\star$, we have*

$$\mathrm{Lev}(x,y) \geq \max\{|x|,|y|\} - \mathrm{lcs}(\{x,y\})$$

Proof. Let $((x_1,y_1),(x_2,y_2),\ldots,(x_n,y_n))$ be an alignment of x and y with cost $\mathrm{Lev}(x,y)$. Remember that $x[i]$ and not x_i denotes the ith symbol of x. We have:

- for all $i \in \mathbb{N}_n^*$, we have $(x_i,y_i) \in ((\Sigma \cup \{\varepsilon\}) \times \Sigma) \cup (\Sigma \times (\Sigma \cup \{\varepsilon\}))$, i.e., in other words a symbol in the alignment can be a single letter or the empty word,
- $x = x_1 x_2 \ldots x_n$,
- $y = y_1 y_2 \ldots y_n$,
- denoting by J the set of all $i \in \mathbb{N}_n^*$ such that $x_i \neq y_i$, we have $\mathrm{Lev}(x,y) = \#J$.

As any alignment symbol can be the empty word, we have $n \geq |x_1 x_2 \ldots x_n| = |x|$ and $n \geq |y_1 y_2 \ldots y_n| = |y|$, and thus:

$$n \geq \max\{|x|,|y|\}.$$

On the other hand, denote $k := \#(\mathbb{N}_n^* \setminus J) = n - \mathrm{Lev}(x,y)$ and let i_1, i_2, $\ldots$, i_k be indexes such that: $\mathbb{N}_n^* \setminus J = \{i_1,i_2,\ldots,i_k\}$ and $i_1 < i_2 < \cdots < i_k$. For all $j \in \mathbb{N}_k^*$, $i \notin J$ means that $x_j = y_j$ and therefore $x_{i_1} x_{i_2} \ldots x_{i_k} = y_{i_1} y_{i_2} \ldots y_{i_k}$ is a subword of x and of y. From that we deduce:

$$\mathrm{lcs}(\{x,y\}) \geq k = n - \mathrm{Lev}(x,y) \geq \max\{|x|,|y|\} - \mathrm{Lev}(x,y).$$

The inequality stated in our lemma follows. $\qquad\square$

The inequality stated in Lemma 3 involved only two words. In order to generalize it to many words (Lemma 5), we need the following lemma.

Lemma 4. *For all $\mu \in \Sigma^\star$ and for all X, $Y \subseteq \Sigma^\star$, we have*

$$\mathrm{lcs}(\{\mu\} \cup X \cup Y) \geq \mathrm{lcs}(\{\mu\} \cup X) + \mathrm{lcs}(\{\mu\} \cup Y) - |\mu| \qquad (2)$$

Proof. Let $p := \mathrm{lcs}(\{\mu\} \cup X)$ and $q := \mathrm{lcs}(\{\mu\} \cup Y)$. By hypothesis for $\{\mu\} \cup X$, there exist indexes i_1, i_2, $\ldots$, i_p satisfying $1 \leq i_1 < i_2 < \cdots < i_p \leq |\mu|$ such that $u := \mu[i_1]\mu[i_2]\ldots\mu[i_p] \in \mathrm{CSub}(\{\mu\} \cup X)$. Similarly, for $\{\mu\} \cup Y$, there exist indexes j_1, j_2, $\ldots j_q$ satisfying $1 \leq j_1 < j_2 < \cdots < j_q \leq |\mu|$ such that $v := \mu[j_1]\mu[j_2]\ldots\mu[j_q] \in \mathrm{CSub}(\{\mu\} \cup Y)$.

Setting $I := \{i_1,i_2,\ldots,i_p\}$ and $J := \{j_1,j_2,\ldots,j_q\}$, we see that u and v share a common subword of length $\#(I \cap J)$. It is also a common subword of all words in $\{\mu\} \cup X \cup Y$. From which we deduce

$$\mathrm{lcs}(\{\mu\} \cup X \cup Y) \geq \#(I \cap J) \qquad (3)$$

On the other hand, since I and J are subsets of $\mathbb{N}_{|\mu|}^*$, we have $\#(I \cup J) \leq |\mu|$ and therefore

$$\#(I \cap J) = p + q - \#(I \cup J) \geq p + q - |\mu| \qquad (4)$$

Combining (3) and (4) gives (2) and concludes the proof. $\qquad\square$

Lemma 5. *For all $\mu \in \Sigma^\star$ and for all finite languages X over Σ, we have:*

$$\sum_{x \in X} \mathrm{Lev}(\mu, x) + (\#X - 1)\,|\mu| \geq \sum_{x \in X} |x| - \mathrm{lcs}(\{\mu\} \cup X)$$

Proof. We proceed by induction on $\#X$. Assume $\#X = 0$; the inequality holds since both members are equal to $-|\mu|$. When $\#X = 1$, the statement follows from Lemma 3.

Now suppose that $\#X \geq 1$. Let $x_0 \in X$ and let $X' := X \setminus \{x_0\}$. We have

$$\mathrm{Lev}(\mu, x_0) \geq |x_0| - \mathrm{lcs}(\{\mu, x_0\})\,, \tag{5}$$

$$\sum_{x' \in X'} \mathrm{Lev}(\mu, x') + (\#X' - 1)\,|\mu| \geq \sum_{x' \in X'} |x'| - \mathrm{lcs}(\{\mu\} \cup X')\,, \tag{6}$$

$$\mathrm{lcs}(\{\mu\} \cup X) \geq \mathrm{lcs}(\{\mu\} \cup X') + \mathrm{lcs}(\{\mu, x_0\}) - |\mu|\,. \tag{7}$$

The inequalities (5) and (6) result respectively from Lemma 3 and from the induction hypothesis. Lemma 4 applied with $(X, Y) := (X', \{x_0\})$ yields (7). Adding (5), (6) and the trivial inequality $|\mu| \geq |\mu|$ we obtain

$$\sum_{x \in X} \mathrm{Lev}(\mu, x) + (\#X')\,|\mu| \geq \sum_{x \in X} |x| - \mathrm{lcs}(\{\mu\} \cup X') - \mathrm{lcs}(\{\mu, x_0\}) + |\mu|$$

$$\geq \sum_{x \in X} |x| - \mathrm{lcs}(\{\mu\} \cup X)$$

where the last inequality deduces from (7). Since $\#X' = \#X - 1$, this concludes the proof.   □

We can now prove the main theorem of this section.

Theorem 5. *The MEDIAN STRING problem is NP-complete even if Σ is binary. Moreover, the MEDIAN STRING problem parameterized in $\#W$ is $W[1]$-hard.*

Proof. Since the edit distance can be computed in polynomial time, it is easy to check that MEDIAN STRING is NP. Now, suppose Σ is the binary alphabet $\{0, 1\}$. The schema of the proof is the following: we reduce LCS to MEDIAN STRING in order to apply Theorem 2 and conclude. The $W[1]$-hardness of MEDIAN STRING parameterized in $\#W$ is deduced from the one of LCS parameterized in $\#L$ since our reduction is parameter preserving.

Let (L, k) be an instance of LCS, L being a non empty finite language over $\{0, 1\}$ and k a positive integer. We transform (L, k) into the instance (W, K) of MEDIAN STRING, as described below. Let

$$n := \#L\,, \quad C := \sum_{x \in L} |x|\,, \quad N := \max\left\{C + \frac{n(n-1)}{2} - k, n - 1\right\}\,,$$

$$K := C + (n-1)N - k - \frac{n(n-1)}{2}\,, \quad W := L0^N \cup \{0^i : i \in \mathbb{N}^\star_{n-1}\}\,.$$

This transformation is polynomial and parameter preserving since $\#W = 2(\#L) - 1$. Hence, it remains to prove

$$\mathrm{lcs}(L) \geq k \iff \mathcal{S}(W) \leq K.$$

($\Rightarrow$) Suppose that $\mathrm{lcs}(L) \geq k$. We want to prove that $\mathcal{S}(W) \leq K$. By hypothesis, it exists $s \in \mathrm{CSub}(L)$ such that $|s| = k$. Let $\mu := s0^N$.

For all $i \in \mathbb{N}^*_{n-1}$, we have $i \leq n-1 \leq N \leq |\mu|_0$, so 0^i is a subword of μ. Hence, by Lemma 2, we have

$$\mathrm{Lev}(\mu, 0^i) = |\mu| - |0^i| = k + N - i$$

Moreover, for all $x \in L$, μ is a subword of $x0^N$; again Lemma 2 applies and we obtain

$$\mathrm{Lev}(\mu, x0^N) = |x0^N| - |\mu| = |x| - k.$$

Using these equalities, we compute the sum

$$\sum_{w \in W} \mathrm{Lev}(\mu, w) = \sum_{x \in L} \mathrm{Lev}(\mu, x0^N) + \sum_{i=1}^{n-1} \mathrm{Lev}(\mu, 0^i)$$

$$= \sum_{x \in L} (|x| - k) + \sum_{i=1}^{n-1} (k + N - i)$$

$$= C - nk + (n-1)k + (n-1)N - \frac{n(n-1)}{2}$$

$$= K$$

and we obtain $\mathcal{S}(W) \leq K$.

($\Leftarrow$) Conversely, assume $\mathcal{S}(W) \leq K$. We will show that $\mathrm{lcs}(L) \geq k$. By hypothesis, it exists $\mu \in \{0, 1\}^*$ such that $\sum_{w \in W} \mathrm{Lev}(\mu, w) \leq K$. First, we prove that $|\mu|_0 \geq n-1$. For this, we note that for all words u, v, $\mathrm{Lev}(u, v)$ is greater or equal to $|v|_0 - |u|_0$. Hence, for all $x' \in L0^N$, we have

$$N - |\mu|_0 \leq |x'|_0 - |\mu|_0 \leq \mathrm{Lev}(\mu, x')$$

so by summing over $x' \in L0^N$, we get

$$nN - n|\mu|_0 \leq \sum_{x' \in L0^N} \mathrm{Lev}(\mu, x') \leq K$$

and:

$$n|\mu|_0 \geq nN - K \geq n(n-1)$$

which is equivalent to $|\mu|_0 \geq n-1$. This implies that for all $i \in \mathbb{N}^*_{n-1}$, 0^i is a subword of μ, and so $\mathrm{Lev}(\mu, 0^i) = |\mu| - i$. Thus,

$$\sum_{i=1}^{n-1} \mathrm{Lev}(\mu, 0^i) = \sum_{i=1}^{n-1} |\mu| - \sum_{i=1}^{n-1} i = (n-1)|\mu| - \frac{n(n-1)}{2}.$$

We can now write

$$\sum_{w \in W} \mathrm{Lev}(\mu, w) = \sum_{x' \in L0^N} \mathrm{Lev}(\mu, x') + (n-1)|\mu| - \frac{n(n-1)}{2}$$

$$\geq \sum_{x' \in L0^N} |x'| - \mathrm{lcs}\left(\{\mu\} \cup L0^N\right) - \frac{n(n-1)}{2} \qquad (8)$$

where the application of Lemma 3 with $X := L0^N$ yields the last inequality.

On the other hand, we have:

$$\sum_{x' \in L0^N} |x'| = \sum_{x \in L} |x0^N| = \sum_{x \in L} (|x| + N) = C + nN \qquad (9)$$

and by Lemma 1

$$\mathrm{lcs}(\{\mu\} \cup L0^N) \leq \mathrm{lcs}(L0^N) = \mathrm{lcs}(L) + N . \qquad (10)$$

By hypothesis, $K \geq \mathcal{S}(W) \geq \sum_{w \in W} \mathrm{Lev}(\mu, w)$; combining this with (8), (9) and (10) yields

$$K \geq \sum_{w \in W} \mathrm{Lev}(\mu, w) \geq (C + nN) - (\mathrm{lcs}(L) + N) - \tfrac{n(n-1)}{2}$$
$$\Rightarrow \mathrm{lcs}(L) \geq C + (n-1)N - \tfrac{n(n-1)}{2} - K = k .$$

This concludes the proof. □

4   Open Problems

Possible improvements of our results. We prove in this paper that CENTRE STRING and MEDIAN STRING are NP-complete, according to the "non weighted" edit distance. Of course, in general for any distance function, the problem remains NP-complete since the edit distance is a particular case. Now, for a fixed alphabet-weighted edit distance the problem may not be NP-complete. To prove the NP-completeness in such setup is not trivial since, since, if $\Sigma = \{0, 1\}$ and if the scores of insertions/deletions of 0 and of 1 are not equal then our reductions do not hold.

Approximation. Although CENTRE STRING and MEDIAN STRING are NP-complete, there exist approximation algorithms with bounded errors [7] and heuristic algorithms [9], [18]. For example, given a finite language W over Σ, *set center* (resp. set median) of W can be found in polynomial time and is an approximate centre (resp. median) of W with performance ratio 2 (resp. $2 - \frac{2}{\#W}$). Note that we call set center of W any word that minimizes the maximum of the distances to strings in the set W *and* belongs to W. An open question subsists: do these problems admit Polynomial Time Approximation Schemes?

Acknowledgment. The authors thank Colin De La Higuera for reading a preliminary version of this work and Jens Gramm for introducing us to parameterized complexity. This is supported by the CNRS STIC Specific Action "Algorithmes et Séquences", by the ACI Jeunes Chercheurs "Combinatoire des mots multidimensionnels, pavages et numération" and by the Math STIC project 2001 "Mots : de la combinatoire à la dynamique symbolique".

References

1. C. de la Higuera and F. Casacuberta. Topology of strings: Median string is NP-complete. *Theoretical Computer Science*, 230:39–48, 2000.
2. X. Deng, G. Li, Z. Li, B. Ma, and L. Wang. A ptas for distinguishing (sub)string selection. In *ICALP*, pages 740–751, 2002.
3. R. G. Downey and M. R. Fellows. *Parameterized Complexity*. Springer, 1999.
4. Michael R. Fellows, Jens Gramm, and Rolf Niedermeier. On the parameterized intractability of CLOSEST SUBSTRING and related problems. In *Symposium on Theoretical Aspects of Computer Science*, pages 262–273, 2002.
5. Jens Gramm, Rolf Niedermeier, and Peter Rossmanith. Exact solutions for CLOSEST STRING and related problems. In *ISAAC*, volume 2223 of *LCNS*, pages 441–453, 2001.
6. Dan Gusfield. Efficient methods for multiple sequence alignment with guaranteed error bounds. *Bull. Math. Biol.*, 55:141–154, 1993.
7. Dan Gusfield. *Algorithms on strings, trees, and sequences: computer science and computational biology*. Cambridge University Press, 1997.
8. Tao Jiang, Eugene L. Lawler, and Lusheng Wang. Approximation algorithms for tree alignment with a given phylogeny. *Algorithmica*, 16(3):302–315, 1996.
9. T. Kohonen. Median strings. *Pattern Recognition Letters*, 3:309–313, 1985.
10. J. Lanctot, M. Li, B. Ma, S. Wang, and L. Zhang. Distinguishing string selection problems. In *SODA: ACM-SIAM Symposium on Discrete Algorithms*, 1999.
11. V. I. Levenshtein. Binary codes capable of correcting deletions, insertions and Reverseals. *Cybernetics and Control Theory*, 10(8):707–710, 1966.
12. M. Li, B. Ma, and L. Wang. On the closest string and substing problems. *Journal of the ACM*, 49(2):157–171, 2002.
13. Ming Li, Bin Ma, and Lusheng Wang. Finding similar regions in many strings. In *Proceedings of the 31st Annual ACM Symposium on Theory of Computing (STOC'99)*, pages 473–482, 1999.
14. Bin Ma. A polynomial time approximation scheme for the closest substring problem. In *CPM*, volume 1848 of *LNCS*, pages 99–107, 2000.
15. D. Maier. The complexity of some problems on subsequences and supersequences. *Journal of the Association for Computing Machinery*, 25:322–336, 1978.
16. L. Marsan and M. F. Sagot. Algorithms for extracting structured motifs using a suffix tree with an application to promoter and regulatory site consensus identification. *J Comput Biol*, 7(3-4):345–62, 2000.
17. C. D. Martinez, A. Juan, and F. Casacuberta. Improving classification using median string and nn rules. In *Spanish Symp. on Pattern Recognition and Image Analysis*, pages 391–395, 2001.
18. C. D. Martinez-Hinarejos, A. Juan, and F. Casacuberta. Use of median string for classification. In *15th International Conference on Pattern Recognition*, volume 2, pages 907–910, september 2000.
19. Pavel Pevzner. *Computational Molecular Biology*. MIT Press, 2000.
20. Krzysztof Pietrzak. On the parameterized complexity of the fixed alphabet shortest common supersequence and longest common subsequence problems. *Journal of Computer and System Sciences*, 2003. to appear.
21. J. S. Sim and K. Park. The consensus string problem for a metric is NP-complete. In R. Raman and J. Simpson, editors, *Proceedings of the 10th Australasian Workshop On Combinatorial Algorithms*, pages 107–113, Perth, WA, Australia, 1999.

22. David J. States and Pankaj Agarwal. Compact encoding strategies for DNA sequence similarity search. In *Proceedings of the Fourth International Conference on Intelligent Systems for Molecular Biology*, pages 211–217. AAAI Press, 1996.
23. Robert A. Wagner and Michael J. Fischer. The string-to-string correction problem. *Journal of the ACM (JACM)*, 21(1):168–173, 1974.
24. L. Wang and D. Gusfield. Improved approximation algorithms for tree alignment. *J. Algorithms*, 25(2):255–273, 1997.

Extracting Approximate Patterns

(Extended Abstract)

Johann Pelfrêne[1,3*], Saïd Abdeddaïm[2], and Joël Alexandre[3]

[1] ExonHit Therapeutics – 65, Boulevard Masséna – 75013 Paris
[2] ABISS,LIFAR – Université de Rouen – 76821 Mont Saint Aignan
[3] ABISS, UMR CNRS 6037 - Université de Rouen – 76821 Mont Saint Aignan
{johann.pelfrene,said.abdeddaim,joel.alexandre}@univ-rouen.fr

Abstract. In a sequence, approximate patterns are exponential in number. In this paper, we present a new notion of basis for the patterns with don't cares occurring in a given text (sequence). The primitive patterns are of interest since their number is lower than previous known definitions (and in a case, sub-linear in the size of the text), and these patterns can be used to extract all the patterns of a text.
We present an incremental algorithm that computes the primitive patterns occurring at least q times in a text of length n, given the N primitive patterns occurring at least $q-1$ times, in time $O(|\Sigma|Nn^2\log^2 n\log\log n)$. In the particular case where $q = 2$, the complexity in time is only $O(|\Sigma|n^2\log^2 n\log\log n)$. We also give an algorithm that decides if a given pattern is primitive in a given text.

1 Introduction

Pattern discovery plays a great role in 'knowledge' extraction [16]. Patterns are repeated structured objects in a text (sequence) or in a set of texts. Extraction algorithms are widely used in bioinformatics for protein motif discovery [6], gene prediction (promotor consensus identification [8], detection of splice sites), repeated patterns [4], sequence alignment [17,15,7,9]. Patterns may be used in other fields such as data compression [2].

We consider a pattern with don't cares: a new symbol is considered to be a joker, matching every other letter of the alphabet. A problem of extracting patterns with don't cares (or simply patterns in the following) from a given text is that they are potentially exponential in number in the size of the sequence from which they are extracted. This is the main problem which arises for an interesting application of patterns: sequence indexing. Since indexing texts has been widely covered in the case of exact patterns [3], due to the exponential number of patterns, indexing approximate patterns had never been considered.

* Supported by the Ministry of Research (France) with the "Bio-ingénierie 2000" program, and the CIFRE convention, in a collaboration between ExonHit Therapeutics and ABISS – http://johann.jalix.org/research

R. Baeza-Yates et al. (Eds.): CPM 2003, LNCS 2676, pp. 328–347, 2003.

A recent work [10] present an idea that tends to reduce the number of patterns, by introducing the maximal non redundant patterns that form a basis for all the patterns appearing at least q times in a sequence (q-patterns).

In [11,12], we have proposed a new notion of primitive q-patterns. The primitive q-patterns form a basis for the patterns that occur at least q times, which is smaller than the maximal non redundant q-patterns basis. The number of primitive 2-patterns is bounded by n [12]. We also proposed an algorithm that computes the primitive 2-patterns in time $O(n^3)$. In this paper, we show the bound for the primitive 2-patterns, and we give a new algorithm, which is both an improvement of the one in [12], and a generalization to every q-pattern. The algorithm computes incrementally the primitive q-patterns given the N primitive (q-1)-patterns in time $O(|\Sigma|Nn^2 \log^2 n \log\log n)$. As there is only one primitive 1-pattern (the text itself), the complexity in time in the case where $q = 2$ is simply $O(|\Sigma|n^2 \log^2 n \log\log n)$. The number of primitive q-patterns is lower than the number of maximal non redundant q-patterns defined by [10]. We also give an algorithm, that decides if a pattern is primitive given the text where it occurs and the considered quorum.

Very recently, Pisanti et al. have independently presented the same basis they call 'tilling motifs', and the algorithm for the case $q = 2$ [13].

First, we will present some general definitions, followed by the definition of a primitive pattern, the new algorithm for extracting these primitive patterns and a new algorithm for deciding whether a given pattern is primitive or not.

2   Definitions

We denote by Σ the alphabet (set of elements called letters) on which are written texts (or sequences or words) $w = w[1]w[2] \cdots w[n] : w[i] \in \Sigma$. The length of w is n, denoted by $|w|$. In the following, we will work on a text t of length $|t| = n$.

In $t = uwv$, the integer $i = |u| + 1$ is a position of the word w in the word t. w is a factor of t at the position i, also denoted by $t[i, |w|]$. If $v = \varepsilon$ (the empty word), w is a suffix of t, denoted $suff(i,t)$.

A pattern is a word written on the alphabet $\Sigma \cup \{\_\}$: the letter $\_$ is to be thought as a don't care (we will name it 'joker' sometimes). Let $\Sigma = \{A, B, C, D\}$, a pattern $m = A\_C$ corresponds to the language $L_m = \{AAC, ABC, ACC, ADC\}$.

The set $pos(m,t)$ is the set of all the positions where the pattern m occurs in the text t. The shift δ of a position list P is a position list such that if $P = \{p_1, p_2, \cdots, p_k\}, P + \delta = \{p + \delta / p \in P, 1 \le p + \delta \le |t|\}$.

A q-pattern is a pattern that occurs at least q times (such that $|pos(m,t)| \ge q$). In the sequence AAAAAXAAAAA, AA_AA_AA is a 2-pattern.

2.1   Maximal Patterns

The idea is to take patterns as long as possible, using the least amount of don't cares as possible. We first create groups of motifs that occur on the same posi-

tions, with eventually a shift : $m_1 \equiv_t m_2 \Leftrightarrow \exists \delta \in \mathbb{Z} : pos(m_1, t) = pos(m_2, t) + \delta$
This is clearly an equivalence relation.

For example, in the text
t=GAC<u>ATG</u>ATTATTGATGCCTAACCGGCTT<u>A</u>TGG<u>ATA</u>GC, the following patterns are in the
same equivalence class : TG__TA, ATG__T, AT___TA et T___TA, ATG__TA.

Next, we define the weight of a pattern m as $w(m)$, the number of solid
letters in the pattern. For example $w(A\_C) = 2$.

Definition 1 (Maximal Pattern) *A maximal pattern m is the shortest pattern (in length) that has the highest weight among the patterns m_i such that
$m_i \equiv_t m$.*

In the previous example, ATG__TA is a maximal pattern.

By this definition, a pattern belongs to the language $\Sigma(\Sigma \cup \{\_\})^* \Sigma \cup \Sigma$.

Lemma 1 (Unicity) *There is a unique maximal pattern in each equivalence
class of $\equiv_t$.*

2.2   Non-redundancy in a Set of Patterns

The notion of redundancy will be used in the negative sense in the following:

Definition 2 (Non-Redundant pattern) *Let M be a set of patterns occurring in a text t. The pattern $m \in M$ is non-redundant in M if there do not exist
patterns $m_i \in M$, $m_i \neq m$, such that $pos(m, t) = \cup_{i>1} pos(m_i, t)$.*

2.3   Non-redundant Maximal q-Patterns

Parida et al. [10] consider the patterns that are non-redundant in the set of all
the maximal q-patterns.

Definition 3 (Irredundant maximal q-pattern [10]) *A maximal q-pattern
m is redundant if there exist maximal q-patterns m_i, $m_i \neq m$, such that
$pos(m, t) = \cup_{i>1} pos(m_i, t)$.*

In the text $t = \text{AAAAAXAAAAA}$, the maximal non-redundant 2-patterns are:
AA, AAA, AAAA, AAAAA, A_AAA_A, AA_AA_AA, AAA_A_AAA, and AAAA__AAAA. The maximal pattern AAA___AAA is redundant because $pos(AAA\_\_\_AAA, t) = \{1, 2, 3\}$ and
$pos(AAAA\_\_AAAA, t) \cup pos(AAA\_A\_AAA, t) = \{1, 2\} \cup \{1, 3\}$.

2.4   Non-redundant Maximal Patterns

As Parida et al. define the non-redundant maximal q-patterns, Apostolico [1]
defines the non-redundant maximal patterns. In that case, it can be remarked
that the definition corresponds to that of Parida in the case $q = 1$.

3   Primitive Patterns

Using the idea of redundancy, we give a new definition, the primitive patterns,
which is more restrictive than the Parida et al. maximal non-redundant definition. The primitive patterns are the maximal non-redundant q-patterns.

3.1   Definition

Definition 4 (Primitive q-pattern [12]) *A maximal q-pattern m is primitive if and only if it is non-redundant in the set of <u>all</u> the q-patterns occurring in the text t.*

In other words, m is primitive if and only if there is no patterns m_i, $m_i \neq m$ and m_i not necessarily maximal, such that $pos(m,t) = \cup_{i>1} pos(m_i, t)$.

In the previous example, there are only five primitive 2-patterns :
AAAAA, A_AAA_A, AA_AA_AA, AAA_A_AAA, and AAAA__AAAA. Using that definition, we may see that the pattern AAA___AAA that is redundant using Parida's definition is, also, not primitive (as we kept the same patterns that make this pattern redundant).

Lemma 2 *Let m be a q-pattern in a text t. If m is primitive, then m is maximal non redundant.*

The proof is straightforward : let m be a primitive q-pattern, then there are no m_i such that $pos(m,t) = \cup_{i>1} pos(m_i, t)$, which means, of course, that there are no maximal q-patterns m'_i such that $pos(m,t) = \cup_{i>1} pos(m'_i, t)$.

We denote the set of all the primitive q-patterns in the text t by $\mathcal{P}(t,q)$.

3.2   Pattern Fusion

The pattern fusion ($\oplus$) is an operation on two patterns that watches the patterns letter by letter, keeping only common letters, and replacing a difference by a don't care. For example, $ABB\_CD \oplus AC\_\_CC = A\_\_\_C\_$.

Definition 5 *Let m_1, m_2 be two patterns. The pattern $m = m_1 \oplus m_2$ is such that for all $i, 1 \leq i \leq min\{|m_1|, |m_2|\}$, $m[i] = m_1[i]$ if $m_1[i] = m_2[i]$, and $m[i] = \_$ otherwise.*

That operation is trivially commutative and associative. The obtained pattern 'tends' to be maximal : all the jokers are present because there is a conflict between two letters on a same position. Some jokers may be present at the beginning of the pattern, or at the end. The operation $\oplus^{max}$ is an extension of the operation $\oplus$, but returns a pattern with no heading or tailing jokers.

Definition 6 *Let m_1, m_2 be two patterns. The pattern $m = m_1 \oplus^{max} m_2$ is such that $m[1], m[|m|] \in \Sigma$, and there exist $u, v \in \{\_\}^*$ such that $umv = m_1 \oplus m_2$.*

These operations are on 'abstract' patterns. We now focus on the text, from which the patterns are extracted, by considering positions of the text, and the corresponding suffix.

Given a set of positions P, we denote by $P_{norm} = P - min\{p_i \in P\} + 1$ the 'normalized' of P (remark that $|P| = |P_{norm}|$, and $1 \in P_{norm}$).

Definition 7 *Let $P = \{p_1, p_2, \cdots p_k\}$ be a set of positions from the text. The pattern $m = \bigoplus_t P$ is such that $m = \oplus_{p_i \in P_{norm}} suff(p_i, t)$.*

And as for the $\oplus$ operation that is extended by $\oplus^{max}$, we define $\bigoplus_t^{max}$ from $\bigoplus_t$:

Definition 8 *Let $P = \{p_1, p_2, \cdots p_k\}$ be a set of positions from the text. The pattern $m = \bigoplus_t^{max} P$ is such that $m[1], m[|m|] \in \Sigma$, and there exist $u, v \in \{\_\}^*$ such that $umv = \bigoplus_t P$*

Theorem 1 *Let t be a text. m is a maximal pattern if and only if there exists a set of positions P, such that $m = \bigoplus_t^{max} P$.*

Proof 1 *Let m be a maximal pattern in t, and $P = pos(m, t)$. Let $m' = \bigoplus_t P$. As $P = P_{norm} + \delta$, m' can be written as $m' = umv$, with $u, v \in \{\_\}^*$. Then is it clear that there exists P such that $m = \bigoplus_t^{max} P$.*

Let P be a set of positions in t, and $m' = \bigoplus_t P$. As m is build using the normalized set of positions, there is no pattern m'', $m'' \equiv_t m'$, such that $w(m'') > w(m')$. Once we remove from m' the tailing and heading jokers, there is still no pattern m'', $m'' \equiv_t m'$, such that $w(m'') > w(m')$. But in that case, the pattern is such that its first and last characters are solid letters, i.e. it is the shortest with the highest weight in its equivalence class : the pattern $m = \bigoplus_t^{max} P$ is maximal.

□

3.3   Characterization

In the following, we give two definitions in order to characterize the primitive patterns.

Covering Pattern

Definition 9 (Covering pattern) *Given a text t and two maximal patterns m and m', m' is said to cover m if and only if there exists $\delta \geq 0$ such that $pos(m', t) + \delta \subseteq pos(m, t)$.*

We will say that a pattern m' covers m on a position p if $p \in pos(m, t)$ and p belongs to the subset of positions covered by m'.

It is clear that the covering relation is reflexive, transitive and antisymmetric (except if $m = m'$), but is not defined for every maximal pattern m and m'.

Hearts

A pattern may have a lot of positions, but only some of them may be sufficient to retrieve the letters composing the pattern. We call heart a subset of the position list of a given pattern such that this subset allows to build the pattern.

Definition 10 (Heart) *Let m be a pattern, and $P \subseteq pos(m, t)$. P is a heart of m, $heart(m, t)$, if and only if $m = \oplus_t^{max} P$.*

A natural consequence is that m must be a maximal pattern, since there exists a set of positions P such that $m = \bigoplus_t^{max} P$ (see theorem 1).

Lemma 3 *Let m be a pattern, and $P \subset pos(m,t)$. If P is a heart of m, then for all $P' \subset pos(m,t)$, $P \cap P' = \emptyset$, we have $P \cup P'$ is a heart of m.*

Characterization

Theorem 2 (Primitive pattern characterization) *Let t be a text, and $q \geq 1$. Let M be the set of all maximal q-patterns of the text t. The five following propositions are equivalent:*

1. *$m \in M$ is a primitive q-pattern;*
2. *there exists a position p^*, $p^* \in pos(m,t)$, such that no pattern $m' \in M, m \neq m'$, covers m on p^*;*
3. *there exists a position p^*, $p^* \in pos(m,t)$, such that $\forall P$, $P \subseteq pos(m,t)$ with $p^* \in P$ and $|P| = q$, we have $m = \bigoplus_t^{max} P$;*
4. *there exists a position p^*, $p^* \in pos(m,t)$, such that $\forall P$, $P \subseteq pos(m,t)$, $p^* \in P$ and $|pos(\bigoplus_t^{max} P, t))| \geq q$, we have $m = \bigoplus_t^{max} P$;*
5. *there do not exist patterns $m_i \in M$, $m_i \neq m$, and associated shifts $\delta_i \geq 0$ such that $pos(m,t) = \cup_{i>1}(pos(m_i,t) + \delta_i)$.*

We call such a position p^* a characteristic position for the primitive pattern m.

Proof 2 $1 \Rightarrow 2$ *Let m be a primitive q-pattern. Then there do not exist patterns m_i (even not maximal) such that $pos(m,t) = \cup_{i>1}pos(m_i,t)$. It means that there exists at least one position not covered by any other pattern.*

$2 \Rightarrow 3$ *Let $p^* \in pos(m,t)$ such that no pattern $m' \in M, m \neq m'$, covers m on p^*. Assume that there exists $P \subseteq pos(m,t)$, $|P| = q$, and $p^* \in P$: it is clear that $m' = \bigoplus_t^{max} P$ is a maximal pattern, with $|pos(m',t)| \geq q$, and m' covers m on p, which is a contradiction.*

$3 \Rightarrow 4$ *Let $p^* \in pos(m,t)$ such that $\forall P \subseteq pos(m,t)$, $p^* \in P$, $|P| = q$, we have $m = \bigoplus_t^{max} P$. Then it is clear that $\forall P \subseteq pos(m,t)$, $p^* \in P$ and $|P| > q$, we have $\bigoplus_t^{max} P = m$ (all the set containing p^*, and having exactly q positions are hearts of m, then adding positions of m cannot form another pattern than m (cf. lemma 3)). If $|P| < q$ and $|pos(\bigoplus_t^{max} P, t)| \geq q$, then as every $P' \subseteq pos(m',t)$, $|P'| = q$, is such that $m = \bigoplus_t^{max} P'$, we have clearly $m' = m$ (cf. 3). Consequently, there exists $p^* \in pos(m,t)$ such that $\forall P \subseteq pos(m,t)$, $p^* \in P$ and $|pos(\bigoplus_t^{max} P, t))| \geq q$, we have $m = \bigoplus_t^{max} P$.*

$4 \Rightarrow 5$ *Assume there exists $p^* \in pos(m,t)$ such that $\forall P \subseteq pos(m,t)$, $p^* \in P$ and $|pos(\bigoplus_t^{max} P, t))| \geq q$, we have $m = \bigoplus_t^{max} P$. It means that no pattern $m' \in M$ is such that $pos(m',t) + \delta \subset pos(m,t)$, with $p - \delta \in pos(m',t)$. Then, clearly, we cannot find maximal patterns m_i and $\delta_i \geq 0$ such that $pos(m,t) = \cup_{i>1}(pos(m_i,t) + \delta_i)$.*

$5 \Rightarrow 1$ *Assume there do not exist patterns $m_i \in M$ and associated shifts $\delta_i \geq 0$ such that $pos(m,t) = \cup_{i>1}(pos(m_i,t) + \delta_i)$. Then, there do not exist non-maximal patterns m'_i such that $pos(m,t) = \cup_{i>1}pos(m'_i,t)$: Assume we can find such patterns m'_i, then for each non-maximal pattern m'_i, there exists a maximal pattern m_i, $m_i \equiv_t m'_i$, and $\delta \geq 0$, which is a contradiction with the hypothesis.*

□

It can be remarked that there can exist more than one characteristic position for a given pattern m.

The example below shows a 2-pattern, `G__TA`, that is not primitive since it is covered by two 2-patterns, `ATG__TA` and `GC_TA`, because we have $pos(G\_\_TA, t) = (pos(ATG\_\_TA, t) + 2) \cup pos(GC\_TA, t)$.

```
GACATGATTATTGATGCCTAACCGGCTTATGGATAGC
   ATG__TA     ATG__TA        ATG__TA
               GC_TA      GC_TA
   G__TA       G__TA      G__TA G__TA
```

3.4 Patterns and Quorum

As the primitive patterns are now defined, it is important to discuss the role of the quorum. Looking for a pattern in a set of sequence leeds naturally to define a thresold that fixes the minimal number of sequence in which the pattern shall be found to be taken into account. However, in a single sequence, the interest of the quorum may not be clear.

In fact, depending on the value of the quorum q, a maximal pattern can be considered as primitive or not.

Lemma 4 (Pattern's life) *Let t be a text, and m a maximal pattern that occurs in q_m positions. There exists an integer $q_p \leq q_m$ such that :*

- *for all quorum $q < q_p$, the q-pattern m is non-primitive,*
- *for all quorum q, $q_p \leq q < q_m$, the q-pattern m is primitive,*
- *and for all quorum q, $q \geq q_m$ the pattern m is not a q-pattern.*

Proof 3 *When we consider a quorum q greater than a position list size, the pattern is clearly not a q-pattern. A maximal pattern is either primitive or not, for a given quorum. It is clear that for $q_p = |pos(m,t)|$, the pattern is primitive. It is also clear that q_p may be lower than $|pos(m,t)|$. We denote by q_p the lowest quorum such that m is primitive. What we may show is that once a pattern becomes primitive for a quorum q_p, it may not be non-primitive for any greater quorum q, $q_p \leq q \leq q_m$.*

The proof is simple: assume there exists $q > q_p$ such that m is not primitive. By theorem 2, all the positions of m are covered by patterns of $\mathcal{P}(t,q)$. As m is primitive for q_p, then by the same proposition, it means there exists a position $p^* \in pos(m,t)$ such that no other pattern covers m on p^*. Let $m' \in \mathcal{P}(t,q)$ cover m on p^* for quorum q and have a look at quorum q_p. As m is primitive, *m' may not be primitive (otherwise, m would not be primitive). But if m' is not primitive for quorum q_p, it means that there exist patterns such that every position is covered, and then a pattern that covers m' should cover m too. This is a contradiction.*

□

This lemma could apply to the non-redundant patterns with minor modifications. It shows in particular the interest of the quorum, since the patterns occuring a small number of times hide the other patterns. As an example, the pattern t (the text itself), hides every other pattern (we recall that $\mathcal{P}(t,1) = \{t\}$ [12]). That is why the first interesting quorum is 2.

3.5 Pattern Basis

Every q-pattern can be generated by the primitive q-patterns. Interesting patterns are, in fact, the ones which are maximal in composition : patterns that are factors of maximal patterns (patterns such that no don't care can be replaced by just one letter). As their number is huge, we cannot keep them in memory, but we want to be able to generate them from the primitive patterns.

A pattern m is generated by $\oplus$ operation.

For a given integer $q > 1$ and a text t, every pattern m, that is maximal in composition, can be written as $m = \oplus_{i>0} m_i[k_i, l_i]$, where m_i are primitive q-patterns. The primitive q-patterns are free, that is to say no primitive q-pattern can be generated using other ones. Then, the primitive q-patterns form a basis for the maximal in composition patterns that occur at least q times in t.

3.6 Number of Primitive Patterns

Given $\delta \geq 1$, $t \oplus^{max} t[\delta]$ is the unique maximal pattern that occurs at least in two positions p and $p + \delta$.

This leads to a bound of the primitive 2-pattern number :

Lemma 5 *The set $\cup_{\delta=2,\cdots|t|} t \oplus^{max} t[\delta]$ contains all the primitive 2-patterns.*

Proof 4 *Let E be the list of the patterns $t \oplus^{max} t[\delta]$. By theorem 1, every pattern in E is maximal.*

Assume now that there exists a pattern m, primitive, but not in E. Let us denote by p and p' two positions of m, and $\delta = |p - p'|$, then there exists $m' = t \oplus^{max} t[\delta]$, $m' \in E$. Then as m' is maximal, there exists $s \geq 0$ such that $p - s$ and $p' - s$ belong to pos(m', t). That means that m' covers m on p and p'.

Then, either $|pos(m,t)| = 2$ and then m is not maximal, or for all $p, p' \in pos(m, t)$, there exists a pattern m_i such that m_i covers m on p and p', which means that m is not primitive.

□

Corollary 1 (Bound on the primitive 2-patterns) *The number of primitive 2-patterns is bounded by $n - 1$.*

In [10], Parida et al. give a linear bound for the non-redundant q-patterns in a text. Since the number of primitive patterns is lower than the number of non-redundant patterns, we could think that the number of primitive patterns are bounded by $3n$. However the following example shows that this is not the case. In the text $A^k X A^k$, with $k = 9$ and $q = 5$, we have 112 primitive 5-patterns, and as k grows, the number of primitive patterns grows too. In section 5, we show there is a strong relation between the number of primitive patterns and C_{n-1}^{q-1}.

Lemma 6 (Upper bound) *Given a text t and a quorum q, there exist at most $C_{|t|-1}^{q-1}$ primitive q-patterns in t.*

Proof 5 *The idea is a consequence of our theorem 2, and particularly the third proposition: a pattern m is q-primitive if and only if there exists a set $P \subseteq pos(m, t)$ such that $|P| = q$ and P contains a characteristic position. Thus, if all the normalized sets containing q positions are build, we have all the q-primitive patterns. There are exactly $C_{|t|-1}^{q-1}$ distinct sets (the normalized sets contain the position 1, we have to choose $q - 1$ among $|t| - 1$).*

□

An other example is given by Pisanti et al. [13] and shows that the $3n$ bound is not correct. In this paper, Pisanti et al. show that there exists an infinite family of strings for which the number of primitive patterns is at least $C_{\frac{n-1}{2}-1}^{q-1}$, and give the same upper bound.

4   Algorithm for Primitive Pattern Extraction

The primitive q-patterns may be obtained as follow :

Lemma 7 (From $\mathcal{P}(t, q-1)$ to $\mathcal{P}(t, q)$) *Let m be a primitive pattern for q, $q \geq 2$ ($m \in \mathcal{P}(t, q)$). If $m \notin \mathcal{P}(t, q-1)$, then there exists a maximal pattern m', $|pos(m', t)| = q - 1$ and a position $p \notin pos(m', t)$ such that $m = m' \oplus^{max} suff(p, t)$.*

Proof 6 *If $m \notin \mathcal{P}(t, q-1)$, then m is not primitive and there exist patterns $m_i \in \mathcal{P}(t, q-1)$ such that m is covered by m_i. As $m \in \mathcal{P}(t, q)$, it means that one position of m that was covered for $\mathcal{P}(t, q-1)$ is not anymore : there exists a pattern m' from $\mathcal{P}(t, q-1)$ that is not taken into account ($|pos(m', t)| = q - 1$). $m' \notin \mathcal{P}(t, q)$, then $|pos(m', t)| < |pos(m, t)|$: there exists a position p such that m is not covered by m' on p. Let $\delta \geq 0$ such that $pos(m', t) + \delta \subset pos(m, t)$. Let P be the set of positions of m such that m is not covered by m' on $p \in P$.*
Let us have a look at the patterns $m_p = m' \oplus^{max} suff(p, t)$, $\forall p \in P - \delta$. Either there exists $p \in P - \delta$ such that $m_p = m$, or $\forall p \in P - \delta, m_p \neq m$ (that is

to say, adding just one position was not enough to build m). Assume no $p \in P - \delta$ is such that $m_p = m$. In this case, there is more than one position in P, and $\forall p \in P - \delta, m_p$ belongs to the basis $\mathcal{P}(t, q)$. It means that m is non-primitive since $pos(m, t) = \cup_{p \in P - \delta}(pos(m_p, t))$.

□

Note that a pattern m can be obtained from a pattern m', $m' \in \mathcal{P}(t, q - 1)$ and $m' \in \mathcal{P}(t, q)$. But in the case where m is primitive, it can be easily shown that there exists a pattern m'', $|pos(m'', t)| = q - 1$, such that m is also obtained by adding a position.

The lemma justifies the incremental approach for computing the basis for a given quorum. Informally speaking, the idea is:

1. generate hearts of possible primitive q-patterns, by taking the primitive (q-1)-patterns that have exactly $q - 1$ positions, and by adding a new position that is not already in the position list of the given pattern.
2. all the primitive (q-1)-patterns, that have more than $q - 1$ positions, are immediately primitive;
3. for each generated pattern, compute its position list : if the size is exactly q, the pattern is primitive, otherwise test the primitivity.

The algorithm EXTRACTPRIMITIVEQPATTERNS $(t, q, \mathcal{P}(t, q - 1))$ computes incrementaly the primitive patterns for a quorum q, given the primitive patterns for the quorum $q - 1$:

```
EXTRACTPRIMITIVEQPATTERNS(text, q, P(t, q − 1))
 1   ▷ Input :
     ▷     the text, the quorum q, the primitive (q-1)-patterns
 2   ▷ Output : result, a trie containing all the primitive q-patterns
 3   result ← NEWTRIE()
 4   for each pattern m ∈ P(t, q − 1), |pos(m, t)| = q − 1 do
 5           for shift ← 1 to |text| do
 6                   if shift ∉ pos(m, t) then
 7                           pattern ← m ⊕^max suff(shift, text)
 8                           PUSH(stack, pattern)
 9                           INSERT(result, pattern)
10   while stack ≠ ∅ do
11           pattern ← POP(stack)
12           pos ← COMPUTEPATTERNPOSITIONS(text, pattern)
13           if TESTPRIMITIVITY(pattern, t, q) then
14                   ▷ The pattern is primitive
15                   Affect pos to the final state
16           else    REMOVE(pattern, result)
17                   ▷ The pattern is not primitive
18   return result
```

In order to store the created patterns, we use a trie, in which the insertion/removal of a word of size n costs $O(n)$. Once the position list is computed, it is associated to the corresponding final state.

Note that for quorum $q = 2$, it is as if we computed all the $t \oplus^{max} suff(\delta, t)$ patterns, for all δ between 2 and $|t|$;

In order to compute the position list of pattern build by a given heart, we need to obtain the pattern itself (the letters composing the pattern): the operation $\oplus^{max}$ is done in time $O(n)$.

Two approaches are possible for testing the primitivity: First, we may use the way patterns are extracted to decide if they are primitive or not (valid for the case $q = 2$, this does not generalize easily). That idea is described in the next section. Second, we may use the way position lists can be encoded as integers, and then use multiplications to discard covered positions. The section 6 describes the algorithm that allows to check the primitivity of a pattern in time $O(|\Sigma|n \log^2 n \log \log n)$.

5   Using the Hearts

Removing non-primitive patterns is done using the hearts that have been studied during the extraction.

We will use a simple counting process for the removal step : a counter $c(p, m)$ is associated to each position for each pattern. Using the theorem 2, and particularly the idea of the third proposition, we have the following accounting scheme:

Lemma 8 *If a pattern m is primitive for a quorum q, a characteristic position p^* can be found in $C^{q-1}_{|pos(m,t)|-1}$ sets P such that $|P| = q$ and $p^* \in P$.*

The number is obtained by fixing the first element of P with p^*, and then we just have to find $q - 1$ positions among the remaining $|pos(m, t)| - 1$.

As for the case $q = 2$, we generate all the possible 2-uplets, and $C^{q-1}_{|pos(m,t)|-1} = |pos(m, t)| - 1$, we just have to count a position in $|pos(m, t)| - 1$ hearts to say that a 2-pattern is primitive. The counters are updated each time a pattern is extracted, in constant time. Still for the case $q = 2$, checking the counters is done in time $O(n)$ for each pattern, that is in time $O(n^2)$ at most: this is the same complexity as the generation step (it does not dominate the time complexity).

In that special case, the algorithm EXTRACTPRIMITIVEQPATTERNS $(t, q, \mathcal{P}(t, q - 1))$ becomes EXTRACTPRIMITIVE2PATTERNS (t). We associate to each pattern the counters, that are automatically updated as a new heart of the pattern is found.

EXTRACTPRIMITIVE2PATTERNS($text$)

```
 1  ▷ Input : the text
 2  ▷ Output : result, a trie containing all the primitive q-patterns
 3  result ← NEWTRIE()
 4  for shift ← 2 to |text| do
 5        pattern, δ ← m ⊕^{max} suff(shift, text)
 6        ▷ δ ≥ 0 is the shift between
           ▷   m ⊕^{max} suff(shift, text) and m ⊕ suff(shift, text)
 7        hearts(m) ← hearts(m) ∪ {1 + δ, shift + δ}
 8        PUSH(stack, pattern)
 9        INSERT(result, pattern)
10  while stack ≠ ∅ do
11        pattern ← POP(stack)
12        pos ← COMPUTEPATTERNPOSITIONS(text, pattern)
13        if MAXCOUNT(hearts(m)) = q − 1 then
14              ▷ The pattern is primitive
15              Affect pos to the final state
16        else    REMOVE(pattern, result)
17              ▷ The pattern is not primitive
18  return result
```

That idea does not generalize trivially, since two possibilities do not appear in the case $q = 2$: we may have patterns that are primitive in $\mathcal{P}(t, q)$ and in $\mathcal{P}(t, q - 1)$ (which complicates the incrementation of the counter); and we may have patterns extracted for $\mathcal{P}(t, q - 1)$, and declared to be non primitive (and then, some of the q-uplets are not formed). In the present version of this extended abstract, we prefered to describe a more simpler algorithm (given in the next section) for doing the primitivity test.

The lemma given above, and the lemma 6 lead to an other approach to compute the primitive q-patterns. It is possible to generate all the sets P, with $1 \in P$ and $|P| = q$. Then for each set, we compute the corresponding maximal pattern, its position list, and we decide whether it is primitive or not. Considering a set, the two steps (computing the maximal pattern, and its position list) can be done in time $O(|\Sigma|n \log^2 n \log \log n)$ using the technique we describe in the following section. The decision of the primitivity is a counting process given by the previous lemma. Such an algorithm would have a complexity in time $O(C_{|t|-1}^{q-1}|\Sigma|n \log^2 n \log \log n)$.

6   Algorithm for Deciding the Primitivity of a Pattern

In this section, we introduce a new operation on sets that can be done using a simple translation of the sets into integers that are to be multiplied. This is the main idea of the Fisher and Paterson's algorithm [5] for computing the position list of a pattern with don't cares in time $O(|\Sigma|n \log^2 n \log \log n)$.

Then, we explain how it is possible to decide, given only a text, a pattern and a quorum, whether the pattern is primitive or not.

6.1 Multisets and Products

A finite set of integers is classically represented with an array of booleans, whose indices belong to a range of integers, where the value at index i is true if and only if the value i is in the set. Integers may also be used to represent a set. For example, the set $E = \{1, 4, 5\}$, can be represented by an array a, in which the indices range from 1 to 6, $a = [true; false; false; true; true; false]$, and an integer 011001.

We introduce the notion of multiset, in which elements are valued. For example, in the multiset $E = \{(1, 5), (4, 2), (5, 1)\}$, the element 4 is valued 2. We use the value as the number of time the element is present in the set. A multiset can also be represented with an array, of integer this time. As an example, we may have $E = \{(1, 5), (4, 2), (5; 1)\}$ represented as $a = [5; 0; 0; 2; 1; 0]$, and the corresponding integer 012005.

The operation TOINTEGER(E, i, j) translates the set E, in which the values are bounded by i and j into an integer where the rightmost coefficient stands for the value of the bound i. The reverse operation, TOMULTISET(k, i), translates the integer k into a multiset with lower bound i.

Adding a constant to a multiset consists in adding the constant to each value: $E + c = \{x + c / x \in E\}$. Realized on an integer, that operation consists in multiplying the integer by b^c, where b is tha basis of the integer.

In the following, we need to produce the union/intersection of a multiset with itself, with a shift of its elements. For example, $E \cup (E + 1) \cup (E + 4) = \{(1, 5), (4, 2), (5, 1)\} \cup \{(2, 5), (5, 2), (6, 1)\} \cup \{(5, 5), (8, 2), (9, 1)\}$. That is $E \cup (E + 1) \cup (E + 4) = \{(1, 5), (2, 5), (4, 2), (5, 8), (6, 1), (8, 2), (9, 1)\}$. We denote by $\odot$ the operation that takes a multiset in the left side, and a set of shifts in the rightside, and computes the union of the multiset with itself at the different shifts: $E \odot \{0, 1, 4\} = E \cup (E + 1) \cup (E + 4)$.

As we use multisets, the result contains both the union and the intersection: the value associated to the element contains the number of time the element is present in the sets. If we set the values to 1 before doing the $\odot$ operation, for each element in the result, we know in how many shifts the element was present. On the example, $E \odot \{0, 1, 4\} = \{(1, 1), (4, 1), (5, 1)\} \odot \{0, 1, 4\} = \{(1, 1), (2, 1), (4, 1), (5, 3), (6, 1), (8, 1), (9, 1)\}$. The element 1 is present in exactly one shift. The element 5 is present in all the three shifts, and thus in the intersection of the three sets.

Since the number of shifts can be equal to the number of elements in the set, this operation can be done with a $O(n^2)$ time algorithm with n the size of the set, and the number of shifts/union to do.

But it is possible to write the operations with integers: $E \odot \{0, 1, 4\}$ can be translated into TOINTEGER$(E, 1, 6)$ + TOINTEGER$(E, 1, 6) \times b^1$ + TOINTEGER$(E, 1, 6) \times b^4 = 011001 + 0110010 + 0110010000 = 0110131011$, where b is the basis used to encode the integers. Which is in fact a simple product of 011001 by 10011.

It is known that a product of integers of k bits can be done with a $O(k \log k)$ time algorithm, using for example the Schönhage-Strassen algorithm [14] (there are numerous fast product algorithms. A technique commonly used is the Fast Fourier Transform).

The basis in which integers are written has to be carrefully chosen, since no retenue shall be propagated in the product. A good choice is to have the basis $\log n$, where n is the maximal integer encoded. Thus, the integer is encoded on $n \log n$ bits, and the product is computed in time $O(n \log^2 n \log \log n)$.

6.2   Fisher and Paterson's Algorithm

The algorithm [5] consists in computing the positions of a pattern with don't cares in a given text t.

In exact pattern matching, the positions of the occurrences of a pattern m in a text t can be seen as an intersection of shifted letter positions: $pos(m, t) = \cap_{i=1}^{|m|}(pos(m[i], t) - i - 1)$. This gives a naive time $O(|m||t|)$ algorithm for computing all the positions of m in t. For example, with $t = AAATAAGTAGAT$, and $m = AA$, we have $pos(m, t) = pos(A) \cap (pos(A) - 1) = \{1, 2, 3, 5, 6, 9, 11\} \cap \{1, 2, 4, 5, 8, 10\} = \{1, 2, 5\}$.

Adapted to pattern matching with don't cares, this could be written as $pos(m, t) = \cap_{i=1, m[i] \neq \_}^{|m|}(pos(m[i], t) - i - 1)$, which is still a quadratic algorithm. With the same text, and the pattern $m = AA\_\_AG$, we have $pos(m, t) = pos(A, t) \cap (pos(A, t) - 1) \cap (pos(A, t) - 4) \cap (pos(G, t) - 5) = \{1, 2, 3, 5, 6, 9, 11\} \cap \{1, 2, 4, 5, 8, 10\} \cap \{1, 2, 5, 6\} \cap \{2, 5\} = \{2, 5\}$.

Therefore, it is possible to do only $|\Sigma|$ intersections, by factorising with the letters of the alphabet : $pos(m, t) = \cap_{c \in \Sigma}(pos(c, t) \odot -(pos(c, m) - 1))$. On the example, we have $pos(AA\_\_AG, t) = (pos(A, t) \odot \{-0, -1, -4\}) \cap (pos(G, t) \odot \{-5\})$.

As $\odot$ can be done by a simple product, we just need to encode the position lists. A recurrent problem is to address: positions in a text are numbered from left to right, and digits in an integer are numbered from right to left. Thus, the notation $\widehat{pos}(m, t)$ will be the set of all the position of m in t, where the numbering of the positions of t begin from the end of the text. As the numbering changes, the negative sign for the shifts disappears. The operation to do becomes $\widehat{pos}(m, t) = \cap_{c \in \Sigma}(\widehat{pos}(c, t) \odot (pos(c, m) - 1))$. We compute $\widehat{pos}(m, t)$ instead of $pos(m, t)$, but it is easy to retrieve $pos(m, t)$.

The algorithm is the following:

1. choose a basis, for example equal to $\log n$ if n is the maximal integer to be represented (the length of the text);
2. for each letter $c \in \Sigma$, compute $P_c = \widehat{pos}(c, t) \odot (pos(c, m) - 1)$ using the integer conversion;
3. do the sum of the Σ products $S = \Sigma_{c \in \Sigma} P_c$;
4. look for every coefficient in S that is equal to the number of fixed letters in the pattern to find the positions.

The time complexity is clearly $O(|\Sigma| n \log^2 n \log \log n)$ (dominated by the step 2).

For example, with the text $t = AAATAAGTAGAT$, and the pattern $m = AA\_\_AG$: $\widehat{pos}(m,t) = (\widehat{pos}(A,t) \odot (pos(A,m)-1)) \cap (\widehat{pos}(G,t) \odot (pos(G,m)-1))$. We do the products $111011001010 \times 10011 = 1111331131111110$ and $100100 \times 100000 = 10010000000$. Then we do the sum, obtaining 1111341141111110. There are two values equal to 4, the number of fixed letters in m, which are at positions 2 and 5 (see figure 1).

t	A	A	A	T	A	A	G	T	A	G	A	T
P_A	1	1	1	3	3	1	1	3	1	1	1	1	1	0
P_G				1	0	0	1	0	0	0	0	0	0	0
S	1	1	1	3	4	1	1	4	1	1	1	1	1	0
$pos(m,t)$					1			1						

Fig. 1. The positions of $m = AA\_\_AG$ in $t = AAATAAGTAGAT$.

6.3   Decision Algorithm

We propose in the following a method that takes a text t, a maximal pattern m and an integer q for the quorum, and decides whether the pattern is primitive or not. With minor changes it is possible to test for maximality, or non-redundancy.

A pattern can be extended with jokers at the both sides. Each joker, inside and outside the pattern, can be replaced by a letter in each occurrence of the pattern in the text. The idea is to count, for each joker, how many times each letter of the alphabet replaces the joker in the text. If a joker can be replaced more than q times by a same letter c, it means that there exists a pattern that has all the fixed letters of the pattern of interest, plus one, the letter c : it is a covering pattern. In that case, the corresponding positions of m are covered, and thus have to be marked. If all the positions of m in t are marked, it means that all the positions are covered, and then m is not primitive (with theorem 2). Informally speaking, the algorithm consists in two steps:

- step 1: compute the counters $count_c(z,m,t)$. $\forall z \in [-|t|;|t|]$, $count_c(z,m,t) = |\{p/p \in pos(m,t)+z, t[p]=c\}|$;
- step 2: compute the covered positions. $p \in pos(m,t)$ is covered if there exists z and a letter $c \in \Sigma$ such that $count_c(z,m,t) \geq q$.

In figure 2, we see the text $t = AAATAAGTAGAT$ aligned with it self three times, using each occurrence of the pattern $m = A\_\_T$ as anchor for the alignment. The counters are represented on the same figure. For example, we may remark that the column standing for the A in the pattern counts three times that letter. We also see that the first joker inside the pattern can be replaced twice by a A: there exists a covering pattern (if we consider $q = 2$). There is not enough information to tell what pattern it is, but there is one that look like

z	-8	-7	-6	-5	-4	-3	-2	-1	0	1	2	3	4	5	6	7	8	9	10	11
m									A	_	_	T								
$p_1 = 1$									A	A	A	T	A	A	G	T	A	G	A	T
$p_2 = 5$					A	A	A	T	A	A	G	T	A	G	A	T				
$p_3 = 9$	A	A	A	T	A	A	G	T	A	G	A	T								
$count_A$	1	1	1	0	2	2	1	0	3	2	2	0	2	1	1	0	1	0	1	0
$count_T$	0	0	0	1	0	0	0	2	0	0	0	3	0	0	0	2	0	0	0	1
$count_G$	0	0	0	0	0	0	1	0	0	1	1	0	0	1	1	0	0	1	0	0

Fig. 2. All the occurrences of the pattern aligned, with the associated counters.

AA_T. It is not interesting to retrieve the pattern, but just to mark the positions p_1 and p_2.

If $q = 2$, the column -3 allows to mark positions p_2 and p_3, and the column 7 let positions p_1 and p_2 marked. As all the positions are marked, the pattern is not primitive: we have $pos(A\_\_T, t) = (pos(A\_\_A\_\_T, t) + 3) \cup (pos(A\_\_T\_\_\_T, t))$. (we do not show here all the possible covering patterns, since two are sufficient to show that m is not primitive.)

It is worth to remark that this sketch of algorithm takes a maximal pattern in input (all the patterns we have to test are maximal), but we could easily check if the pattern is maximal because if a joker can be replaced by a single letter in each position of the pattern, it is clear that the pattern is not maximal.

Step 1: Compute the Counters

First, we will show that if a counter for the letter c is higher than the quorum, it means that there exists a covering pattern. Then, we show how to compute efficiently the counters.

Proposition 1 *If there is a value in the counter of the letter c that higher than q and lower than $pos(m, t)$, there exists a covering pattern for the positions where, in the text, the corresponding joker is replaced by the letter c.*

The proof is, in fact, trivial: in such a case, a pattern m' would occur at some of the positions of m (with eventually a shift), and thus having all the fixed letters of m, plus one other fixed letter. We have clearly $pos(m', t) + \delta \subsetneq pos(m, t)$ and $|pos(m', t)| \geq q$.

Computing the counters is done by $count_c = \widehat{pos}(c, t) \odot (pos(m, t) - 1)$, for a letter $c \in \Sigma$.

Step 2: Compute the Covered Positions

On the example (fig. 2), we see that the first joker in the pattern $A\_\_T$ can be replaced twice by a A. In order to discover what positions are covered, we could build the pattern AA_T, and look for its positions. As the second joker is also replaced twice by a A, we could build A_AT, and look for its positions, and so on for each joker that can be replace by a A. This method would give all the covered positions, but with a bad time algorithm.

Instead of computing the positions of the pattern AA_T, we can build the pattern $\_A\_$. The positions we find are at least the ones that AA_T has, and others. We can do so for each pattern (see fig 3), and then build an aggregate pattern, m_A.

$z = -4$	A _ _ _ A _ _ T
$z = -3$	_ A _ _ A _ _ T
$z = 1$	A A _ T
$z = 2$	A _ A T
$z = 4$	A _ _ _ T A
m_A	A A _ _ _ A A _ A

Fig. 3. the pattern m_A

Then it is possible to apply the algorithm in order to find its positions, $\widehat{pos}(A, t) \odot (pos(A, m_A) - 1)$, but using a different interpretation of the result. In the normal case, we look for values equal to the number of fixed letters, since we want all the letters of the pattern to appear. In our case, we just want that there exists one letter, thus, we are just looking for non null values.

On the example, we build the pattern $m_A = AA\_\_\_AA\_A$, and the pattern $m_T = \_\_\_T_____T$ relatively to $A\_\_T$ (see fig. 4). Note that as the positions of m_A cover the positions of m with a shift equal to 4, we have to divide the integer by b^4, that is remove the last four values (or in terms of multiset, we have to add the value 4 to each element).

m	A _ _ T
m_A	A A _ _ _ A A _ A
m_T	_ _ _ T _ _ _ _ _ _ T

Fig. 4. Step 2: building patterns

The result of the products are presented in figure 5. Note that to simulate the division, we add four 0s to the integer representing the positions of m (the whole integers obtained after the products are given here).

In figure 5, the underlined values stands for positions of m, and show that all of them are covered.

For each counter, we can easily know what letters belong to covering patterns. We denote by $CovPat(m, c, count_c)$ the operation that build the pattern m_c, in which all the jokers that can be replaced by the letter c are really replaced by it, and the fixed letters in m are removed. We denote by k_c the maximal distance

$pos(\_\_\_\_m, t)$	0 0 0 0 $\underline{1}$ 0 0 0 $\underline{1}$ 0 0 0 $\underline{1}$ 0 0 0 0 0 0 0
$\widehat{pos}(A,t) \odot pos(A, m_A) - 1$	1 1 2 2 $\underline{3}$ 2 1 3 $\underline{4}$ 2 3 2 $\underline{3}$ 2 0 1 1 1 1 0
$\widehat{pos}(T,t) \odot pos(T, m_T) - 1$	1 0 0 0 $\underline{1}$ 0 0 0 $\underline{2}$ 0 0 0 $\underline{1}$ 0 0 0 1 0 0 0

Fig. 5. Step 2: products and covered positions

from the first letter of the pattern to the first joker replaced by a letter c, such that the replaced joker is present *before* the first letter of the pattern, $k_c = 0$ if no joker before the first letter of m is replaced.

Such patterns m_c may not occur in the text, but when computing $P_c = (\widehat{pos}(c,t) \odot pos(c, m_c) - 1) + k_c$, if an element is present in the set, whatever is the corresponding value, it means that there is a covering pattern at the corresponding position. The operation $\text{ToSet}(E)$ converts a multiset into a set.

Proposition 2 *Let m be a maximal pattern in a text t. We consider the letters $c \in \Sigma$ such that there exists z with $count_c(z, m, t) \geq q$. Let m_c be the patterns $m_c = CovPat(m, c, count_c)$, and $k \geq 0$ the shifts of m_c relatively to m. All the covered positions of m are in the set:*

$$\text{ToSet}(\widehat{pos}(m,t)) \cap \text{ToSet}(\bigcup_{c \in \Sigma} (\widehat{pos}(c,t) \odot (pos(c, m_c) - 1)) + k_c)$$

Algorithm and Complexity

Finally, the algorithm $\text{IsPrimitive}(m, t, q)$, where m is a maximal pattern in the text t, is the following:

1. Step 1: for $c \in \Sigma$, compute $count_c = \widehat{pos}(c, t) \odot (pos(m, t) - 1)$;
2. Step 2:
   a) for $c \in \Sigma$, compute $m_c = CovPat(m, c, count_c)$, and k_c;
   b) for $c \in \Sigma$, compute $E_c = \widehat{pos}(c, t) \odot (pos(c, m_c) - 1)$;
   c) convert the multisets E_c in sets S_c (forget the values associated to each element);
   d) compute $I = \widehat{pos}(m, t) \cap (\cup_{c \in \Sigma}(S_c + k_c))$;
   e) if $|I| = |pos(m, t)|$, the pattern is not primitive, it is primitive otherwise.

Theorem 3 (Complexity) *The complexity in time for deciding whether a pattern is primitive or not in a text t of size n with quorum q is $O(|\Sigma| n \log^2 n \log \log n)$.*

Proof 7 *The first step consists in Σ products. The step (2.a) allows to build patterns of size $O(n)$ in linear time. The step (2.b) computes the product of each pattern. Finally, as the $|\Sigma| + 1$ intersections are computed using additions of integers, it is clear that the complexity is dominated by the $|\Sigma|$ products.*

$\square$

A slightly modified version of this algorithm can be used to find the letters composing a pattern, given a text and a set of positions : once the position list if computed, we can either keep the letters of the pattern, or the position list (or just an heart of the pattern, to reduce memory space), since one can be computed from the other.

7   Complexity

Theorem 4 *For a quorum $q \geq 2$, the algorithm* EXTRACTPRIMITIVEQPAT-TERNS *for extracting the primitive q-patterns from a text t of size n takes a time $O(N|\Sigma|n^2 \log^2 n \log \log n)$, where N is the number of primitive (q-1)-patterns.*

Proof 8 *The first step of* EXTRACTPRIMITIVEQPATTERNS *for generating the pattern generation takes $O(Nn^2)$ time as for each if the N patterns that are $(q-1)$ primitive, we study the n positions, and we need to compute the letters compising the pattern, and to insert it into our data structure.*

The second step, which consists in computing the set of all the positions of all the extracted patterns, dominates the overall complexity. Computing the position list of each extracted pattern takes $O(|\Sigma|n \log^2 n \log \log n)$ time using Fischer and Paterson's [5] algorithm. As we have $O(Nn)$ patterns to study, this step takes $O(N|\Sigma|n^2 \log^2 n \log \log n)$.

The last step consists in deciding whether a pattern is primitive or not: this can be done in time $O(|\Sigma|n \log^2 n \log \log n)$. This has to be done for each pattern having more than q positions, that is $O(N|\Sigma|n^2 \log^2 n \log \log n)$.

□

Corollary 2 *For the case where $q = 2$, the algorithm* EXTRACTPRIMI-TIVE2PATTERNS *for extracting the primitive 2-patterns from a text t of size n takes a time $O(|\Sigma|n^2 \log^2 n \log \log n)$.*

8   Conclusion and Perspectives

We give an algorithm to compute incrementally the primitive patterns in time $O(N|\Sigma|n^2 \log^2 n \log \log n)$, with N the number of primitive patterns for the previous quorum. As we seen, the number N of primitive patterns, at worst, is not linear (a bound is given in [13]). Therefore, it could be useful to study the average case, and the properties of the texts for which there is not such an explosion. Preliminary tests show that this approach could be of interest.

Acknowledgments. The authors would like to thank the reviewers for their remarks and corrections. We would like to thank the bio-informatics team at Exonhit. Thanks are given also to Laxmi Parida, who answered to a lot of questions and requests for examples, and to Gérard Duchamp for the useful talk we had about the product.

References

1. A. Apostolico. Pattern discovery and the algorithmics of surprise. In P. Frasconi and R. Shamir, editors, *Proceedings of the NATO ASI on Artificial Intelligence and Heuristic Methods for Bioinformatics*, October 2001.
2. A. Apostolico and L. Parida. Compression and the wheel of fortune. In *Proceedings of Data Compression Conference (DCC)*, Snowbird, Utah, March 2003.
3. M. Crochemore, C. Hancart, and T. Lecroq. *Algorithmique du Texte*. Vuibert, 2001.
4. M. Crochemore and M.-F. Sagot. Motifs in sequences: localization and extraction. In A. Konopka and al., editors, *Handbook of Computational Chemistry*. Marcel Dekker, Inc, 2001.
5. M. J. Fischer and M. S. Paterson. String matching and other products. *SIAM-AMS proceedings*, pages 113–125, 1974.
6. I. Jonassen, J. Collins, and D. Higgins. Finding flexible Patterns in unaligned protein sequences. *Protein Science*, pages 1587–1595, 1995.
7. C. Lawrence, S. Altschul, M. Boguski, J. Liu, A. Neuwald, and J. Wootton. *Science*, volume 262, page 208. 1993.
8. L. Marsan and M.-F. Sagot. Extracting structured motifs using a suffix tree – Algorithms and application to consensus identification. In S. Minoru and R. Shamir, editors, *Proceedings of the 4th Annual International Conference on Computational Molecular Biology (RECOMB)*, Tokyo, Japan, 2000. ACM Press.
9. B. Morgenstern, A. Dress, and T. Werner. Multiple DNA and protein sequence alignment based on segment-to-segment comparison. In *Proceedings of the National Academy of Sciences USA*, pages 1209–12103, 1996.
10. L. Parida, I. Rigoutsos, A. Floratos, D. Platt, and Y. Gao. Pattern discovery on character sets and real-valued data: linear bound on irredundant motifs and an efficient polynomial time algorithm. In *Proceedings of the 11th Symposium on Discrete Algorithms*, pages 297–308, 2000.
11. J. Pelfrêne. Indexation de motifs approches. *Rapport de DÉA*, September 2000.
12. J. Pelfrêne, S. Abdeddaïm, and J. Alexandre. Un algorithme d'indexation de motifs approchés (poster and short talk). In *Journées Ouvertes Biologie Informatique Mathématiques*, Saint-Malo, pages 263–264, June 2002.
13. N. Pisanti, M. Crochemore, R. Grossi, and M.-F. Sagot. Bases of motifs for generating repeated patterns with don't cares. Technical report, Università di Pisa, February 2003.
14. A. Schönhage and V. Strassen. Schnelle Multiplikation grosser Zahlen. *Computer (Arch. Elektron. Rechnen)*,7:281–292, 1971.
15. G. Schuler, S. Altschul, and D. Lipman. *Proteins: Structure, Function, and Genetics*, volume 9, pages 180–190. 1991.
16. J. Wang, B. Shapiro, and D. Shasha. *Pattern Discovery in Biomolecular Data*. Oxford University Press, 1999.
17. M. Waterman and R. Jones. *Methods in enzymology*, page 221. Academic Press, London, 1990.

A Fully Linear-Time Approximation Algorithm for Grammar-Based Compression

Hiroshi Sakamoto

Department of Informatics, Kyushu University
Fukuoka 812-8581, Japan
hiroshi@i.kyushu-u.ac.jp

Abstract. A linear-time approximation algorithm for the *grammar-based compression*, which is an optimization problem to minimize the size of a context-free grammar deriving a given string, is presented. For each string of length n over unbounded alphabet, the algorithm guarantees $O(\log^2 n)$ approximation ratio without suffix tree and runs in $O(n)$ time in the sense of randomized model.

1 Introduction

The optimization problem handled in this paper is the *grammar-based compression* to find a *small* context-free grammar which generates an input string. This problem is known to be NP-hard and not approximable within a constant factor [2,11,16]. Moreover, due to a relation with an algebraic problem [8], it is unlikely that an $O(\log n / \log \log n)$-approximation algorithm is found. The framework of the grammar-based compression can uniformly describe the dictionary-based coding schemes which are widely presented for real world text compression. For example, LZ78 [20] (including LZW [17]) and BISECTION [7] encodings are considered as algorithms to find a straight-line program, which is a very restricted CFG. Lehman and Shelat [11] also showed the lower bounds of the approximation ratio of almost dictionary-based encodings to the smallest CFG. However, these lower bounds are relatively large to $O(\log n)$ ratio.

The first polynomial-time $\log n$-approximation algorithms were produced by Charikar, Lehman, Liu, et al. [3], and Rytter [15], independently. In particular, the latter algorithm based on suffix tree construction is attracted by its simplicity in view of the implementation for large text data.

Rytter's algorithm runs in $O(n \log |\Sigma|)$ time for unbounded alphabet Σ and in linear time for any constant alphabet. This gap is caused by the construction of a suffix tree in the algorithm to retrieve whether a string appear in the input in linear time. The edges labeled by characters leaving a node are lexicographically sorted. Thus, in this representation, sorting is a lower bound for suffix tree construction in case of unbounded alphabets.

In this paper we propose another method for the grammar-based compression without suffix tree construction. The starting point of this study is RE-PAIR encoding by Larsson and Moffat [9] which recursively replaces all pairs like ab in

R. Baeza-Yates et al. (Eds.): CPM 2003, LNCS 2676, pp. 348–360, 2003.

an input string according to the frequency. This encoding scheme is also included in the framework of the grammar-based compression, while only the lower bound $O(\sqrt{\log n})$ of its approximation ratio is known [10]. Its nontrivial upper bound is still an important open problem.

As was shown in [9], RE-PAIR encoding is simple and space-efficient. Thus, we develop this strategy in our approximation algorithm. Our algorithm is not RE-PAIR in itself but is partially based on the recursive replacement of pairs. The main idea is explained as follows. Assume that a string contains nonoverlapping intervals X and Y which represent a same substring. The aim of our algorithm is to compress them into some intervals which have a common substring as long as possible. More precisely, X and Y are compressed into $X' = \alpha\beta\gamma$ and $Y' = \alpha'\beta\gamma'$ such that the length of disagreement string $\alpha\gamma$ is bounded by a constant. If this encoding is realized for all such intervals, then the input is expected to be compressed in a sufficiently short string by successively applying this process to the resulting intervals X' and Y'.

In case X and Y are partitioned by some delimiter characters on their both sides, it is easy to compress them into a same string by RE-PAIR like algorithm. However X and possibly Y are generally overlapping with other intervals which represent other different substrings. The main goal is to construct an algorithm for the required encoding without suffix tree.

We call our algorithm LEVELWISE-REPAIR since the replacement of pairs is restricted by the level in which the pairs exist. More precisely, if an interval is replaced by a nonterminal, any interval containing the nonterminal is not replaced again within the same loop.

In this paper, we assume a standard RAM model for reading any $O(\log n)$ bit integer in constant-time. We additionally assume three data structures, *doubly-linked list, hash table,* and *priority queue* to gain constant-time access to any occurrence of a pair ab. The construction of such data structures for input string is basically presented in [9]. Using these structures, the running time of LEVELWISE-REPAIR is reduced to linear-time for unbounded alphabets in the sense of randomized computation.

The approximation ratio is obtained from the comparison with the size of the output grammar and the size of the *LZ-factorization* [19] for an input string. Since a logarithmic relation between *LZ*-factorizations and minimum grammars was already shown in [15], a polylogarithmic approximation ratio of LEVELWISE-REPAIR is concluded.

2   Preliminaries

We assume the following standard notations and definitions concerned with strings. An *alphabet* is a finite set of symbols. Let A be an alphabet. The set of all strings of length i over A is denoted by A^i and the length of a string w is denoted by $|w|$.

The ith symbol of w is denoted by $w[i]$ and $w[i,j]$ denotes the interval from $w[i]$ to $w[j]$. If $w[i,j]$ and $w[i',j']$ represent a same substring, it is denoted by

$w[i, j] = w[i', j']$. An expression $\sharp(\alpha, \beta)$ denotes the number of occurrences of a string α in a string β. For example, $\sharp(ab, ababbba) = 3$ and $\sharp(bab, ababbba) = 2$.

A substring $w[i, j] = x^k$ for a symbol x is called a *repetition*. In particular, in case $w[i-1], w[j+1] \neq x$, we may write $w[i, j] = x^+$ if we do not need to specify the length k. Intervals $w[i, j]$ and $w[i', j']$ ($i < i'$) are called to be *overlapping* if $i' \leq j < j'$ and to be *independent* if $j < i'$.

A substring ab of length two in a string w is called a *pair* in w. Similarly, an interval $w[i, i+1]$ is called a *segment* of ab if $w[i, i+1] = ab$. For a segment $w[i, i+1]$, two segments $w[i-1, i]$ and $w[i+1, i+2]$ are called the *left* and *right* segments of $w[i, j]$, respectively.

A *context-free grammar* (CFG) is a 4-tuple $G = (\Sigma, N, P, S)$, where Σ and N are disjoint alphabets, P is a finite set of binary relations called *production rules* between N and the set of strings over $\Sigma \cup N$, and $S \in N$ is called the *start symbol*. Elements in N are called *nonterminals*. A production rule in P represents a replacement rule written by $A \to B_1 \cdots B_k$ for some $A \in N$ and $B_i \in \Sigma \cup N$.

We assume that any CFG considered in this paper is *deterministic*, that is, for each $A \in N$, exactly one production $A \to \alpha$ exists in P. Thus, the language $L(G)$ defined by G is a singleton set.

The *size* of G, denoted by $|G|$, is the total length of right sides of all production rules. In particular, we obtain $|G| = 2|N|$ in case of Chomsky normal form. The grammar-based compression problem is then defined as follows.

Problem. (Grammar-Based Compression)
INSTANCE: A string w
SOLUTION: A deterministic CFG G for w
MEASURE: The size $|G|$ of G

3   Approximation Algorithm

We present the approximation algorithm, named by LEVELWISE-REPAIR, for the grammar-based compression in Fig 1. This algorithm calls two procedures *repetition(,)* in Fig. 2 and *arrangement(,)* in Fig. 3.

3.1   Outline of the Algorithm

We begin with the outline of the procedures. *repetition(,)* receives a string w and replaces all repetitions $w[i, j] = x^+$ of length k in w by the nonterminal $A_{(x,k)}$. The production $A_{(x,k)} \to BC$ is then added to P and nonterminals B, C are defined recursively as $B = C = A_{(x,k/2)}$ provided k is even and $BC = A_{(x,k-1)}x$ otherwise. Consequently the interval $w[i, j]$ is replaced by a nonterminal corresponding to the root of a binary derivation tree for x^+ of depth at most $O(\log k)$.

Next *arrangement(,)* receives w and counts the frequency of all pairs in w. All such pairs are stored in a priority queue in the frequent order, where two different pairs in a same frequency are ordered by FIFO manner. This queue is indicated by *list* in line 3 of Fig. 3 and this order is fixed until all elements are popped according to the following process.

In *arrangement*(,), a dictionary D is initialized and a unique index $id = \{d_1, d_2\}$ is created for each pair ab. The aim of the procedure is, for each segment $w[i, i + 1] = ab$, to decide whether $w[i, i + 1]$ is added to D and assign d_1 or d_2 to $w[i, i+1]$ by a certain decision rule. The decision rule is explained in the next subsection. All segments in D are finally replaced by appropriate nonterminals.

After all pairs are popped from the priority queue, the algorithm actually replaces all segments in D; If $w[i, i + 1] = w[i', i' + 1] = ab$ and they are in D, then they are replaced a same nonterminal. The resulting string is then given to *repetition*(,) as a next input and the algorithm continues this process until there is no more pair ab such that $\sharp(ab, w) \geq 2$. An example run of *repetition*(,) is shown in Example 1.

Example 1. Let us consider the sample string $-a^7 - a^4 - a^5-$, where $a \in \Sigma$ and each $-$ is a symbol not equal to a. *repetition*(,) replaces all repetitions as follows. The first repetition a^7 is replaced by A_7 and a^7 is recursively parsed by the production rules $A_7 \to A_6 a$, $A_6 \to A_3^2$, $A_3 \to A_2 a$, and $A_2 \to a^2$. Similarly, a^4, a^5 are replaced by A_4, A_5 and $A_4 \to A_2^2, A_5 \to A_4 a$ are additionally defined.

By contrast, we must prepare several notions to explain the details of the other procedure *arrangement*(,). These notions are given in the next subsection.

```
1  Algorithm LEVELWISE-REPAIR(w)
2      initialize P = N = ∅;
3      while( ∃ab[♯(ab, w) ≥ 2] ) do{
4          P ← repetition(w, N);        (replacing all repetitions)
5          P ← arrangement(w, N);       (replacing frequent pairs)
6      }
7      if(|w| = 1) return P;
8      else return P ∪ {S → w};
9  end.
```

notation: $X \leftarrow Y$ denotes the addition of the set Y to X.

Fig. 1. The approximation algorithm for Grammar-Based Compression. An input is a string and an output is a set of production rule of a deterministic CFG for w.

3.2   Decision Rule for Assignment

In *arrangement*(,), a unique index $id = \{d_1, d_2\}$ of integers is set for each pair ab and segments of ab are assigned d_1 or d_2 as well as some of them are added to a

```
1  procedure repetition(w, N)
2      initialize P = ∅;
3      while( ∃w[i, i + j] = a⁺ )do{
4          replace w[i, i + j] by A₍ₐ,ⱼ₎;
5          P ← {A₍ₐ,ⱼ₎ → BC} and N ← {A₍ₐ,ⱼ₎, B, C} recursively;
6      }
7      return P;
8  end.
```

$$BC = \begin{cases} A^2_{(a,j/2)}, & \text{if } j \geq 4 \text{ is even} \\ A_{(a,j-1)} \cdot a, & \text{if } j \geq 3 \text{ is odd} \\ a^2, & \text{otherwise} \end{cases}$$

Fig. 2. The procedure *repetition*(,). An input is a string and a current alphabet. An output is a set of production rules deriving all repetitions in the input.

current dictionary D. The index assigned for a segment and a current D are the factors for deciding whether the segment is finally replaced by a nonterminal. This process is explained by the following notions.

Definition 1. *A set of segments of a pair ab is called a group if all segments are assigned by the index $id = \{d_1, d_2\}$ for ab. A group is divided into at most two disjoint subsets whose elements are assigned only d_1 and d_2, respectively. Such subsets are said to be subgroups of the group. Moreover subgroups are categorized into the following three types depending on a current dictionary D. A subgroup is said to be selected if all segments in the subgroup are in D, unselected if all segments in the subgroup are not in D, and irregular otherwise.*

Definition 2. *A segment is called free if the left and right segments of it are not assigned, and is called left-fixed (right-fixed) if only the left (right) segment of it is assigned, respectively.*

The assignment for segments are decided as the following manner. Let ab be a current pair popped from the priority queue. At first, the sets F_{ab}, L_{ab}, and R_{ab} are computed based on the set C_{ab} of all segments $w[i, i+1] = ab$.

F_{ab} is the set of free segments, that is, the both sides of each $w[i, i+1] \in F_{ab}$ are not assigned. In this case all segments in F_{ab} are assigned d_1 and added to the dictionary D.

L_{ab} is the set of the left-fixed segments, that is, the left side of each segment in L_{ab} is assigned and the other is not. Let L be the set of such assigned segments.

Since all segments in L are assigned, L is divided into some disjoint groups like $L = L_1 \cup L_2 \cdots \cup L_k$ such that L_ℓ is assigned by a unique $id = \{d_1, d_2\}$ and each group L_ℓ consists of some subgroups. Given L_{ab} and L, *arrangement*(,)

```
1    procedure arrangement(w, N)
2        initialize D = ∅;
3        make list: the frequency list of all pairs in w;
4        while( list is not empty )do{
5            pop the top pair ab in list;
6            set the unique id = {d₁, d₂} for ab;
7            compute the following sets based on C_ab = {w[i, i + 1] = ab}:
8                F_ab = {s ∈ C_ab | s is free },
9                L_ab = {s ∈ C_ab | s is left-fixed },
10               R_ab = {s ∈ C_ab | s is right-fixed };
11           D ← assignment(F_ab) ∪ assignment(L_ab) ∪ assignment(R_ab);
12       }
13       replace all segments in D by appropriate nonterminals;
14       return the set P of production rules computed by D and update N by P;
15   end.
```

```
16       subprocedure assignment(X)
17           in case( X = F_ab ){ D ← F_ab and set id(s) = d₁ for all s ∈ F_ab;}
18           in case( X = L_ab   (resp. X = R_ab))do{
19               compute the set Y of all left (resp. right) segments of X;
20               for each( yx ∈ YX   (resp. xy ∈ XY ))do{
21                   in case (1): y is a member of an irregular subgroup,
22                       set id(x) = d₂;
23                   in case (2): y is a member of an unselected subgroup,
24                       set id(x) = d₁ and D ← {x};
25                   in case (3): y is a member of a selected subgroup,
26                       if the group has an irregular subgroup,
27                           set id(x) = d₂;
28                       else if the group has an unselected subgroup,
29                           set id(x) = d₁;
30                       else if Y contains an irregular subgroup,
31                           set id(x) = d₂;
32                       else set id(x) = d₁;
33               }
34           }
35           return D;
36       end.
```

notation: $yx \in YX$ in line 20 denotes $y = w[i - 1, i] \in Y$ and $x = w[i, i + 1] \in X$.

Fig. 3. The procedure $arrangement(,)$ and its subprocedure $assignment()$. An input is a string and a current alphabet. The output is a set of production rules.

finds all $w[i - 1, i] \in L$ belonging to an unselected subgroup and then adds their all right segments $w[i, i + 1] \in L_{ab}$ to D.

Next the assignments for L_{ab} are decided as follows. Each $w[i, i+1] \in L_{ab}$ is assigned d_2 if the left segment $w[i-1, i] \in L$ is in an irregular subgroup and each $w[i, i+1] \in L_{ab}$ is assigned d_1 if $w[i-1, i] \in L$ is in an unselected subgroup.

Any remained segment is $w[i, i+1] \in L_{ab}$ such that its left segment $w[i-1, i] \in L$ belongs to a selected subgroups of a group. In this case, the procedure checks whether the group also contains an unselected or irregular subgroup. If it contains an irregular subgroup, $w[i, i+1]$ is assigned d_2, else if it contains an unselected subgroup, $w[i, i+1]$ is assigned d_1, and otherwise, the procedure checks whether there is other group in containing an irregular subgroup; If so, $w[i, i+1]$ is assigned d_2 and else $w[i, i+1]$ is assigned d_1. Consequently, a single group for L_{ab} assigned d_1 or d_2 is constructed from k groups $L = L_1 \cup L_2 \cdots \cup L_k$.

The case of R_{ab} is symmetric, that is, by the set R of the right assigned segments for R_{ab}, the assignments for R_{ab} and the dictionary are decided by R. The segments in $C_{ab} \setminus F_{ab} \cup L_{ab} \cup R_{ab}$ are skipped since both sides of any segments in C'_{ab} are already assigned.

Example 2. Fig. 4 illustrates how the set X of left-fixed segments are assigned from their left segments. Y is the set of left segments of the set X of current segments. Assume that X is left-fixed, e.g. all segments in Y are already assigned and classified into the groups p and q. The group g for X is obtained from group p and q. The indices of group p, q, and g are denoted by $id = \{1, 2\}$, $\{3\}$, $\{4, 5\}$, respectively. The mark '+' denotes that the marked segments are added to a current dictionary D. On the first figure of (1), since Y contains an *unselected* subgroup, then the corresponding segments in X are added to D. On the other hand, the assignments for X is decided as follows. There are an *irregular* subgroup and an *unselected* subgroup, then the corresponding segments in X are assigned the different indices 4 and 5, respectively. Finally, the remained segments are members of a *selected* subgroup. The case of a selected subgroup is synchronized with the assignment for a subgroup contained in the same group. In this case the selected subgroup is contained in group p. This group also contains the *irregular* subgroup. Thus, the assignment for the corresponding segments in X are are assigned 5. Fig. 4 shows only the case that q consists of a single subgroup; nevertheless this figure is sufficiently general since the assignment for X is invariable even if q contains other subgroups.

4   Approximation Ratio and Running Time

In this section, we analyze the approximation ratio and running time of our algorithm. We first show that the running time is bounded by $O(n^2)$ time. This bound is finally reduced to $O(n)$ time in this section.

Lemma 1. LEVELWISE-REPAIR runs in at most $O(n^2)$ time in the length of input.

Proof. Using a counter, for each repetition x^k in w, we can construct all nonterminals in the binary derivation for x^k in $O(k)$ time. Thus, the required time for

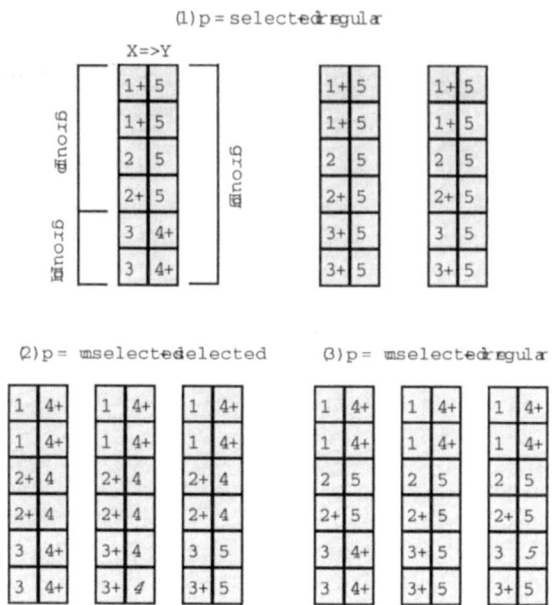

Fig. 4. Deciding the assignment for a set X of current segments from the set Y of their left segments and a current dictionary D.

$repetition(w, N)$ is $O(n)$. For other computation, we initially construct a doubly-linked list for w to gain constant-time access to any occurrence of a pair ab. Since this technique was already implemented in [9], we briefly explain this idea.

The length of the linked-list, that is, the number of nodes is n such that the ith node n_i contains at most five pointers $a(i)$, $suc(i)$, $pre(i)$, $latter(i)$, and $former(i)$, where $a(i)$ is $w(i)$, $suc(i)$ and $pre(i)$ are pointers for the nodes n_{i-1} and n_{i+1}, respectively, $latter(i)$ is the pointer for the next occurrence of ab for $w[i, i+1] = ab$, and the $former(i)$ is similar. The time to construct this linked-list is $O(n)$.

The priority list of all pairs in w is simultaneously constructed. Whenever the top of the priority list, say ab, is popped, the total length traced by the algorithm to compute the set C_{ab}, F_{ab}, L_{ab}, and R_{ab} is at most $O(k)$ for the number k of all occurrences of ab. Similarly, the sets L for F_{ab} and R for R_{ab} can be computed in $O(k)$ time.

Using hash table, for each $w[i, i+1] \in L_{ab}$ we can decide the group of $w[i-1, i] \in L$ in $O(1)$ time. The other conditions can be also computed in $O(1)$ time. Thus the running time of $arrangement(,)$ for a pair ab is also $O(k)$. Since an output string by $arrangement(,)$ is shorter than its input (if not, the algorithm terminates), the number of execution of the outer-loop is at most n times. Hence the running time of $Levelwise\text{-}repair$ is at most $O(n^2)$.   □

Lemma 2. Let w be an input string for $repetition(,)$, $w[i_1, j_1] = w[i_2, j_2]$ be nonoverlapping intervals of a same substring in w, and w' be the output string. Let I_1 be the shortest interval in w' satisfying that I_1 corresponds to an interval in w which contains $w[i_1, j_1]$. The other interval I_2 is similarly defined. Then it holds that $I_1[2, |I_1| - 1] = I_2[2, |I_2| - 1]$.

Proof. We can assume $w[i_1, j_1] = w[i_1, j_1] = usv$ such that $u = a^+$ and $v = b^+$ for some $a, b \in N$. The intervals $w[i_1 + |u|, i_1 + |us| - 1] = w[i_2 + |u|, i_2 + |us| - 1] = s$ are compressed into a same string $\tilde{s}$. There exist $i \leq i_1$ and $i' \leq i_2$ such that $w[i, i_1] = w[i', i_2] = a^+$ are compressed into some symbols A_1 and A_2, and such indices exist also for j_1 and j_2. Thus, the strings represented by the intervals in w' corresponding to $w[i_1, j_1]$ and $w[i_2, j_2]$ are of the form $A_1 \tilde{s} B_1$ and $A_2 \tilde{s} B_2$, respectively. Hence these intervals satisfies the statement.   □

Lemma 2 shows that any intervals represented by a same substring are compressed by $repetition(,)$ into intervals which have a long common substring each other. The main purpose of this section is to show that the same property is satisfied by $arrangement(,)$. We prepare several notions prior to the proof.

Let p and q be pairs in a priority queue constructed in $arrangement(,)$. A pair p is said to be *more frequent* than q if p is former element than q in the queue. Similarly, a segment s is also said to be *more frequent* than s' if the pair of s is more frequent than that of s'.

Definition 3. An interval $w[i, j]$ is said to be decreasing if $w[k, k + 1]$ is more frequent than $w[k + 1, k + 2]$ for all $i \leq k \leq j - 2$, and conversely, is said to be increasing if $w[k+1, k+2]$ is more frequent than $w[k, k+1]$ for all $i \leq k \leq j-2$. A segment $w[i, i+1]$ is said to be local maximum if $w[i, i+1]$ is more frequent than $w[i-1, i], w[i+1, i+2]$ and is said to be local minimum if $w[i-1, i], w[i+1, i+2]$ are more frequent than $w[i, i+1]$.

Here we note that any repetition like a^+ is replaced by a nonterminal. Thus for any two segments representing different strings, one of them is more frequent than the other.

Definition 4. Let $w[i, j]$ and $w[i', j']$ be independent occurrences of a substring and D be a current dictionary. Let s_k and s'_k be the kth segments in $w[i, j]$ and $w[i', j']$ from the left most segments, respectively. Then s_k, s'_k are said to agree with D if either $s_k, s'_k \in D$ or $s_k, s'_k \notin D$, and are said to disagree with D otherwise.

Lemma 3. Let w be an input for $arrangement(,)$, $w[i, i + j] = w[i', i' + j]$ be independent occurrences of a same substring in w, and D be the computed dictionary. Then the following two conditions hold: (1) for each $6 \leq k \leq j - 6$, $w[i + k, i + k + 1]$ and $w[i' + k, i' + k + 1]$ agree with D and (2) for each $w[\ell, \ell+3]$ contained in $w[i, i + j]$, at least one segment in $w[\ell, \ell + 3]$ is in D.

Proof. We first show Condition (1). If $w[i, i + j]$ contains a local maximum segment $s = w[i + k, i + k + 1]$, then s is the first segment chosen from $w[i + k -$

$1, i + k + 2]$. Thus, s and the corresponding segment s' in $w[i', i' + j]$ are added to D and assigned a same index.

Similarly it is easy to see that any segments agree with D between two adjacent local maximum segments. Thus, the remained intervals are a long decreasing prefix and a long increasing suffix of $w[i, i + j]$ and $w[i', i' + j]$. In order to prove this case, we need the following claim:

Claim. Any group computed by *arrangement*$(,)$ consists of at most two different subgroups of selected, unselected, and irregular.

This claim is directly obtained from Definition 1 (See Fig. 4 which illustrates all the cases). Let $w[i, i + j]$ contains a decreasing prefix of length at least six. The first segment chosen from the prefix of $w[i, i+j]$ is $w[i, i+1]$, and $w[i', i'+1]$ is also chosen simultaneously. They are then classified into some groups. Since the prefix is decreasing, succeedingly chosen segments are the right segments of $w[i, i + 1]$ and of $w[i', i' + 1]$. They are indicated by s and s' respectively. Since s and s' are both left-fixed and represent a same pair, they are classified into a same group g.

Case 1: The group g consists of a single subgroup. In this case, s and s' are both contained in one of (a) selected, (b) unselected, and (c) irregular subgroup. Case (a) satisfies that s and s' are assigned a same index and are both added to D. Thus, from the segments, no disagreement happens within the prefix. Case (b) and (c) converge to (a) within at least two right segments from s are chosen.

Case 2: The group g containing s and s' consists of two different subgroups. The right segments of s and s' are assigned some indices according to the types of the groups in which s and s' are contained. All the combinations of two different subgroups are (i) selected and unselected, (ii) selected and irregular, and (iii) unselected and irregular. In the first two cases, the right segments are all classified into a single subgroup. In the last case, any segment are classified into a selected or unselected subgroup, that is, this case converges to case (i). Thus, all case of (i), (ii), and (iii) converge to Case 1 within further two right segment from s and s' are chosen.

Consequently, it is guaranteed that two segments $w[i + k, i + k + 1]$ and $w[i' + k, i' + k + 1]$ are assigned a same index and they are added to D within four right segments from s and s' are chosen. It follows that any disagreement of $w[i, i+j]$ and $w[i', i' + j]$ in the decreasing prefix happens within only the range $w[i, i + 6]$ and $w[i', i' + 6]$. The increasing suffix case can be similarly shown.

We next show Condition (2). Since all local maximum segments are added to D, the possibility for unsatisfying Condition (2) is only the cases of a decreasing prefix and increasing suffix of $w[i, i + j]$. As is already shown in the above, any segment is classified into one of a selected, unselected, and irregular subgroup. Moreover, the last two subgroups converge to a selected subgroup within two segments. Thus, $w[i, i + j]$ and $w[i', i' + j]$ has no three consecutive segments which are not added to D.   □

Example 3. Fig. 5 illustrates the convergence of a long prefix case. Let a string w contain 8 independent intervals which have the same prefix '*abcdefg*' and this prefix be decreasing. The 1-8 rows represent such 8 intervals. Assume that the set of segments of ab are already assigned and classified into two group p and q. The last 4 rows correspond to other 4 intervals in w which have the same prefix '$bcdefg$'. This figure shows that the assignments for all 12 rows converge on the column of cd in a same group. All segments of this group are set to a same *selected* subgroup on this column. We note that the convergence of 1-8 rows is guaranteed regardless of the last 4 rows since for each group g', the assignments for right segments of g' are not affected by other groups as long as g' contains 2 subgroups (See Example 2). Finally each interval is compressed in the string shown in its right side. Nonterminals $B, C, D, E,$ and F are associated with the production rules $B \rightarrow bc, C \rightarrow cd, D \rightarrow de, E \rightarrow ef,$ and $F \rightarrow fg$, respectively. The left most '$-$' denotes an indefinite character since they depend on their left sides.

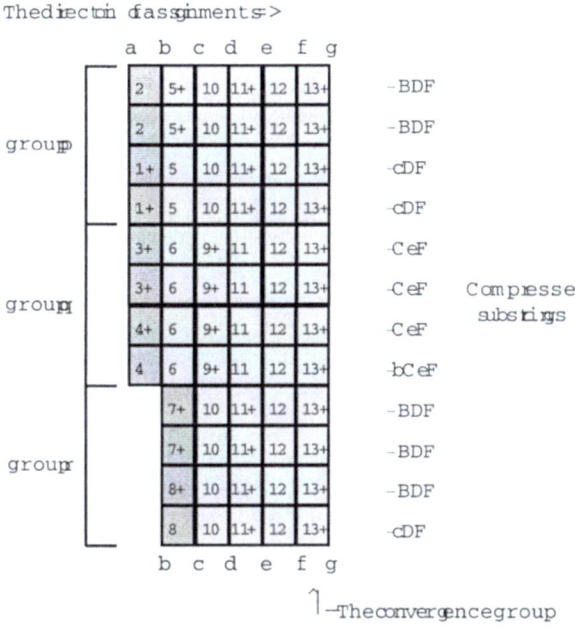

Fig. 5. The convergence of assignments for a long *decreasing* prefix case.

Finally, we show the main result of this paper by comparing the size of output grammar G with the *LZ-factorization* [19] of w. Here we recall its definition: The *LZ-factorization* of w denoted by $LZ(w)$ is the decomposition $w = f_1 \cdots f_k$, where $f_1 = w[1]$ and for each $1 \leq \ell \leq k$, f_ℓ is the longest prefix of $f_\ell \cdots f_k$ which occurs in $f_1 \cdots f_{\ell-1}$. Each f_ℓ is called a *factor*. The size of $LZ(w)$, denoted by $|LZ(w)|$, is the number of its factors.

Theorem 1 ([15]). For each string w and its deterministic CFG G, $|LZ(w)| \leq |G|$.

Theorem 2. The approximation ratio of LEVELWISE-REPAIR is $O(\log^2 n)$ and the running time is $O(n)$ for each string of length n.

Proof. By Theorem 1, it is sufficient to prove $|G|/|LZ(w)| = O(\log^2 n)$. For each factor f_ℓ, the prefix $f_1 \cdots f_{\ell-1}$ contains at least one occurrence of f_ℓ. We denote f_ℓ by $w[i, i+j]$ and other occurrence by $w[i', i'+j]$, respectively. By Lemma 2 and 3, after one loop of the algorithm is executed, the substrings represented by $w[i, i+j]$ and $w[i', i'+j]$ are compressed into some strings $\alpha\beta\gamma$ and $\alpha'\beta\gamma'$, respectively, where $|\alpha|, |\gamma| \leq 4$. By Lemma 3, we obtain $|\beta| \leq \frac{3}{4}j$. Since β occurs in the compressed string at least twice, Lemma 2 and 3 can be applied to the strings until these strings are sufficiently short strings.

Thus, the interval $w[i, i+j]$ corresponding to f_ℓ is compressed into a string of length at most $O(\log j)$. It follows that w is compressed into a string of length at most $O(k \log n)$, where $k = |LZ(w)|$. Hence, we can estimate $|G| = 2|N| + c \cdot k \log n$ with a constant c and the set N of all nonterminals of G.

The number of different nonterminals in the final string is at most $c \cdot k \log n$. If a nonterminal $A \in N$ occurs in this string and $A \rightarrow BC \in P$, then the pair BC must occur in the prior string at least twice. Thus, the number of different nonterminals in the prior string is also bounded by $c \cdot k \log n$. Since the depth of the loop of the algorithm is $O(\log n)$, we obtain $|N| \leq ck \log^2 n$. Therefore $|G|/|LZ(w)| = O(\log^2 n)$.

The running time of LEVELWISE-REPAIR can be reduced in linear time in n since the number of execution of the outer loop is $O(\log n)$ and $|\beta| \leq \frac{3}{4} \cdot j$. $\qquad \square$

5   Conclusion

For the grammar-based compression problem, we presented a fully linear time algorithm which guarantees $O(\log^2 n)$ approximation ratio for unbounded alphabets under randomized model. There are several open problems. An important problem is an upper bound of the approximation ratio of RE-PAIR algorithm [9]. Other problem is to construct a space-economic approximation algorithm preserving the approximation ratio.

Acknowledgement. The author would like to thank Alistair Moffat for his helpful comments. The author also thank all anonymous referees for their careful proof reading and comments on the draft.

References

1. G. Ausiello, P. Crescenzi, G. Gambosi, V. Kann, A. Marchetti-Spaccamela, and M. Protasi. *Complexity and Approximation: Combinatorial Optimization Problems and Their Approximability Properties.* Springer, 1999.

2. S. De Agostino and J. A. Storer. On-Line versus Off-Line Computation in Dynamic Text Compression. *Inform. Process. Lett.*, 59:169–174, 1996.

3. M. Charikar, E. Lehman, D. Liu, R. Panigrahy, M. Prabhakaran, A. Rasala, A. Sahai, and A. Shelat. Approximating the Smallest Grammar: Kolmogorov Complexity in Natural Models. In *Proc. 29th Ann. Sympo. on Theory of Computing*, 792–801, 2002.

4. D. Gusfield. *Algorithms on Strings, Trees, and Sequences*. Computer Science and Computational Biology. Cambridge University Press, 1997.

5. T. Kida, Y. Shibata, M. Takeda, A. Shinohara, and S. Arikawa. Collage System: a Unifying Framework for Compressed Pattern Matching. *Theoret. Comput. Sci.* (to appear).

6. J. C. Kieffer and E.-H. Yang. Grammar-Based Codes: a New Class of Universal Lossless Source Codes. *IEEE Trans. on Inform. Theory*, 46(3):737–754, 2000.

7. J. C. Kieffer, E.-H. Yang, G. Nelson, and P. Cosman. Universal Lossless Compression via Multilevel Pattern Matching. *IEEE Trans. Inform. Theory*, IT-46(4), 1227–1245, 2000.

8. D. Knuth. Seminumerical Algorithms. Addison-Wesley, 441–462, 1981.

9. N. J. Larsson and A. Moffat. Offline Dictionary-Based Compression. *Proceedings of the IEEE*, 88(11):1722–1732, 2000.

10. E. Lehman. Approximation Algorithms for Grammar-Based Compression. PhD thesis, MIT, 2002.

11. E. Lehman and A. Shelat. Approximation Algorithms for Grammar-Based Compression. In *Proc. 20th Ann. ACM-SIAM Sympo. on Discrete Algorithms*, 205–212, 2002.

12. M. Lothaire. *Combinatorics on Words*, volume 17 of *Encyclopedia of Mathematics and Its Applications*. Addison-Wesley, 1983.

13. M. Farach. Optimal Suffix Tree Construction with Large Alphabets. In *Proc. 38th Ann. Sympo. on Foundations of Computer Science*, 137–143, 1997.

14. C. Nevill-Manning and I. Witten. Compression and Explanation Using Hierarchical Grammars. *Computer Journal*, 40(2/3):103–116, 1997.

15. W. Rytter. Application of Lempel-Ziv Factorization to the Approximation of Grammar-Based Compression. In *Proc. 13th Ann. Sympo. Combinatorial Pattern Matching*, 20–31, 2002.

16. J. A. Storer and T. G. Szymanski. The Macro Model for Data Compression. In *Proc. 10th Ann. Sympo. on Theory of Computing*, pages 30–39, San Diego, California, 1978. ACM Press.

17. T. A. Welch. A Technique for High Performance Data Compression. *IEEE Comput.*, 17:8–19, 1984.

18. E.-H. Yang and J. C. Kieffer. Efficient Universal Lossless Data Compression Algorithms Based on a Greedy Sequential Grammar Transform–Part One: without Context Models. *IEEE Trans. on Inform. Theory*, 46(3):755–777, 2000.

19. J. Ziv and A. Lempel. A Universal Algorithm for Sequential Data Compression. *IEEE Trans. on Inform. Theory*, IT-23(3):337–349, 1977.

20. J. Ziv and A. Lempel. Compression of Individual Sequences via Variable-Rate Coding. *IEEE Trans. on Inform. Theory*, 24(5):530–536, 1978.

Constrained Tree Inclusion

Gabriel Valiente[*]

Department of Software, Technical University of Catalonia, E-08034 Barcelona,
Spain, Fax 00 34 934 017 014, `valiente@lsi.upc.es`

Abstract. The tree matching problem is considered of given labeled
trees P and T, determining if the pattern tree P can be obtained
from the text tree T by deleting degree-one and degree-two nodes
and, in the case of unordered trees, by also permuting siblings. The
constrained tree inclusion problem is more sensitive to the structure
of the pattern tree than the general tree inclusion problem. Further,
it can be solved in polynomial time for both unordered and ordered
trees. Algorithms based on the subtree homeomorphism algorithm of
(Chung, 1987) are presented that solve the constrained tree inclusion
problem in $O(m^{1.5}n)$ time on unordered trees with m and n nodes,
and in $O(mn)$ time on ordered trees, using $O(mn)$ additional space.
These algorithms can be improved using results of (Shamir and Tsur,
1999) to run in $O((m^{1.5}/\log m)n)$ and $O((m/\log m)n)$ time, respectively.

Keywords. Tree inclusion, tree pattern matching, subtree homeomorphism, noncrossing bipartite matching

1 Introduction

The tree inclusion problem was introduced in [12, Ex. 2.3.2-22] and further
studied in [11], motivated by the study of query languages for structured text
databases, and has since received much attention, from the theoretical side and
also as a primitive for querying collections of XML documents [1,10,20]. Given a
pattern tree P and a text tree T, both with labels on the nodes, the tree inclusion
problem consists in locating the smallest subtrees of T that include P, where a
tree is included in another tree if can be obtained from the latter by deleting
nodes and, in the case of unordered trees, by also permuting siblings. Deleting
a node v from a tree entails deleting all edges incident to v and inserting edges
connecting the parent of v (for nonroot nodes) with the children (if any) of v.

Tree inclusion has two main drawbacks. First, the solution to a tree inclusion
query of a pattern tree in a text tree is not sensitive to the structure of the
query, in the sense that even for ordered trees, many structural forms of the
same pattern (that is, many different pattern trees with the same labeling) may
be included in the same text tree. As a matter of fact, any smallest supertree
under minor containment [17] of a set of pattern trees includes all of the pattern
trees (although not necessarily in a minimal way).

[*] Partially supported by Spanish CICYT project MAVERISH (TIC2001-2476-C03-01)

R. Baeza-Yates et al. (Eds.): CPM 2003, LNCS 2676, pp. 361–371, 2003
© Springer-Verlag Berlin Heidelberg 2003

Example 1. Three forms of the same query, shown to the left of the following picture, are all included at the node labeled A in text tree, shown to the right of the picture.

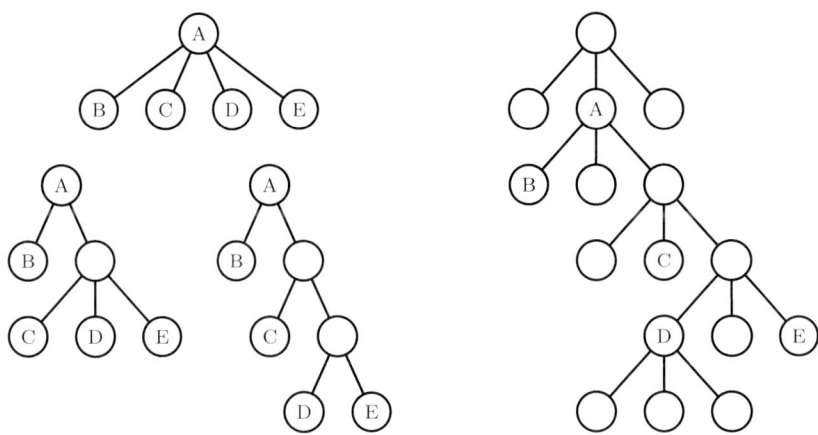

The second drawback is the complexity of tree inclusion. The tree inclusion problem on unordered trees is NP-hard [11,15], although it can be solved in polynomial time on ordered trees. A dynamic programming algorithm was given in [11] that solves the ordered tree inclusion problem in $O(mn)$ time and space in the worst case and also on the average, for a pattern tree with m nodes and a text tree with n nodes. Several improvements were since proposed [2,4,19].

These drawbacks stem from the generality of tree inclusion. In this paper, a constrained form of the tree inclusion problem is presented which is more sensitive to the structure of the pattern tree and can be solved in polynomial time, on both unordered and ordered trees. In the constrained formulation, a tree is included in another tree if can be obtained from the latter by deleting degree-one and degree-two nodes and, in the case of unordered trees, by also permuting siblings. Therefore, it is related to the problem of finding homeomorphic subtrees in a tree. The solution method used is based on the subtree homeomorphism algorithm of [5], which involves solving a series of maximum bipartite matching problems. The constrained tree inclusion algorithm takes $O(m^{1.5}n)$ time using $O(mn)$ additional space on unordered trees with m and n nodes, and can be improved using a result of [21] to run in $O((m^{1.5}/\log m)n)$ time.

For ordered trees, a simple algorithm for the noncrossing bipartite matching problem is also presented that takes time linear in the number of vertices of the bipartite graph. The constrained tree inclusion algorithm takes thus $O(mn)$ time using $O(mn)$ additional space on ordered trees with m and n nodes, and can also be improved using the result of [21] to run in $O((m/\log m)n)$ time.

Constrained tree inclusion is related to the tree edit problem. As a matter of fact, in tree inclusion the elementary edit operation of insertion is forbidden and thus, constrained tree inclusion is polynomially equivalent to degree-two tree edit distance, in which a node can be deleted only when it is a leaf or has one child. The degree-two tree edit distance problem can be solved in

$O(mn \min(\deg(P), \deg(T)))$ time on unordered trees, and in $O(mn)$ time on ordered trees P and T with m and n nodes, respectively [27,28], using $O(mn)$ additional space.

The rest of the paper is organized as follows. The constrained tree inclusion problem is defined in Section 2, where the necessary notation is also introduced. The solution of constrained tree inclusion problems on unordered trees in addressed in Section 3, where an algorithm based on the subtree homeomorphism algorithm of [5] is described that solves the constrained tree inclusion problem in $O(m^{1.5}n)$ time using $O(mn)$ additional space, on unordered trees with m and n nodes. The algorithm, which involves the solution of a series of small maximum bipartite matching problems, is extended in Section 4 to ordered trees by means of a simple algorithm to find a noncrossing matching covering one of the bipartite sets in an ordered bipartite graph. The constrained tree inclusion problem is thus solved in $O(mn)$ time and space, on ordered trees with m and n nodes. Finally, some conclusions are outlined in Section 5.

2   Constrained Tree Inclusion

In the tree inclusion problem, a pattern tree P is included in a text tree T if P can be obtained by deleting some nodes from T and, in the case of unordered trees, by also permuting sibling nodes [11]. In constrained tree inclusion, these deletion operations are allowed on degree-one nodes (leaves) and degree-two nodes (with one child) only.

The following notation will be used. The set of nodes and the set of edges of a tree T are denoted by $V(T)$ and $E(T)$, respectively. The number of children of node v in tree T is denoted by $outdeg(T, v)$ or just $outdeg(v)$, if T is clear from the context. The subtree of T rooted at node $v \in V(T)$ is denoted by $T[v]$. The label of node $v \in V(T)$ is denoted by $label(v)$. Further, $root(T)$ denotes the root of tree T, and $P \cong T$ denotes that trees P and T are isomorphic.

Definition 1. *A tree P is included in a tree T, denoted by $P \sqsubseteq T$, if there is a sequence of nodes $v_1, v_2, \ldots, v_k$ in $V(T)$ such that*

- $T_{i+1} \cong delete(T_i, v_{i+1})$
- $outdeg(T_i, v_{i+1}) \leqslant 1$

for $1 \leqslant i \leqslant k - 1$, with $T_0 = T$ and $T_k = P$.

Example 2. A tree P, shown to the left, is included in a tree T, shown to the right of the following picture. The pattern tree P can be obtained from the text tree T by deleting degree-one and degree-two nodes, as shown from right to left: $T_1 \cong delete(T_0, w)$ and $T_2 \cong delete(T_1, y)$.

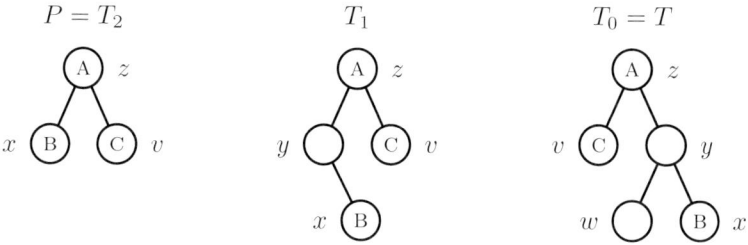

$$P = T_2 \qquad\qquad T_1 \qquad\qquad T_0 = T$$

3   Solving Constrained Tree Inclusion Problems on Unordered Trees

The number of pattern trees that are included in a text tree is exponential in the number of nodes of the text tree.

Example 3. The ordered tree represented by $a(b, c(a, b), d)$ has 25 nonempty included subtrees, represented by a, b, c, d, $a(a)$, $a(b)$, $a(c)$, $a(d)$, $c(b)$, $c(b)$, $a(a, d)$, $a(b, a)$, $a(b, b)$, $a(b, c)$, $a(b, d)$, $a(c, d)$, $c(a, b)$, $a(b, c(a))$, $a(b, c(b))$, $a(c(a), d)$, $a(c(b), d)$, $a(b, c, d)$, $a(b, c(a), d)$, $a(b, c(b), d)$, and $a(b, c(a, b), d)$, but does not include the 4 subtrees represented by $a(a, b)$, $a(a, b, d)$, $a(b, a, b)$, and $a(b, a, b, d)$, which would require the deletion of a degree-three node (with two children). Further, there are $25 + 36 = 61$ nonempty subtrees included in the unordered tree represented by the same term: the previous 25 subtrees, and $a(a, b)$, $a(b, a)$, $a(c, b)$, $a(d, a)$, $a(d, b)$, $a(d, c)$, $a(d, c(a))$, $a(d, c(b))$, $a(c(a), b)$, $a(c(b), b)$, $a(b, c, d)$, $a(c, b, d)$, $a(c, d, b)$, $a(d, b, c)$, $a(d, c, b)$, $a(b, d, c(a))$, $a(c(a), b, d)$, $a(c(a), d, b)$, $a(d, b, c(a))$, $a(d, c(a), b)$, $a(b, d, c(b))$, $a(c(b), b, d)$, $a(c(b), d, b)$, $a(d, b, c(b))$, $a(d, c(b), b)$, $a(b, c(b, a), d)$, $a(b, d, c(a, b))$, $a(b, d, c(b, a))$, $a(c(a, b), b, d)$, $a(c(b, a), b, d)$, $a(c(a, b), d, b)$, $a(c(b, a), d, b)$, $a(d, b, c(a, b))$, $a(d, b, c(b, a))$, $a(d, c(a, b), b)$, and $a(d, c(b, a), b)$. Again, the 19 subtrees represented by $a(b, a)$, $a(a, d, b)$, $a(b, a, d)$, $a(b, d, a)$, $a(d, a, b)$, $a(d, b, a)$, $a(a, b, b)$, $a(b, b, a)$, $a(a, b, b, d)$, $a(a, b, d, b)$, $a(a, d, b, b)$, $a(b, a, d, b)$, $a(b, b, a, d)$, $a(b, b, d, a)$, $a(b, d, a, b)$, $a(b, d, b, a)$, $a(d, a, b, b)$, $a(d, b, a, b)$, and $a(d, b, b, a)$ are not included in the previous unordered tree, because that would require the deletion of a degree-three node.

The key to an efficient solution lies in the fact that a constrained tree inclusion problem instance can be decomposed into a series of smaller, independent problem instances. As a matter of fact, it will be shown that in order to determine whether or not $P[v] \sqsubseteq T[w]$, where node $v \in V(P)$ has children $v_1, v_2, \ldots, v_p$ and node $w \in V(T)$ has children $w_1, w_2, \ldots, w_t$, it suffices to know if $P[x] \sqsubseteq T[y]$ for all $x \in \{v, v_1, v_2, \ldots, v_p\}$ and $y \in \{w_1, w_2, \ldots, w_t\}$.

Definition 2. *Let P and T be unordered trees, and let $w \in V(T)$. The included subtrees at node w, denoted by $S(w)$, are the set*

$$S(w) = \{v \in V(P) \mid P[v] \sqsubseteq T[w]\}$$

That is, $S(w)$ is the set of roots of those subtrees of P that are included in $T[w]$. Two direct consequences of the previous definition are given next.

Fact 1 *For all nodes $v \in V(P)$ and $w \in V(T)$, $P[v] \sqsubseteq T[w]$ if and only if $v \in S(w)$.*

Fact 2 *$P \sqsubseteq T$ if and only if $\{w \in V(T) \mid root(P) \in S(w)\} \neq \emptyset$.*

The next result assures correctness of decomposing a constrained tree inclusion problem in independent subproblems.

Lemma 1 (Chung). *Let $v \in V(P)$ have children $v_1, v_2, \ldots, v_p$, and let $w \in V(T)$ have children $w_1, w_2, \ldots, w_t$. Then, $P[v] \sqsubseteq T[w]$ if and only if either there is a child w_j of w such that $P[v] \sqsubseteq T[w_j]$, or $label(v) = label(w)$ and there is a subset of p different nodes $\{u_1, u_2, \ldots, u_p\} \subseteq \{w_1, w_2, \ldots, w_t\}$ such that $P[v_i] \sqsubseteq T[u_i]$ for $1 \leqslant i \leqslant p$.*

Lemma 2. *Let T be a tree. There is a sequence of node deletion operations that transform T into the tree T' with $V(T') = \{root(T)\}$ and $E(T') = \emptyset$.*

Proof. By deleting all nonroot nodes of T in postorder, the children (if any) of a node will have already been deleted when the node is considered for deletion, meaning the node has become a degree-one node (a leaf), which can thus be deleted. □

Corollary 1. *$P[v] \sqsubseteq T[parent(w)]$ if $P[v] \sqsubseteq T[w]$, for all nodes $v \in V(P)$ and all nonroot nodes $w \in V(T)$.*

Proof. $P[v] \sqsubseteq T[w]$, and $T[w]$ can be obtained from $T[parent(w)]$ by deleting $T[x]$ for all siblings x of node w and, then, deleting node $parent(w)$, which has become either a degree-one or a degree-two node. □

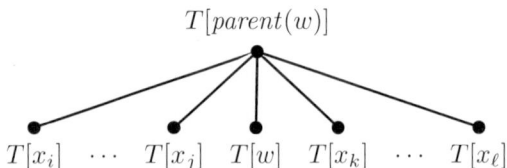

$$T[parent(w)]$$

$$T[x_i] \quad \cdots \quad T[x_j] \quad T[w] \quad T[x_k] \quad \cdots \quad T[x_\ell]$$

Proof (Lemma 1). (If.) Immediate. (Only if.) In the first case, $P[v]$ can be obtained from $T[w]$ by deleting $T[w_1], T[w_2], \ldots, T[w_{j-1}], T[w_{j+1}], \ldots, T[w_t]$ and, then, deleting node w, which has become either a degree-one or a degree-two node. In the second case, $P[v]$ can be obtained from $T[w]$ by deleting $T[w_i]$ for all $w_i \in \{w_1, w_2, \ldots, w_t\} \setminus \{u_1, u_2, \ldots, u_p\}$. □

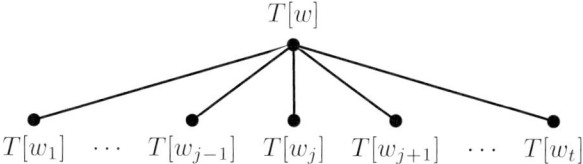

$$T[w]$$

$$T[w_1] \quad \cdots \quad T[w_{j-1}] \quad T[w_j] \quad T[w_{j+1}] \quad \cdots \quad T[w_t]$$

Remark 1. A result similar to Lemma 1 was enunciated without proof in [5, Lemma 1] for the subtree homeomorphism problem but does not carry over to constrained tree inclusion (it does not even hold for subtree homeomorphism) because deletion of degree-one nodes, not only of degree-two nodes, is required for Lemma 2 to hold.

Now, the set of included subtrees $S(w)$ can be computed for each node $w \in V(T)$ in a bottom-up way, as follows.

Algorithm 1 *A procedure call of the form included(P, T, S), where P and T are unordered trees, computes $S = \{w \in V(T) \mid root(P) \in S(w)\}$. Thus, $P \sqsubseteq T$ if and only if $S \neq \emptyset$.*

> **procedure** included (P, T, S)
> $S := \emptyset$
>   unmark all nodes $w \in V(T)$
>   **for all** nodes $w \in V(T)$ in postorder **do**
> $S(w) := \{v \in V(P) \mid outdeg(v) = 0 \wedge label(v) = label(w)\}$
>   **if** $outdeg(w) \neq 0$ **then**
>   let $w_1, w_2, \ldots, w_t$ be the children of w
> $S(w) := S(w) \cup S(w_1) \cup S(w_2) \cup \cdots \cup S(w_t)$
>   **for all** nonleaf nodes $v \in V(P)$ with $outdeg(v) \leqslant t$ and
> $label(v) = label(w)$ in postorder **do**
>   **if** $v \notin S(w)$ **then**
>   let $v_1, v_2, \ldots, v_p$ be the children of v
>   construct a bipartite graph
> $G = (\{w_1, w_2, \ldots, w_t\} \cup \{v_1, v_2, \ldots, v_p\}, E)$
>   with $(w_j, v_i) \in E$ if and only if $v_i \in S(w_j)$
>   for $1 \leqslant j \leqslant t$ and $1 \leqslant i \leqslant p$
>   **if** there is a matching in G with p edges **then**
> $S(w) := S(w) \cup \{v\}$
>   **end if**
>   **end if**
>   **end for**
>   **end if**
>   **if** node w is unmarked and $root(P) \in S(w)$ **then**
> $S := S \cup \{v\}$
>   mark w and all ancestors of w
>   **end if**
>   **end for**
> **end procedure**

Lemma 3. *The constrained tree inclusion algorithm is correct.*

Proof. Follows from Lemma 1. Notice that, all leaves $v \in V(P)$ with $label(v) = label(w)$ are added to $S(w)$ for all nodes $w \in V(T)$, both leaves and nonleaf nodes, in postorder. A leaf $v \in V(P)$ with $v \in S(w)$ for a nonleaf node $w \in V(T)$ corresponds to a constrained tree inclusion $P[v] \sqsubseteq T[w]$ in which $T[w_j]$ is deleted for all children w_j of node w. Note also that by marking all ancestors of a node $w \in V(T)$ with $root(P) \in S(w)$, only the (roots of) smallest subtrees of T that include P are collected in S. $\square$

Lemma 4. *The constrained tree inclusion algorithm takes $O(m^{1.5}n)$ time using $O(mn)$ additional space, on unordered trees with m and n nodes.*

Proof. Let P and T be unordered trees with nodes $\{v_1, v_2, \ldots, v_m\}$ and $\{w_1, w_2, \ldots, w_n\}$, respectively. The time complexity of the algorithm is dominated by the solution of a series of small maximum bipartite matching problems. Since a maximum matching problem on a bipartite graph with r and s vertices can be solved in $O(r^{1.5}s)$ time [9], the time complexity of the algorithm is

$$\sum_{i=1}^{n}\sum_{j=1}^{m} O(outdeg(w_i)\,outdeg(v_j)^{1.5})$$

$$= \sum_{i=1}^{n} O(m^{1.5}outdeg(w_i)) \quad \left(\text{because} \sum_{j=1}^{m} outdeg(v_j) = m-1\right)$$

$$= O(m^{1.5}n) \quad \left(\text{because} \sum_{i=1}^{n} outdeg(w_i) = n-1\right)$$

Regarding space complexity, the collection of sets $S(w)$ for $w \in V(T)$ is stored in a two-dimensional array S of m by n integers, indexed by the postorder number of the nodes in the trees and with the representation invariant that $S(order(v), order(w))$ if and only if $v \in S(w)$, for all nodes $v \in V(P)$ and $w \in V(T)$. The algorithm thus uses $O(mn)$ additional space. $\square$

Corollary 2. *The constrained tree inclusion algorithm can be improved to take $O((m^{1.5}/\log m)n)$ time, on unordered trees with m and n nodes.*

4   Solving Constrained Tree Inclusion Problems on Ordered Trees

The previous algorithm for constrained tree inclusion on unordered trees can be used to solve the constrained tree inclusion problem on ordered trees as well. The series of maximum bipartite matching problems become now matching problems on *ordered bipartite graphs*.

Definition 3. *An ordered bipartite graph is a bipartite graph $G = (V \cup W, E)$ with orderings $V = (v_1, v_2, \ldots, v_p)$ and $W = (w_1, w_2, \ldots, w_q)$.*

A matching in an ordered bipartite graph is a noncrossing matching in the bipartite graph, with respect to the given orderings on the bipartite sets of vertices.

Definition 4. *A noncrossing matching M in an ordered bipartite graph $G = (V \cup W, E)$ is a subset of edges $M \subseteq E$ such that no two edges are incident to the same vertex and no two edges are crossing, that is, for all edges (v_i, w_k) and (v_j, w_ℓ) in M, $i < j$ if and only if $k < \ell$.*

The problem of finding a noncrossing matching of a bipartite graph with n vertices can be solved in $O(n \log \log)$ time [14,26]. However, the decision problem of whether an ordered bipartite graph $(V \cup W, E)$ has a noncrossing matching of size $|W|$ can be solved in $O(n)$ time, where $n = |V| + |W|$.

As a matter of fact, given an ordered bipartite graph $(V \cup W, E)$, the greedy strategy of always choosing the first noncrossing edge joining some vertex $v_i \in V$ with vertex $w_j \in W$, for $1 \leqslant j \leqslant |W|$, gives a noncrossing matching with $|W|$ edges, as long as such a matching does exist. The following algorithm is an efficient implementation.

Algorithm 2 *A function call of the form matching(V, W, E) returns true if and only if there is a noncrossing matching of the ordered bipartite graph $(V \cup W, E)$ covering all vertices of W.*

    function matching (V, W, E)
      $i := 1$
      for $j := 1$ **to** $|W|$ **do**
        while $i \leqslant |V| - |W| + j$ **and** $(v_i, w_j) \notin E$ **do**
          $i := i + 1$
        end while
        if $i \leqslant |V| - |W| + j$ **then**
          do nothing: edge (v_i, w_j) belongs to the matching
          $i := i + 1$
        else
          return false
        end if
      end for
      return true
    end function

Remark 2. Note that the noncrossing bipartite matching problem is polynomially equivalent to the sequence inclusion problem. As a matter of fact, let $G = (V \cup W, E)$ be an ordered bipartite graph with $V = (v_1, v_2, \ldots, v_p)$ and $W = (w_1, w_2, \ldots, w_q)$, where $p \geqslant q$. Then, there is a noncrossing matching in G covering all vertices of W if and only if there is a subsequence $v_{i_1} v_{i_2} \ldots v_{i_q}$ of $v_1 v_2 \ldots v_p$ such that $(v_{i_j}, w_j) \in E$ for $1 \leqslant j \leqslant q$.

Lemma 5. *The noncrossing bipartite matching algorithm is correct.*

Proof. Let $V = (v_1, v_2, \ldots, v_p)$ and $W = (w_1, w_2, \ldots, w_q)$. For $1 \leqslant j \leqslant q$, the edge (v_i, w_j) with the smallest i for $i \leqslant p - (q - j)$ is added to the matching. Choosing the smallest such i guarantees that the matching is noncrossing, and the upper bound $p - (q - j)$ guarantees that the noncrossing matching can be completed to size q with the remaining $q - j$ edges (if they exist). Now, as soon as no such noncrossing edge (v_i, w_j) exists for a vertex w_j, no noncrossing matching of size q exists and the matching procedure fails. Otherwise, one noncrossing edge belongs to the matching for each j with $1 \leqslant j \leqslant q$, and the matching procedure is thus successful.   □

Lemma 6. *The noncrossing bipartite matching algorithm takes time linear in the number of vertices of the bipartite graph.*

Proof. On an ordered bipartite graph $G = (V \cup W, E)$, the inner loop is executed at most $|V| + |W|$ times and, thus, the number of edge existence tests made is linear in the number of vertices of G.   □

Corollary 3. *The constrained tree inclusion algorithm takes $O(mn)$ time using $O(mn)$ additional space, on ordered trees with m and n nodes.*

Proof. As in the proof of Lemma 4,

$$\sum_{i=1}^{n}\sum_{j=1}^{m} O(outdeg(w_i)\, outdeg(v_j)) = \sum_{i=1}^{n} O(m \cdot outdeg(w_i)) = O(mn).$$

                                                            □

Corollary 4. *The constrained tree inclusion algorithm can be improved to take $O((m/\log m)n)$ time, on ordered trees with m and n nodes.*

5   Conclusions

A constrained form of the tree inclusion problem is addressed in this paper in which a pattern tree can be obtained from a text tree by deleting degree-one and degree-two nodes and, in the case of unordered trees, by also permuting siblings. The constrained tree inclusion problem is more sensitive to the structure of the pattern tree than the general tree inclusion problem, in which there is no restriction of node degree for deletion and, unlike the latter, can be solved in polynomial time for both unordered and ordered trees.

Based on the subtree homeomorphism algorithm of [5], an algorithm is given that solves the constrained tree inclusion problem in $O(m^{1.5}n)$ time using $O(mn)$ additional space, on unordered trees with m and n nodes. The algorithm, which involves the solution of a series of small maximum bipartite matching problems, is extended to ordered trees by means of a simple algorithm to find a noncrossing matching covering one of the bipartite sets in an ordered bipartite graph, a problem that is polynomially equivalent to the sequence inclusion problem, solving thus the constrained tree inclusion problem in $O(mn)$ time and space, on ordered trees with m and n nodes.

The constrained tree inclusion algorithm also solves, with a minor modification, the subtree isomorphism problem. While there are efficient algorithms for unordered subtree isomorphism [16,18,21,25], known ordered subtree isomorphism algorithms solve the problem in a restricted form, in which a subtree is either a prefix of the tree [3, Sect. 5d] or the whole tree rooted at a node [6,8, 7,13,22,24]. Further details about unordered and ordered subtree isomorphism can be found in [23, Sect. 4.2].

Acknowledgment. The author would like to acknowledge with thanks the anonymous referees, whose comments and criticism have led to a substantial improvement of this paper.

References

1. S. Abiteboul, P. Buneman, and D. Suciu. *Data on the Web: From Relations to Semistructured Data and XML.* Morgan Kaufmann, 2000.
2. L. Alonso and R. Schott. On the tree inclusion problem. *Acta Informatica*, 37(9):653–670, 2001.
3. A. T. Berztiss. *Data Structures: Theory and Practice.* Academic Press, New York, 2nd edition, 1975.
4. W. Chen. More efficient algorithm for ordered tree inclusion. *Journal of Algorithms*, 26(2):370–385, 1998.
5. M.-J. Chung. $O(n^{2.5})$ time algorithms for the subgraph homeomorphism problem on trees. *Journal of Algorithms*, 8(1):106–112, 1987.
6. P. Dublish. Some comments on the subtree isomorphism problem for ordered trees. *Information Processing Letters*, 36(5):273–275, 1990.
7. R. Grossi. Further comments on the subtree isomorphism for ordered trees. *Information Processing Letters*, 40(5):255–256, 1991.
8. R. Grossi. A note on the subtree isomorphism for ordered trees and related problems. *Information Processing Letters*, 39(2):81–84, 1991.
9. J. E. Hopcroft and R. M. Karp. An $n^{5/2}$ algorithm for maximum matchings in bipartite graphs. *SIAM Journal on Computing*, 2(4):225–231, 1973.
10. P. Kilpeläinen and H. Mannila. Retrieval from hierarchical texts by partial patterns. In *Proc. 16th Annual Int. ACM SIGIR Conf. Research and Development in Information Retrieval*, pages 214–222. ACM Press, 1993.
11. P. Kilpeläinen and H. Mannila. Ordered and unordered tree inclusion. *SIAM Journal on Computing*, 24(2):340–356, 1995.
12. D. E. Knuth. *Fundamental Algorithms*, volume 1 of *The Art of Computer Programming*. Addison-Wesley, Reading MA, 3rd edition, 1997.
13. E. Mäkinen. On the subtree isomorphism problem for ordered trees. *Information Processing Letters*, 32(5):271–273, 1989.

14. F. Malucelli, T. Ottmann, and D. Pretolani. Efficient labelling algorithms for the maximum noncrossing matching problem. *Discrete Applied Mathematics*, 47(2):175–179, 1993.

15. J. Matoušek and R. Thomas. On the complexity of finding isomorphisms and other morphisms for partial k-trees. *Discrete Mathematics*, 108(1–3):343–364, 1992.

16. D. W. Matula. Subtree isomorphism in $O(n^{5/2})$. *Annals of Discrete Mathematics*, 2(1):91–106, 1978.

17. N. Nishimura, P. Ragde, and D. M. Thilikos. Finding smallest supertrees under minor containment. *Int. Journal of Foundations of Computer Science*, 11(3):445–465, 2000.

18. S. W. Reyner. An analysis of a good algorithm for the subtree problem. *SIAM Journal on Computing*, 6(4):730–732, 1977.

19. T. Richter. A new algorithm for the ordered tree inclusion problem. In *Proc. 8th Annual Symp. Combinatorial Pattern Matching*, volume 1264 of *Lecture Notes in Computer Science*, pages 150–166. Springer-Verlag, 1997.

20. T. Schlieder and H. Meuss. Querying and ranking XML documents. *Journal of the American Society for Information Science and Technology*, 53(6):489–503, 2002.

21. R. Shamir and D. Tsur. Faster subtree isomorphism. *Journal of Algorithms*, 33(2):267–280, 1999.

22. G. Valiente. An efficient bottom-up distance between trees. In *Proc. 8th Int. Symp. String Processing and Information Retrieval*, pages 212–219, Piscataway NJ, 2001. IEEE Computer Science Press.

23. G. Valiente. *Algorithms on Trees and Graphs*. Springer-Verlag, Berlin, 2002.

24. R. M. Verma. Strings, trees, and patterns. *Information Processing Letters*, 41(3):157–161, 1992.

25. R. M. Verma and S. W. Reyner. An analysis of a good algorithm for the subtree problem, corrected. *SIAM Journal on Computing*, 18(5):906–908, 1989.

26. M.-S. Yu, L. Y. Tseng, and S.-J. Chang. Sequential and parallel algorithms for the maximum-weight independent set problem on permutation graphs. *Information Processing Letters*, 46(1):7–11, 1993.

27. K. Zhang. Efficient parallel algorithms for tree editing problems. In *Proc. 7th Annual Symp. Combinatorial Pattern Matching*, volume 1075 of *Lecture Notes in Computer Science*, pages 361–372, Berlin Heidelberg, 1996. Springer-Verlag.

28. K. Zhang, J. T.-L. Wang, and D. Shasha. On the editing distance between undirected acyclic graphs. *International Journal of Foundations of Computer Science*, 7(1):43–57, 1996.

Working on the Problem of Sorting by Transpositions on Genome Rearrangements

Maria Emilia M.T. Walter, Luiz Reginaldo A.F. Curado, and
Adilton G. Oliveira

University of Brasilia, Department of Computer Science, Brasilia, Brasil
mia@cic.unb.br, lutitus@ig.com.br, adilton@hotmail.com

Abstract. In computational biology, genome rearrangements is a field in which we investigate the combinatorial problem of sorting by transpositions. This problem consists in finding the minimum number of transpositions (mutational event) that transform a chromosome into another. In this work, we implement the 1.5-approximation algorithm proposed by Christie [2] for solving this problem, introducing modifications to reduce its time complexity, and we also propose heuristics to further improve its performance. Comparing our experimental results with the best known results, we had better performance. This work targets to contribute for discovering the complexity of the problem of sorting by transpositions, which remains open.

1 Introduction

Genome rearrangements, in computational biology, is an area where we study problems of rearrangement distance, which basically consists in finding the minimal number of mutations, affecting large portions of genomes, that transform an organism into another. Particularly, transposition is a mutational event moving one block of genes from a region to another inside a chromosome (Fig. 1).

$$1 \quad 4 \quad 3 \quad 2 \quad 7 \quad 8 \quad 5 \quad 6 \quad 9$$

$$1 \quad 4 \quad 5 \quad 6 \quad 3 \quad 2 \quad 7 \quad 8 \quad 9$$

Fig. 1. An example of the transposition event, where each number represent a block of genes.

The *problem of transposition distance* consists in finding the minimal number of transpositions transforming a chromosome into another. For this problem, Bafna and Pevzner [1] proposed an approximation algorithm with factor 1.5, based on some properties of a structure named *cycle graph*. Guyer, Heath and

R. Baeza-Yates et al. (Eds.): CPM 2003, LNCS 2676, pp. 372–383, 2003.

Vergara [4] and Vergara [7] developed algorithms based on heuristics. Christie [2] proposed and implemented a different approximation algorithm, also with factor 1.5, using new properties of the cycle graph. Walter, Dias and Meidanis [8] proposed and implemented another approximation algorithm with factor 2.25, but simpler than the others. Oliveira [6], extending the theory of Bafna and Pevzner [1], implemented their algorithm. Christie [2] and independently Meidanis, Walter and Dias [5] have computed the exact transposition distance of a permutation and its inverse. Recently, Dias and Meidanis [3] presented a polynomial algorithm for sorting by prefix transpositions.

The objectives of this work were to implement the 1.5-approximation algorithm of Christie [2] for the problem of transposition distance, reducing the time complexity of the original algorithm, and to propose heuristics to further improve the performance of this algorithm. In Section 2 we make basic definitions. In Section 3 we present results that permitted us to lower the complexity of Christie's algorithm. In Section 4 we describe heuristics to improve the experimental results of Christie's algorithm. In Section 5, we present some experiments and a comparison from our implementation with the results of Bafna and Pevzner [1] (implemented in Oliveira [6]) and the results of Walter, Dias and Meidanis [8], noting that our heuristic leaded to better results. Finally in Section 6 we present some conclusions and suggest future works.

2   Definitions

A permutation π represents a sequence of n blocks $\pi = [\pi(0)\pi(1)\ldots\pi(n)\pi(n+1)]$, where $\pi(0) = 0$, $\pi(n+1) = (n+1)$ and $\pi(i), 1 \leq i \leq n$, is an integer in $1..n$. The identity permutation is defined as $\iota = [0\ 1\ 2\ \ldots\ n\ n+1]$. The inverse permutation is defined as $R = [0\ n\ n-1\ \ldots\ 2\ 1\ n+1]$.

The transposition $\tau(i,j,k), 1 \leq i < j < k \leq n+1$, is defined as the event transforming the permutation $\pi = [\pi(0)\ldots\pi(i-1)\pi(i)\pi(i+1)\ldots\pi(j-1)\pi(j)\pi(j+1)\ldots\pi(n+1)]$ in the permutation $\pi \cdot \tau(i,j,k) = [\pi(0)\ldots\pi(i-1)\pi(j)\ldots\pi(k-1)\pi(i)\ldots\pi(j-1)\pi(k)\ldots\pi(n+1)]$. For example, if we apply the transposition $\tau(2,4,6)$ to the permutation $\pi = [0\ 1\ 5\ 6\ 2\ 4\ 3\ 8\ 7\ 9]$, then $\pi \cdot \tau(2,4,6) = [0\ 1\ 2\ 4\ 5\ 6\ 3\ 8\ 7\ 9]$.

Now we define the *transposition distance problem*. Given two permutations π and σ, we want to compute a shortest sequence of transpositions $\tau_1, \tau_2, \ldots, \tau_t$ such that $\pi \cdot \tau_1 \cdot \tau_2 \cdot \ldots \cdot \tau_t = \sigma$ and t is minimum. We call t the *transposition distance* of π and σ and denote it by $d(\pi, \sigma)$. Without loss of generality, we can fix $\sigma = \iota$, and have the equivalent *problem of sorting by transpositions* (Fig. 2). So, our developments will be done on the last problem. To simplify, we will denote $d(\pi, \iota)$ by $d(\pi)$.

The cycle graph $G(\pi)$, $|\pi| = n$, is a directed edge-colored graph, defined by a set of nodes $\{0, 1, \ldots, n+1\}$, a set of gray edges $\{(i, i+1), 0 \leq i \leq n\}$ and a set of black edges $\{(\pi(i), \pi(i-1)), 1 \leq i \leq n+1\}$ (Fig. 3).

In each node of $G(\pi)$, the edge entering this node is followed by an edge with the other color leaving this node. This matching of edges decompose the

$$\pi \, = 0 \quad 4 \quad 5 \quad 2 \quad 3 \quad 1 \quad 6$$

$$\pi_1 \, = \, \pi \cdot \tau \, (1,3,6) \, = \, 0 \quad 2 \quad 3 \quad 1 \quad 4 \quad 5 \quad 6$$

$$\pi_2 \, = \, \pi_1 \cdot \tau \, (1,3,4) \, = \, 0 \quad 1 \quad 2 \quad 3 \quad 4 \quad 5 \quad 6$$

Fig. 2. Sorting the permutation $\pi = [0\ 4\ 5\ 2\ 3\ 1\ 6]$ with two transpositions. In this example $d(\pi) = 2$.

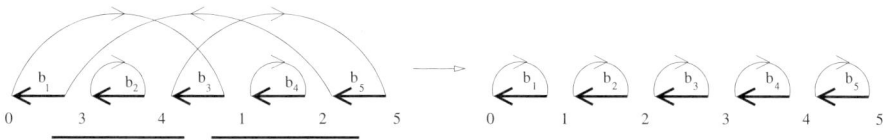

Fig. 3. The cycle graph $G(\pi)$, for $\pi = [0\ 3\ 4\ 1\ 2\ 5]$ and a 2-transposition, indicated by the gray bars.

graph in alternate cycles, and this decomposition is unique, because each node is connected with exactly two edges, each one with a different color, entering or leaving this node. The black edge $(\pi(i), \pi(i-1))$ is indicated as b_i. The cycle graph is drawn in such a way that the black edges occur in the order $b_1, \ldots, b_{n+1}$, from left to right in $G(\pi)$ (Fig. 3).

The number of cycles in $G(\pi)$ is denoted by $c(\pi)$. In Fig. 3 the first cycle graph has $c(\pi) = 3$ and the second one has $c(\pi) = 5$. The basic idea to sort a permutation π_n, $|\pi_n| = n$, is to increase $c(\pi_n)$ in $G(\pi_n)$ to $n + 1$. This can be explained by the fact that $G(\iota_n)$ has exactly $n + 1$ cycles, being the only permutation with this number of cycles.

Bafna and Pevzner [1] defined the *length* of a cycle C as the number of black edges in C; an *even cycle* as an even length cycle, and an *odd cycle* as an odd length cycle. They presented the following result, where $\Delta c(\pi \cdot \tau)$ is the variation of the number of cycles in $G(\pi)$ when transposition τ is applied.

Theorem 1. $\Delta c(\pi \cdot \tau) \in \{-2, 0, +2\}$

So the best we can do in each transposition is to increase the number of cycles in a cycle graph by 2. Bafna and Pevzner [1] proved the following lower bound, using Theorem 1 and the fact that all $n + 1$ cycles of $G(\iota_n)$ have length 1. In this equation, $c_{odd}(\pi_n)$ denotes the number of odd cycles in $G(\pi_n)$.

Theorem 2.

$$\frac{(n+1) - c_{odd}(\pi_n)}{2} \le d(\pi_n)$$

Bafna and Pevzner [1] and independently Christie [2] have demonstrated the following upper bound.

Theorem 3.

$$d(\pi_n) \leq \frac{3((n+1) - c_{odd}(\pi_n))}{4}$$

In the proof of the last theorem, Christie [2] used some properties of the cycles in the cycle graph of a permutation, based on the definitions below.

Firstly, we can observe from Theorems 2 and 3 that to sort a permutation we have to increase the number of odd cycles as fast as we can. So, Christie [2] defined a 2-*transposition* as a transposition generating two new odd cycles in $G(\pi)$, and a 0-*transposition* as a transposition not increasing the number of odd cycles in $G(\pi)$. A *triple* of black edges x, y, z from $G(\pi)$ is *oriented* if the transposition applied in these edges increases by two $c(\pi)$ in $G(\pi)$. A *triple* of black edges is *non-oriented* if the transposition applied on these edges does not increase $c(\pi)$ in $G(\pi)$. A cycle is *oriented* if it has at least one oriented triple. A *cycle* is *non-oriented* if it is not oriented. A *triple* of black edges x, y, z of $G(\pi)$ is *fully oriented* if the transposition $\tau(x, y, z)$ is a 2-transposition, that is, if the transposition applied on these edges generates two new odd cycles in $G(\pi)$. A cycle containing a fully oriented triple is a *fully oriented cycle*. For example, in Fig. 3, the cycle in which the transposition acts is a fully oriented cycle, because it has the fully oriented triple (b_1, b_3, b_5).

Two *pairs of black edges* (b_i, b_j) and (b_k, b_l) *intersect* if $i < k < j < l$, or $k < i < l < j$. A *cycle intersect with two black edges* x and y if the cycle has two black edges that intersect with x and y. Two *cycles* C and D *intersect* if C has two black edges that intersect with two black edges of D.

Two *triples of black edges* (b_i, b_j, b_k) and (b_l, b_m, b_n) are *interleaving* if $i < l < j < m < k < n$ or $l < i < m < j < n < k$. Two *cycles* C and D are *interleaving* if C has a triple of black edges interleaving with a triple of black edges of D. A *transposition* is *interleaving* with a cycle C if the three edges on which the transposition acts are interleaving with three black edges of C.

The *canonical labeling of a cycle* $C, |C| = L$, is obtained associating new labels $1, 2, \ldots, L$ to the black edges of C, preserving the relative position of the edges. We define it associating label 1 to the edge $b_{C_{min}}$, label 2 to edge b_i of C with i minimum, but larger than C_{min}, and so forth, until label L associated with C_{max}. Then, the canonical labeling of C is the permutation of the set $\{1, 2, \ldots, L\}$ obtained from this labeling, beginning in edge 1 and traversing the other edges in the order they appear in the cycle. For example, in Fig. 3, the cycle with black edges b_1, b_3, b_5, has canonical labeling $(1, 2, 3)$.

3   Lowering the Time Complexity of Christie's Algorithm

Christie [2] noted that fully oriented triples could be obtained in time $O(n^3)$, and the other steps of his algorithm could be executed in time $O(n^2)$. So, the time complexity for obtaining fully oriented triples determines the time complexity of his algorithm. In this section we present a $O(n^2)$ time complexity algorithm for obtaining fully oriented triples, which allowed us to reduce the time complexity of Christie's algorithm.

The canonical labeling of a cycle can be divided in two subsequences. The *outer sequence* $O(C)$ of the canonical labeling is defined taking the first element, the third, and so on. The *inner sequence* $I(C)$ of the canonical labeling is defined taking the second element, the fourth, and so on. Christie [2] proved the following four results.

Proposition 1. *Let* $[x\ y\ z]$ *be a subsequence of the canonical labeling of a cycle. Then* (x, y, z) *is an oriented triple if and only if* $x < y < z$, $y < z < x$ *or* $z < x < y$.

Theorem 4. *If C is an oriented even cycle, then C is also a fully oriented cycle.*

Lemma 1. *Let* $[x\ y\ z]$ *be a subsequence of the canonical labeling of an even cycle C, such that* $x < y < z$, $y < z < x$ *or* $z < x < y$. *Then* (x, y, z) *is a fully oriented triple if and only if* $[x\ y\ z]$ *is neither a subsequence of $I(C)$ nor a subsequence of $O(C)$.*

Lemma 2. *Let* $[x\ y\ z]$ *be a subsequence of the canonical labeling of an odd cycle C. Then* (x, y, z) *is a fully oriented triple if and only if one of the following six conditions occurs:*

1. $x < y < z$, x *and* z *are in* $O(C)$ *and* y *is in* $I(C)$
2. $x < y < z$, x *and* z *are in* $I(C)$ *and* y *is in* $O(C)$
3. $y < z < x$, x *and* z *are in* $O(C)$ *and* y *is in* $I(C)$
4. $y < z < x$, x *and* z *are in* $I(C)$ *and* y *is in* $O(C)$
5. $z < x < y$, x *and* z *are in* $O(C)$ *and* y *is in* $I(C)$
6. $z < x < y$, x *and* z *are in* $I(C)$ *and* y *is in* $O(C)$

We will show now that it is possible to find a fully oriented triple in a cycle C without using sets $I(C)$ and $O(C)$. We will denote $p(C, x)$ as the position of edge x in cycle C, according to the canonical labeling of C. The proof of next result is simple and will be omitted.

Lemma 3. *Let x and y be two black edges of a cycle C in $G(\pi)$, $|C| = L$. Then:*

1. x *and* y *belong both to* $I(C)$, *or* $O(C)$, *if and only if* $|p(C, x) - p(C, y)| \equiv 0 \bmod 2$
2. $x \in I(C)$ *and* $y \in O(C)$, *or vice-versa, if and only if* $|p(C, x) - p(C, y)| \equiv 1 \bmod 2$

The above lemma will be used in the following two results.

Lemma 4. *Let* $[x\ y\ z]$ *be a subsequence of the canonical labeling of an even cycle C, $|C| > 2$, such that* $x < y < z$, $y < z < x$ *or* $z < x < y$. *Then* (x, y, z) *is a fully oriented triple if and only if* $|p(C, x) - p(C, y)| \equiv 1 \bmod 2$ *or* $|p(C, x) - p(C, z)| \equiv 1 \bmod 2$ *or* $|p(C, y) - p(C, z)| \equiv 1 \bmod 2$.

Proof. ($\Rightarrow$)From Lemma 1, as $[x\ y\ z]$ is a subsequence of the canonical labeling of an even cycle C such that $x < y < z$, $y < z < x$ or $z < x < y$, the triple (x, y, z) is a fully oriented triple of C if and only if $[x\ y\ z]$ is not a subsequence of $O(C)$ nor a subsequence of $I(C)$. This occurs when two of the three edges x, y, z are related by one of the following ways:

- $x \in O(C)$ and $y \in I(C)$ or
- $x \in I(C)$ and $y \in O(C)$ or
- $x \in O(C)$ and $z \in I(C)$ or
- $x \in I(C)$ and $z \in O(C)$ or
- $y \in O(C)$ and $z \in I(C)$ or
- $y \in I(C)$ and $z \in O(C)$

Then, using Lemma 3:

1. First and second conditions are equivalent to $|p(C, x) - p(C, y)| \equiv 1 \bmod 2$.
2. Third and fourth conditions are equivalent to $|p(C, x) - p(C, z)| \equiv 1 \bmod 2$.
3. Fifth and sixth conditions are equivalent to $|p(C, y) - p(C, z)| \equiv 1 \bmod 2$.

($\Leftarrow$) From Lemma 3 (b), if $|p(C, x) - p(C, y)| \equiv 1 \bmod 2$, $x \in O(C)$ and $y \in I(C)$, or vice-versa. If $|p(C, x) - p(C, z)| \equiv 1 \bmod 2$, then $x \in O(C)$ and $z \in I(C)$, or vice-versa. If $|p(C, y) - p(C, z)| \equiv 1 \bmod 2$, then $y \in O(C)$ and $z \in I(C)$, or vice-versa. In all of these cases the x, y, and z do not belong simultaneously to $O(C)$ or to $I(C)$. Then taking a subsequence of the canonical label of an even cycle C, $[x, y, z]$ such that $x < y < z$, $y < z < x$ or $z < x < y$, it follows from Lemma 1 that (x, y, z) is a fully oriented triple of C. □

The proof of next lemma, that uses Lemmas 2 and 3, is analogous to that of Lemma 4, and will be omitted.

Lemma 5. *Let $[x\ y\ z]$ be a subsequence of the canonical labeling of an odd cycle C, $|C| > 2$. Then (x, y, z) is a fully oriented triple in C if and only if one of the following three conditions occurs:*

1. $x < y < z$, $|p(C, x) - p(C, z)| \equiv 0 \bmod 2$ *and* $|p(C, x) - p(C, y)| \equiv 1 \bmod 2$
2. $y < z < x$, $|p(C, x) - p(C, z)| \equiv 0 \bmod 2$ *and* $|p(C, x) - p(C, y)| \equiv 1 \bmod 2$
3. $z < x < y$, $|p(C, x) - p(C, z)| \equiv 0 \bmod 2$ *and* $|p(C, x) - p(C, y)| \equiv 1 \bmod 2$

Lemmas 4 and 5 gave another characterization for fully oriented triples in even and odd cycles, different from the conditions presented by Christie [2] on Lemmas 1 and 2.

Next, we will demonstrate the correctness and the $O(n^2)$ time complexity of an algorithm to find fully oriented triples in odd cycles.

Lemma 6. *Given the canonical labeling of an odd cycle C, $|C| > 2$, algorithm FindTripleOddCycleCase1 returns a fully oriented triple (x, y, z), with $x < y < z$, if such a triple exists. If it does not, the algorithm will indicate it.*

Proof. The algorithm tries to find three edges x, y, z, in cycle C such that $x < y < z$ and $|p(C, x) - p(C, z)| \equiv 0 \bmod 2$ and $|p(C, x) - p(C, y)| \equiv 1 \bmod 2$, because if it exists, case (1) of Lemma 5 says that (x, y, z) is a fully oriented triple.

In the first *for*, we associate with edge x all possible values in the canonical labeling of C. For each x: in the first *while*, the algorithm choose the first edge y such that $y > x$ and $|p(C, x) - p(C, y)| \equiv 1 \bmod 2$. For these edges x and y, the algorithm tries to find edge z, such that $z > y$ and $|p(C, x) - p(C, z)| \equiv 0 \bmod 2$:

- If edge z following edge y in the canonical labeling of C is such that $z > y$ (we note that $z > x$, because $y > x$) and $|p(C, x) - p(C, z)| \equiv 0 \bmod 2$ then (x, y, z) is a fully oriented triple, by case (1) of Lemma 5.
- If edge z following edge y in the canonical labeling of C is such that $z < y$, $z > x$ and $|p(C, x) - p(C, z)| \equiv 1 \bmod 2$ then we change current edge y by edge z, noting two facts. First, for this new edge y, we have the relation $|p(C, x) - p(C, y)| \equiv 1 \bmod 2$. Second, taking the old value of edge y, if there was an edge $z' > y$, with $|p(C, x) - p(C, z')| \equiv 0 \bmod 2$, $|p(C, x) - p(C, y)| \equiv 1 \bmod 2$ (so (x, y, z') would be a fully oriented triple) the relation between edges x, y and z' will also be preserved between x, y' (y changed to z) and z'. This shows that we are not "losing" a fully oriented triple when changing edge y. In other words, a change in edge y occurs if the algorithm finds another edge y' following edge y in the canonical labeling of C such that $x < y' < y$, and the conditions between edges x and y are preserved between edges x and y'.
- If edge z following edge y in the canonical labeling of C is such that $z > y$ and $z > x$ and $|p(C, x) - p(C, z)| \equiv 1 \bmod 2$, or $z < y$ and $z < x$, or $z < y$, $z > x$ and $|p(C, x) - p(C, z)| \equiv 0 \bmod 2$ or $z < x$, then the algorithm just tries another value for z, following the current value of z in the canonical labeling of C.

These are all possible relations among edges z, x and y.   □

We note that algorithm *FindTripleOddCycleCase*1 (Fig. 4), which implements case (1) of Lemma 5, although not testing all possibilities for triples (x, y, z), finds a fully oriented triple, if it exists.

Lemma 7. *Algorithm FindTripleOddCycleCase1 (with input $[c_1, c_2, \ldots, c_L]$, the canonical labeling of an odd cycle C, $|C| = L > 2$) executes in $O(L^2)$ steps.*

Proof. Command *for* is executed $L - 2$ times. Inside this command, steps 2 to 6 are time constant. Command *while* on the step 7 is executed at most $(L - 1)$ times. Inside this command, steps 8 to 11 are time constant. Steps 13, 14 and 15 are also time constant. Command *while* in step 16 is executed at most L times. Inside this command, steps 17 to 24 are all time constant. Step 25 is time constant. The *while* commands are executed one after the other, and therefore, both execute at most in L steps. Thus, the algorithm has time complexity of $O(L^2)$.   □

```
algorithm FindTripleOddCycleCase1
input: Canonical labeling [c_1, c_2, ..., c_L] of an odd cycle C, |C| = L > 2
output: (x, y, z), x < y < z, a fully oriented triple in C;
          NO, if such a triple does not exist in C.
begin
    1. for i = 1 to L − 2 do
    2.        x = c_i
    3.      { Figuring out edge y }
    4.        j = i + 1
    5.        y = c_j { |p(C, x) − p(C, y)| ≡ 1 mod 2 }
    6.        foundY = false
    7.        while not foundY and j < L − 1 do
    8.            if y > x then
    9.                foundY = true
   10.            else
   11.                j = j + 2
   12.      { Figuring out edge z }
   13.        if foundY then
   14.            k = j + 1
   15.            z = c_k { |p(C, x) − p(C, z)| ≡ 0 mod 2 }
   16.            while k ≤ L do
   17.                if z > y and |i − k| ≡ 0 mod 2 then
   18.                    return (x, y, z) { fully oriented triple found }
   19.                else if z < y and z > x and |i − k| ≡ 1 mod 2 then
   20.                    j = k
   21.                    y = c_k
   22.                    k = k + 1 {tries next edge z in the canonical labeling}
   23.                else
   24.                    k = k + 1 {tries next edge z in the canonical labeling}
   25. Return NO { A fully oriented triple does not exist in C}
end.
```

Fig. 4. The algorithm to find fully oriented triples in odd cycles.

We developed analogous results to find fully oriented triples in odd cycles for the other two cases of Lemma 5, $y < z < x$ and $z < x < y$. The corresponding algorithms are $FindTripleOddCycleCase2$ and $FindTripleOddCycleCase3$, executed in sequence. The proof of the following theorem is based on these results.

Theorem 5. *Given an odd cycle C, $|C| > 2$, algorithms*
$FindTripleOddCycleCase1$, $FindTripleOddCycleCase2$ and
$FindTripleOddCycleCase3$ return a fully oriented triple (x, y, z), with $x < y < z$,
$y < z < x$ or $z < x < y$, if such a triple exists. If such a triple does not exist, it
will be indicated at the end of the execution of the algorithms.

Next theorem follows from Lemma 7 and the results for the other cases.

Theorem 6. *Algorithms $FindTripleOddCycleCase1$, $FindTripleOddCycleCase2$*
and $FindTripleOddCycleCase3$ executes in $O(n^2)$ steps, where n is the length of
the permutation being sorted.

The following results show the correctness and $O(n^2)$ time complexity for an algorithm to find fully oriented triples in even cycles. Initially, let us take R, a sequence with $|R| = L$. Define R_i, $0 \le i \le L − 1$, as the sequence in which $R_i(j) = R(i + j \bmod L)$, $1 \le j \le L$. Then R_i is the sequence obtained from R rotating its elements from left to right, such that $R(i)$ will be the first element of R_i. R_{L-i} is the inverse sequence of R_i, because when applying R_{L-i} to R_i we obtain $R_L = R_0 = R$. For example, if we take $R = [1\ 2\ 3\ 4\ 5]$, then $R_1 = [5\ 1\ 2\ 3\ 4]$, $R_2 = [4\ 5\ 1\ 2\ 3]$, $R_0 = R_5 = R$.

Now, we enunciate some technical lemmas.

Lemma 8. *Let us take R, $|R| = L \geq 3$. If there are three elements x, y and z such that $x < y < z$ in one of the sequences R_i from R, $0 \leq i \leq L - 1$, and such that $[x\ y\ z]$ is a subsequence of R_i, then there is a triple (x', y', z') in R, such that $[x'y'z']$ is a subsequence of R, and $x' < y' < z'$, $y' < z' < x'$ or $z' < x' < y'$.*

Proof. Let us take x, y and z three elements of one of the sequences R_i from R, $0 \leq i \leq L - 1$, such that $[x\ y\ z]$ is a subsequence of R_i and $x < y < z$. Taking a, b, c as the positions of x, y, z in R_i we have:

- $x = R_i(a) = R(i + a \bmod L)$
- $y = R_i(b) = R(i + b \bmod L)$
- $z = R_i(c) = R(i + c \bmod L)$

When we apply the inverse sequence R_{L-i} to R_i, $0 \leq i \leq L - 1$, we will obtain sequence R. Let us take now x', y', z' such that $x' = R_{L-i}(R_i(a)) = R(a)$, $y' = R_{L-i}(Ri(b)) = R(b)$ and $z' = R_{L-i}(R_i(c)) = R(c)$. Then, in R, $R(a) < R(b) < R(c)$ or $R(b) < R(c) < R(a)$ or $R(c) < R(a) < R(b)$, in which case $x' < y' < z'$ or $y' < z' < x'$ or $z' < x' < y'$. □

Lemma 9. *Let us take R as the canonical labeling of a cycle C, $|C| = L > 2$, L even. Then, if there are three elements x, y and z such that $x < y < z$ in one of the sequences R_i of R, $0 \leq i \leq L - 1$, and such that $[x\ y\ z]$ is a subsequence of R_i, then the triple (x, y, z) is oriented in R.*

Proof. From Lemma 8, if there are three elements x, y and z in one of the subsequences R_i of R, $0 \leq i \leq L - 1$, such that $x < y < z$ and $[x\ y\ z]$ is a subsequence of R_i, then there are elements x', y' and z' in R such that $[x'\ y'\ z']$ is a subsequence of R, and $x' < y' < z'$, $y' < z' < x'$ or $z' < x' < y'$. Following Proposition 1, one of these conditions define an oriented triple (x', y', z') in C. □

Lemmas 4 and 9, and Theorem 4, prove the following lemma.

Lemma 10. *Let us take C, $|C| = L > 2$, L even, R the canonical labeling of C. If there are three elements x, y and z in one of the sequences R_i from R such that $x < y < z$ and $[x\ y\ z]$ is a subsequence of R_i, $0 \leq i \leq L - 1$, and $|p(C, x) - p(C, y)| \equiv 1 \bmod 2$ or $|p(C, x) - p(C, z)| \equiv 1 \bmod 2$ or $|p(C, y) - p(C, z)| \equiv 1 \bmod 2$ then (x, y, z) is a fully oriented triple in R.*

Then, if we find a triple in any sequence R_i from the canonical labeling R of an even cycle, following the conditions of Lemma 10, it will be fully oriented. Besides, we can prove that if a fully oriented triple (x, y, z) exists in R, it can be found in one of the sequences R_i from R, analyzing only the case $x < y < z$ in all R_i. Then the three cases of Lemma 4 can be reduced to only one case. The following lemma proves the correctness of the algorithm for searching fully oriented triples in even cycles.

Lemma 11. *Given an even cycle C, $|C| > 2$, algorithm FindTripleEvenCycle returns a fully oriented triple (x, y, z), with $x < y < z$, if such a triple exists. If such a triple does not exist, the algorithm will indicate it.*

Algorithm *FindTripleEvenCycle* receives as input the canonical labeling of an even cycle C, $|C| = L > 2$ and returns a fully oriented triple in C if it exists or NO, if such a triple does not exist in C. It builds all sequences R_i, $0 \leq i \leq L - 1$, and searches in each R_i a triple (x, y, z), such that $x < y < z$ and $|p(C, x) - p(C, y)| \equiv 1 \bmod 2$ or $|p(C, x) - p(C, z)| \equiv 1 \bmod 2$ or $|p(C, y) - p(C, z)| \equiv 1 \bmod 2$. These cases are analysed sequentially and within each case x is fixed and the loops for y and z are independent from each other. This algorithm is very similar to *FindTripleOddCycleCase*1, and will not be described here.

The proofs of the next results are very similar to those for the cases of odd cycles.

Lemma 12. *Algorithm FindTripleEvenCycle (with input $[c_1, c_2, \ldots, c_L]$, the canonical labeling of an even cycle C, $|C| = L > 2$) executes in at the most $O(L^2)$ steps.*

The above lemma is used in the proof of the following theorem, that shows the correctness of the algorithm that finds a fully oriented triple for even cycles.

Theorem 7. *Given an even cycle C, $|C| > 2$, algorithm FindTripleEvenCycle returns a fully oriented triple (x, y, z), with $x < y < z$, $y < z < x$ or $z < x < y$, if such a triple exists. If such a triple does not exist, it will be indicated by the algorithm.*

The next result shows the time complexity of the algorithm.

Theorem 8. *Algorithm FindTripleEvenCycle executes in $O(n^2)$ steps, where n is the length of the permutation being sorted.*

4   Analyzing Not Intersecting Non-oriented Even Cycles

In this section, we introduce heuristics that improved the results obtained by the algorithm of Christie [2]. The new idea here is to investigate even cycles, and not only odd cycles, as Christie [2] and Bafna and Pevzner [1] did.

Lemma 13. *If $G(\pi)$ has an even oriented cycle C then we can apply two 2-transpositions in C.*

Lemma 14. *Let us take C and D two not intersecting non-oriented even cycles in $G(\pi)$, C with the more left black edge and D the other cycle. Then if $G(\pi)$ has a non-oriented cycle E, $|E| \geq 3$, and three edges x, y, z from E such that x, y, z interleave with three from the four edges C_{min}, C_{max}, D_{min}, D_{max}, then it is possible to apply a sequence of two 2-transpositions in $G(\pi)$.*

The proofs of these two lemmas indicate which two 2-transpositions we have to apply. Fig. 5 shows an example for Lemma 14. Christie's algorithm does not verify these conditions, and his algorithm can "lose" these two 2-transpositions.

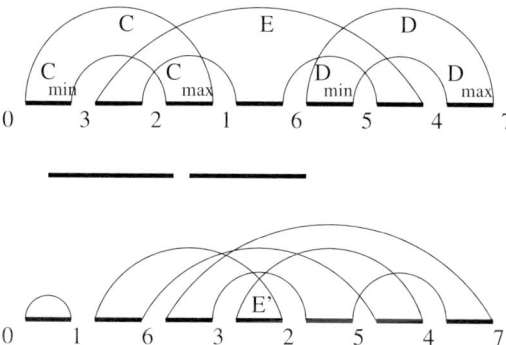

Fig. 5. This is an example of a non-oriented cycle E having interleaving edges with two not intersecting non-oriented even cycles, C and D. Note that $\tau(1,3,5)$ applied to the first cycle graph, and $\tau(2,4,6)$ applied to the second cycle graph, are 2-transpositions.

5   Experiments

In Table 1 we show the results from the implementations of the algorithms of Christie [2], Bafna and Pevzner [1] (Oliveira [6]), Walter, Dias and Meidanis [8], besides the results obtained by Christie's algorithm with the heuristic of not intersecting non-oriented even cycles. In this table, *length* indicates the length of the analyzed permutations, *permutation number* indicates the number of permutations for each length, and the other columns show the number of permutations in which $d(\pi)$ is different from the value computed by the indicated algorithm.

Table 1. Comparing the results of the algorithms.

length	permutation number	Christie	Bafna and Pevzner	Walter, Dias and Meidanis	Christie with the heuristic
2	2	0	0	0	0
3	6	0	0	0	0
4	24	0	0	0	0
5	120	0	0	0	0
6	720	3	0	6	0
7	5,040	37	1	72	0
8	40,320	588	135	1,167	40
9	362,360	7,970	4,361	14,327	1,182

We can see from this table that the heuristic on even cycles significantly reduced the number of cases in which $d(\pi)$ is different from the value computed by the Bafna and Pevzner [1] algorithm, implemented by Oliveira [6], the best we got.

6   Conclusions and Future Works

In this work we implemented the approximation algorithm with factor 1.5 proposed by Christie [2] for the problem of sorting by transpositions, reducing its time complexity and proposing a heuristic that further improved its performance. We also compared the results obtained in this implementation with the results of Bafna and Pevzner [1] (implemented in Oliveira [6]), and with the results Walter, Dias and Meidanis [8], obtaining better results.

As future works, we suggest the analysis of non-oriented cycles interleaving with a triple of edges of not intersecting even cycles, because we conjecture that transpositions applied in these interleaving edges would improve the results, when comparing to the transposition distance. The analysis of the permutations of lengths 8 and 9 with values of distance (computed by the algorithm of Christie with the heuristic) different from the transposition distance would also suggest new interesting properties of the cycles in the cycle graph. Finally, another important study is to discover the transposition diameter, yet not known.

References

1. V. Bafna and P. Pevzner. Sorting by transpositions. In *Proceedings of the 6th Annual ACM-SIAM Symposium on Discrete Algorithms*, pages 614–623, 1995.
2. D. A. Christie. *Genome rearrangements problems*. PhD thesis, Glasgow University, Scotland, 1998.
3. Z. Dias and J. Meidanis. Sorting by prefix transpositions. In *String Processing and Information Retrieval – SPIRE 2002*, 2002. Lecture Notes in Computer Science, v. 2476, 2002.
4. S. A. Guyer, L. S. Heath, and J. P. C. Vergara. Subsequences and run heuristics for sorting by transpositions. In *4th Dimacs International Algorithm Implementation Challenge*, 1995.
5. J. Meidanis, M. E. M. T. Walter, and Z. Dias. Transposition distance of strictly decreasing sequences. In *Workshop on String Processing-WSP* 97, 1997.
6. E. T. G. Oliveira. Implementations of algorithms to the problem of sorting by transpositions. Master's thesis, Department of Computer Science, University of Brasilia, 2001.
7. J. P. C. Vergara. *Sorting by Bounded Permutations*. PhD thesis, Virginia Polytechnic Institute and State University, 1997.
8. M. E. M. T. Walter, Z. Dias, and J. Meidanis. A new approach for approximating the transposition distance. In *String Processing and Information Retrieval - SPIRE 2000*, pages 199–208, 2000.

Efficient Selection of Unique and Popular Oligos for Large EST Databases

Jie Zheng[1], Timothy J. Close[2], Tao Jiang[1], and Stefano Lonardi[1]

[1] Dept. of Computer Science & Engineering
[2] Department of Botany & Plant Sciences
University of California
Riverside, CA 92521

Abstract. EST databases have grown exponentially in recent years and now represent the largest collection of genetic sequences. An important application of these databases is that they contain information useful for the design of gene-specific oligonucleotides (or simply, oligos) that can be used in PCR primer design, microarray experiments, and genomic library screening. In this paper, we study two complementary problems concerning the selection of short oligos, *e.g.*, 20–50 bases, from a large database of tens of thousands of EST sequences: (i) selection of oligos each of which appears (exactly) in one EST sequence but does not appear (exactly or approximately) in any other EST sequence and (ii) selection of oligos that appear (exactly or approximately) in many ESTs. The first problem is called the *unique oligo* problem and has applications in PCR primer and microarray probe designs. The second is called the *popular oligo* problem and is useful in screening genomic libraries (such as BAC libraries) for gene-rich regions. We present an efficient algorithm to identify all unique oligos in the ESTs and an efficient heuristic algorithm to enumerate the most popular oligos. By taking into account the distribution of the frequencies of the words in the EST database, the algorithms have been carefully engineered to achieve remarkable running times on regular PCs. Each of the algorithms takes only a couple of hours (on a 1.2 GHz CPU, 1 GB RAM machine) to run on a dataset 28 Mbases of barley ESTs from the HARVEST database. We present simulation results on synthetic data and a preliminary analysis of the barley EST database.

1   Introduction

Expressed sequence tags (ESTs) are partial sequences of expressed genes, usually 200–700 bases long, which are generated by sequencing from one or both ends of cDNAs. The information in an EST allows researchers to infer functions of the gene based on similarity to genes of known functions, source of tissue and timing of expression, and genetic map position. EST sequences have become widely accepted as a cost-effective method to gather information about the majority of expressed genes in a number of systems. They can be used to accelerate various research activities, including map-based cloning of genes that control

R. Baeza-Yates et al. (Eds.): CPM 2003, LNCS 2676, pp. 384–401, 2003.

traits, comparative genome analysis, protein identification, and numerous methods that rely on gene-specific oligonucleotides (or *oligos*, for short) such as the DNA microarray technology.

Due to their utility, speed with which they may be obtained, and the low cost associated with this technology, many individual scientists and large genome sequencing centers have been generating hundreds of thousands of ESTs for public use. EST databases have been growing exponentially fast since the first few hundreds sequences obtained in the early nineties by Adams *et al.* [1], and now they represent the largest collection of genetic sequences. As of September 2002, the number of sequences deposited in NCBI's dbEST [2] has reached 13 million sequences out of the 18 million sequences which composes the entire GenBank.

With the advent of whole genome sequencing, it may appear that ESTs have lost some of its appeal. However, the genomes of many organisms that are important to society, including the majority of crop plants, have not yet been fully sequenced, and the prospects for large-scale funding to support the sequencing of any but a few in the immediate future is slim to none. In addition, several of our most important crop plants have genomes that are of daunting sizes and present special computational challenges because they are composed mostly of highly repetitive DNA. For example, the *Triticeae* (wheat, barley and rye) genomes, each with a size of about 5×10^9 base pairs per haploid genome (this is about twice the size of maize, 12 times the size of rice, and 35 times the size of the Arabidopsis genomes), are too large for us to seriously consider whole genome sequencing at the present time.

Of the many EST databases, we will be especially interested in the dataset of barley (*Hordeum vulgare*). Barley is premiere model for Triticeae plants due to its diploid genome and a rich legacy of mutant collections, germplasm diversity, mapping populations (see http://www.css.orst.edu/barley/nabgmp/nabgmp.htm), and the recent accumulation of other genomics resources such as BAC [3] and cDNA libraries [4,5]. Nearly 300,000 publicly available ESTs derived from barley cDNA libraries are currently present in dbEST. These sequences have been quality-trimmed, cleaned of vector and other contaminating sequences, pre-clustered using the software TGICL (http://www.tigr.org/tdb/tgi/software/) and clustered into final assemblies of "contigs" (*i.e.*, overlapping EST sequences) and "singletons" (*i.e.*, non-overlapping EST sequences) using CAP3 [6]. The collection of the singletons and consensus sequences of the contigs, called *unigenes*, form our main dataset. As of July 23, 2002, the collection has 46,145 unigene ESTs of a total of 28,475,017 bases. This dataset can be obtained from http://harvest.ucr.edu/ using the HARVEST viewer.

In this paper, we study two computational problems arising in the selection of short oligos (*e.g.*, 20–50 bases) from a large EST database. One is to identify oligos that are *unique* to each EST in the database. The other is to identify oligos that are popular among the ESTs. More precisely, the *unique oligo* problem asks for the set of all oligos each of which appears (exactly) in one EST sequence but does not appear (exactly or approximately) in any other EST sequence, whereas the *popular oligo* problem asks for a list of oligos that appear (exactly or approximately) in the largest number of ESTs. [1]

A unique oligo can be thought of as a "signature" that distinguishes an EST from all the others. Unique oligos are particularly valuable as locus-specific PCR primers for placement of ESTs at single positions on a genetic linkage map, on microarrays for studies of the expression of specific genes without signal interference from other genes, and to probe genomic libraries [7] in search of specific genes.

Popular oligos can be used to screen efficiently large genomic library. They could allow one to simultaneously identify a large number of genomic clones that carry expressed genes using a relatively small number of (popular) probes and thus save considerable amounts of money. In particular for the database under analysis, it has been shown previously by a number of independent methods that the expressed genes in Triticeae are concentrated in a small fraction of the total genome. In barley, this portion of the genome, often referred to as the *gene-space*, has been estimated to be only 12% of the total genome [8]. If this is indeed true, then at most 12% of the clones in a typical BAC library would carry expressed genes, and therefore also the vast majority of barley genes could be sequenced by focusing only on this 12% of the genome. An efficient method to reveal the small portion of BAC clones derived from the gene-space has the potential for tremendous cost savings in the context of obtaining the sequences of the vast majority of barley genes. The most commonly used barley BAC library has a 6.3 fold genome coverage, 17-filter set with a total of 313,344 clones [3]. This number of filters is inconvenient and costly to handle, and the total number of BAC clones is intractable for whole genome physical mapping or sequencing. However, a reduction of this library to a gene-space of only 12% of the total would make it fit onto two filters that would comprise only about 600 Mb. This is about the same size as the rice genome, which has been recently sequenced. A solution for the popular oligo problem should make it possible to develop an effective greedy approach to BAC library screening, enabling a very inexpensive method of identifying a large portion of the BAC clones from the gene-space. This would also likely accelerate progress in many crop plant and other systems that are not being considered for whole genome sequencing.

Our Contribution. In this paper, we present an efficient algorithm to identify all unique oligos in the ESTs and an efficient heuristic algorithm to enumerate the most popular oligos. Although the unique and popular oligos problems are complementary in some sense, the two algorithms are very different because unique oligos are required to appear in the ESTs while the popular oligos are not. In particular, the heuristic algorithm for popular oligos is much more involved than that for unique oligos, although their (average) running times are similar. The algorithms combine well-established algorithmic and data structuring techniques such as hashing, approximate string matching, and clustering, and take

---

[1] Note that, a popular oligo does not necessarily have to appear exactly in any EST.

advantage of the facts that (i) the number of mismatches allowed in these problems is usually small and (ii) we usually require a pair of approximately matched strings to share a long common substring (called a *common factor* in [9]). These algorithms have been carefully engineered to achieve satisfactory speeds on PCs, by taking into account the distribution of the frequencies of the words in the input EST dataset. For example, running each of the algorithms for the barley EST dataset from HARVEST takes only a couple of hours (on a 1.2 GHz AMD machine). This is a great improvement over other brute-force methods, like the ones based on BLAST. [2] Simulations results show that the number of missed positives by the heuristic algorithm for popular oligos is very limited and can be controlled very effectively by adjusting the parameters.

Previous related work. The problem of finding infrequent and frequent patterns in sequences is a common task in pattern discovery. A quite large family of pattern discovery algorithms has been proposed in the literature and implemented in software tools. Without pretending to be exhaustive, we mention MEME [10], PRATT [11,12], TEIRESIAS [13], CONSENSUS [14], GIBBS SAMPLER [15,16], WINNOWER [17,18], PROJECTION [19,20], VERBUMCULUS [21], MITRA [22], among others. Although these tools have been demonstrated to perform very well on small and medium-size datasets, they cannot handle large datasets such as the barley EST dataset that we are interested in. In particular, some of these tools were designed to attack the "challenge" posed by Pevzner and Sze [17], which is in the order of a few Kbases. Among the more general and efficient tools, we tried to run TEIRESIAS on the 28 Mbases barley EST dataset on an 1.2GHz Athlon CPU with 1GB of RAM, without being able to obtain any result (probably due to lack of memory).

The unique oligo problem has been studied in the context of *probe design* [23,9,24]. The algorithms in [23,24] consider physical and structural properties of oligos and are very time consuming. (The algorithm in [24] also uses BLAST.) A very recent algorithm by Rahman [9] is, on the other hand, purely combinatorial. It uses suffix arrays instead of hash tables, and requires approximately 50 hours for a dataset of 40 Mb on a high-performance Compaq Alpha machine with 16 Gb of RAM. However, his definition of unique oligos is slightly different from ours (to be given in the next section).

The rest of the paper is organized as follows. Section 2 defines the unique and popular oligo problems formally. The algorithms are presented in Section 3. Experimental results on the barley EST dataset and simulation results can be found in Section 4. In Section 5, we draw some concluding remarks. The Appendix B explains the popular oligo algorithm with an example.

---

[2] For example, one can identify unique oligos by repeatedly running BLAST for each EST sequence against the entire dataset. This was the strategy previously employed by the HARVEST researchers.

2   The Unique and Popular Oligo Problems

We denote the input dataset as $X = \{x_1, x_2, \ldots x_k\}$, where the generic string x_i is an EST sequence over the alphabet $\Sigma = \{A, C, G, T\}$ and k is the cardinality of the set. Let n_i denote the length of the i-th sequence, $1 \leq i \leq k$. We set $n = \sum_{i=1}^{k} n_i$, which represents the total size of the input. A string (or oligo) from Σ is called an l-mer if its length is l.

Given a string x, we write $x_{[i]}$, $1 \leq i \leq |x|$, to indicate the i-th symbol in x. We use $x_{[i,j]}$ as a shorthand for the substring $x_{[i]}x_{[i+1]} \ldots x_{[j]}$ where $1 \leq i \leq j \leq n$, with the convention that $x_{[i,i]} = x_{[i]}$. Substrings in the form $x_{[1,j]}$ correspond to the *prefixes* of x, and substrings in the form $x_{[i,n]}$ to the *suffixes* of x. A string y *occurs* at position i of another string x if $y_{[1]} = x_{[i]}, \ldots, y_{[m]} = x_{[i+m-1]}$, where $m = |y|$. For any substring y of x, we denote by $f_x(y)$ the number of occurrences of y in x. $f_X(y)$ denotes the total number of occurrences of y in $x_1, \ldots, x_k$.

The *color-set* of y in the set $X = \{x_1, x_2, \ldots, x_k\}$ is a subset $col(y) = \{i_1, i_2, \ldots, i_l\}$ of $\{1, 2, \ldots, k\}$ such that if $i_j \in col(y)$ then y occurs at least once in x_{i_j}. We also say that y has *colors* $i_1, i_2, \ldots, i_l$. The number of colors, l, of y is denoted as $c_X(w)$. Clearly $f_X(y) \geq c_X(y)$.

Given two strings x and y of the same length, we denote by $H(x, y)$ the Hamming distance between x and y, that is, the number of mismatches between x and y. If $H(x, y) \leq d$, we say that x d-*matches* y and x is a d-*mutant* of y. The set of all the strings that d-match x is called the d-*neighborhood* of x. The notion of occurrences and colors can be extended to d-occurrence and d-colors by allowing up to d mismatches. If a string y has d-mutants at j distinct positions in a string x, we say that y has j d-occurrences in x. If a string y has at least one d-occurrence in each of j sequences in X, we say that y has j d-colors in X.

In the context of DNA hybridization, most papers define the specificity of an l-mer in terms of its mismatches to the length-l substrings of target sequences, although some also consider its physical and structural characteristics such as melting temperature, free-energy, GC-content, and secondary structure [23,24]. In [9], Rahman took a more optimistic approach and used the length of the longest common substring (called the longest common factor or LCF) as a measure of unspecificity. Given the nature of our target applications, we will take a conservative approach in the definitions of unique and popular oligos.

Definition 1. *Given the set $X = \{x_1, x_2, \ldots x_k\}$ of ESTs and integers l and d, a* unique oligo *is an l-mer y such that y occurs in at least one EST and the number of d-colors of y in X is exactly one. In other words, y appears exactly in some EST but does not appear approximately in any other EST.*

Suppose that strings x and y have the same length. For any given constants c, d, we say that string x (c, d)-*matches* string y if x and y can be partitioned into substrings as $x = x_1 x_2 x_3$ and $y = y_1 y_2 y_3$ with $|x_i| = |y_i|$ such that (i) $|x_2| = c$, (ii) $x_2 = y_2$, and (iii) the string $x_1 x_3$ d-matches the string $y_1 y_3$. In the above partition, we call x_2 a *core* in the (c, d)-match between x and y. (Note that a (c, d)-match may have many cores). The notion of d-occurrences and d-colors can be easily extended to (c, d)-occurrences and (c, d)-colors.

Definition 2. *Given the set* $X = \{x_1, x_2, \ldots x_k\}$ *of ESTs and integers* l, d, c *and* T, *a* popular oligo *is an* l-mer y *such that the number of* (c, d)-colors *of* y *in* X *is greater than or equal to* T. *In other words, the* l-mer y *appears approximately in at least* T *ESTs.*

The use of pooled oligo probes for BAC library screening [7] generally have lengths from 24 to 40 bases. Given this range and based on discussion with researchers from the Triticeae community, we consider $l = 33$ and $d = 5$ in the unique oligo problem. In the popular oligo problem, we consider $d = 1, 2, 3$ and $c = 20$.

3   The Algorithms

Our goal is to determine unique and popular oligos for a given set X of EST sequences. Although the objectives of the two problems seem complementary, our algorithms are quite different. The algorithm for the popular oligo problem turns out to be much more involved because popular oligos are not necessarily contained in the ESTs. Nevertheless, both algorithms share some common strategies such as the idea of separating dissimilar strings as early as possible to reduce the search space. To achieve this in the popular oligo problem, we first find cores (*i.e.*, c-mers) that appear exactly in at least two ESTs. Then we cluster all length-l substrings of the ESTs by their cores of length c using a hash table, and then, within each cluster, we cluster again the l-mers based on Hamming distance between regions flanking the cores. Candidate popular oligos are then enumerated from the small groups resulted from the two clusterings, and their number of colors are counted. For the unique oligo problem, the notion of cores, however, does not exist. On the other hand, due to the small number of mismatches allowed (relative to l), two l-mers that d-match each other must contain substrings that 1-match each other. Such substrings are called *seeds*. We can thus cluster the ESTs by their seeds using a dictionary. For each cluster, we compare the ESTs in the cluster by counting mismatches in the regions flanking the seeds. The details are given below.

3.1   Unique Oligos

Recall that the unique oligo problem is to identify length-l substrings of the ESTs that have exactly one d-color in the dataset X, for a given value of d. Our strategy is first to eliminate all the those l-mers that cannot be unique oligos. The algorithm is based on the following observation. Assume that x and y are two l-mers such that $H(x, y) \leq d$. Divide x and y into $t = \lfloor d/2 \rfloor + 1$ substrings. That is $x = x_1 x_2 \cdots x_t$ and $y = y_1 y_2 \cdots y_t$, where the length of each substring is $q = \lceil l/t \rceil$, except possibly for the last one. In practice, one can always choose l and d so that l is a multiple of t and hence x and y can be decomposed into t substrings of length q, which we call *seeds*. It is easy to see that since $H(x, y) \leq d$, at least one of the seeds of x has at most one mismatch with the corresponding seed of y.

UNIQUE-OLIGO-SELECTION(X, l, m)
Input: EST sequences $X = \{x_1, x_2, \ldots, x_k\}$
                l: length of the oligos to be reported
                d: maximum number of mismatches for non-unique oligos
Output: Mark each unique l-mers in X
1   $t, q \leftarrow \lfloor d/2 \rfloor + 1, \lceil l/t \rceil$
2   $table \leftarrow$ HIT(X, q)
3   EXTENSION$(X, t, q, table, d)$

COMPARE$(table[i][j], table[i][k], d)$
1   $q_1 \leftarrow$ the q-mer located at $table[i][j]$
2   $q_2 \leftarrow$ the q-mer located at $table[i][k]$
3   **for** each pair of l-mers that contain q_1 and q_2 as seeds respectively **do**
4           $c \leftarrow H(q_1, q_2)$
5   **if** $c \le d$ **then**
6   mark the two l-mers as *"non-unique"*

HIT(X, q)
1   **for** $i \leftarrow 1$ **to** 4^q **do**
2   initialize $table[i]$
3       $index[i] \leftarrow 0$
4   **for** $i \leftarrow 1$ **to** k **do**
5   **for** $j \leftarrow 1$ **to** $n_i - q + 1$ **do**
6           $key \leftarrow$ MAP$(x_{i,[j,\ldots,j+q-1]})$
7           $table[key][index[key]] \leftarrow \langle i, j \rangle$
8           $index[key] \leftarrow index[key] + 1$
9   **return** $table$

Fig. 1. The algorithm for identifying unique oligos (continues in Figure 2).

Using this idea, we design an efficient two-phase algorithm. In the first phase, we cluster all the possible seeds from the ESTs into groups such that within each group, a seed has no more than one mismatch with the other seeds. In the second phase, we check whether extending the flanking regions of a seed would result in a d-match with the corresponding extension of any other seed in the same group. If so, the l-mer given by this extension is not a unique oligo.

Phase 1. (HIT) We file all q-mers (seeds) from the input ESTs into a dictionary with 4^q entries. (If 4^q cannot fit in the main memory, one could use a hash table of an appropriate size.) Each entry of the table points to a list of locations where the q-mer occurs in the EST sequences. Using the table we can immediately locate identical seeds in the ESTs.

EXTENSION($X, t, q, table, m$)
1   **for** $i \leftarrow 1$ **to** 4^q **do**
2   **List** $mut \leftarrow mutant$ $list$ of $table[i]$
3       $len \leftarrow$ # of records in $table[i]$
4   **for** $j \leftarrow 1$ **to** len **do**
5   **for** $k \leftarrow j + 1$ **to** len **do**
6   COMPARE($table[i][j]$, $table[i][k]$, m)
7   **for** $h \leftarrow 1$ **to** # of records in mut **do**
8               $mutlen \leftarrow$ # of records in $table[mut[h]]$
9   **for** $k \leftarrow 1$ **to** $mutlen$ **do**
10   COMPARE($table[i][j]$, $table[mut[h]][k]$, m)

MAP(**string** S)
1   map S into an integer X with function $f : \{$A,C,G,T$\} \rightarrow \{0, 1, 2, 3\}$
2   **return** X

Fig. 2. The algorithm for identifying unique oligos (continued).

Phase 2. (EXTENSION) We compare the corresponding flanking regions of each pair of matching seeds to determine whether they can be extended to a pair of l-mers that d-match each other. Here, we also collect seeds that have exactly one mismatch with each other as follows. For each table entry corresponding to a seed y, we record a list of other seeds that have exactly one mismatch with y, by looking up table entries that correspond to all the 1-mutants of y. This list is called a *mutant list* of y. We examine all the seeds in the mutant list, and compare the flanking regions of the q-mers and that of y in the same way as we did for identical seeds, except that now the cutoff for the number of mismatches in the flanking regions is $d - 1$.

The algorithm is summarized in Figures 1 and 2.

Time complexity. Suppose that the total number of bases in X is n. The time complexity of phase one is simply $\Theta(qn)$. The time complexity of phase two depends on the distribution of the number of seeds filed into each table entry. Simply speaking, if the distribution is more or less uniform (which is the case in our experiment) and each table entry contains $r \approx n/4^q$ identical seeds, the number of comparisons within the table entry is $O(r^2)$. The number for comparisons for each mutant lists of size $3q$ is $O(qr^2)$. Each comparison requires extension of the seeds and takes $2(l - q)$ time. Since there are 4^q entries in the table, the overall time complexity is $O((l - q)qr^2 4^q)$. Given the exponential dependency on q, one needs to make sure that q is not too large before using the algorithm. (Again, in our experiment on the barley dataset, $l = 33, d = 5$ and $q = 11$.)

In the practice of EST data analysis, we need also consider the reverse complementary strand of each EST, which implies more stringency in the choice of unique oligos. The above algorithm can be easily modified to take into account reverse complementary EST strands without a significant increase of complexity.

3.2   Popular Oligos

Recall that the objective is to find all l-mers that have sufficiently large number of (c, d)-colors in X. Since popular oligos are not required to appear exactly in the EST sequences, we cannot solve the problem by exhaustive enumerate all the substrings of length l in the EST sequences and count their (c, d)-colors. In fact, one can easily show that the problem is NP-hard in general.

A straightforward algorithm is to consider all l-mers occuring in the ESTs and for each l-mer, enumerate all its (c, d)-mutants and count their number of (c, d)-colors. However, the number of (c, d)-mutants of an l-mer over the DNA alphabet is more than $\binom{l-c}{d}3^d$. Hence, the "brute-force" method becomes compuitionally impractical due to its memory requirement as soon as the input size reaches the order of hundreds of thousands of bases (like the barley dataset). [3]

We can reduce the search space using the same idea as in the algorithm for unique oligos, except that here the role of seeds is played by cores. Observe that, if a (popular) oligo has (c, d)-occurrences in many ESTs, many of these ESTs must contain length-l substring that share common cores. Based on this observation, we propose a heuristic strategy that first clusters the l-mers in the EST sequences into groups by their cores, and then enumerates candidate l-mers by comparing the members of each cluster in a hierarchical way.

An outline of the algorithm is illustrated in Figure 3. Here, we determine the popularity of the cores (i.e., length-c substrings) from the ESTs in the first step. For each popular core, we consider extension of the cores into l-mers by including flanking regions and cluster them using a well-known hierarchical clustering method, called *unweighted pair group method with arithmetic mean* (UP-GMA) [25]. We recall that UPGMA builds the tree bottom-up in a greedy fashion by merging groups (or subtrees) of data points that have the smallest average distance. Based on the clustering tree, we compute the common oligos shared by the l-mers by performing set intersection. These common oligos shared by many l-mers become candidate popular oligos. Finally, we count the number of colors of these candidates, and output the oligos with at least T colors. A more detailed description is given below. A complete example of the algorithm on a toy dataset is also given in the appendix.

Phase 1. We compute the number of colors for all c-mers in the ESTs to determine whether they could be candidate cores for popular l-mers, using a hash table. According to our definition, a popular oligo should have a popular core. We therefore set a threshold T_c on the minimum number of colors of each

---

[3] When $d = 3$ and $c = 20$, $\binom{l-c}{d}3^d = \binom{13}{3}3^3 = 7,722$ for the barley dataset. Hence, the straightforward algorithm would have to count the number of colors for about $7,722 \cdot 28 \times 10^6 = 217 \times 10^9$ l-mers.

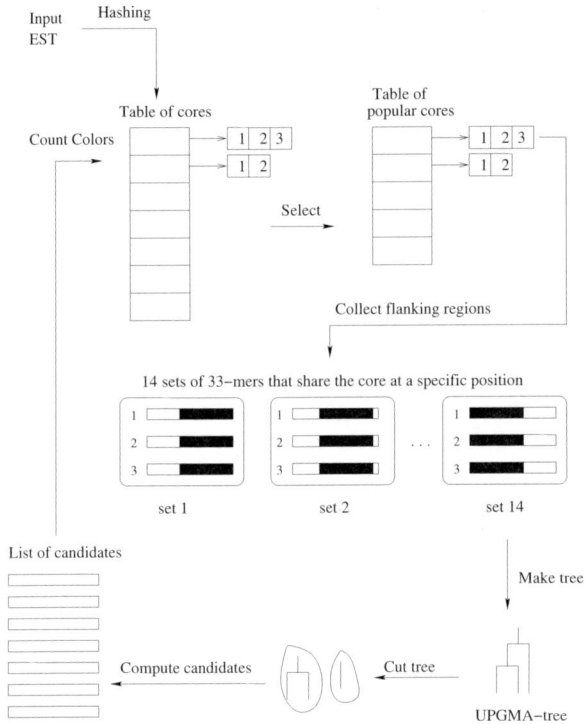

Fig. 3. An overview of the algorithm for selecting popular oligos. For convenience of illustration, the length of the oligos is assumed to be $l = 33$, and the length of the cores is assumed to be $c = 20$.

popular core, depending on T, c, l and X. All cores that have a number of colors below T_c are filtered out, and considered "unpopular". However, since an l-mer can (c, d)-match another l-mer with any of its $l - d + 1$ cores, it is possible that we might miss some popular oligos that critically depend on unpopular core. The parameter T_c represents a tradeoff between precision and efficiency. We will show in Section 4 the effect of changing T_c on the output. We will see that in practice we might miss only a negligible number of popular oligos.

Phase 2. Here we collect the substrings flanking the popular cores. For each popular core, we construct $l - c + 1$ sets of substrings, one for each possible extension of the core into an l-mer. Each set contains substrings of length $l - c$ constructed by concatenating the left and right flanking regions.

Phase 3. For each set of flanking substrings, we would like to identify all $(l - c)$-mers that have d-occurrences in many of these substrings. In order to achieve this efficiently, we first cluster the substrings according to their mutual Hamming distance using the well-known hierarchical clustering method UPGMA. In the process of building the clustering tree, whenever the Hamming distance between some leaves in the tree is zero we compress the distance matrix by combining the identical strings into one entry. This significantly reduces the running time not only because the tree becomes smaller, but also because the number of common d-mutants of two different l-mers is much less than that of two identical ones. As we can see later, a significant proportion of the running time is spent on the intersection of the sets of d-mutants. Compressing the distance matrices avoids intersecting identical sets of d-mutants, which is expensive and also useless. We then create a set of d-mutants for each substring represented at the leaves and traverse the tree bottom-up. At each internal node u, we compute the intersection of the two sets attached to the children, using a hash table based on the hash function described in [26]. This intersection represents all the $(l - c)$-mers that have d-occurrences in all the leaves (substrings) under the node u. As soon as the intersetion becomes empty, we prune the tree. At the end of this process, we obtain a collection of sets of $(l - c)$-mers, each of which, together with the popular core, represents a candidate popular oligo.

Phase 4. Given the candidate popular oligos, we need to count their number of (c, d)-colors. Before counting, we radix-sort the candidates and remove duplicates. More precisely, due to the possibly very large number of candidates and duplicates (as in the barley case), we sort the candidates in several stages as follows. For each core, we radix-sort all the candidates derived from the core and remove duplicates. Then we merge the sorted list into the sorted list containing all the candidates from those cores that have already been processed.

Time complexity. Phase 1 costs time $O(cn)$. In phase 2, if the number of popular cores selected in the first step is p and the average number of occurrence of the cores is r, this phase costs $O(nr(l - c))$. For phase 3, the time for building a UPGMA tree, including the computation of the distance matrix, is $O((l - c)r^2)$, where r stands for the number of strings to be clustered. Since a (binary) UPGMA tree with r leaves has $2r - 1$ nodes, the time for traversing (and pruning) the tree is $O(r\binom{l-c}{d}3^d)$, where $\binom{l-c}{d}3^d$ is the number of d-mutants at each leaf. Finally for phase 4, if the total number of candidates is m, counting the colors for the candidates, excluding the time for radix-sort, costs time $O(rm(l - c))$.

4   Implementation and Results

We have implemented both algorithms in C and tested the programs on a desktop PC with a 1.2GHz AMD Athlon CPU and 1GB RAM, under Linux. The main dataset is a collection barley ESTs from HARVEST containing $k = 46,145$ EST sequences with a total of $n = 28,475,017$ bases. Before doing the searches, we first cleaned the dataset by removing PolyT and PolyA repeats.

As mentioned above, our first task was to search for unique oligos of length $l = 33$ with a minimum number of mismatches $d = 5$. Based on these parameters, each oligo was divided into three seeds of length $q = 11$. Hence, our dictionary table had $4^{11} \approx 4$ million entries. The efficiency of our algorithm critically depends on the statistical distribution of the seeds in the dictionary. The statistics of the seeds in our experiment (before the extension phase) is shown in table in the left of Figure 4. Clearly, most seeds occur less than 20 times in the ESTs

# of occurrences	# of seeds
0	242399
1-9	3063288
10-19	708745
20-29	120698
30-39	31637
40-49	11908
50-5049	15629

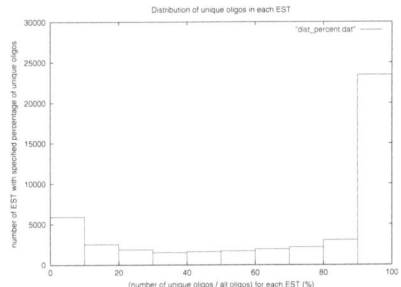

Fig. 4. LEFT: Distribution of frequencies of seeds in barley ESTs. RIGHT: Distribution of unique oligos. The horizontal axis stands for the percentage of unique oligos over all 33-mers in an EST, and the vertical axis stands for the number of ESTs whose unique oligos are at a certain percentage of all its 33-mers.

and this is the main reason why our algorithm was able to solve the dataset efficiently. The final distribution of unique oligos is shown in the right of Figure 4. Note that, there are many ESTs (slightly more than half of the entire dataset) whose length-33 substrings are almost all unique oligos. In particular, there are 13,430 ESTs whose length-33 substrings are all unique oligos and there are 2,159 ESTs that contain no unique oligos. The whole computation took 2 hours and 26 minutes and used about 200 MB of memory.

Table 1. Distribution of the number of colors of the cores. The left column is the range of the number of colors. The right column is the number of cores with a certain number of color.

colors	number of cores
1	22523412
2-10	2128677
11-20	5148
21-30	1131
31-40	492
41-50	346
51-60	242
61-70	77
71-80	34
81-90	29
91-100	43
101-176	19

Our second task was to search for popular oligos with length $l = 33$ and core length $c = 20$. We considered different choices for the maximum number of mismatches d ourside the core and the threshold T_c on the minimum number of

Table 2. Running time for enumerating popular oligos (in seconds).

	$T_c = 30$	$T_c = 40$	$T_c = 50$	$T_c = 60$
$d = 1$	184	166	156	153
$d = 2$	278	219	177	338
$d = 3$	3808	1788	730	359

colors for the popular cores. The distribution of the number of colors of the cores is shown in Figure 1. From the table we can see that the number of cores decreases almost exponentially as the number of colors increases. On the other hand, cores with low colors are unlikely to contribute to popular oligos. Therefore, it is important to filter them out to increase the efficiency.

The running time of this program varies with the parameters d and T_c, as shown in the Figure 2. The memory used in the program was mainly for storing the candidate popular oligos. In general, about 64 MB suffices since the program reuses the memory frequently.

4.1   Simulations

To evaluate the performance of our heuristics for selecting popular oligos we also have run a few simulations as follows. We generated a set of artificial ESTs by creating first a set of k random sequences and then injecting a controlled number of approximate occurrences of a given set of oligos. The initial set of oligos, denoted by $I_1, \ldots, I_s$, was also generated randomly over Σ. Each oligo I_i was assigned a predetermined number of colors C_i. We decided that the distribution of the C_i should be Gaussian, that is, we defined $C_i = C \, e^{-i^2/2} / \sqrt{2\pi}$ where C is a fixed constant which determines the maximum number of colors. As said, the positions in-between the oligos were filled with random symbols over the DNA alphabet.

We then ran our program for popular oligos on the artificial ESTs dataset and output a set of candidate oligos $O_1, \ldots, O_t$ with their respective colors $C'_1, \ldots, C'_t$. The output oligos were sorted by colors, that is $C'_i \geq C'_j$, if $i < j$. Since the output contained redudant candidates that came from the mutations of the original popular oligos, we removed those candidates that were a d-mutant of another oligo with an higher number of colors. More precisely, if O_i was a d-mutant of O_j, and $1 \leq i < j \leq t$, then O_j was discarded. This "compression step" did not eliminate good candidates for the following reason. Since the input oligos $I_1, \ldots, I_s$ were generated randomly they were very unlikely to be similar. As a consequence, the corresponding output oligos were also unlikely to be eliminated.

Finally, we compared the pair (I, C) with (O, C'). The more similar (O, C') is to (I, C), the better is the heuristic of our algorithm. Recall that I and O were sorted by decreasing number of of colors. We compared the entries in (I, C) with the ones in (O, C'), position by position. For each $1 \leq i \leq u$, where $u = \min(s, t)$, we computed the average difference between C and C'

Table 3. The average relative errors between the number of colors in the input and the number of colors in output for a simulated experiment ($n = 1,440,000$, $k = 720$, $c = 20$, $C = 100$, $s = 100$).

	$d = 2$	$d = 3$
$T_c = 10$	0.019	0.022
$T_c = 15$	0.000	0.002
$T_c = 20$	0.018	0.000
$T_c = 25$	0.000	0.001
$T_c = 30$	0.000	0.000

as $E = (1/u) \sum_{i=1}^{u} \frac{|C(i)-C'(i)|}{C'(i)}$. If we assume that I and O contain the same set of oligos, the smaller is E, the more similar is (I, C) to (O, C'). To validate this assumption, we also searched the list of oligos I in O, to determine whether we missed completely some oligos.

Figure 3 shows the value of E for one run of the program on a dataset of $n = 1,440,000$ bases composed by $k = 2,000$ sequences of size 720. We generated a set of $s = 100$ oligos with a maximum number of colors $C = 100$. In the analysis, we fixed the size of the core to be $c = 20$, whereas the maximum number of mismatches d outside the core and the threshold T_c were varied. The results show that the average relative error is below 2%. In the final version of the paper, we plan to repeat these simulations a few times to get more stable and reliable results. We also compared the list of input oligos with the list of output oligos and we found that sometimes the program misses one or two oligos out of 100. However, the number of colors of these missed oligos is always near the threshold T_c. We never miss an oligo whose number of color is above $T_c + 10$.

5   Conclusion

We have proposed two algorithms to find unique and popular oligos in large EST databases. The size of our dataset, in the order of tens of millions of bases, was the real challenge due to the limitation in the size of main memory in common 32-bits architectures. Our algorithms were able to produce a solution in a reasonable amount of time on a regular PC with a modest amount of memory. As far as we know, no other existing tools are capable of handling such a dataset with such limited resources. Simulations show that the number of missed oligos by the heuristic algorithm for popular oligos is negligible and can be controlled very effectively by adjusting the parameters. Although the algorithms were initially designed to address the challenges from the barley EST dataset, the methods can be easily adapted to solve similar problems concerning infrequent and frequent oligos on other large datasets. The software will be released in the public domain in the near future.

Acknowledgments. The authors would like to thank Steve Wanamaker for producing the sequence assemblies and unigene datasets that were used in this work.

References

1. Adams, M.D., Kelley, J.M., Gocayne, J.D., Dubnick, M., Polymeropoulos, M.H., Xiao, H., Merril, C.R., Wu, A., Olde, B., Moreno, R.F., Kerlavage, A.R., McCombie, W.R., Venter, J.C.: Complementary DNA sequencing: Expressed sequence tags and human genome project. Science **252** (1991) 1651–1656
2. Boguski, M., Lowe, T., Tolstoshev, C.: dbEST–database for "expressed sequence tags". Nat. Genet. **4** (1993) 332–3
3. Yu, Y., Tomkins, J.P., Waugh, R., Frisch, D.A., Kudrna, D., Kleinhofs, A., Brueggeman, R.S., Muehlbauer, G.J., Wise, R.P., Wing, R.A.: A bacterial artificial chromosome library for barley (*Hordeum vulgare L.*) and the identification of clones containing putative resistance genes. Theor Appl Genet **101** (2000) 1093–1099
4. Close, T., Wing, R., Kleinhofs, A., Wise, R.: Genetically and physically anchored EST resources for barley genomics. Barley Genetics Newsletter **31** (2001) 29–30
5. Michalek, W., Weschke, W., Pleissner, K., Graner, A.: Est analysis in barley defines a unigene set comprising 4,000 genes. Theor Appl Genet **104** (2002) 97–103
6. Huang, X., Madan, A.: CAP3: A DNA sequence assembly program. Genome Research **9** (1999) 868–877
7. Han, C., Sutherland, R., Jewett, P., Campbell, M., Meincke, L., Tesmer, J., Mundt, M., Fawcett, J., Kim, U., Deaven, L., Doggett, N.: Construction of a BAC contig map of chromosome 16q by two-dimensional overgo hybridization. Genome Research **104** (2000) 714–721
8. Barakat, A., Carels, N., Bernardi, G.: The distribution of genes in the genomes of gramineae. Proc. Natl. Acad. Sci. U.S.A. **94** (1997) 6857–6861
9. Rahmann, S.: Rapid large-scale oligonucleotide selection for microarrays. In: Proceedings of the First IEEE Computer Society Bioinformatics Conference (CSB'02), IEEE Press (2002)
10. Bailey, T.L., Elkan, C.: Unsupervised learning of multiple motifs in biopolymers using expectation maximization. Machine Learning **21** (1995) 51–80
11. Jonassen, I., Collins, J.F., Higgins, D.G.: Finding flexible patterns in unaligned protein sequences. Protein Science **4** (1995) 1587–1595
12. Jonassen, I.: Efficient discovery of conserved patterns using a pattern graph. Comput. Appl. Biosci. **13** (1997) 509–522
13. Rigoutsos, I., Floratos, A.: Combinatorial pattern discovery in biological sequences: The TEIRESIAS algorithm. Bioinformatics **14** (1998) 55–67
14. Hertz, G., Stormo, G.: Identifying DNA and protein patterns with statistically significant alignments of multiple sequences. Bioinformatics **15** (1999) 563–577
15. Lawrence, C.E., Altschul, S.F., Boguski, M.S., Liu, J.S., Neuwald, A.F., Wootton, J.C.: Detecting subtle sequence signals: A Gibbs sampling strategy for multiple alignment. Science **262** (1993) 208–214
16. Neuwald, A., Liu, J., Lawrence, C.: Gibbs motif sampling: Detecting bacterial outer membrane protein repeats. Protein Science **4** (1995) 1618–1632
17. Pevzner, P.A., Sze, S.H.: Combinatorial approaches to finding subtle signals in DNA sequences. In: Proc. of the International Conference on Intelligent Systems for Molecular Biology, AAAI press, Menlo Park, CA (2000) 269–278

18. Keich, Pevzner: Finding motifs in the twilight zone. In: Annual International Conference on Computational Molecular Biology, Washington, DC (2002) 195–204
19. Tompa, M., Buhler, J.: Finding motifs using random projections. In: Annual International Conference on Computational Molecular Biology, Montreal, Canada (2001) 67–74
20. Buhler, J., Tompa, M.: Finding motifs using random projections. J. Comput. Bio. **9** (2002) 225–242
21. Apostolico, A., Bock, M.E., Lonardi, S.: Monotony of surprise and large-scale quest for unusual words (extended abstract). In Myers, G., Hannenhalli, S., Istrail, S., Pevzner, P., Waterman, M., eds.: Proc. of Research in Computational Molecular Biology (RECOMB), Washington, DC (2002)
22. Eskin, E., Pevzner, P.A.: Finding composite regulatory patterns in DNA sequences. In: Proc. of the International Conference on Intelligent Systems for Molecular Biology, AAAI press, Menlo Park, CA (2002) Bioinformatics S181–S188
23. Li, F., Stormo, G.D.: Selection of optimal DNA oligos for gene expression arrays. Bioinformatics **17** (2001) 1067–1076
24. Rouillard, J.M., Herbert, C.J., Zuker, M.: Oligoarray: Genome-scale oligonucleotide design for microarrays. Bioinformatics **18** (2002) 486–487
25. Swofford, D. In: PAUP: Phylogenetic Analysis Using Parsimony version 4.0 beta 10. Sinauer Associates, Sunderland, Massachusetts (2002)
26. Wesselink, J.J., de la Iglesia, B., James, S.A., Dicks, J.L., Roberts, I.N., Rayward-Smith, V.J.: Determining a unique defining DNA sequence for yeast species using hashing techniques. Bionformatics **18** (2002) 1004–1010
27. Ito, M., Shimizu, K., Nakanishi, M., Hashimoto, A.: Polynomial-time algorithms for computing characteristic strings. In Crochemore, M., Gusfield, D., eds.: Proceedings of the 5th Annual Symposium on Combinatorial Pattern Matching. Number 807 in Lecture Notes in Computer Science, Asilomar, CA, Springer-Verlag, Berlin (1994) 274–288

Appendix: An Example of the Popular Oligo Algorithm

We show a small example to illustrate each step of the algorithm. Again, let l denote the length of oligos, c the length of cores, d the maximum number of mismatches between two oligos, and T_c the threshold on the minimum number of colors of popular cores. In this toy example, $l = 8, c = 5, d = 1, T_c = 3$. The input is composed of four artificial EST sequences as shown in Figures 5. EST1 and EST3 are highly similar, which represents the similarity between some of the EST sequences in real data.

Phase 1. Seven cores out of 148 possible cores are selected as popular cores. Each entry of the hash table points to a list of positions of the occurrences of the core in the input ESTs.

Phase 2. We collect the flanking regions for the seven cores. Figure 6 shows the four sets of flanking regions for AAGGC. Note that the fourth set has one fewer element than the other 3 sets. The reason is that the core AAGGC occurs at the right boundary of EST0, and therefore has a shorter flanking region.

Phase 3. We cluster the flanking regions using UPGMA. In Figure 8, we show the clusters for set 2 of the core AAGGC. Observe that the Hamming distance

>EST0
TGGAGTCCTCGGACACGATCACATCGACAATGTGAA
GGCGA

>EST1
GTGAAGGAGGTAGATCAAATAGAGCCTGCCCTAAAA
AGGCAGCTTATAATCTCCACTGCT

>EST2
TCCGACTACTGCACCCCGAGCGGATCACACAATGGAA
GGCCCGTGCGC

>EST3
GTGAAGGAGGTAGATACTCGTATACGATCACTGCCTA
AAAAGGCAGCTTATAATCTCCATATCGCTG

Fig. 5. The example dataset. The table of popular cores.

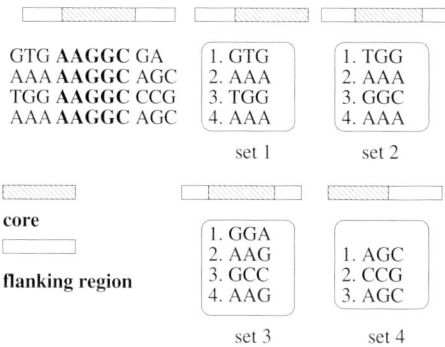

Fig. 6. Collecting flanking regions for the core AAGGC.

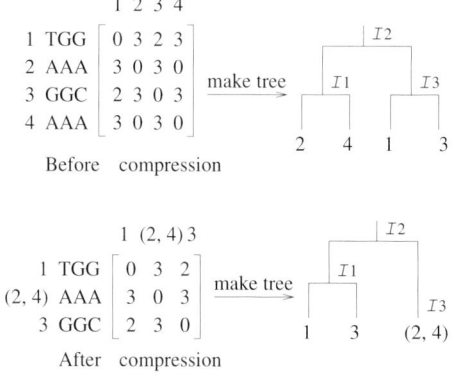

Fig. 7. UPGMA tree construction for set 2 of the core AAGGC.

between entry 2 and entry 4 is zero and therefore distance matrix is compressed by combining the identical strings into one entry. We then need to enumerate all the 1-mutants of the strings denoted by the leaves of the trees, that is AAA, TGG

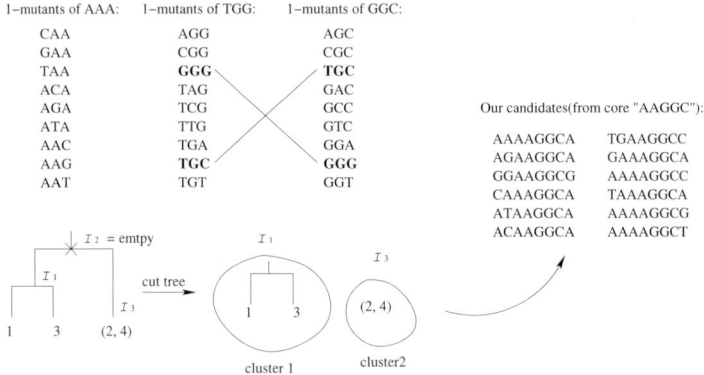

Fig. 8. Clustering set 2 of the core AAGGC

and GGC. Because leaf 1 and leaf 3 share the same parent, we apply intersection on their sets of 1-mutants and get the set $I_1 = \{$GGG, TGC$\}$. Then we apply intersection between I_1 and I_3, where I_3 represents the set of the 1-mutants of leaf AAA. Since the resulting intersection I_2 is empty, we prune the tree at I_2 and separately output the strings in I_1 and I_3 as flanking regions of candidate popular oligos. Note that, we output the 1-mutants of I_3 even if it is represented only by one node, because it is actually the intersection of two occurrences of AAA, and therefore all elements of I_3 have at least two d-matches in the EST sequences.

Author Index